D1713575

MANUAL OF
SEDIMENTARY PETROGRAPHY

SEPM REPRINT SERIES NUMBER 13
Classic facsimile edition of
The Century Earth Science Series
KIRTLEY F. MATHER, *Editor*

MANUAL OF SEDIMENTARY PETROGRAPHY

I. Sampling, Preparation for Analysis, Mechanical Analysis, and Statistical Analysis

BY

W. C. KRUMBEIN

UNIVERSITY OF CHICAGO

II. Shape Analysis, Mineralogical Analysis, Chemical Analysis, and Mass Properties

BY

F. J. PETTIJOHN

UNIVERSITY OF CHICAGO

originally published by

APPLETON-CENTURY-CROFTS, INC.

NEW YORK

reprinted by

SOCIETY OF ECONOMIC PALEONTOLOGISTS AND MINERALOGISTS

TULSA, OKLAHOMA

ISBN 0-918985-78-1
© 1988 by
Society of Economic Paleontologists and Mineralogists
P.O. Box 4756
Tulsa, OK 74159-0756

Original COPYRIGHT, 1938, BY
D. APPLETON-CENTURY COMPANY, INC.

PRINTED IN THE UNITED STATES OF AMERICA

FOREWORD

THE BOOK

W.C. Krumbein and F.J. Pettijohn's *Manual of Sedimentary Petrography* was published in 1938, exactly one-half century ago. It was the first comprehensive survey of methods of laboratory analysis for sediments and sedimentary rocks to be published in the United States and, overall, it was the third such book. The two earlier books were Henry B. Milner's *Sedimentary Petrography* (1922 and 1929, Thomas Murby & Co.), published in Great Britian, and Lucien Cayeux's *Introduction à l'Étude Petrographique des Roches Sedimentaires* (1931, Imprimerie Nationale), published in France. In the years following 1938 and, in particular beginning in the mid-1960s, several other methodology books were published. Krumbein and Pettijohn's manual, however, remained the primary reference source on sedimentary petrography methodologies for nearly 30 years and consequently had a large and lingering impact on the development of sedimentology in North America. Although never revised, it was reprinted a number of times. The original printer's plates were destroyed in the 1970s, and so the current facsimile edition is a photo-reproduction of one of the surviving copies.

With the many reference books and journal articles on laboratory techniques available, the reader might well wonder why Krumbein and Pettijohn's manual is being republished now. There are several good reasons. First, it is one of the first books of its kind and so is of considerable historical interest. Second, unlike more recent works, it describes the initial development of many of the methodologies still in use today. In this connection, the authors' citation of a surprisingly extensive early literature is especially useful. Third, it describes in some detail the identification of detrital (and, in particular, heavy) minerals in grain mounts, a technique not included in more recent works and one which is fast becoming a lost art. Fourth, and most importantly, it describes many simple, low-cost techniques that are effective and reliable alternatives to some of the more elaborate and expensive methodologies now in vogue. In this regard, Krumbein and Pettijohn's manual is especially valuable as a compendium of still useful techniques that have been forgotten or overlooked by modern petrographers. Some of these techniques may even prove to be better, in certain circumstances, than the more conventional approaches that we tend to use out of habit.

In re-reading Krumbein and Pettijohn's manual, I was constantly struck by the timelessness of many of the sedimentary petrography methodologies. The book has lost little of its usefulness in the 50 years since it was first published. The clarity of writing, the heuristic treatment of the subject, and the many excellent line drawings combine to produce a text that is both informative and easy to follow. By any measure, the manual is still an important reference.

In the years following the publication of their manual, both Krumbein and Pettijohn went on the have long and illustrious careers in geology. Sadly, Krumbein passed away in 1979 at the age of 77, but Pettijohn, who is now 84, is still very active. He has recently published three books: *Memoirs of an Unrepentant Field Geologist* (1984, University of Chicago Press), the second edition of *Sand and Sandstone* (1987, with P.E. Potter and R. Siever, Springer-Verlag) and *A Century of Geology at Johns Hopkins University* (1988, Earth and Planetary Science Department at Johns Hopkins University and Gateway Press). Pettijohn has also served in recent years on several National Research Council panels. Perhaps the reader has wondered, as I have, about the circumstances surrounding the remarkable careers of W.C. Krumbein and F.J. Pettijohn. What deliberate decisions, fortuitous accidents, and external influences and events caused them to become the scientists and educators that they were? In what follows, I will try to provide some answers to these questions.

THE FIELD GEOLOGIST AND THE METHODOLOGIST

Francis John Pettijohn was born in 1904 to schoolteacher parents in Waterford, Wisconsin. He was the first of six children. His father had a highly varied and successful career as a school administrator, primarily in universities, and as a result of this, Pettijohn grew up in widely separated parts of the country. His family lived briefly in North Dakota and Washington, D.C., but he spent most of his formative years in Wisconsin, Indiana, and eventually Minnesota. The young Pettijohn developed an enthusiasm for outdoor activities, and particularly enjoyed exploring caves and quarries, and collecting fossils. When he entered the University of Minnesota in Minneapolis as a freshman in 1921, he had already decided on a career in geology.

FOREWORD

Pettijohn received his bachelor's and master's degrees in geology from the University of Minnesota in 1924 and 1925, respectively. His M.S. thesis was supervised by the renowned petrologist Frank F. Grout and dealt with phosphatized fossils and nodules in the Platteville Limestone in the Minneapolis area. During the summers of 1924 and 1925, Pettijohn worked as a field assistant with the Minnesota Geological Survey mapping Precambrian igneous and metamorphic rocks. It was from this experience that he acquired his lifelong interest in the Precambrian metasediments of the Canadian Shield and also his love of geological field work.

Pettijohn spent the next two years (1925-1927) as an instructor at Oberlin College in Ohio. He found teaching geology to be an especially rewarding experience and two years later would make it his life's work. In 1927 he went to the University of California at Berkeley to resume his graduate studies. After one year, Pettijohn returned to the University of Minnesota and in 1930 received his Ph.D. in geology. Professor Grout again supervised his research, this time on the Precambrian metasediments and associated rocks of northwestern Ontario, Canada.

In 1929 Pettijohn accepted a faculty position at the University of Chicago, a position he would hold for the next 23 years. He was hired as the Geology Department's "sedimentationist" and thus, quite by accident, by his own account, embarked on a career in sedimentology. Up until that time Pettijohn had considered himself simply a petrologist who was equally interested in all rock types. It was, of course, his work on Precambrian metasediments that qualified him for the position at Chicago. At the time of Pettijohn's appointment, sedimentology was just beginning to be recognized as a distinct subdiscipline of geology. The impetus for this development was the rapidly expanding role of geology in the petroleum industry. The Society of Economic Paleontologists and Mineralogists had just been organized in 1926, and in 1931 the inaugural issue of the Journal of Sedimentary Petrology would appear. Pettijohn had gotten into sedimentology on the ground floor, and he was destined to influence the course of its future development.

It 1930 Pettijohn taught his first course in "sedimentation." One of the students in the class was W.C. Krumbein.

William Christian Krumbein was born in 1902 in Beaver Falls, Pennsylvania. He was orphaned at an early age and thereafter was raised by his uncle who lived in the Chicago area. German was spoken at home, and so Krumbein grew up fluent in the language. Regrettably, I have been unable to discover anything more about his childhood years.

Krumbein received his bachelor's degree in 1926 from the University of Chicago's School of Business. As an undergraduate, he took a geology course as an elective and in so doing discovered that he had an interest in geology. During the next four years, Krumbein worked for a collection agency as an auditor, comptroller, and eventually its director, but at the same time continued at the University of Chicago as a part-time student in geology. His interest in advanced statistical theory apparently began in 1927 as a result of sharing an office with M. King Hubbert. Krumbein took his first course in probability theory at Hubbert's urging after a conversation they had about why some parts of pebbles abrade faster than other parts. The course had a profound effect on him because he discovered that statistics were useful not only for summarizing data but also, in some cases, for investigating physical processes. In 1930 Krumbein quit his job at the collection agency (or perhaps lost it because of the stock market crash) and pursued his graduate studies in geology full time. He received his master's and doctoral degrees in geology in 1930 and 1932, respectively. His M.S. thesis, which was supervised by the mineralogist D. Jerome Fisher, was basically a review of the various methods of mineral identification. Professor J. Harlen Bretz, a glacial geologist, supervised his Ph.D. dissertation on the textural analysis of glacial tills in northern Illinois. Pettijohn, who was untenured at the time, served as an informal coadvisor on the dissertation.

It was as a student in Pettijohn's sedimentation course in 1930 that Krumbein first recognized the geological potential of the statistical procedures he had learned as a business major and applied as a businessman. Nowhere in geology at the time was the potential greater than for sedimentologic data, especially grain size data. It was in Pettijohn's course then that the whole future direction of Krumbein's professional career was set: the statistical analysis and mathematical treatment of sedimentologic data.

In 1931 Krumbein was given a part-time instructorship in the University of Chicago's experimental "general college," where he taught introductory geology to non-majors. The appointment was upgraded to full-time in 1933, but it was not until 1938 that Krumbein became formally attached to the Geology Department. During the intervening years, he worked summers mapping glacial deposits for the Illinois Geological Survey, he continued to take mathematics and physics courses to develop his quantitative abilities further, and he aggressively pursued his research on grain size analysis, publishing several papers, including one on his now famous phi size scale. During this same period, his office was next door to Pettijohn's, and the two men developed a close working relationship and a shared interest in sedimentology. This led to their collaboration on the *Manual of Sedimentary Petrography*. Krumbein wrote the first part of the book, which deals with sampling, grain size analysis, and graphical and statistical procedures; and Pettijohn wrote the second part on the analysis of grain shape, mineralogy, and bulk properties. The original price of the book in 1938 was $6.50. The sales were initially quite small, but, beginning with World War II, they dropped off to nearly zero. It was not until after the war, when students returned to the universities and basic research was once again a luxury that could be afforded, that large numbers of the book were sold.

With the coming of the war, Krumbein and Pettijohn's paths began to diverge. In 1941 Krumbein was awarded a Guggenheim fellowship and so took a leave of absence from the University of Chicago. He spent a year at the Universities of Iowa and California (Berkeley) studying fluid hydraulics and sediment transport under flume and wave tank conditions, and then in 1942 joined the Landing Beach Intelligence Group attached to the Beach Erosion Board of the Army Corps of Engineers. In 1943 Pettijohn also took a leave of absence and began working for the United States Geological Survey as a field geologist mapping the Precambrian iron deposits of northern Michigan. Shortly after the war, Krumbein and Pettijohn had agreed to collaborate on another book which was to be titled *Sedimentary Rocks*. A contract was signed with Harper and Row in 1948, but Krumbein lost interest and dropped out of the project. Pettijohn pressed ahead and in 1949 published the book. During the same period, Krumbein collaborated with Larry L. Sloss and in 1951 published *Stratigraphy and Sedimentation* (W.F. Freeman and Co.).

Pettijohn continued his association with the U.S. Geological Survey on a part-time basis for several years after the war. In 1952 he left Chicago and joined the geology faculty at Johns Hopkins University in Baltimore, Maryland, where he has remained to this day. He retired in 1973 and is now a Professor Emeritus. During the 1950s and early 1960s, Pettijohn was a consultant to the Shell Oil Company's research laboratory in Bellaire, Texas. It was here that he developed his intense interest in paleocurrent and basin analyses. Krumbein had a similar enlightening experience with an oil company. Immediately after the war, he spent a year at Gulf Oil's research laboratory in Pittsburgh, Pennsylvania, where he was introduced to quantitative facies mapping using subsurface data. In 1946 Krumbein joined the geology faculty at Northwestern University in Evanston, Illinois, where he remained through his retirement in 1970 and until his death in 1979. In the interim he served as a consultant to the Brookhaven National Laboratory and the Beach Erosion Board of the Army Corps of Engineers.

During the post-war years, Krumbein and Pettijohn each went on to publish a multitude of papers and several books, and to reap many honors. Both men were exemplary geologists, who were highly respected researchers as well as gifted and much loved teachers. The modern field of sedimentology owes much of its spirit and direction to their contributions.

Although reprinted many times, Krumbein and Pettijohn's *Manual of Sedimentary Petrography* was never revised, primarily because of Pettijohn's disillusionment with the value of quantitative textural analyses as a means of interpreting sedimentary processes. The original premise of the book was that such analyses would provide many important insights into the transportation and deposition of ancient sediments. Shortly after publication of the book, Pettijohn returned to his earlier belief in the supremacy of field observations in solving geological problems. Laboratory techniques were in his view only occasionally useful adjuncts to field work. Pettijohn had become a die-hard field geologist and was no longer comfortable with the emphasis on laboratory techniques that a revision of the book would entail. Krumbein, on the other hand, was becoming more, not less, enamored with the original premise of the book. Increasingly, he was devoting himself to statistical models, and to the quantitative analysis of sediment textures and facies, both from the standpoint of interpreting the physical dynamics of sedimentary processes and of developing new techniques for studying these processes. Krumbein had become a methodologist, and as such, he was usually more interested in the means by which geological problems were solved than in the problems themselves. A philosophical schism had opened between the two geologists and their *Manual of Sedimentary Petrology* was to be their first and only collaborative effort.

JAMES A. HARRELL
The University of Toledo
July 1988

FOREWORD

SOURCES OF INFORMATION

1. F.J. Pettijohn, 1984, Memoirs of an Unrepentant Field Geologist: The University of Chicago Press, Chicago, 260 p.
2. A.L. Howland, 1975, W.C. Krumbein — the Making of a Methodologist: Geological Society of America Memoir 142, p. xi-xiii.
3. D.J. Fisher, 1963, The Seventy Years of the Department of Geology, University of Chicago: The University of Chicago Press, Chicago, 147 p.
4. Who Was Who in America with World Notables, Vol. VII, 1977-1981: Marquis Who's Who, Inc., 1981 — article on W.C. Krumbein, p. 331.
5. American Association of Petroleum Geologists Bulletin, v. 38, p. 2437 (1954) — biographical data on W.C. Krumbein in a section on "Nominees for Editor, 1955-56."
6. conversations with:
    a. Francis J. Pettijohn
    b. Paul E. Potter
    c. Larry L. Sloss
    d. M. King Hubbert

EDITOR'S PREFACE

WHEN Sorby and his contemporaries were laying the foundations for modern petrographic research, back in the middle of the nineteenth century, the importance of microscopic examination of sedimentary rocks was stressed on at least one occasion. Nevertheless, for more than half a century the scientific study of that type of rocks received relatively slight attention. Many petrographers peered through their microscopes at thin sections of igneous rocks; many paleontologists measured minutely the shape of fossils; many stratigraphers debated the systematic position and correlation of sedimentary formations; but few geologists gave any really serious thought to the accurate and detailed study of the mineralogic characteristics of sediments and the rocks formed therefrom.

During the last twenty years, however, there has been a notable increase in interest in this phase of geology, with a very gratifying expansion of knowledge as a result of extraordinary improvements in the techniques of study. This is in part a by-product of the application of geology in the petroleum industry, in part a result of the discovery that sedimentary rocks provide a field for pure research unexcelled by any other field within the broad area of earth science.

The closely related sciences of sedimentary petrography and sedimentary petrology are to-day established on a firm base of technical procedure and deductive theory. They have become worthy of a lifetime of specialization which will well repay the devotion of a considerable minority of geologists.

Most of the data which such specialists should use are widely scattered through a large number of memoirs and periodicals which have been published within the last fifteen years. Many of these are to be found only in the journals devoted to physics, soil science, statistical method, and colloidal chemistry, and in other sources which like those journals are not ordinarily available in geologic laboratories. No adequate handbook of sedimentary petrography has hitherto been published in this country. Drs. Krumbein and Pettijohn have therefore rendered a signal service to the student and worker in this field by preparing this very useful volume.

It is primarily concerned with the methods of petrographic analysis of the sedimentary rocks, including the unconsolidated sediments. It covers every step of the process, from the field sampling to the final graphic and statistical analysis, with due regard for theory as well as method. It will serve admirably not only as a textbook for students but also as an indispensable aid for the professional worker dealing as in petroleum geology with sedimentary rocks and the valuable resources which they contain.

KIRTLEY F. MATHER.

PREFACE

I often say that if you measure that of which you speak, you know something of your subject; but if you cannot measure it, your knowledge is meager and unsatisfactory.—Lord Kelvin.

The recognition, long overdue, of the value of laboratory analysis in the study of sediments is beginning to be apparent. From the colloidal chemist, the ceramist, the ore-dressing engineer, the pedologist, the mineralogist, the statistician, and others the geologist has adapted methods which will eventually go far toward making the study of sediments a more exact science.

The writers of this volume do not depreciate field study. Such studies are a necessary prelude to the laboratory work and are fundamental to the science of geology. Realization, however, that sediments, like all other rocks, are a product of definite physical and chemical processes and are capable of definite analysis on the basis of carefully gathered quantitative data has been very slow indeed among geologists as a group. We believe that any consideration of the origin of a deposit which neglects the analysis of the material itself is quite incomplete.

It has been pointed out by some that our ability to interpret analytical data lags greatly behind our ability to make the analyses and that therefore further refinement in technique and greater accuracy in description are superfluous. It is even argued that laboratory analyses only confirm what the field geologist already knows, and such work is therefore regarded only as a refinement and not as a new contribution. The authors are not in sympathy with this view. It may be evident, even to the naked eye, that the sand along a beach decreases in size in the direction of transport, but whether the rate of decrease in size is exponential or conforms to some other law is not evident. To discover some underlying law or relationship introduces a new element into geological theory and opens up new avenues of thought.

In addition to the establishment of new principles and the interpretation of rock origins, laboratory study of sediments has important economic applications. Well known is the study of "heavy minerals," which has proved its worth in the correlation of sedimentary formations. The technologist has long recognized the necessity for physical analysis of the materials with which he is concerned. The geologist is often called upon to prospect for and estimate the worth of pottery clays, brick earths, fire-clay, fullers' earth, molding sand, etc. He must therefore be able to use such methods of analysis as will serve to indicate the usefulness of a deposit for the purpose intended. In fact, any one engaged in the study of the particulate substances, natural or artificial (cement, paint pigments, etc.), will find valuable the methods of study of particle size and particle shape and the optical methods of identification.

In so far as the geologist is involved in problems of petroleum production and reserves, or engaged in mapping where the soils are the only clue to the nature of the subjacent formations, or engaged in prospecting for or estimating the worth of alluvial deposits, or involved in a study of the problems of soil erosion and reservoir silting, he will find the methods described herein pertinent.

The science of sedimentary petrology, or sedimentology, has now reached a stage of development which involves a large number of techniques unique to this science and distinct from those employed in the study of the igneous and metamorphic rocks. These techniques are described in a widely scattered literature—in the literature of ceramics, pedology, petroleum technology, hydrology, etc. Growing interest in sedimentology, as evidenced by the increasing number of courses devoted to the subject, the establishment of a journal devoted exclusively to this field, and the increasing use of its methods and principles in the exploitation of petroleum and other mineral resources, has, we believe, justified the attempt to bring together, for the benefit of the geologist and other students of sedimentary materials, methods of analysis applicable to these substances.

The compilation of material from an extensive literature has raised numerous perplexing problems. Sciences vary widely in their terminology and in their use of mathematics. Fundamental principles common to one field are largely unknown to other fields. The authors have accordingly decided to write this book primarily from the point of view of the geologist, with the hope that it will be of value to workers in other fields, at least to the extent of marshaling some of the literature for them. Geologists as a group are not mathematically inclined, but among geologists are many who have a command of mathematics and physics. The problem of writing a volume of interest to both extreme groups has been difficult.

No pretense is made of making the volume complete or exhaustive. It is inevitable that, in a work as broad as the present one, the authors should give most space to those fields and methods with which they are most familiar. Nevertheless we believe the allotment of space to the various techniques and fields reflects fairly well present-day interests and needs. Where methods are well established and generally familiar, as are the optical methods, summaries suffice; where newer and less well known procedures are involved, more detail has been supplied.

It is perhaps inevitable also that there should be some omissions of important material, owing to the wide literature involved. The authors would appreciate advices concerning such omissions. Some selection has had to be made by the authors, but as far as possible references are given to further details elsewhere. No apology is made for using personal material for illustrations and examples; the greater convenience of working with familiar material is its own justification.

The present volume is largely the joint effort of the two authors, but fortunately their fields of specialization adapt themselves to a division of the book into two parts. This division is more apparent than real. Individually the authors assume responsibility for their separate portions; jointly they assume responsibility for the apportionment of space and the thread of continuity which runs through the book.

This manual is written for a person of average training in the methods of science. It is assumed only that the worker has had an elementary training in laboratory technique—such that he can handle a chemical balance intelligently—and that he has a working knowledge of elementary physical and mathematical theory and some knowledge of crystallography.

We believe the book will be found suitable as a textbook for courses in sedimentary petrography. We have therefore attempted to explain both principles and objectives of the various techniques of analysis and to raise in the student's mind a critical attitude toward the purposes and methods of sedimentary analysis.

The authors are indebted to many writers and workers for the final design and content of this book. Gessner's excellent treatise, *Der Schlämmanalyse,* Johannsen's *Manual of Petrographic Methods,* Larsen and Berman's *Microscopic Determination of the Non-opaque Minerals,* and Boswell's *Mineralogy of Sedimentary Rocks* have been of inestimable value, and numerous essays and comprehensive articles have furnished inspiration and information. Credit has been given in many of these cases.

The authors are indebted to numerous individuals for advice and criticism. Dr. Carl Eckart of the Department of Physics of the University of Chicago has helped in the mathematical treatment of the theoretical parts of mechanical analysis; Dr. M. W. Richardson of the

Department of Psychology has critically read the chapters on statistics; and Mr. Paul Reiner has criticized several portions of the text. Many of the illustrations were prepared by Messrs. H. Holloway, A. Lundahl, and W. C. Rasmussen, of the University of Chicago. Among our colleagues in the Department of Geology, Drs. J. H. Bretz, Carey Croneis, and A. Johannsen have made valuable suggestions as to style and content. Among other geologists and sedimentary petrologists who have read portions of the text are Dr. W. W. Rubey of the United States Geological Survey, J. L. Hough and Dr. Gordon Rittenhouse of the United States Soil Conservation Service, and Mr. G. H. Otto of the Soil Conservation Laboratory, Pasadena, California. Dr. Kirtley Mather, Editor of the Century Earth Science Series, has been unfailing in his encouragement during the preparation of the text. Messrs. D. H. Ferrin and F. S. Pease, Jr., of D. Appleton-Century Company have smoothed many difficulties in the arduous task of seeing the book through the press.

W. C. KRUMBEIN.
F. J. PETTIJOHN.

Chicago, Illinois

CONTENTS

PART I

SAMPLING, PREPARATION FOR ANALYSIS, MECHANICAL ANALYSIS, AND STATISTICAL ANALYSIS

PAGE

CHAPTER 1. INTRODUCTION 3
Definitions. Properties of component grains. Attributes of grains in the aggregate. Properties of the aggregate. Preliminary field and laboratory schedules. Field observations during sampling.

CHAPTER 2. THE COLLECTION OF SEDIMENTARY SAMPLES . . 11
Introduction. Purposes of sampling. Outcrop samples, discrete, serial, channel, and compound. Sub-surface samples. Bottom samples. The problem of weathering. The problem of induration. The collection of oriented samples. Size of samples. Containers for samples. Capacities of sample containers. Labeling and numbering of samples. Theory of sampling sediments.

CHAPTER 3. PREPARATION OF SAMPLES FOR ANALYSIS . . . 43
Introduction. Preliminary disaggregation. Sample splitting. Preparation for mechanical, mineralogical, shape, and surface texture analysis. Physical dispersion procedures. Chemical dispersion procedures. Theory of coagulation. General critique of dispersion. Generalized dispersion routine.

CHAPTER 4. THE CONCEPT OF A GRADE SCALE 76
Introduction. Modern grade scales. Problems of unequal class intervals. Functions of grade scales, descriptive and analytic. Choice of a grade scale.

CHAPTER 5. PRINCIPLES OF MECHANICAL ANALYSIS . . . 91
Introduction. Classification of disperse systems. Concept of size in irregular solids. Settling velocities of small particles. Stokes' law and its assump-

CONTENTS

tions. Other laws of settling velocities. Theory of sedimenting systems. Odén's general theory. Principles of modern methods. Principles of older methods. Theory of sieving. Theory of microscopic methods of analysis. Summary.

CHAPTER 6. METHODS OF MECHANICAL ANALYSIS 135
Introduction. Sieving methods. Direct measurement of large particles. Decantation methods. Rising current elutriation. Air elutriation. The sedimentation balance. Continuous sedimentation cylinders. The pipette method. The hydrometer method. Photocell method. Microscopic methods of analysis. Comparisons of methods of mechanical analysis.

CHAPTER 7. GRAPHIC PRESENTATION OF ANALYTICAL DATA . 182
Introduction. General principles of graphs. Choice of dependent and independent variables. Graphs involving two variables. Histograms, cumulative curves, and frequency curves. Graphs with distance or time as independent variable. Scatter diagrams. Graphs involving three or more variables. Isopleth maps and triangle diagrams. Mathematical analysis of graphic data. Linear, power, and exponential functions.

CHAPTER 8. ELEMENTS OF STATISTICAL ANALYSIS 212
Introduction. The concept of a frequency distribution. Histograms and cumulative curves as statistical devices. Introduction to statistical measures of the central tendency, the degree of scatter, and degree of asymmetry. Arithmetic and logarithmic frequency distributions. Quartile and moment measures. The question of frequency.

CHAPTER 9. APPLICATION OF STATISTICAL MEASURES TO SEDIMENTS 228
Introduction. Quartile measures, arithmetic geometric, and logarithmic. Moment measures, arithmetic, geometric, and logarithmic. Special statistical measures. Sorting indices. Choice of statistical devices. Statistical correlation. Chi-square test. Theory of control. The probable error.

CONTENTS

CHAPTER 10. ORIENTATION ANALYSIS OF SEDIMENTARY PARTICLES 268
Introduction. Collection of oriented samples. Laboratory analysis of particle orientation. Presentation of analytical data. Statistical analysis.

PART II

SHAPE ANALYSIS, MINERALOGICAL ANALYSIS, CHEMICAL ANALYSIS, AND MASS PROPERTIES

CHAPTER 11. SHAPE AND ROUNDNESS 277
Introduction. Review of quantitative methods. Choice of method. Procedure of analysis. Method for large fragments. Wadell's method for sand grains.

CHAPTER 12. SURFACE TEXTURES OF SEDIMENTARY FRAGMENTS AND PARTICLES 303
Introduction. Surface textures of large fragments. Surface textures of small fragments.

CHAPTER 13. PREPARATION OF SAMPLE FOR MINERAL ANALYSIS 309
Introduction. Disaggregation. Clarification of grains. Special preparation problems.

CHAPTER 14. SEPARATION METHODS 319
Preliminary concentration of heavy minerals. Separation on basis of specific gravity. Heavy liquids. Standardization of heavy liquids. Separation apparatus. Use of centrifuge. Analytical procedure. Separation on basis of magnetic permeability. Separation on basis of dielectric properties. Separation on basis of electrical conductivity. Separation on basis of visual properties. Separation on basis of shape. Separation on basis of surface tension. Separation on basis of chemical properties. Errors in separation. Systematic schemes of separation.

CHAPTER 15. MOUNTING FOR MICROSCOPIC STUDY 357
Splitting. Mounting. Preparation of thin sections. Film method of study.

CONTENTS

CHAPTER 16. OPTICAL METHODS OF IDENTIFICATION OF MINERALS 366
Introduction. The polarizing microscope. Measurements of small particles. Fundamental optical constants. Observations in ordinary light. Observations in plane polarized light (crossed nicols). Observations in convergent light. Special methods for the study of clays. Preparation. Identification.

CHAPTER 17. DESCRIPTION OF MINERALS OF SEDIMENTARY ROCKS 412
Introduction. Mineral descriptions. Determinative tables. Miscellaneous tables. Record forms.

CHAPTER 18. MINERAL FREQUENCIES AND COMPUTATION . . 465
Pebble counts. Thin-section analysis. Mineral frequencies. Presentation of results. Calculation of mineral frequencies based on analysis of several fractions. Statistical methods. Mineral variations. Statistical correlation.

CHAPTER 19. CHEMICAL METHODS OF STUDY 490
Introduction. Quantitative analysis. Methods. Computations based on quantitative analysis. Microchemical methods. Organic content. Insoluble residues. Staining methods.

CHAPTER 20. MASS PROPERTIES OF SEDIMENTS 498
Introduction. Color of sedimentary materials. Specific gravity of mineral grains and of sedimentary rocks. Porosity. Definitions. Determination of porosity. Methods of porosity measurement. Permeability. Plasticity. Definitions. Methods of measurement. Hygroscopicity. Miscellaneous mass properties.

CHAPTER 21. THE LABORATORY, EQUIPMENT, AND ORGANIZATION OF WORK 522
The laboratory. Apparatus. Reference books. Organization.

AUTHOR INDEX 533
SUBJECT INDEX 539

PART I

SAMPLING, PREPARATION FOR ANALYSIS,
MECHANICAL ANALYSIS, AND
STATISTICAL ANALYSIS

CHAPTER 1

INTRODUCTION

SCOPE OF SUBJECT

THE study of sediments is concerned with (1) the physical conditions of deposition of a sediment, whether glacial, fluvial, marine, etc.; (2) the time of formation or age of the deposit; and (3) the *provenance*, or area of denudation that furnished the material composing the sediment. All of the analytical methods described in this volume have as their common aim the elucidation of these points.

Various names have been applied to the detailed study of sediments, ranging from *sedimentation* through *sedimentary petrology* to *sedimentology*. The latter word has not come into general use, despite its conciseness and clear meaning; it may be said that usage favors the second term. Whatever name may be ultimately chosen, there is no doubt that the subject involves a complete study of sediments from the point of view and with the methods of pure science. Here are included not only geological methods of study, as typified by field work, but also the methods of the chemist, the physicist, and the statistician. In short, the complete study of sediments must make use of any and all devices which lead to an understanding of the nature and origin of the sediment in question.

This broad viewpoint means that the study of sediments may be approached from various angles. From one angle it may be a study of the size attributes of sediments as physical mixtures of particles; from another it may be a study of mineral suites which by depositional conditions have been united into a single deposit; or, the sediment may be considered as a composite of sizes, shapes, and minerals controlled by complex environmental conditions, and the investigation may seek to evaluate the conditions of that environment. All of these points of view are related, and in their ultimate end are directed to the elucidation of geological problems, many in direct connection with historical geology.

Whatever the point of view applied to sedimentary investigations, laboratory studies will become an increasingly important source of data.

Not alone do laboratory analyses supplement and refine field observations, but often they afford data which cannot be gleaned by field methods alone. Criticisms are often leveled against the application of refined methods of analysis to geological problems, either on the ground that they give a specious air of preciseness to fundamentally approximate data or on the ground that geology is completely studied in the field and laboratory studies should be left to chemists and physicists. The first criticism has been more pertinent in the past than it is now, because even the rather poor data afforded by early laboratory studies of sediments have paved the way for improvements in technique and interpretation, as well as for tests to determine the degree of accuracy of the data. The second criticism needs no answer: the world is the geologist's domain, and he is justified in using whatever techniques he requires to solve problems fundamentally geological. True, much remains to be improved in the laboratory study of sediments, but there can be no doubt that interest is growing in the subject, and will continue at an accelerated pace during the next decade at least.

Among soil scientists there is a fairly standardized routine of analysis, but this stage of development has by no means been reached in sedimentary studies. One finds in a single year papers prepared on methods of analysis or presentation of data as remote as the poles, and it is no small problem to determine the relative dependability of the methods or the degree to which the results are comparable. It is perhaps too early to advocate the adoption of standardized routines for sedimentary analysis, inasmuch as it is not clear in all cases whether current data are the most valuable for the ends toward which they are directed. Natural phenomena are exceedingly complex when examined in detail, and analytical procedures must be developed which do not destroy the very data being sought. It is at least fortunate that much current work is done with modern techniques, based on sound theory, but there still are numerous aspects of the subject where a state bordering on chaos prevails.

The scientific study of sediments may be divided into two broad divisions. The first of these is the field and laboratory investigation of sediments, which yields data that lead to their description and classification. The second part of the subject is concerned with the laws of sedimentation and the origin of sedimentary deposits. To the first aspect may be applied the term *sedimentary petrography* or *sedimentography*. The second division is properly designated as *sedimentary petrology* or *sedimentology*.

The distinction between petrography and petrology is, according to

INTRODUCTION 5

Tyrrell,[1] that petrography is the study of rocks as specimens, whereas petrology is the science of rocks, that is, of the more or less definite units of which the earth is built. These general terms may apply equally well to igneous, sedimentary, or metamorphic rocks. Specifically, Milner[2] has defined sedimentary petrology as follows:

> [Sedimentary] petrology connotes something more than mere description of rock-types based on microscopical analysis, and in its wider sense embraces comprehensive investigations of their nature, origin, mode of deposition, inherent structures, mineralogical composition, mechanical constitution, textural analysis, various chemical and physical properties, in short, all data leading to an understanding of the natural history of the formations under review.

In practice, one seldom distinguishes between sedimentary petrography and sedimentary petrology. Most studies of sediments, perhaps, are directed toward the petrological aspects of the problem: the clarification of details of origin, transportation, deposition, or diagenesis. Actually, of course, the petrographic aspects precede the petrological, because it is first necessary to assemble facts about the sediments, both from the field and the laboratory.

It is only with the methods of petrographic analysis that this volume is concerned. The purpose of the book is to present theories and methods of examining sediments, from the field sampling to the final graphic and statistical analysis. Petrological aspects are only touched upon as they apply to certain details of analytical methods and to indicate the underlying purposes of the laboratory investigations.

The point of view which this book presents is based on the premise that every sediment is a response to a definite set of environmental conditions. Whatever the conditions may be, there are characteristics of the sediment which may be measured in the laboratory and which reflect the environmental factors that produced them. A change in the environment (pressure, temperature, chemical associations) results in a corresponding adjustment of the rock to its new conditions. Owing to incomplete adjustments, however, the rock materials may have some characteristics inherited from previous states and some due to conditions existing at the moment. Refinements of technique and interpretation, however, make possible the unraveling of even such complexities. This point of view requires a careful consideration of all techniques used, a critical evaluation of data, and a reliance on the methods of pure science.

[1] G. W. Tyrrell, *The Principles of Petrology* (London, 1926), p. 1.
[2] H. B. Milner, Report of research on sedimentary rocks by British petrologists for the year 1927: *Rep. Com. Sed.*, Nat. Research Council, 1928, p. 9.

Mineralogical, as well as size and shape studies, throw light on the physical conditions of deposition, but it is largely on minerals alone that we have to rely for the determination of age (petrographic correlation) and for knowledge of the provenance of the deposit. The problem of provenance or of source of the sediments involves a knowledge of the composition of the parent rock types, a knowledge of mineral stability in respect both to climate and to mechanical wear; and an understanding of the relations between mineral frequencies and the transportation of the sediment.

Twenhofel [1] has defined sedimentation as follows:

Sedimentation includes that portion of the metamorphic cycle from the destruction of the parent rock, no matter what its origin or constitution, to the consolidation of the products derived from that destruction (with any additions from other sources) into another rock.

The term *sedimentation* thus connotes a process and is to be distinguished from the products of that process, the sedimentary rocks. Twenhofel [2] has also defined a sediment as

...a deposit of solid material (or material in transportation which may be deposited) made from any medium on the earth's surface, or in its outer crust under conditions of temperature approximating those normal to the surface.

PROPERTIES OF SEDIMENTS

Every sedimentary deposit has certain fundamental characteristics or properties, some of which are associated with the individual particles and others with the aggregate of all the particles. In some cases there is an overlap, but the following classification indicates the principal characteristics.

Properties of component grains. The fundamental properties of the component grains of a sediment are (1) sizes, (2) shapes, (3) surface textures, and (4) mineralogical composition. The last characteristic determines such attributes as density, hardness, color, and the like, of each grain. Each of these four fundamental properties may be examined in the laboratory.

The fundamental properties of the particles are important because they reflect either directly or indirectly many of the vicissitudes through which the sediment has passed. Size is related to the medium of transportation

---

[1] W. H. Twenhofel, *Treatise on Sedimentation*, 2nd ed. (Baltimore, 1932), p. xxvii.
[2] *Loc. cit.*

INTRODUCTION

and its velocity; shape is related in part to the medium of transportation and to the distance and rigor of transport; surface texture may reflect subsequent changes due to solution, or it may furnish clues to the method of transportation. Finally, the mineralogical composition indicates possible source rocks, as well as any post-depositional changes that may have occurred.

Attributes of component grains in the aggregate. Interest in the component grains of a sediment often involves the frequency distributions of grain properties in the aggregate; for example, size is expressed in terms of a size frequency distribution (mechanical analysis) rather than by cataloging the individual size of each particle. In similar manner, shape, mineral composition, and other properties may be considered statistically as distributions of grain properties. Each of these distributions may then be studied in terms of their own characteristics, such as average size, average density, average degree of sorting or sizing, and the like.

Another important attribute of the component grains in terms of their aggregate properties is the orientation of the particles in space (the "fabric" of the rock). The orientation of the particles, considered statistically, may indicate among other things whether deposition was subaqueous or subaërial.

Properties of the aggregate. In addition to the attributes of the individual particles, there are various aggregate properties of the sediment which are important. These include (1) the cementation of the particles in the specimen, (2) structures, such as bedding, concretions, and the like, and (3) the color of the sediment. These properties also furnish information about the history of the sediment. The color of the sediment in the aggregate, including the nature of the cement, may help determine conditions of deposition, or post-depositional changes. Some of these aggregate properties may be controlled in large part by properties of the component grains. The orientation of the particles helps determine such structures as bedding and in addition may be a factor in such aggregate properties as porosity and permeability.

To a large extent the aggregate properties, as they are defined above, may be studied in the field, whereas the properties of the component grains and their distribution in the sediment may best be considered in the laboratory. A thorough examination of sediments therefore involves a combination of field and laboratory work, and in the modern development of the science neither is complete without the other. Laboratory methods must be quantitative, inasmuch as quantitative data are necessary to the development of complete theories of sediment transportation and deposition. In the future development of the science there can be little doubt that this quantitative and theoretical aspect of the science will be increasingly emphasized.

PRELIMINARY FIELD AND LABORATORY SCHEDULE

It is appropriate that a schedule outlining the examination of sediments be given here as a preliminary outline by which the scope of the present book may be indicated. The schedule is divided into two parts: the first lists the features of the rock that may be observed in the field by ordinary geological methods, and the second includes those characteristics which are best determined quantitatively in the laboratory. This volume is directly concerned with the details of the second section of the schedule.

The following organization of data on sedimentary rocks is adapted, with some changes, from the excellent report on the field description of sedimentary rocks by Goldman and Hewett.[1]

FIELD SCHEDULE

External form of the rock unit
  Dimensions, persistence, regularity

Color
  Wet or dry, on basis of accepted color scheme [2]

Bedding
  Sharp or transitional
  Plane, undulatory, or ripple-marked
  Thickness
    Constant or variable
    Rhythmic or random
  Attitude and direction of bedding surfaces
    Horizontal, inclined, or curved
    Parallel, intersecting, or tangential to other beds
    Relation of particle properties to attitude and direction [3]
  Markings of bedding surfaces
    Mudcracks, rain prints, footprints, etc.
  Disturbances of bedding
    Folding or crumpling
    Intraformational conglomerates

[1] M. I. Goldman and D. F. Hewett, Schedule for field description of sedimentary rocks: National Research Council, Committee on Sedimentation, Washington, D. C.

[2] M. I. Goldman and H. E. Merwin, Color chart and explanation of the color chart for the description of sedimentary rocks, prepared under the auspices of the Division of Geology and Geography, of the National Research Council, Washington, D. C., 1928.

[3] Such properties as porosity and permeability may be determined in the laboratory from oriented field samples.

INTRODUCTION

Concretions
  Kinds, size
  Condition and distribution
  Orientation with respect to bedding
  Form, size, composition
  Internal structure
  Boundary against country rock
    Sharp or transitional
  Relation to bedding
  Distribution
    Random or regular

Organic constituents [1]
  Kinds, size
  Condition
    Whole or broken
  Distribution
  Orientation with respect to bedding

Laboratory Schedule [2]

Preparation of sample for analysis
  Sample splitting
  Disaggregation and dispersion

Particle size analysis

Shape analysis
  Roundness
  Sphericity

Surface texture analysis

Mineralogical analysis
  Separation of heavy minerals
  Microscopic examination

Orientation of particles in sample

Mass properties of sediment
  Porosity and permeability
  Specific gravity

Chemical analysis

Graphic presentation of data

Statistical analysis of data

It may be mentioned that although the above schedules present little overlap, in actual practice some of the quantitative data are obtained

[1] This refers only to the megascopic remains. Microfossils demand specialized laboratory techniques not included in this volume.
[2] This outline presupposes that samples have been collected in the field. For a field outline of sampling routine, see below; the subject of sampling forms Chapter 2 of this volume.

directly in the field and part of the field data are secured in the laboratory. For example, the study of pebble orientation may be conducted at the exposure, and if the rock being investigated is consolidated, some field observations, such as bedding and other structures, may be observed from the sample. On the other hand, if the material is incoherent, such features as bedding, grain orientation, and the like are not preserved during sampling, and steps must be taken to complete such observations in the field. It may be said as a general rule that too much data cannot be collected. This is true especially in a field such as sedimentary petrology, where research has not yet advanced to the stage where it may be predicted whether a given set of observational data are pertinent to the study or not.

Schedule of field observations during sampling. In addition to the general observations to be made on the formation as a whole, as outlined above, there are several specific observations to be made in the field at the time a sample is collected. These specific data include:

1. Location of the sampling locality, either as a point on a map or with reference to some easily located landmark.
2. Nature of the sampling point, as an outcrop, roadcut, or ditch.
3. Nature of the material sampled, including type of rock, portion of bed sampled, and so on.
4. Nature of the sample, as from a single point, a composite sample from several parts of the bed, as a channel through the bed, etc.
5. Relation of sample to surrounding rock, as, for example, from just beneath a stained zone, whether cut by joints, and the like.
6. Topography of sampling site, as river bottom, terrace, top of hill, etc.
7. Depth of sample beneath immediate surface at point of sampling.
8. Zone of weathering from which sample is taken, if this can be determined.
9. A field evaluation of the total condition of the sample for the purpose desired, as excellent, good, fair, poor. This is desirable when many samples are collected and the laboratory work may involve using scattered samples to outline the scope of the study.

It cannot be emphasized too strongly that the detailed investigation of sediments should not be a hurried process. The investigator should spend adequate time in the field, examining the general set-up of the problem, locating sampling sites, measuring sections, and in general accumulating sufficient data so that work need not be delayed during the ensuing laboratory season owing to failure to observe adequately in the field. As a general rule it is better to acquire too many samples and field observations than not enough of either.

CHAPTER 2

THE COLLECTION OF SEDIMENTARY SAMPLES

INTRODUCTION

THE physical impossibility of analyzing an entire sedimentary formation, or even an appreciable part of one, renders it necessary to work with samples. A sample is assumed to be a representative part of the formation at the point of sampling, or sometimes of the entire formation. How nearly it is representative determines in large measure the validity of the final conclusions, assuming, of course, that the methods of analysis correctly describe the sample.

Interest in a sediment may arise in a variety of ways. It may be a question of the economic exploitation of a limestone or a fire-clay; it may be merely a desire to supplement general geological field work with some quantitative data. On the other hand, the study may involve consideration of the conditions of sedimentation, the agents that formed the deposit, the possible source rocks, and the like. One may thus argue that the process of sampling may be either casual or precise, depending upon the ends in view. This is an attitude with which the authors cannot wholly concur. It seems reasonable that if a sample is worth collecting, it is worth collecting well.

PURPOSES OF SAMPLING

Samples for display. In unconsolidated material a display sample may consist merely of a small vial of the material—sand, silt, or clay—or it may consist of a selection of pebbles in a tray. The collection of such samples affords no particular difficulties unless structures, such as bedding and grain orientation are to be preserved.

If the structure of unconsolidated sediments is to be preserved, and if the material is sufficiently fine-grained to be cohesive, an undisturbed sample may be collected by a routine procedure. Antevs [1] describes the process for collecting unconsolidated varved clay samples as follows:

[1] E. Antevs, Retreat of the last ice-sheet in eastern Canada: *Canadian Geol. Survey,* Memoir 146, p. 12, 1925.

"The samples are taken in tight troughs of zinc plate, conveniently 19⅝ inches long, 2 inches wide, and ¾ inch high. The face of the clay bank is carefully smoothed and the trough is cautiously pressed in, a knife being used to cut away the clay just outside the edges, until the trough is entirely filled with clay. The troughs are then cut out from the bank, and the projecting clay is removed."

If the rock is indurated, the display sample may consist of a chip or a trimmed hand specimen. The hand specimen should be about 3 x 4 in. in size and from 1 to 1½ in. thick. The smallest dimension is usually chosen at right angles to the bedding. The corners of the specimen should be rectangular and not rounded,[1] so that they conform to the standards set for hand specimens of igneous rocks.

Samples for commercial analysis. Samples of sediments collected for commercial analysis present a number of problems peculiar to the purposes for which they are used. In general, however, the methods of sampling are similar to those used for the detailed laboratory study of sediments for scientific purposes.

Commercial analyses may be made for such diverse purposes as the determination of the CaO or MgO content of limestone; the fuel value or the determination of special constituents of coal; the value of gravel for use as road materials; or the value of silica sand for glass-making. Regardless of the purpose of the analysis or the state of induration of the material, the prime requisite is that the sample must be representative of the formation. A specialized aspect of commercial sampling is the prospecting of economically useful deposits. This topic does not properly come within the scope of the volume, and interested readers are referred to standard texts on the subject.[2]

Samples for detailed laboratory investigations. A critical choice of samples, necessary in any detailed study of sediments, should take into consideration as many elements of the problem as may be evaluated, so that the final results are not weakened by poor samples, collected without regard to the purposes of the study.

Sediments may vary in terms of the coarseness of their particles, in the degree of sorting or homogeneity, in their manner of bedding or arrangement of particles, in their degree of induration, and in their degree of alteration. In any given formation, one must also consider the vertical and lateral variations in size of the formation, the presence or absence of bedding, changes in the thickness of the formation or its individual beds, and changes in the shape, size, and arrangement of its particles. Further, some sediments are exposed to view in extensive outcrops, and others are hidden within drilled wells or are covered by bodies of water. Each of these cases presents its own problems, some of which are far from being solved.

[1] A. Johannsen, *Manual of Petrographic Methods,* 2nd ed. (New York, 1918), p. 607.
[2] C. Raeburn and H. B. Milner, *Alluvial Prospecting* (London, 1927).

COLLECTION OF SAMPLES

Unfortunately, there is at present no general mathematical theory of sampling sediments which enables one in every case to determine the technique of sampling *a priori;* the science of sampling is still in the stage where "rule of thumb" procedures predominate. These practical rules are based on experience and thus are satisfactory in an empirical sense; happily, they are supported by favorable results, and to some extent they may be checked by statistical theory. In a later section of the chapter some of the elementary aspects of sampling theory will be discussed.

OUTCROP SAMPLES

Sedimentary formations exposed in outcrops are the most convenient to sample because the sampling site may be examined in detail and some judgment may be used in choosing the particular point of sampling. Given such an outcrop, the problem involves the number of samples to be taken, the size of the samples, and the desirability of preserving structures or particle orientations.

Spot samples.[1] An isolated sample taken at a particular point on the outcrop may be termed a spot sample, or a discrete sample. Such samples are collected separately and kept separately, being thus distinguished from composite samples.

The decision to collect a spot sample may be based on the apparent homogeneity of the deposit as exposed to view. If the outcrop represents a bank of unbedded sand or silt, or even glacial till, with no changes in composition detectable by eye, a single sample may be taken from any convenient point along the outcrop. Unless the object of the study is the investigation of weathering, the principal precaution to be followed is that no weathered or altered phases of the formation be included. This necessitates taking the sample at some distance below the soil horizon, and beneath the surface of the outcrop face.

An area on the outcrop face is first cleaned or scraped, and a sample taken by scooping out a limited amount from a square or circular zone. In general it may be desirable to have the depth of penetration about as great as the width of the face sampled, so that a roughly cubical or cylindrical volume is obtained. The sediment may be removed with a scoop, the point or chisel of a hammer, or a small pick. A bag or other receptacle should be at hand so that none of the sample is spilled or lost.

[1] The term *grab sample* is often used for individual samples collected at a given point. The term may imply a degree of carelessness in the collection of the sample, and the substitute term *spot sample* is used here.

When the exposure is horizontal, such as the surface of a dune or a beach, the sample may be collected in various ways. A simple method is to dig a shallow hole with vertical walls, and to take the sample from one of the walls so exposed. A more convenient method is to have a short section of downspout pipe with fluted sides,[1] which may be forced down into the sand for a distance of six inches or so. The sand around the tube is then dug away, and by inserting the hand beneath the tube, the entire sample may be preserved, even in dry sand.

A spot sample is strictly valid only for the point being sampled. A single sample, used to generalize about the material exposed in the outcrop, should be relied upon only when the sediment is quite homogeneous, or when a limited part of the formation is to be studied. If variations occur vertically or laterally, or if an extensive area is to be studied, it is better to rely on a series of spot samples.

Serial samples. Spot samples which are part of a related set of samples may be called serial samples. They are collected in accordance with some predetermined plan, involving an arbitrary but usually equal interval of spacing. Each spot sample is kept in a separate container and is usually handled as a unit during the study.

Serial samples may be arranged along a line of traverse across a formation, or they may represent a set of samples collected at intervals along a river or beach. Likewise, the serial set may extend vertically across the thickness of a formation. When the series is arranged along a line in this manner, the series is linear, and it is not necessary that the line be straight.

In contrast to linear series of samples, either horizontal or vertical, is a grid series of samples collected over an area, or over the face of a vertical outcrop. The grid may represent a square pattern of lines superimposed over the area, spaced according to the detail with which the work is to be carried on. Samples are collected at the points of intersection of the crossing lines. In some cases section line roads or even township lines may be used for grid patterns. From this extreme the grid spacing may range down to a foot or so for very detailed studies. As a first approach one chooses a grid interval having sampling points spaced equally over the area, at such distances as the detailed nature of the study suggests. As samples are collected at the grid intersections, attention is paid to any noticeable variations that occur from one sample to the next. If occasional changes are noted, it may be well to collect an intermediate sample at half-grid interval, to cover the transition; on the other hand, if prominent changes occur between each succeeding sample, it may be desirable to halve the interval over the entire area.

The primary function of the grid method of sampling is to assure oneself of random samples, distributed more or less evenly over the area

[1] This device was developed by G. H. Otto; its use is described in W. C. Krumbein, The probable error of sampling sediments for mechanical analysis: *Am. Jour. Sci.*, vol. 27, pp. 204-214, 1934.

considered. If a grid pattern is not used, and samples are taken at scattered intervals, personal bias may influence one in the location of the samples.

In most cases a simple rectangular grid, as illustrated in Figure 1, is probably to be preferred, although other patterns will suggest themselves. In the study of an alluvial fan, for example, one may choose a series of concentric arcs of circles to space the samples equally from the apex of the fan; the cross lines may be radial from the apex or may cross

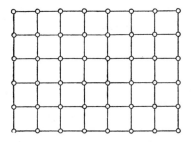

FIG. 1.—Simple rectangular grid for sample control. The distance between sampling points may be chosen as a function of the size of the area and the number of samples to be collected.

FIG. 2.—Radial logarithmic sampling grid, designed to furnish more detailed data near source of sediment. Grids of this type may apply to studies of alluvial fans.

the arcs at equal intervals. In some sampling problems it may be logical to space the grid lines on logarithmic intervals, as shown in Figure 2. For example, if it is suspected that some property of a sediment varies exponentially from its source outward, a closer spacing near the source will afford more critical data in the steep part of the exponential curve. Such exponential functions (see Chapter 7) may be expected in some cases of average size or thickness of deposits.

The collection of serial samples, whether linear or grid, assumes that the formation will outcrop at the grid points. If this is not so, one must choose between taking a sample from the nearest outcrop or drilling at the exact sampling point.

Some of the situations in which serial samples, either linear or grid, are indicated may be mentioned. If a beach is to be studied, the samples may be arranged as a linear series near the strand line. If, however, only a limited stretch of the beach is being investigated, a grid may be used involving one set of samples along the strand, another in the shallow off-shore water, and a third higher up on the beach. Rivers afford another instance. A linear set may be taken to study the changes in the sediment downstream, or a grid may be laid over a terrace or river bar to study detailed variations. Vertical series of samples are indicated where there are several formations in the study but each formation is itself a more or less homogeneous unit. In general, serial samples are indicated whenever the variations from point to point along or over a deposit are the subject of study. Soil surveys have long used the principle of grid sampling, in which the data from numerous closely spaced samples are used to prepare maps of soil types.[1]

When collecting serial samples it may be difficult to decide whether the interval between samples should be relatively long or short. One solution is to collect the samples at the shorter intervals, but in the laboratory to analyze only alternate samples. If these suffice to bring out the variations adequately, the intermediate samples may be discarded. Where abrupt changes occur, the intermediate samples may be included to cover the transitions.

The same principles of laying out grids apply to samples collected from bodies of water. The grid may be made to conform with the configuration of a bay, or it may be based simply on a series of sailing courses, more or less parallel, and extending at right angles from the shore. By an appropriate spacing of the sampling intervals in relation to the sailing courses, any type of grid may be followed.

As far as the authors are aware, the first clear statement of the value of grid patterns in the control of sampling was made by Pratje in 1932.[2] He pointed out that by laying a closely spaced network over the region to be studied a sufficient number of related samples could be obtained to evaluate the environment. He designated his approach as the regional-statistical method of studying sediments.

Channel samples. A channel sample may be defined as an elongated sample taken from a relatively narrow zone of an outcrop. The channels involve a continuous strip of the material from top to bottom of the channel zone. Channel samples are important whenever the average characteristics of the formation are to be determined. Consequently such samples are widely used in commercial sampling, as for road gravel, fire-clays, molding sand, and the like. For scientific studies channel samples are to be avoided if they extend through zones of weathering or alteration or in any other manner introduce complexities into the sample collected. There are cases, however, in which channel samples

---

[1] Samples for soil surveys are usually collected from the center point of the rectangular grid pattern, instead of from the points of intersection. Either approach is equally logical.

[2] O. Pratje, Die marinen Sedimente als Abildung ihrer Umwelt und ihre Auswertung durch regional-statistische Methode: *Fortschritte der Geol. u. Paläon.*, vol. II, pp. 220-245, 1932.

COLLECTION OF SAMPLES 17

are indicated for detailed work on sediments, and these will be considered below.

The channel sampling method may be illustrated by a common procedure used in obtaining samples of road gravels. The more or less vertical face of the gravel pit is cleared of surface material for a width of about a foot or eighteen inches, extending from top to bottom of the exposure. The fallen material is cleared from the base of the wall, and a shallow indentation is dug at the base of the cleared strip, for the insertion of the edge of a tarpaulin beneath the channel zone. After these preliminaries, a small sugar scoop is used to scrape the material from the channel, allowing it to fall on the tarpaulin below. The depth of the channel is made equal to the diameter of the largest pebble in the sampling zone. This insures a representative sample, on the whole.

After the material has been removed from the face of the outcrop, there will be a channel extending from top to bottom, about a foot wide and two or three inches deep. The material from this strip lies on the tarpaulin below. The tarpaulin with its pile of gravel is pulled away from the face, and the gravel is thoroughly mixed with a shovel, and spread into a roughly conical heap. This heap may then be divided into four quarters with the shovel, and two alternate quarters discarded, if the size of the sample is too large.

Channel samples for general testing purposes are usually collected normal to the bedding of the deposit. The purpose is to obtain a wide range of the material to be tested, so that both its average composition and its extremes of size may be known. For the detailed study of sediments, however, the inclusion of separate beds in a single sample may be a distinct disadvantage. Channel samples tend to mask details, because they furnish no data on the range of sizes in individual beds; they yield composite data made up of several sets of individual data, and furnish no information whatever on the degree of sorting, or mineralogical composition of individual beds. Thus where the sediment is composed of numerous thin members having a wide range of characteristics (a glacial outwash deposit is an example), it is doubtful whether a channel sample taken through the deposit will yield any detailed knowledge about the conditions of deposition.

On the other hand, where the material is unbedded and apparently homogeneous, as in loess, it may be advantageous to choose channel samples instead of spot samples. In the absence of any evidence of heterogeneity, it may be argued that a channel sample is more representative than a sample from a point in the outcrop. Samples intermediate between a spot sample and a channel sample may also be considered. That is, instead of taking a single channel from top to bottom of the deposit, two or three separate shorter channels may be used, each collected separately. In this manner hidden variations are disclosed in the analyses, which would remain masked in a single channel.

The decision between spot samples, short channels, and long channels must depend partly on the judgment of the collector. It is not possible to make rigid rules, inasmuch as the purposes of the study influence the decision. The important point, perhaps, is that the collector should be aware of the choice at his disposal in a given case, so that his judgment may be sound.

Compound samples.[1] A compound sample is a mixture of a number of spot samples combined to form an aggregate single sample. For example, a number of small pieces of limestone are collected at various points within a quarry and combined into a single composite. This sample is analyzed for, say, its MgO content, and the value obtained is taken as an average value for the quarry. Compound samples find wide application in commercial sampling, because, like the channel sample, they afford average values.

An advantage of compound samples over channel samples for furnishing average data arises in some cases from the fact that the sampling localities may be spread widely over the formation studied, instead of being confined to a single vertical cut. The same effect may be obtained by taking a number of channel samples and mixing them all into one composite. Again the nature of the material and the purposes of the study will control the choice of method to be followed.

If the detailed variation of sediment characteristics is being studied, the same disadvantages which apply to channel samples may be extended to compound samples, which afford average values and merge variations into a single value. There is at least one important type of compound sample, however, which is of importance in detailed studies. It was mentioned earlier that any single spot sample is rigorously valid only for the exact point of sampling. There is a possibility that this single sample may actually represent a deviation or departure from the composition round about rather than represent the general characteristics of the sediment in the sampling vicinity. Such possible deviation of a sample from the average is referred to as the *probable error*[2] of the sample.

---

[1] Both compound samples and channel samples belong to the general group of *composite samples,* in which more than a single set of characteristics may be combined.

[2] The probable error is a statistical measure of the chance deviation of a given sample from the average value of the material being sampled. It is that error which will not be exceeded by half the samples collected, and hence it serves as a measure of the reliability of the sample.

COLLECTION OF SAMPLES

The subject of probable errors in sampling is discussed later (page 41), but an introduction to the topic is necessary here to clear the discussion. The general concept of the probable error may be illustrated with a specific example. It is desired to sample an exposure of unbedded sand, about 10 ft. thick and 30 ft. long. A single sample is to be collected from the bank, to represent that locality in a set of serial samples. One decides to take the sample from the center point of the exposure. How would his results have differed if he had taken the sample from one of the sides, at the top, or at the bottom of the cut? For simplicity it will be assumed that the sand appears to be homogeneous to the eye and that it is equally convenient to take the sample anywhere in the exposure.

The only answer that can be given to the question at this point is a general one: it depends on the actual variation of the sand from point to point, regardless of how the eye may appraise it. Studies have been made of the probable error in a few cases, and it is found that even the most apparently homogeneous sands actually do vary slightly from point to point. In beach sands, for example, the variation in the average grain size ranges from 0.8 to 4.1 per cent over distances of a few hundred feet.[1] An important point that emerges from a study of the probable error is that, regardless of its magnitude, the error may be reduced to any desired value by mixing a number of discrete samples into a single compound sample. Actually a mixture of four discrete spot samples into one compound sample will reduce the error by about half. Ten samples, on the other hand, will reduce it to about 0.3 of itself. Thus the maximum reduction for practical purposes is obtained by mixing four discrete samples into one composite. In the example given above, then, any concern about the effect of the precise sampling point may be halved by taking four samples scattered over the exposure and mixing them into a single compound sample.

From the point of view of the probable error of sampling, compound samples may deserve consideration in any studies where apparently homogeneous sediments are involved. Serial samples along a beach, for example, may be composed of a series of compound samples, each made by combining four spot samples collected within a short distance of each other at each of the main sampling points. In this manner the accidental sampling of an unsuspected deviation in the sand may to a large extent be avoided. The precaution to be followed in this type of sampling is to make certain that the distance between the individual samples which make up each composite is small compared to the distance between the main serial points. The need for this precaution is that each composite of four samples must represent essentially a point on the scale at which the successive serial samples are spaced, so that complexities are not introduced into the study by averaging actual sediment variations rather than random deviations within small areas.

[1] W. C. Krumbein, *loc. cit.*, 1934.

Single vs. composite samples. From the preceding sections it may be noted that there are no fixed rules regarding the relative merits of the several types of samples. In every case qualifications must be included, and these qualifications are partly a reflection of the lack of quantitative data on the problem. There is a pressing need for more research on the problems of sampling, so that quantitative data may be available for the development of a general theory of sampling sediments. One function of composite samples (either compound or channel samples) is to reduce the non-systematic variations that may be present at any given sampling point, so that when a series of samples is considered as a whole, the error due to sampling may be kept smaller than the actual range of sediment variation from point to point along the series. Another function of composite samples is to reduce the extremes of sediment variation at any given locality by combining the properties of all the material at that point. The functions of discrete or spot samples include the preservation of individual differences among adjacent samples. Chance may in some cases introduce an appreciable error in these samples, but at least the individuality is preserved and important data are not obscured by a general average composition. It is obvious that if sediment variations are to be studied within a single outcrop, discrete samples will bring the variations to light, whereas composite samples will tend to suppress them. The decision to use compound samples as against discrete samples thus depends in part on the scale of the investigation.

An attempt to establish sound principles for sampling sediments was made very recently by Otto,[1] who classified sampling techniques into four groups in terms of the purposes to be fulfilled. These groups are samples for engineering uses, for descriptive purposes, for environmental studies, and for correlation studies. His discussion of environmental sampling is of especial interest. For this purpose he developed the concept of a *sedimentation unit,* defined as that thickness of sediment, at a given sampling point, which was deposited under essentially constant physical conditions. Chance deviations about average values may be present, but these deviations should themselves form a unimodal distribution. Otto's classification and analysis offer a basis for a generalized theory of sediment sampling, which may be applied in the field to a variety of problems.

---

[1] G. H. Otto, The sedimentation unit and its use in field sampling: *Jour. Geology,* vol. 41, pp. 569-582, 1938. Through Mr. Otto's kindness the authors were privileged to read the manuscript.

THE PROBLEM OF WEATHERING

Up to the present the assumption has been tacitly made that only unweathered sediments were involved in the sampling process. In general, sediments collected for studies of the conditions of deposition or the nature of source rocks should be unweathered. For other studies, involving the alteration history of the sediments, it may be necessary to collect samples from the weathered zones. Sample collectors should be able to recognize weathered sediments in the field and should understand some of the changes which are introduced into the sediment by weathering changes. Leighton and MacClintock,[1] and more recently Grim, Bray, and Leighton,[2] have shown that weathering involves the development of four horizons, each of which is characterized by certain features. The studies were conducted on glacial till and loess, but the principles found are of general application.

The first change that occurs is oxidation, which affects mainly the iron-bearing minerals. The net result is a change in color inclining toward brown. Following the oxidation comes a leaching of the more soluble minerals, such as the carbonates, notably calcite. The third stage is the decomposition of the silicates, during which feldspars and similar minerals are decomposed. Finally, near the surface, the soil zone proper is evolved, with only the more resistant minerals remaining, notably quartz. The chemical changes are accompanied by changes in the size distribution and other physical attributes of the sediment. Thus a calcareous sediment, which includes primary grains of calcite, has a different size distribution after it has been subjected to leaching. Similarly the breakdown of the feldspars into clays and colloids involves significant changes in the physical properties. The drainage conditions at the site of weathering also influence the process of decomposition, so that different end-products result from well and poorly drained situations.

The authors of the present volume have found that in some sediments, such as sand, loess, and glacial till, there are no significant differences in the general properties of unweathered samples and oxidized samples as far as routine size analysis is concerned. Usually changes become noticeable in the leached zone and are generally striking in the silicate-decomposition zone. Quartz sand, with relatively few and resistant heavy minerals, appears to remain essentially unaffected by weathering conditions. For most present-day studies it seems possible to use material from the oxidized zone if unweathered samples are not available. This does not imply that oxidation is negligible, but rather that present-day ex-

[1] M. M. Leighton and P. MacClintock, Weathered zones of the drift sheets of Illinois: *Jour. Geology*, vol. 38, pp. 28-53, 1930.
[2] R. E. Grim, R. H. Bray and M. M. Leighton, Weathering of loess in Illinois: *Geol. Soc. America, Proceedings*, 1936, p. 76.

perimental errors appear to be at least of the order of magnitude of the oxidation changes.

When studies are undertaken specifically on the effects of weathering, the sampling procedure should involve first the identification of the several weathering zones, if they are fully developed, and a collection of samples from each zone as well as from the transitions from one zone to the next. The authors have relied on vertical series of spot samples rather closely spaced, rather than a channel sample through each zone. The use of discrete samples here offers an opportunity for a more detailed picture of the changes from point to point in the weathering profile.

The Problem of Induration

Just as weathering introduces complexities into the sedimentary picture, induration complicates the study by virtue of the physical and chemical changes involved. If the induration is due simply to the cementation of non-calcareous grains by calcite, no special problems are introduced, because presumably the original material may be recovered by leaching in acid. Where secondary material has been introduced intimately into the rock, or where changes have taken place in the original material itself, the problem of original constitution may be quite complex. For the most part, however, the problem belongs to the laboratory, where the successful breaking-down of the rock may be a difficult problem.

The sampling process in the case of indurated rock differs from that with unconsolidated rocks mainly in the greater difficulty of obtaining representative samples. Channel samples, for instance, may have to be literally chiseled from the outcrop. Grout [1] discussed the sampling of igneous rocks for chemical analysis, and the principles he developed appear to apply also to indurated sediments.

The Collection of Oriented Samples

For certain types of investigations it is important that the exact orientation of rock specimens or sedimentary particles be known. If the rock is a sandstone or other consolidated sediment, the dip and strike may be painted on the rock face with a quick-drying enamel before the specimen is broken off. With these lines of reference it is possible to prepare sections at any given orientation for studies of the rock fabric.

---

[1] F. Grout, Rock sampling for chemical analysis: *Am. Jour. Sci.*, vol. 24, pp. 394-404, 1932.

COLLECTION OF SAMPLES 23

If the sediment is loose sand, it may be impregnated with paraffin in some instances, or with a dilute bakelite varnish.

Among coarse sediments like gravel, the individual pebbles may be large enough to have their orientation marked directly on them. Wadell [1] developed a technique in which horizontal and vertical lines were drawn on the pebbles with red and black enamel respectively, so that in the laboratory the exact orientation of each pebble in space may be reproduced. The detailed procedure used by Wadell is described in Chapter 10 of this volume.

SUB-SURFACE SAMPLES

The sampling procedures outlined in the preceding sections assumed complete exposure of the deposits or, at worst, sediments concealed beneath a thin veneer of surface materials, so that shallow pits expose the unaltered sediment. Beyond a depth of several feet, reliance must be placed on mechanical devices for obtaining the samples. Moreover, different methods apply with depth, because of the mechanical difficulty of penetrating very far beneath the surface with hand-operated tools.

Hand auger samples. Hand augers may be used for obtaining samples down to a depth which seldom exceeds 20 or 30 ft.; for most practical purposes the labor involved excludes this device for depths much greater than 10 ft.

Hand augers were developed in connection with soil sampling, where the sampling depth is often limited to about 30 in. For such shallow depths augers are excellent, and many types have been devised for general and special purposes. A simple hand auger of general utility for sedimentary purposes [2] may be constructed from an ordinary steel bit about 2 in. in diameter, of the type used for drilling wood. The screw at the end of the bit and the small flanges on the first whorl are filed off. The bit is welded to a hollow steel tube $\frac{1}{2}$ in. in diameter and about 3 ft. long. Additional lengths of tubing with threaded connections are also prepared, so that the assembled auger is about 12 or 14 ft. long. The handle is made from an 18-in. length of the same tubing, with a threaded connection at the middle, so that it acts as a crosspiece.

---

[1] H. Wadell, Volume, shape, and shape position of rock fragments in openwork gravel: *Geografiska Annaler,* 1936, pp. 74-92.
[2] This device was developed by the Illinois State Geological Survey. It has been extensively used for a variety of purposes, such as test holes, collecting samples, and the like.

In using the auger, the bit is turned down to about its own depth in the soil and withdrawn. The coiled sample is unwound and laid on a square of canvas provided for the purpose. The process is repeated, in each case turning the bit only so far that the labor of withdrawing it is not excessive. The coiled samples of sediment are placed end to end on the canvas, in the order of their removal.

Certain precautions should be followed in using any auger, to avoid too great a contamination of the sample. The surface material about the drill hole should be scraped away so that loose fragments do not fall into the hole. Also, in dropping the auger into the hole, some material is invariably scraped from the walls. This is tamped down, and when the next sample is withdrawn, it will be found at the top of the bit. By examining the unwound coil it is usually possible to determine how much should be discarded.

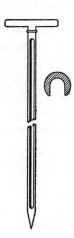

FIG. 3.—Auger for small samples of soil or clay, after Mitscherlich.

Augers of this general type are most effective with silt or clay; also such deposits as loess, lake clay, and glacial till may be quite conveniently sampled. Sand generally does not form a coil around the bit, but fine sand, if moist, can often be sampled. Water may be poured into the hole to facilitate the sampling. Very wet material like quicksand, on the other hand, merely flows from the bit.

Sampling devices which permit the collection of undisturbed samples at comparatively shallow depths have also been described. A recent device, suitable for soil samples, was developed by Heyward.[1]

Mitscherlich [2] described a special type of boring tube for small samples at shallow depths. The instrument is a hollow tube with a slot along one side (Figure 3), so that in cross section it is like the letter C. The diameter of the tube is about an inch. The lower end is pointed, and at the upper end is a crossbar for a handle. The tube is forced into the ground, twisted through a single turn, and withdrawn. The sample so obtained is a small core of the sediment.

Other auger-like devices, suitable for shallow depths, are post-hole diggers and golf-hole drills. The former is a common farm implement consisting of two handles with semicylindrical blades. The instrument is twisted into the soil and a cylindrical sample withdrawn. It may be used for depths up to about 5 ft. The golf-hole

---

[1] F. Heyward, Soil sampling tubes for shallow depths: *Soil Science,* vol. 41, pp. 357-360, 1936.

[2] E. A. Mitscherlich, *Bodenkunde für Land- und Forstwirte* (Berlin, 1905), pp. 314-315.

COLLECTION OF SAMPLES

drill is a hollow steel cup about 3 in. in diameter, with a plunger within it for expelling the plug obtained when the drill is forced into the ground.

Augers suitable for loose sand have been described by Veatch.[1] Such devices consist essentially of an auger or cutting tool surrounded by a metal cylinder which retains the loose material picked up by the bit.

Drive-pipe samples. A device commonly used for sampling clays and other fine-grained sediments is an ordinary iron pipe about an inch or two in diameter, which is driven vertically into the deposit. A heavy metal collar should be fastened to the upper end of such pipes to prevent spreading, and also to facilitate subsequent removal.

For shallow depths, up to about 5 ft., the pipe may be set upright in a small pit and driven down with a sledge hammer. A wooden box is provided for standing upon until the pipe is down far enough to be struck from ground level. After the pipe has been driven in, it is raised by a jack or with block and tackle, and the core removed.

When depths up to 15 or 20 ft. are involved, a tripod and heavy weight may be used for driving the pipe. The tripod, also made of pipe, is set up above the sampling site, and the pipe itself is set into a shallow pit in an upright position. An iron weight of about 50 lbs. is used as the driver. The weight is a solid cylinder with a ring on one end and a thin metal rod about 4 ft. long on the other. A rope is tied to the ring and run over a pulley in the tripod. The thin rod is inserted into the drive pipe and the weight is lifted about 3 ft. above the pipe. The weight is released, and the rod guides it onto the drive-pipe. By repeating the blows, the pipe is driven into the ground. To remove the pipe, a block and tackle is used.

The core in the pipe is removed with a pressure screw, which is mounted on a rigid frame. The screw is turned into the pipe, forcing the core out at the other end.

Drive-pipes may also consist of an outer driving pipe with an inner collar, in which two halves of a tin cylinder, split lengthwise, are inserted. After withdrawing the drive-pipe, the inner tube is removed and the sample readily obtained.

A modification of the drive-pipe is the so-called sampling rod, described by Simpson.[2] It consists of a pipe about 2 in. in diameter and 7 ft. long, with a narrow vertical slit extending from the bottom nearly to the top of the pipe. The slit allows the pipe to yield slightly as it is forced down and enables it to hold the sample by tightening when it is withdrawn.

[1] A. C. Veatch, Geology and underground water resources of northern Louisiana and southern Arkansas: *U. S. Geol. Survey, Prof. Paper 46,* pp. 93 ff., 1906.

[2] D. Simpson, Sand sampling in cyanide works: *Trans. Inst. Min. and Met.,* vol. 16, pp. 30-41, 1906-1907.

Drilled Well Samples

Samples from drilled wells differ considerably among themselves depending upon the method of drilling employed. Three methods are used, and because of the wide variation in the quality of samples obtained the discussion will be based on the method employed.

Diamond core drilling. This method affords the best type of samples, because a solid core of the material is preserved during drilling. The drilling is accomplished by means of a bit made of a hollow steel cylinder, along the lower edge of which are set black diamonds to act as the cutting edge. The bit is attached to a core shell and core barrel, the latter in turn being connected to a series of hollow steel rods which extend to the surface. The rods and tools are rotated, and a stream of clear water is run down through the rods and core barrel. The water serves to keep the cutting edge cool, and as it flows upward between the rods and the wall of the bore hole, it carries away the rock cuttings.

As the bit penetrates into the rock, a core is cut out, which gradually fills the bit and the core barrel. When the core barrel is filled, the entire mechanism is pulled to the surface and the core removed. In this manner a continuous record is had of the rocks penetrated during the drilling. The cores may be taken to the laboratory for a complete examination, or portions of each type of rock may be removed as samples. In general, the core is excellently preserved, although occasionally soft materials may be washed away by the circulating water.

Percussion drilling. In this method of drilling a series of rock cuttings are obtained which in general are satisfactory for laboratory study. Certain characteristics of the particles, such as size and shape, may suffer, but mineral content and microfossils may be examined.

For drilling, a steel cutting tool is attached to a string of heavy steel cylinders which are suspended from the drilling rig with a rope or cable. The string of tools is alternately lifted and dropped, and the repeated blows of the bit serve to cut into the rock. The product is a granular or powdery material, which is kept wet either by seepage into the hole or by pouring water into it. At intervals, the string of tools is withdrawn and a bailing tube is lowered to withdraw the cuttings. The bailer is a hollow cylinder with a valve at the bottom. The contents of the bailer are dumped at the side of the rig, and the rock debris which constitutes the sample is left behind as the water runs off.

There is considerable danger of contamination of the sample from percussion drilling, due to material caving from the hole, or by the mixture of two formations if the bailer is not used often enough. With

COLLECTION OF SAMPLES

reasonable care, however, fair samples may be obtained. A sample taken from the bailer contents is roughly equivalent to a channel sample taken through the depth penetrated between two successive bailings. It thus represents an average sample from a given thickness. There is also some danger due to selective losses, inasmuch as the finer material is swept away with the washings from the bailer.

Rotary drilling. From the point of view of sampling, rotary drilling furnishes the poorest kind of well samples.

In this method of drilling, a bit, shaped like a fishtail, is rotated at the end of a stem of hollow rods. A stream of mud-laden water is circulated downward through the hollow stem, both to lubricate the cutting tool and to cool it. As the mud returns to the surface, it carries with it cuttings of the rock penetrated. It is from the mixture of mud and cuttings that the sample is taken.

The continued use of the same drilling mud results in considerable contamination of samples in rotary drilling. Various methods have been devised for obtaining satisfactory samples. The mud may be passed over screens to collect the cuttings, or flumes may be fitted with wiers to allow the cuttings to settle. A detailed discussion is given by Whiteside.[1]

BOTTOM SAMPLES

The collection of sedimentary samples from the bottom of bodies of standing water is a somewhat specialized procedure which requires apparatus of one sort or another, depending upon several factors, such as the depth of water, the nature of the bottom, whether an undisturbed sample is required, and whether the sample should be large or small. Numerous devices have been developed,[2] but many of them represent slight modifications of a few fundamental types. These types include apparatus which is dragged along the bottom, tubes which are driven vertically into the bottom deposits, or mechanical devices which snap a sample of the sediment between spring-operated jaws. Of these types, the tube samplers are perhaps the most extensively used.

Bottom dredges and drag buckets. Dredges and drag buckets are of several types. Among the earlier forms, of which modified versions are still in use, are dredges of the *Challenger* type.[3] The Challenger dredge consists of two

[1] R. M. Whiteside, Geologic interpretations from rotary well cuttings: *Bull. A. A. P. G.*, vol. 16, 1932, pp. 653-674.
[2] F. M. Soule, Oceanographic instruments and methods: Nat. Research Council, Bull. 85, pp. 411-454, 1932.
[3] J. Murray, *Report on the Scientific Results of the Voyage of H.M.S. Challenger* (London, 1885), vol. 1, pp. 73 ff.

parts, the iron framework which skims the surface of the deposit, and a bag or sack which collects and retains the skimmings. Another type of sampler which is pulled along the bottom is the Gilson sampler,[1] developed in 1906. It consists of a hemispherical bowl attached in the center to an iron rod. The Mann sampler, according to Trask,[2] consists of a cylindrical iron tube about 4 in. in diameter and 6 in. long. It is closed at one end, and at the open end is attached to a sounding line. The sampler is dragged along the bottom until filled.

A sampling device similar to the bucket type is the cup lead described in the *Challenger* report.[3] This consists of a hollow cone, fitted with a sliding lid, and fastened to a weighted spike. The lid prevents loss of the sample in the cone.

A bottom sampler devised by Lugn[4] for collecting sediments from the Mississippi River belongs in this classification. The device consists of two weights, rigidly attached to a central stem, and a loose-fitting cup which rests around a shoulder on the lower weight. The dimensions are such that, when the instrument lies on its side, the cup inclines easily without falling from the shoulder, and it slips back to vertical when the instrument is pulled up to the surface. In use, the instrument is dragged along the bottom; when it is hauled up, the cup fits tightly enough about the lower collar to prevent losses.

Bottom sampling tubes. These bottom sampling devices consist essentially of a tube of varying length, with weights attached. As it strikes the bottom, the tube settles into the deposit and fills with a core of the sediment. Many variations of this instrument have been used, and it is perhaps the most widely used of bottom sampling devices.

Among the earliest of such instruments was the Baillie Rod.[5] It was made of an iron pipe about 2½ in. in diameter, beveled at the bottom and fitted with a butterfly valve. A more effective device, fitted with a comb-valve to prevent the loss of sediment, was the Buchanan Combined Water Bottle and Sampling Tube.[6]

Among modern apparatus the Ekman Sampler[7] and its modifications are the most important. Trask[8] used an instrument essentially like the

[1] Stina Gripenberg, A study of the sediments of the North Baltic and adjoining seas: *Fennia*, vol. 60, no. 3, p. 11, 1934.

[2] P. D. Trask, *Origin and Environment of Source Sediments of Petroleum* (Houston, Texas, 1932), p. 14.

[3] J. Murray, *op. cit.*, 1885, p. 69.

[4] A. L. Lugn, Sedimentation in the Mississippi River: *Augustana Library Publications*, no. 11, 1927.

[5] J. Murray, *op. cit.*, 1885, pp. 59 ff.

[6] *Ibid.*, pp. 117 ff.

[7] V. W. Ekman, An apparatus for the collection of bottom samples: *Publications de Circonstances*, no. 27, Copenhagen, 1905. (Reference from Gripenberg, *loc. cit.*, 1934.)

[8] P. D. Trask, *op. cit.*, 1932, pp. 11-13.

COLLECTION OF SAMPLES

original, and its description will serve to indicate the general pattern. The apparatus (Figure 4) consists of a galvanized iron pipe, 3 ft. long and 1½ in. in diameter. The lower end of the pipe is open, and attached to the top is a vertical check valve with a perforated reducer, which in turn is attached to a stem with lead weights. The stem is fastened to a sounding line, and the apparatus is allowed to fall to the bottom, where the tube is driven into the sediment a distance dependent upon the softness of the deposit. After hauling the sampler to the surface, the collected sediment is driven from the pipe into a container. If it is desired, cardboard cylinders may be placed inside the tube, to act as receptacles for the sediment.

The Ekman type sampler is very effective in water to any depth, especially with silt or clay. Cores up to 120 cm. in length may be obtained. It is unsuccessful with coarser sediments, however, because of the absence of a valve at the base of the tube.

The most recent modification of the Ekman type of sampler is the Piggot sampler,[1] which introduces a new principle into the design of such instruments. This device was designed to apply an impulse to the sampling tube when it strikes the bottom, so that the tube will be driven farther into the mud than is the case with gravity-settling alone. Essentially the sampler consists of a sampling tube which is attached to a heavy mass acting as a gun. Within the gun is a charge of powder and a cap, set off by the impact of the device on the bottom. The tube is driven into the mud by the force of the explosion and is prevented from escape by an auxiliary cable, which permits the two parts to be withdrawn to the surface.

In detail the assembled instrument is about 15 ft. long and weighs about 400 lbs. Adequate hoisting equipment is accordingly necessary for its use. However, the instrument has been successful in depths greater than 15,000 ft., and it is capable of taking cores 10 ft. long. The core is collected in a brass tube within the outer casing, to facilitate removal and storage of the sample. A schematic diagram of the sampling

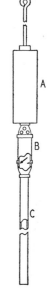

FIG. 4.—Modified Ekman bottom sampler. The valve at B prevents loss of the core in tube C during withdrawal of the instrument. A is a heavy mass to assure penetration of the tube into the sediment. (Adapted from Trask, 1930.)

---

[1] C. S. Piggot, Apparatus to secure cores from the ocean bottom: *Geol. Soc. America, Bulletin*, vol. 47, pp. 675-684, 1936.

tube is shown in Figure 5; a more detailed drawing may be found in the original paper.

Hydraulic coring tubes. Varney and Redwine [1] have recently developed a hydraulic coring instrument which successfully applies the principle that the differential pressure due to high water pressure outside the instrument, and low air pressure inside, may be used to drive a coring tube into the sea bottom. The apparatus consists of a coring tube passing through and attached to a piston sliding in a cylinder. The piston is supported near the upper end of the cylinder by trigger arms, which release it when the bottom is struck and permit the piston to move downward under water pressure, driving the core barrel with it. The apparatus was used in water from 50 to 300 ft. deep, with penetration varying from 3 to 7 ft.

Another type of tube sampler, which relies on mechanical force to obtain a longer sample, is the Knudsen sampler, described by Trask.[2] It consists of a tube fastened to a drum, around which latter are a number of turns of sounding line. A catch is released on impact, and as the line is pulled upward the drum rotates and operates a pump which draws the water from the collecting tube. Pressure differences cause the tube to settle farther into the mud. Trask reports difficulty in the use of the instrument.

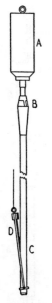

F I G. 5.—Piggot bottom sampler. The gun A drives the core bit C into the sediment on impact. B is a "water exit port" to facilitate penetration and to prevent loss of the sample. The "stirrup" D permits recovery of the core bit.

Clam-shell snappers and other closing types of sampler. Among the more recent sampling devices which are especially suitable for medium-grained sediments are the clam-shell snappers and related devices.[3] These instruments consist of two jaws which are held open by a trigger, as shown in Figure 6. When the device strikes the bottom a spring is released which snaps the jaws shut. Snapping devices may have several jaws which open in orange-peel fashion.

Larger devices of the closing type, which are operated by cables which draw the jaws together, instead of with springs, are represented by the Peterson Dredge.[4] This consists of two jaws hinged at their intersection and closed by means of a chain which pulls them together.

[1] F. M. Varney and L. E. Redwine, A hydraulic coring instrument for submarine geologic investigations: *Rep. Com. Sed.*, Nat. Research Council, 1937, pp. 107-113.
[2] P. D. Trask, *op. cit.*, 1932, p. 15.
[3] F. M. Soule, *loc. cit.*, 1932.
[4] O. Pratje, Die Sedimente des Südaltlantischen Ozeans: *Wiss. Ergeb. d. Deutsch Atlantischen Exped. auf d. Meteor*, vol. 3, part 2, p. 12, 1935.

COLLECTION OF SAMPLES

Miscellaneous devices. A simple instrument which may be used to obtain information about bottom deposits is a small sounding lead with an indentation in the bottom, into which wax is placed. When the lead strikes the bottom, some of the sediment clings to the wax and furnishes information about the nature of the bottom.

Another sampling device used to determine the nature of bottom deposits is the sampling "spud."[1] This consists of a long rod, along which grooves have been machined to form a series of cups with lips directed upward. The cups are spaced a little more than an inch apart along the length of the rod. The rod is forced down into the deposit, and as it is withdrawn each cup catches a small amount of sediment from the depth to which it penetrated. A vertical section is thus disclosed. The spud is operated by hand and is used in relatively shallow water.

F I G. 6.—
C l a m-shell snapper.

SIZE OF COLLECTED SAMPLES

The size of sample to be collected in any given case depends mainly upon two considerations: the coarseness of the sediment and the uses to which the sample is to be put. What is desired is the smallest sample that will adequately represent the material. As a general rule, field samples of medium- and fine-grained sediments are much larger than the amount required for a single laboratory determination, but as the material becomes coarser, the field sample tends to be about sufficient for a single detailed analysis, owing to the labor of transporting very large samples. From the point of view of sample size, it is immaterial whether the sample is discrete or composite; the essential point is that enough material be present to give adequate representation to the largest sizes present.

Wentworth[2] has investigated the relation of coarseness to sample size, and his practical rule is that the samples should be

large enough to include several fragments which fall in the largest grade present in the deposit. Several fragments may be interpreted as a number sufficiently large so that the probability of a serious accidental deviation from

---

[1] H. M. Eakin, Silting of reservoirs: *U. S. Dept. Agric., Tech. Bull. 524*, p. 27, 1936.
[2] C. K. Wentworth, Methods of mechanical analysis of sediments: *Univ. Iowa Studies in Nat. Hist.*, vol. 11, no. 11, 1926.

the normal number of such fragments in a sample collected by a reliable random method is small.

At the other extreme, according to Wentworth, it is hardly advisable to collect less than about 125 g. of any sediment regardless of its fineness.

Wentworth summarized his findings in a table, a modified form of which is given in Table 1. The sediment size is expressed in terms of the coarsest material present, which is indicated in the first column. The

TABLE 1

PRACTICAL SAMPLE WEIGHTS

Diameter of Coarsest Size (mm.)	Suggested Weight of Sample	Approximate Volume of Sample
Cobbles 128-64 mm.	32 kg.	......
Pebbles 64-4 mm.	16 to 2 kg.	......
Granules 4-2 mm.	1 kg.	1 liter
Sand 2-$\frac{1}{16}$ mm.	500 to 125 gm.	500 cc.
Silt $\frac{1}{16}$-$\frac{1}{256}$ mm.	125 gm.	250 cc.
Clay Under $\frac{1}{256}$ mm.	125 gm.	250 cc.

second column shows the range of suggested sample weights for the given sizes, and the third column indicates the approximate volume of material which coincides with the given weight.

The table affords a basis for judging sample size in terms of the coarseness of material. In addition it is necessary to consider the types of analyses to be performed on the sample. Wentworth's values are for mechanical analysis, but the general values hold for most purposes. With coarse sediments, from sand upward, the same sample may be used for size, shape, and mineral analyses, but among the silts and clays some methods of size analysis (the pipette method, Chapter 6) prevent the re-use of the material; however, as only 20 or 30 g. are used for this purpose, the suggested sample size is adequate to cover a number of analyses. Chemical analysis likewise destroys the sample. If storage or display material is to be made of part of the sample, that must be allowed for except in cases where the material analyzed itself becomes the display or storage material.

An interesting approach to the problem of sample size was made by Knight [1] in connection with ceramic materials. The principle introduced involves sample sizes proportional to the square of the diameters of the particles, starting with a half-gram sample of 200-mesh material.

CONTAINERS FOR SAMPLES

Sample containers may be used for transportation, storage, or display. In some cases the same container serves more than a single purpose, but for convenience the treatment is based on function. Indurated specimens may be wrapped in newspaper for shipment, and displayed in trays or boxes. The greater part of the discussion, accordingly, will be devoted to unconsolidated material.

Containers for collecting and shipping. Current usage varies considerably in the choice of containers for field samples. The commonest container is undoubtedly a bag. Wentworth [2] recommends cloth bags for most general use with dry sediments. The authors experimented extensively with brown kraft paper bags, and found them suited to more purposes than has generally been believed. Even wet sand may be transported in them if certain precautions of packing the bags are followed. The most commonly used containers for wet or damp sediments have been the familiar glass Mason jars. Ice-cream cartons of the cylindrical type appear to be increasingly used for the same purpose; their light weight when empty and the absence of a breakage risk suit them admirably for the purpose.

Storage containers. It is common practice to retain parts of each field sample for later reference. The most convenient storage containers are the cylindrical cartons mentioned earlier, because of their compact size and the convenience with which they may be stacked one above the other. The next most convenient storage container is a brown kraft paper bag. These also are compact and may be stacked. These two types of containers apply to medium- and fine-grained sediments. If gravels are to be stored, small, square corrugated paper cartons may be used, or in exceptional cases, cement bags.

If the storage samples are to be used extensively, as in assigning them to classes for analysis, the cylindrical cartons are best. Paper bags do not

[1] F. P. Knight, Jr., The importance of accurate sampling in the production and use of ceramic materials: *Jour. Am. Ceram. Soc.*, vol. 15, pp. 444-451, 1932.
[2] C. K. Wentworth, *loc. cit.*, 1926.

stand much handling, but have proved their worth for dead storage.

When the storage of samples involves the preservation of their original moisture, Mason jars are perhaps the most useful containers.

Display containers. Display containers may be of several types, depending partly on the size of material involved. Ordinary cardboard trays, 2 x 3 x ½ in. or 3 x 4 x 1 in., are convenient for small displays of pebbles. Other sizes of trays are available, some fitted with glass covers to exclude dust. Finer sediments may be displayed very conveniently in small round glass vials of about 2-oz. (liquid) capacity and measuring about an inch in diameter and 2½ in. in height. These may be obtained either with corks or with screw tops. Another convenient display jar is the inverted type of bottle, round-topped, and fitted with a cork at the base. These are available in a variety of sizes.

Capacities of Sample Containers

Table 2 lists the more common types of sample containers and indicates their capacities and dimensions. The following discussion supplements the data in the table.

Cloth bags. Cloth bags are available in many varieties. The mouths may have tying strings attached, or drawstrings. The weave or weight of the bag may be chosen in terms of the fineness of the sediment; a general rule is that the cloth mesh should be finer than the smallest particles. Sized cloth is satisfactory for dry sediments, but wet samples may soften the sizing.

Paper bags. Where cost is a factor, and where most of the samples are medium to fine-grained, paper bags have a wide applicability. The cost is practically negligible, and sizes 1 to 3 may be had for 15 to 20¢ per hundred. The authors have used paper bags with many kinds of sediments in the past several years, and the loss due to breakage or leakage has been negligible. In one instance wet sands were carried 400 mi. by automobile with no losses. Certain precautions, however, must be followed in using paper bags for transportation. The bag is filled about half full of sediment, and is tamped by jostling the bag on a plane surface. The bag is then closed by folding the upper part into a series of underfolds until the package forms a rectangular solid. The bags need not be tied with string. After the bags have been folded, they are placed upside down in a corrugated paper box, and packed tightly together. The reason for packing them upside down is that dampness may loosen the bottom seams, and if the bags are later removed while still damp, no loss is occasioned by the bottoms dropping out. If the bags remain packed until they are dry, the seams reglue themselves. Added strength is gained if the bags are used double.

For fine sediments like silt, loess, or clay, especially when the sediment is dry, the #1 bags are suitable, and they may be packed in larger kraft bags. For example, the #1 bags fit very snugly crosswise in #8 bags, in rows of

COLLECTION OF SAMPLES

TABLE 2

CAPACITIES OF SAMPLE CONTAINERS

CLOTH BAGS

Dimensions (in.)	Capacity (g. of sand)
12 x 18	10 kg.
9 x 14	5 "
7 x 9	2 "
5 x 8	1 "
4 x 6	500 gm.
3 x 4½	200 "
2 x 4	100 "

BROWN KRAFT PAPER BAGS

Number and Capacity (lb.)	Length	Dimensions (in.) Width	Breadth
12 *	13¾	7¼	4¼
10 *	13	6¼	4¼
8 *	12½	6	4
6	11½	6	3½
5	11	5¼	3¼
4	9½	5	3¼
3 **	8½	4¾	3
2 **	8	4¼	2½
1 **	7½	3½	2

* Suitable for packing smaller bags.
** Suitable for sand and fine sediments.

GLASS JARS AND VIALS (CYLINDRICAL)

Dimensions (in.)	Volume	Capacity (g. of sand)
7½ x 4¼	1 quart	1,400
5½ x 3½	1 pint	700
4¼ x 3½	½ pint	400

CYLINDRICAL ICE-CREAM CARTONS

Dimensions (in.)	Volume	Capacity (g. of sand)
7 x 3½	1 quart	1,600
4 x 3½	1 pint	800
2¼ x 3½	½ pint	400

three. Thus about nine of the smaller bags may be assembled in larger bags, a decided convenience when large numbers of samples are collected. Paper bags are not suitable for gravel, because the pebbles wear through. In general, sizes 1, 2, and 3 are used for individual samples, and sizes 8, 10, and 12 for repackaging.

Glass jars. Mason jars have found wide use for collecting wet samples, where there may be as much water as sediment. They are eminently satisfactory for this purpose, but there are inconveniences attached to them. The risk of breakage is present, but may be largely eliminated by carrying them always in their cartons. Present tendencies seem to be to replace these jars with cylindrical cartons, which are less expensive and come in the same sizes.

Cylindrical waterproof cartons. Damp or wet samples are most conveniently carried in these cartons, which are supplied with friction tops to prevent leakage.[1] Modern cartons of this type are free from wax, so that none is rubbed off by the sediment. The largest size is suitable for fine gravel, and for smaller particles the cartons range down in size to a half pint. For greatest all-around adaptability in sampling, it is difficult to choose between cartons and cloth bags; the bags have the advantage of occupying less room when empty, but the cartons have the advantage that they pack more readily when filled. The cartons may be used for either wet or dry sediments, but cloth bags have limitations when used for wet material.

Paper envelopes. Another type of container, not mentioned in the table, is the small paper envelope, fitted with an aluminum strip across the top, so that the edge may be rolled over and the envelope sealed securely against losses. These envelopes are usually too small for field samples, but they are convenient for storing small samples of laboratory materials. For example, the sieve separates obtained during mechanical analysis may be stored in this manner, and during heavy mineral work, the light and heavy separates may be kept in them.

LABELING AND NUMBERING OF SAMPLES

Every sample should be numbered or labeled at the time of sampling. A convenient plan, suggested by Wentworth,[2] and used by the authors as a standard procedure, is to number all samples serially during a given sampling expedition, regardless of their nature or locality. The serial number alone is marked on the sample, and at the same time a notation is made in the field book, giving all necessary data. Suggested field observations to be made at the time of sampling are given elsewhere (Chapter 1). It is also convenient to indicate the location and number of each sample on a map. In this manner samples, notebook, and map are all coördinated.

[1] A satisfactory brand of carton is the "Titelok," manufactured by Sutherland Paper Co., Kalamazoo, Michigan.
[2] C. K. Wentworth, *loc. cit.*, 1926.

COLLECTION OF SAMPLES

If it is necessary to keep two or more projects separate during sampling, an appropriate number of capital letters may be used to designate the several projects, and the samples under each project may be numbered serially, as A1, A2, etc. In some instances, also, it may be desirable to number the samples in accordance with a predetermined grid pattern, regardless of the order in which the samples are collected. This may be accomplished by labeling one set of coördinates with capital letters and the other with numbers. In this manner any sample, as C1 or D3, may at once be located. The authors' experience suggests, however, that the grid keying may just as conveniently be accomplished in the laboratory from serially numbered samples.

Individual practices vary, but in the labeling of specimens the authors have found that large numbers, legibly written directly on the bag or container with an indelible pencil, are adequate. It is often advisable to write the number in more than one place on the container to avoid erasure by friction. Likewise, large figures do not blur into illegibility as smaller ones often do, due to damp samples. If the rocks are consolidated, the number may be written with indelible pencil on a small square of adhesive tape, which is fastened to the specimen. The same number may also be written on the outside of the wrapping, to facilitate sorting in the laboratory.

For storage and other laboratory purposes it is advisable to have some system of distinguishing among samples from different projects. Wentworth[1] suggested a series of ciphers and key digits which could be written in front of the field serial number, to distinguish the projects. The authors follow a plan of entering the samples in an accession catalogue, approximately following the suggestion of Johannsen[2] in connection with igneous rocks. In the accession catalogue a new serial number is given to each sample. The record itself is an ordinary daybook, ruled into the following columns:

Accession Number	Field Number	Type of Sediment	Location	Collector	Date	Remarks

The accession numbers are serial, regardless of the project. The original field number is placed in the second column, and a short descriptive term, as "dune sand," "glacial till," or the like is placed in the third column. The location may be indicated in full detail or roughly by county or locality. The collector's initials are usually sufficient, as is the year of

---

[1] C. K. Wentworth, *loc. cit.*, 1926.
[2] A. Johannsen, *Manual of Petrographic Methods*, 2nd ed. (New York, 1918), pp. 609 ff.

collection. Under "Remarks" a number of items may be included. A given project may be bracketed with a reference to the field notes; if some of the data from the samples are published, that may be indicated with a reference.

The accession catalogue is used for storage purposes. For reference, a card catalogue may be made, cross-referring types of sediments with localities, and the like. The following examples show a sediment-type card and a locality card:

BEACH SAND

14   North Chicago, Ill.
21   Benton Harbor, Mich.
42-57 Waverly Beach, Ind.

WISCONSIN
Door County

131   River sand
153   Beach gravel
160-177 Beach gravel

More elaborate numbering systems and cross-reference schemes may be devised, based on a complete classification of sedimentary materials. Milner [1] follows a classification of letters and numbers, but the unsatisfactory state of present sedimentary classifications suggests that a simple descriptive scheme be followed. As genetic classification develops, cards may be prepared with references to the sample numbers that belong in each classification.

THEORY OF SAMPLING SEDIMENTS

Statisticians have paid considerable attention to the theory of sampling, but a direct application of the principles to sedimentary petrology is not apparent without careful consideration. Problems of sampling have received very little formal treatment from a strictly sedimentary point of view. This may be attributed to the difficult nature of the problem and to the fact that virtually no sedimentary petrologists are trained in

[1] H. B. Milner, *Sedimentary Petrography*, 2nd ed. (London, 1929), pp. 266 ff.

COLLECTION OF SAMPLES

mathematical statistics. The present section will not attempt to establish a general theory of sampling sediments but instead will discuss the nature of the problem the types of work that have already been done.

In the discussion of compound samples (page 18) it was mentioned that practically every sampling procedure was subject to some error. This may be illustrated by supposing a population or universe of 10,000 spherical pebbles, from which samples of 100 pebbles each are to be drawn. The pebbles range in size continuously from r_1 to r_2, where r_2 may be taken as about twice r_1. The population has an average radius r_{av}, which is not known. Suppose five samples of 100 pebbles are drawn from the population at random, and the average radius of each sample determined by measurement. By the law of error, the chances are that no two of these averages would be identical. Similarly, it is likely that not one of the individual averages would be identical with r_{av}. However, if additional samples are withdrawn and measured, the average values of the samples would tend to distribute themselves as a symmetrical bell-shaped distribution about the value of r_{av}. Furthermore, the peak of the distribution would, for all practical purposes, coincide with the value of r_{av}.

This general principle suggests a simple definition of a satisfactory sample from any given sedimentary deposit: a random sample may be defined as one in which the characteristics of the sample show no systematic variations from the characteristics of the deposit at the sampling locality.[1] This affords a basis for collecting representative samples in the field. In order that deviations of the sample be random, the material sampled must be homogeneous, and this suggests that individual beds or strata be used as fundamental units in sampling, inasmuch as in a majority of cases a given bed will have approximately the same characteristics throughout its thickness at the sampling site.[2]

On this basis, samples for detailed investigations should not transect more than a single bed. Here, however, the question of defining a single bed arises, and superimposed on that is a question of practicability in terms of the labor and inconvenience involved. By starting out from this fundamental concept, however, the individual worker may form his own judgment as to the sampling unit he should use. The principle of beds or strata as units in sampling has been implied in the body of the chapter,

[1] This discussion illustrates the parallelism of the authors' and Otto's independent approaches to the problem of sampling. By introducing the term *sedimentation unit* (page 20), and by classifying sampling techniques more fully, Otto developed the present notions more rigorously.

[2] Graded bedding is a marked exception, but the simpler case is used here as a first approach.

in connection with the choice of sample in a given case. It applies most strongly in studies directed toward an elucidation of the detailed history of a sediment, rather than to studies involving only average characteristics.

An example may be given here of the point of view involved. Suppose an outwash terrace composed of numerous beds of varying thicknesses and degrees of coarseness. How shall the terrace be sampled to determine its characteristics in detail? One answer is to treat each bed as a homogeneous population and to sample it individually. Each sample could thereupon be analyzed individually, and from the series of average values obtained a frequency distribution of average sizes could be constructed which would reflect the general characteristics of the terrace. Not only average conditions, but the spread of the individual beds would be brought out by such a study. The same reasoning applies not only to size characteristics, but to mineral content, shape attributes, and the like.

There is a further question involved, however, even in the case of a sample from a given stratum: how may one know whether his sample of that bed is actually a random sample? There are several tests by which this may be learned, and all of them rest on the principle that a random sample will, on the average, show no significant systematic deviations from the true characteristics of the bed. These tests include the probable error method, the chi-square test, and the theory of a state of control, all of which have been applied to sedimentary problems and are mentioned in Chapter 9 of this volume. Each of the methods involves technical statistical operations, and hence they are appropriately deferred to the chapter on statistics. The probable error method has been applied specifically to sampling problems by Krumbein,[1] the chi-square test was used by Eisenhart[2] in a discussion of geological correlation, and the theory of control was used by Otto[3] in connection with test samples split from a field sample.

The sampling error is a function of the homogeneity of the sediment, of the precise locality in which the sample is collected, and of the manner in which it is collected. Regardless of the magnitude of the error, however, it was pointed out earlier that the sampling error could be reduced to any value desired by securing compound samples. This arises from the nature of the error function and may be illustrated as follows: suppose a series of numbers, $r_1, r_2, \ldots r_n$, and their average value r_{av}. If any two

---

[1] W. C. Krumbein, *loc. cit.*, 1934.
[2] C. Eisenhart, A test for the significance of lithological variations: *Jour. Sed. Petrology*, vol. 5, pp. 137-145, 1935.
[3] G. H. Otto, The use of statistical methods in effecting improvements on a Jones sample splitter: *Jour. Sed. Petrology*, vol. 7, pp. 101-133, 1937.

COLLECTION OF SAMPLES

of the numbers are chosen at random, they will generally differ from r_{av} by some fixed amount, which may be referred to as their error. However, if the pair of numbers are themselves averaged, the value of their average will never be farther from r_{av} than the extreme of the paired values.

It has been found [1] that the error of the mean of a set of observations varies inversely as the square root of the number of observations. Expressed mathematically, this is $E_m = E/\sqrt{n}$, where E_m is the error of the mean, E is the error of a single observation, and n is the number of observations made. This equation may be expressed as a ratio, $E_m/E = 1/\sqrt{n}$. In this latter case it is possible to study, by means of a simple graph, the behavior of the function as n increases. By choosing values of n from 1 to 10, the corresponding values of E_m/E are found, as shown in Table 3. Figure 7 is a graph of the function, which demonstrates that as n increases the error decreases rapidly at first and then more slowly. The point where the curve begins to flatten out is at about $n = 4$, where $E_m/E = 0.5$. As the curve is followed out to $n = 10$, $E_m/E = 0.316$, so that the rate of change has decreased appreciably.

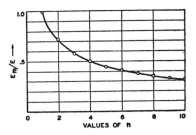

Fig. 7.—Graph of the function $E_m/E = 1/\sqrt{n}$. Data from Table 3.

TABLE 3

VALUES OF PE_m/E AND n FROM $PE_m/E = 1/\sqrt{n}$

n	PE_m/E
1	1.000
2	0.707
3	0.577
4	0.500
5	0.447
6	0.408
7	0.378
8	0.354
9	0.333
10	0.316

[1] H. L. Rietz, *Handbook of Mathematical Statistics* (Boston, 1924), p. 77.

The greater relative reduction in the error when four observations are combined, together with the added effort required to collect a large number of samples, suggests as a first approximation that in sampling sediments where the sampling error may be a factor, four discrete samples be combined into a single composite. It was for this reason that composites of four were suggested in the discussion of compound samples.

It is possible to apply the error equation to the problem of reducing the sampling error to any given value. Suppose the error of the individual sample is 4.5 per cent, as has been found in certain beach sands. It is desired to reduce this error to 0.5 per cent. How many samples must be combined into a composite? The solution is found by first determining the decimal value of the ratio 0.5:4.5. This is 0.111. Hence $E_m/E = 0.111 = 1/\sqrt{n}$. Solving this for $\sqrt{n}$ yields $\sqrt{n} = 1/0.111 = 9.0$, and hence $n = 81$ samples.

The present discussion of sampling theory as applied to sediments is far from exhaustive, and there are numerous problems which have not even been touched upon. For example, an important problem connected with the areal study of sediments involves the determination of the change in sediment characteristics as the formation is followed away from its source. Among the questions which arise is whether the variations observed are due to an actual change in sedimentary characteristics or whether they are due to other causes, such as sampling errors and the like. Problems of this sort have not been investigated in detail from a sedimentary point of view, but analogous problems have been studied in agricultural science. In this instance the problem was whether observed variations in crops were due to actual variations among the species studied or whether they were due to variations in soil fertility over the experimental plot. The nature of the problem and methods of attacking it statistically are given by Fisher.[1] The method applied by Fisher is called *analysis of variance,* and it seems likely that applications of the technique to sedimentary problems will yield significant results.

[1] R. A. Fisher, *The Design of Experiments* (Edinburgh, 1935), Chap. 4.

CHAPTER 3

PREPARATION OF SAMPLES FOR ANALYSIS

INTRODUCTION

THE preparation of sedimentary materials for study is a usual preliminary to their detailed analysis. Field samples commonly are larger than the laboratory test sample, and in addition the sediment may be in a state of aggregation unsuitable for direct analysis. Several steps may be involved in the preparation of the samples, and the nature of the treatment depends upon the sediment and on the study to be made.

Although there is no universal method of treatment for all possible types of sedimentary analysis, there are several steps in the process which have more or less in common. The first treatment is a preliminary disaggregation of the field sample into smaller aggregates suitable for splitting off test samples. The test samples are then further disaggregated or dispersed to a state suitable for the type of analysis to be performed. In mechanical analysis the sample may be treated with mild chemical agents which effect a separation of the aggregates into individual particles and supply each particle with electrical charges which prevent reaggregation during analysis. In mineralogical analysis, on the other hand, strong treatment with acids or bases may be required to clean the surfaces of the grains and to prepare them for microscopic study.

Care should be exercised during splitting to obtain a representative part of the field sample. It is not sufficient to pour a quantity of the field sample from the container, especially with medium- or coarse-grained sediments, because properties such as size, shape, density, magnetic properties, coefficient of friction, and elasticity may cause a selective error.[1] Such errors may have a serious effect on the final results, especially if very small samples are to be split off, as in heavy mineral studies. With very fine-grained material no serious errors may be involved if the material is thoroughly mixed and portions extracted with a spatula, as is often done for chemical analysis, but even here formal methods of splitting are to be preferred.

[1] G. H. Otto, Comparative tests of several methods of sampling heavy mineral concentrates: *Jour. Sed. Petrology*, vol. 3, pp. 30-39, 1933.

If the sediment is composed of loose grains or particles, the splitting process may be undertaken without preliminaries. However, if the material occurs as aggregates, a preliminary disaggregation should be performed, to avoid having too large lumps of material in the test sample.

SAMPLE SPLITTING

To avoid selective errors, the field sample should be separated into individual grains or small aggregates before splitting off a test sample. If the sediment is a loosely cemented sandstone, the preliminary treatment may consist of gently crushing the rock with a rubber pestle or a wooden rolling pin. Clays and silts often harden during drying, and the lumps may be broken in a similar manner, so that no aggregates larger than a pea are present. If the rock is partially indurated, as a shale, some crushing device may be used to obtain pea-sized fragments if it is possible to do so without destroying individual particles. Gravels which are cemented may sometimes be separated into pebbles by acid leaching, but before any chemical methods are used, precautions should be taken to see that no material which may be needed for the analysis is lost. The purpose of preliminary disaggregation is merely to obtain the material in such form that it may be quartered into smaller samples.

Quartering by hand. The simplest method of splitting samples is to pour the field sample into a conical pile on a large sheet of smooth-surfaced paper and, with a spatula or other device (the hands may be used with coarse material), to separate the heap into four quarters by cutting it pie fashion along two normal diameters. Alternate quarters are retained, and the others are laid aside. If the remaining quarters are still too large, they may be recombined into a smaller pile and the process repeated.

An adaptation of hand quartering was made by Pettijohn,[1] who used four rectangular sheets of paper and overlapped them to form a square composed of one-fourth of each sheet. The sample was poured on the center of the square, spread into a circular heap, and the papers pulled apart. Opposite quarters were recombined and the process repeated until a sufficiently small split was obtained.

Knife-edge splitters. Several years ago Krumbein[2] experimented with a common splitting device which consisted of a conical hopper, the lower opening of which was superimposed over two crossed knife edges, which separated

---

[1] F. J. Pettijohn, Petrography of the beach sands of southern Lake Michigan: *Jour. Geology*, vol. 39, pp. 432-455, 1931.
[2] G. H. Otto, *loc. cit.*, 1933.

PREPARATION OF SAMPLES

the stream of grains into four divisions. Otto,[1] in comparing several splitting methods, improved on the original device but found that for certain types of samples the deviation was larger than in other splitting devices.

Jones sample splitter. One of the most widely used devices for splitting samples is the Jones sample splitter, which consists of a series of inclined chutes leading alternately to two pans placed on opposite sides of the apparatus. The sample is poured into a hopper, using a rectangular pan, the width of which is equal to the width of the set of chutes. The sample is split to the desired size by resplitting the right- and left-hand halves alternately.

FIG. 8.—Jones Sample Splitter. (Courtesy W. S. Tyler Company, Cleveland, O.)

The commercial model of the Jones splitter (Figure 8) has splitting compartments about 1 cm. in width, so that the use of the device is limited to particles smaller than about a centimeter in diameter. Wentworth[2] studied the error involved in the use of the Jones splitter, using sandy gravel having a range of sizes from about 8 mm. to $\frac{1}{8}$ mm. diameter. His results showed that the error was larger with the larger sizes, but the relation between particle size and magnitude of the error was not constant. More recently Otto[3] made similar tests, involving a method of statistical control. The method involves a study of the performance characteristics of the instrument in terms of the deviations from expected theoretical results. The study indicated among other things that the personal element affected results, in the manner in which the material was poured into the hopper. Otto thereupon designed a modified Jones-type splitter and applied the control method to it. The new device showed satisfactory agreements between expected and attained results and was not influenced by the personal element.

Otto's modified Jones-type splitter differs from commercial models in several respects. The hopper is so designed that the sample can only be poured in a standard manner; the pans have lugs on them, which eliminates the personal element both in pouring and receiving. The receiving pans also have dust covers to prevent losses during splitting. Complete working drawings for the instrument are given in Otto's paper.

---

[1] G. H. Otto, *loc. cit.*, 1933.
[2] C. K. Wentworth, The accuracy of mechanical analysis: *Am. Jour. Sci.*, vol. 13, pp. 399-408, 1927.
[3] G. H. Otto, The use of statistical methods in effecting improvements on a Jones sample splitter: *Jour. Sed. Petrology*, vol. 7, pp. 101-133, 1937.

At an earlier date Otto also designed a miniature form of the Jones splitter, called a "Microsplit."[1] This device is described more fully in Chapter 15.

Rotary type sample splitters. Wentworth, Wilgus and Koch[2] developed a rotary type of sample splitter which was made in two sizes, one for large samples and the other for small samples. The device consists of a set of cylindrical tubes arranged around the periphery of a rotating table. The sample is fed into a hopper which delivers it through a funnel to a position above the rotating set of cylinders. As the table rotates, the grains are distributed among the tubes. The speed of rotation may be varied and the tubes made to pass as often as desired beneath the funnel. The splits could be used individually as sixteenths of the original amount, or opposite tubes could be combined in various ways. A series of tests were conducted on the relative accuracy of the rotary splitter and a Jones splitter; the rotary splitter showed a marked superiority.

Oscillatory sample splitters. Several years ago J. E. Appel developed an oscillatory type sample splitter at the University of Chicago. The device consists of a vertical brass funnel suspended on an axis and rocked over a knife edge through a small arc. The device is more fully described in Chapter 14.

Significance of statistical tests on sample splitters. In all the investigations made of the relative merits of sample splitters, the data were based on the grades recovered, rather than on the effect of the splitter on the statistical parameters of the frequency curve. In other connections it has been noted that comparative analyses of sediments may show fluctuations in the amounts of material collected by the several sieves, and yet the statistical values of the sediment as a whole will be only slightly affected. It would be instructive to have a study made, either in terms of probable error or in connection with the theory of control, on the size distribution as a whole with various splitting devices. The only data on this question known to the authors are contained in an incidental study conducted by Krumbein[3] in connection with field sampling errors. The probable error of splitting and sieving combined was found to range from 0.75 to 1.42 per cent, as computed with respect to the median grain size of beach sands. The sample splitter was a commercial Jones splitter, and the magnitude of the total laboratory error suggests that for most analyses the error is perhaps not unduly large.

[1] G. H. Otto, *loc. cit.*, 1933.
[2] C. K. Wentworth, W. L. Wilgus and H. L. Koch. A rotary type of sample splitter: *Jour. Sed. Petrology*, vol. 4, pp. 127-138, 1934.
[3] W. C. Krumbein, The probable error of sampling sediments for mechanical analysis: *Am. Jour. Sci.*, vol. 27, pp. 204-214, 1934.

PREPARATION OF SAMPLES

DISAGGREGATION OF TEST SAMPLES

Further disaggregation, after splitting test samples from the original material, must be undertaken with an understanding of the effect of each process on the characteristics of the sediment. Of primary importance is the principle that no method should be used which alters or destroys any of the data to be obtained in the subsequent analysis, without an evaluation of the error introduced. Unconsolidated sediments may receive so little pretreatment that their properties are not appreciably affected. As the degree of cementation or induration increases, however, methods of disaggregation become progressively more "violent" and there is greater likelihood that some of the characteristics of the sediment may be modified. When it appears that disaggregation can only be had at the expense of greatly altered properties, it may be better to rely on thin-section methods of analysis.

It may be pertinent here to contrast the kinds of treatment which may be resorted to in the several types of analysis:

Preparation for mechanical analysis. In disaggregating sediments for mechanical analysis the precautions to be followed are that the grains should not be broken and that none of the primary constituents should be removed by the disaggregation process. Breakage of grains results in an error in the average size and in other statistical constants, because the larger grains are reduced in number and the smaller ones increased. The removal of primary material will yield results which are biased in the degree to which the removed material is an integral part of any given grade size.

The treatment accorded to samples for mechanical analysis depends also on the coarseness of the sediment. Coarser particles are usually sieved, whereas fine-grained sediments are usually separated in terms of their settling velocities in water. For sieving it is only necessary to obtain a state of disaggregation such that each particle is individual; for finer sediments precautions must also be taken to see that the individual grains do not reaggregate during analysis. This second requisite of the fine-grained sediments automatically rules out certain chemical procedures, which may coagulate the particles during the analysis.

Preparation for shape and surface texture analysis. Owing to limitations in the practical application of techniques of shape and surface texture analysis, most work of this kind is limited to coarse- and medium-grained sediments. The principal precautions that must be followed relate to the breakage of the grains, which strongly affects their roundness although it may not seriously change their degree of sphericity. In like manner, methods of disaggregation which alter surface textures, either by solution or abrasion, must be avoided in surface texture analysis. Tests to be applied in such cases may involve the examination of particular grains before and after disaggregation, to determine whether changes have occurred.

Preparation for mineralogical analysis. For certain kinds of mineralogical studies it is not necessary to prevent breakage of grains, and individual surface textures usually need not be preserved. Attention is generally focused on the heavy minerals, and hence a wider choice of disaggregation procedures is available for mineralogical studies. When it is necessary to count mineral frequencies for statistical comparisons, however, the factor of grain breakage must be considered. Inasmuch as mineralogical analysis is a complete subject within itself, further details of the disaggregation processes are given in Chapter 13.

COARSE- AND MEDIUM-GRAINED SEDIMENTS

Unconsolidated gravel and sand present no disaggregation problems. With such materials one may proceed to the analysis as soon as a test sample has been split from the field sample. If on the other hand the sediment is consolidated, several devices are available for disaggregation. Several fragments should be examined under a binocular microscope before treating the rock, so that the proper choice of method may be made.

Removal of cement. Among the more common cementing materials in sedimentary rocks are calcite, iron oxide, quartz (silica in general) and organic materials like bitumen. The simplest of these to eliminate is the calcite cement, which may be removed by gently heating the rock fragments in dilute hydrochloric acid. Before resorting to acid treatment, however, it is well to note whether any primary calcite fragments are present in the sediment. In gravel there may be limestone pebbles, and some sands contain calcite and other carbonates as an integral part of the size frequency distribution. If acid leaching is resorted to indiscriminately, the mechanical analysis will be inaccurate to the extent that primary carbonate particles are present. In the absence of any alternative, the rock may be leached in acid and some method used to correct for lost material. The data for the corrections may be obtained microscopically in many cases.

Iron oxide cement may be removed with stannous chloride. A solution of the salt is added to dilute hydrochloric acid, and the rock fragments are heated in the solution. Tester used 15-18 per cent HCl with about 10 per cent of stannous chloride.[1] Because of the acidic nature of this solvent, the same precautions should be taken about primary carbonate grains as in the case of acid treatment alone.

Silica cements are in general the most difficult to remove. If the cement

---

[1] A. C. Tester, The Dakota stage of the type locality. Appendix A, Laboratory Methods: *Iowa Geol. Survey,* vol. 35, p. 305, 1931.

PREPARATION OF SAMPLES

is quartz and shows secondary enlargement of the primary quartz grains in the sediment, disaggregation is practically hopeless. Thin-section methods of mechanical analysis (Chapter 6) may be used in such cases. When the cement is opal or amorphous silica, the use of concentrated alkalies is sometimes sufficient to remove the material.[1] Such alkalies also affect some of the mineral grains, however, and should be used with recognition of this fact.

Occasionally pyrite occurs as a cementing agent. By boiling the specimen in dilute nitric acid the pyrite may commonly be dissolved.[2]

Organic cements, such as bitumen, are most effectively treated by using such solvents as ether, acetone, benzol, or gasoline.[3] Needless to say, the specimen should not be boiled in such solvents unless provisions are taken to avoid fires or explosions. Reflux condensers attached to Pyrex flasks are usually sufficient.

For the removal of colloidal binding materials in soils, such as organic matter, iron oxide, or colloidal silica, Truog and others [4] recently developed a procedure involving the use of oxalic acid and sodium sulphide. The nascent hydrogen sulphide liberated in the soil suspension dissolved the cementing materials and effected a completed dispersion of the soil.

Disruption of rock specimens. In some cases the cementing material may defy all attempts to remove it, or the rock may be highly indurated. When dealing with conglomerates, individual pebbles may be chiseled from the matrix, but for mechanical analyses care must be exercised to obtain the complete size range of the pebbles, as well as a representative number of the several sizes. If the rock material is not coarse-grained enough for individual treatment of the particles, various disrupting devices may be tried.

The rock may be heated to redness and plunged into water,[5] whereupon some of the grains may be loosened. This method is extremely effective in altering minerals, however, and should not be resorted to as a general rule for mechanical analysis. Grain breakage is also a common accompaniment of such violent treatment. A less destructive method of treatment involves the use of saturated solutions of various chemicals, which

---

[1] G. L. Taylor and N. C. Georgesen, Disaggregation of clastic rocks by use of a pressure chamber: *Jour. Sed. Petrology,* vol. 3, pp. 40-43, 1933.

[2] E. S. Dana, *Textbook of Mineralogy,* 3rd rev. ed. (New York, 1922), p. 377.

[3] E. M. Spieker, Bituminous sandstone near Vernal, Utah: *U. S. Geol. Survey Bull. 822c,* pp. 77-98, 1930.

[4] E. Truog, J. R. Taylor, R. W. Pearson, M. E. Weeks and R. W. Simonson, Procedure for special type of mechanical and mineralogical soil analysis: *Proc. Soil Sci. Soc. America,* vol. 1, pp. 101-112, 1936.

[5] This procedure was used as early as 1863 by R. Ulbricht: Ein Beitrag zur Methode der Boden-analyse, *Landwirts. Versuchs-Stat.,* vol. 5, pp. 200-209, 1863.

are allowed to permeate the rock by prolonged soaking. The fragments are then removed and allowed to dry, whereupon the force of crystallization will sometimes disrupt the rock. Sodium sulphate [1] and sodium hyposulphite (hypo) [2] have been used for this purpose. A somewhat analogous method is to saturate the rock with sodium carbonate solution and then plunge the rock in acid.[3] The pressure of the escaping CO_2 aids in freeing the grains from each other.

Neumaier [4] used ammonium nitrate in an ingenious manner to effect a disaggregation of sedimentary grains. The salt is soluble to the extent of 177 g. per liter at 20° C., and 1,011 g. per liter at 100° C. The sediment was accordingly placed in a saturated solution of the salt at 110° for five minutes, whereupon the solution was rapidly cooled to room temperature. The resulting crystallization of the excess ammonium nitrate forced apart the aggregates of the sediment. The process was repeated, and finally the salt was removed by washing with distilled water.

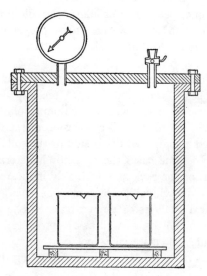

FIG. 9.—Taylor and Georgesen pressure disaggregator.

Pressure-chamber disaggregation. Taylor and Georgesen [5] developed a pressure chamber which proved very effective in disaggregating indurated rocks. The chamber consists of a 12-in. length of 10-in. steel casing, with a plate of half-inch steel welded to one end to form the base. A flange of the same material was welded to the open end, to afford a means of fastening a cover of half-inch steel to it. A stopcock and pressure gauge were added to the cover (Figure 9). Within the chamber was a wooden rack on which several beakers could be set. The specimen to be disaggregated is placed in a beaker and covered with an appropriate solution. Additional

[1] M. Morris, Unsoundness of certain types of rocks: *Iowa Acad. Sci. Proc.,* vol. 38, pp. 175-181, 1931.
[2] I. Tolmachoff, Crystallization of certain salts used for the disintegration of shales: *Science,* vol. 76, pp. 147-148, 1932.
[3] A. Mann, *Proc. U. S. Nat. Mus.,* vol. 60, pp. 1-8, 1932.
[4] F. Neumaier, Über Vorbehandlungsverfahren der Sedimente zur Sclämmanalyse: *Zentr. f. Min.,* Abt. A, pp. 78-95, 1935.
[5] G. L. Taylor and N. C. Georgesen, *loc. cit.,* 1933.

solution is poured into the bottom of the chamber, the lid is fastened on, and the vessel heated with blowtorches until a desired pressure is registered on the gauge. The maximum pressure used was 350 lb. to the square inch. The chamber proved successful with conglomerates, grits, sandstones, siltstones, and shales, cemented with calcium carbonate, iron oxide, silica, or combinations of the three. In most cases the specimens were either completely disaggregated or so weakened that they crumbled under a rubber mallet.

Fine-grained Sediments

Considerable attention has been devoted to the question of disaggregating and dispersing fine-grained sediments for mechanical analysis. Because of the difficulty of determining in all cases the effect of various agents on the extremely small particles, it has been considered safest to avoid the more rigorous methods used with coarse sediments and in general to avoid the use of harsh chemicals. In addition to the actual chemical changes which may accompany drastic treatment, there is the factor that the clay minerals may be so thoroughly coagulated that they cannot be dispersed without considerable effort. Among fine-grained sediments the processes of disaggregation and dispersion[1] are usually carried out simultaneously, either by the use of physical methods alone or, more commonly, by a combination of physical and chemical methods.

Mud and silt require very little treatment and usually offer no difficulties. Partially or completely indurated rocks, however, or sediments with abundant soluble salts, may yield to no methods. Experimentation with small samples is often necessary before suitable techniques are found, and various tests are available to determine whether dispersion is complete or not.

It is of fundamental importance, theoretically, that the dispersive treatment should be vigorous enough to separate the aggregates into individual particles, but should not break the crystal fragments. The development of any universal technique which lies between these limits may be impossible in practice, but one may approach it to varying degrees in given cases.

---

[1] Disaggregation, as the term is used here, refers to the breaking-down of aggregates into smaller clusters or into individual grains. Dispersion refers to the process of actually separating and dispersing the particles throughout some fluid medium, so that each grain acts as an individual when settling.

PHYSICAL DISPERSION PROCEDURES FOR MECHANICAL ANALYSIS

Prolonged soaking in water. The sample is crushed into small lumps and allowed to soak either in water or in dilute solutions of electrolytes.[1] The period of soaking allows each particle to be surrounded by a film of water or fills the pores of the rock and so loosens the grains and aids in their dispersion. The method is particularly suitable for partially indurated sediments. Rubey,[2] using it in his study of Cretaceous shales, reviewed earlier work and emphasized that the ease of disintegration varies with the moisture content of the sample. Rubey soaked his samples in dilute ammonia for eight weeks, but he pointed out that prolonged soaking may dissolve fine particles and hydrate minerals. In general, the period of soaking depends on the degree of consolidation of the sediment. Dragan[3] recommended a 24-hr. period of soaking in distilled water as advantageous to the dispersion of soils.

Rubbing or trituration in water. The sample is made into a paste with water and rubbed with the finger or a stiff brush or triturated with a rubber pestle. Water is added from time to time, and the dispersed material poured into a beaker, until all the aggregates are destroyed. Whittles[4] emphasized the necessity of wetting the samples gradually so that the water penetrates them throughout. The general method has found considerable favor among analysts.

Shaking in water. Shaking the sediment in water is a widely used dispersion procedure. Reciprocating, end-over-end, and rotary shakers are commonly used. Joseph and Snow[5] considered reciprocating shakers preferable. Periods of shaking varying from 1 to 24 hr. have been recommended. Richter[6] observed some breaking of grains during shaking, and Nolte[7] compared the size-reduction effects of shaking and boiling

---

[1] While the use of electrolytes is a chemical procedure, it is mentioned here as a variation which often accompanies this and following physical procedures. In the discussion, however, the effects of the electrolyte are not considered, inasmuch as that subject will be treated under a separate head.

[2] W. W. Rubey, Lithologic studies of fine-grained Upper Cretaceous sedimentary rocks of the Black Hills region: *U. S. Geol. Survey, Prof. Paper 165A*, pp. 1-54, 1930.

[3] I. C. Dragan, Die Vorbehandlung der Bodenproben zur mechanischen Analyse: *Landwirts. Jahrb.*, vol. 74, pp. 27-46, 1931.

[4] C. L. Whittles, Methods for the disintegration of soil aggregates and the preparation of soil suspensions: *Jour. Agric. Sci.*, vol. 14, pp. 346-369, 1924.

[5] A. F. Joseph and O. W. Snow, The dispersion and mechanical analysis of heavy alkaline soils: *Jour. Agric. Sci.*, vol. 19, pp. 106-120, 1929.

[6] G. Richter, Die Ausführung mechanischer und physikalischer Bodenanalysen: *Int. Mitt. für Bodenkunde*, vol. 6, pp. 193-208; 318-346, 1916.

[7] O. Nolte, Der Einfluss des Kochens und des Schuttelns auf seine Mineralteilchen: *Landwirts. Versuchs-Stat.*, vol. 93, pp. 247-258, 1919.

PREPARATION OF SAMPLES

on carefully separated grades. He found less reduction from shaking than from boiling, although he warned against too long a period of shaking. Olmstead, Alexander, and Middleton [1] considered the breaking of particles to be negligible even after 16 hr. of shaking. Hissink [2] contended that rubbing with a brush had as great a grinding effect as shaking.

There are many types of shaking machine on the market, and others may be constructed at little cost. A simple end-over-end shaker, described by Puri and Keen (see below), consists of a wheel to which two bottles are fastened. The wheel is rotated by a motor. A reciprocating shaker of the type used by the United States Bureau of Soils [3] is illustrated in Figure 10. It consists of a box divided into compartments for nursing bottles, which lie lengthwise in the device. A motor drives a gear system which imparts a to-and-fro motion to the box. This type of shaker has been used with considerable success at the laboratories of the University of Chicago. Rotary shakers are available at various laboratory supply houses; they consist of a metal plate with clamps to hold flasks; during the shaking the flasks are swung through an elliptical motion similar to hand shaking.

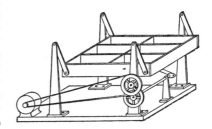

Fig. 10.—Reciprocating shaker, adapted from Briggs, Martin, and Pearce, 1904.

A detailed study of shaking was made by Puri and Keen [4] in 1925. They shook soil samples in water for varying lengths of time and measured the degree of dispersion by the percentage of fine material set free. This percentage, called the "dispersion factor," increased rapidly at first and then more slowly, as shown in Figure 11, adapted from the original paper. The curve was found to agree with the empirical equation $d = a + k \log t$, where d is the dispersion factor, a and k are constants, and t is the time. The test periods extended over intervals as great as 100 hr., but in no case did the dispersion factor reach an upper limit. Dispersion was thus shown to be a continuous function of time, but a 24-hr. period

[1] L. B. Olmstead, L. T. Alexander and H. E. Middleton, A pipette method of mechanical analysis of soils based on improved dispersion procedure: *U. S. Dept. Agric. Tech. Bull. 170*, 1930.
[2] D. J. Hissink, Die Methode der mechanischen Bodenanalyse: *Int. Mitt. für Bodenkunde*, vol. 11, pp. 1-11, 1921.
[3] L. J. Briggs, F. D. Martin and J. R. Pearce, The centrifugal method of mechanical soil analysis: *U. S. Dept. Agric., Bur. of Soils, Bull. 24*, 1904.
[4] A. N. Puri and B. A. Keen, The dispersion of soil in water under various conditions: *Jour. Agric. Sci.*, vol. 15, pp. 147-161, 1925.

of shaking was found sufficient to carry the degree of dispersion over the steep part of the curve. The original moisture content of the sample and the concentration of the suspension were found materially to affect dispersion.

Experiments by Krumbein [1] showed that the effectiveness of shaking also depends on the presence of coarse material in the sediment. A lake clay, having no particles larger than 0.03 mm. in diameter, contained undispersed clay pellets after 6 hr. of shaking, whereas a glacial till with considerable sand was fully disaggregated within an hour. Davis [2] used rubber balls to hasten the dispersion of fine material, and glass beads have also been used. The effect is similar to the sand in till, reducing the time of shaking to a fraction of its previous length. Comparative analyses of two samples of the lake clay, one of which was shaken for 12 hr. without glass beads, and the other 1 hr. with them, showed no differences in the size distribution beyond the limits of experimental error, so that the grinding effect of the glass beads appears to be negligible.

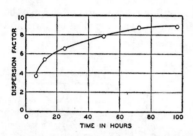

FIG. 11.—Graph of dispersion factor as a function of time. After Puri and Keen, 1925.

Stirring in water. In 1927 Bouyoucos [3] used an electric drink mixer for dispersing soils. Due to the high speed of the stirrer, wire baffles were placed in the cup to prevent circular motion of the suspension. Bouyoucos also [4] compared stirring with shaking and found that 10 min. of the former were more effective than 16 hr. of the latter. In sandy soils he noticed some apparent breaking of the sand grains from more prolonged stirring. The stirrer is one of the most effective physical dispersion devices.

Vibration in water. Among the newer dispersion procedures is the use of rapid vibrations to achieve dispersion. In 1924 Whittles [5] developed a mechanism in which a rapidly vibrating hammer struck the celluloid

---

[1] W. C. Krumbein, The dispersion of fine-grained sediments for mechanical analysis: *Jour. Sed. Petrology*, vol. 3, pp. 121-135, 1933.

[2] R. O. E. Davis, Colloidal determination in mechanical analysis: *Jour. Am. Soc. Agron.*, vol. 17, pp. 275-279, 1935.

[3] G. J. Bouyoucos, The hydrometer as a new and rapid method for determining the colloidal content of soils: *Soil Science*, vol. 23, pp. 319-331, 1927.

[4] G. J. Bouyoucos, Studies on the dispersion procedure used in the hydrometer method for making mechanical analysis of soils: *Soil Science*, vol. 33, pp. 21-26, 1932.

[5] C. L. Whittles, *loc. cit.*, 1924.

bottom of a glass cylinder at controlled frequencies and amplitudes. A fair degree of dispersion was obtained by trituration with a rubber pestle, followed by vibration for one hour at 10,000 vibrations per minute.

In 1931 Olmstead [1] used supersonic waves for dispersion. The vibrations are produced by a piezoelectric quartz crystal immersed in an oil bath and energized by a vacuum tube oscillator. The energy is transmitted through the oil into a flask containing the soil suspension. The sample is vibrated for several 2-min. periods, followed by decantation of the dispersed material. Olmstead compared his method with rubbing and found a close agreement, but the supersonic method was much more rapid.

Ignition. Heating the samples in the dry state has been used by some workers. Nolte [2] mentioned the far-reaching physical and chemical effects of such treatment, and Richter [3] found that among several procedures, ignition gave the least satisfactory results due to the destruction of colloids and the possible fusion of grains.

Boiling in water. The boiling of samples in water or in dilute electrolytes is a procedure about which there has been considerable controversy. In 1919 Nolte [4] furnished a summary of the situation. He concluded that boiling simultaneously reduced the size of the larger particles and coagulated the smaller ones. In the same year Odén [5] observed that boiling destroyed aggregates larger than 10 microns, while particles smaller than 1 micron coagulated to aggregates between 1 and 2 microns in diameter. In 1927 von Hahn [6] referred to boiling as a "barbaric practice" because of its physical and chemical effects on the suspension.

The most important contribution on the subject was made by Wiegner [7] in 1927. He compared the effects of boiling on samples in which the water-soluble salts were either washed out or left in. In the washed samples dispersion was increased by boiling, whereas the salts in the unwashed soils caused coagulation. He thus showed that the same treatment may either prevent or aid dispersion, depending on the presence or absence of appreciable amounts of foreign electrolytes.

Periods of boiling have varied from about 10 min. to more than 40 hr.

[1] L. B. Olmstead, Dispersion of soils by a supersonic method: *Jour. Agric. Research*, vol. 42, pp. 841-852, 1931.
[2] O. Nolte, *loc. cit.*, 1919.
[3] G. Richter, *loc. cit.*, 1916.
[4] O. Nolte, *loc. cit.*, 1919.
[5] S. Odén, Über die Vorbehandlung der Bodenproben zur mechanischen Analyse: *Bull. Geol. Inst. Upsala*, vol. 16, pp. 125-134, 1919.
[6] F.-V. von Hahn, *Dispersoidanalyse* (Leipzig u. Dresden, 1927).
[7] G. Wiegner, Method of preparation of soil suspension and degree of dispersion as measured by the Wiegner-Gessner apparatus: *Soil Science*, vol. 23, pp. 377-390, 1927. (Translated by R. M. Barnette.)

It is doubtful whether boiling for 24 hr. or more aids dispersion, even if no foreign electrolytes are present. Wiegner considered an hour to be sufficient. Krumbein [1] followed the procedure of heating the suspensions to the boiling point but not allowing them to boil. In this manner the agitation due to heating is able to perform its function of dispersion without the disadvantages that may follow a more prolonged application of heat.

A variation of the usual procedure of boiling the samples was used in 1933 by Postel,[2] who used steam agitation to effort dispersion. The clay was placed in a flask, and a copper pipe was inserted through a stopper. The pipe was connected with a boiler which supplied steam at 30 lbs. pressure. The steam condensed in the flask, but in the 10 min. during which it flowed, the clay was sufficiently dispersed to permit elutriation.

Removal of water-soluble salts. The washing-out of foreign electrolytes, as a preliminary treatment of samples, has received increasing attention during the last decade. Although the procedure is essentially physical in nature, detailed discussion is deferred to a later section of this chapter.

CHEMICAL DISPERSION PROCEDURES FOR MECHANICAL ANALYSIS

Leaching in acids or alkalies. Leaching in dilute acids as a preliminary treatment has long been followed by soil scientists. From the viewpoint of sedimentary petrology, however, it cannot be emphasized too strongly that all primary carbonate particles in a sediment should be retained during mechanical analysis, because they are an integral part of the size frequency distribution. When only secondary carbonates are present they may be removed, but when both primary grains and secondary cement are present, the one cannot be removed without destroying the other, and the decision to remove all or none must depend upon the problem at hand.

The digestion of samples in strong alkalies, such as sodium hydroxide, has also been practised. Such treatment is in the same category as acid treatment from the point of view of sedimentary petrology.

Peptization with very dilute electrolytes. The use of small amounts of peptizing electrolytes, such as ammonium hydroxide, sodium car-

---

[1] W. C. Krumbein, *loc. cit.*, 1933.
[2] A. W. Postel, The preparation of clay samples for elutriation by steam agitation: *Jour. Sed. Petrology*, vol. 3, pp. 119-120, 1933.

bonate, or sodium oxalate, is very widespread, and few techniques do not include one or another of them. The purpose of adding these electrolytes is to disperse the sediment into individual particles and to prevent the particles from coagulating during the subsequent analysis. The subject is of such importance in the conduct of mechanical analysis that the following theoretical considerations are included here as an introduction to the process of coagulation.[1]

Stable and Unstable Suspensions

When a single particle settles in water, it does so at a rate which depends in part on its size and shape, and on the nature of the fluid (Chapter 5). When a system of particles settles, they may descend as individuals essentially uninfluenced by their neighbors, or they may coagulate and settle as aggregates. It is obvious that the results of a given analysis are sound only when the former condition holds, since mechanical analysis is an attempt to determine the frequency distribution of the primary particles in the sediment. It is of paramount importance, therefore, that the factors affecting coagulation be known, to guard against introducing errors of considerable magnitude into the results.

Colloidal suspensions differ from true solutions in that the latter are permanently stable, whereas the former are not necessarily so. It is generally accepted that in a stable colloidal suspension each particle has an electric charge arranged about it in a double layer. The charge may be positive or negative, depending in part on the nature of the colloid. The nature of the double layer may be visualized as follows: In a solid particle the atoms are held in a crystalline structure, and within the interior of the particle each atom is balanced, in terms of its valence, with corresponding atoms of other elements. The atoms at the boundary of the particle, however, are only partially satisfied in terms of their valence, and they are therefore capable of attracting a swarm of ions from the surrounding fluid. Of this swarm of ions, either the positive or negative ions (depending upon the nature of the particles and of the ions in solution) arrange themselves alongside the solid particle and constitute the inner layer, which gives the particle its charge. Meanwhile, the oppositely charged ions in solution swarm about the inner layer. Thus is built up a double layer of ions. More technical details of the double layer may be found in standard reference books on colloids.[2]

Under the influence of Brownian movement the charged colloidal particles are brought into the vicinity of others, but as long as the charges are above a critical potential (Figure 12) the particles repulse each other and adherence is prevented. If the charges are below the critical potential, or zero, the particles may adhere when they collide, with the result that aggregates are formed. These aggregates begin to settle and eventually the entire dispersed phase may settle out of suspension as a flocculent precipitate. The rate of

[1] The terms *coagulation* and *flocculation* appear to be used synonymously by many authors. The former has been chosen here for the sake of consistency.

[2] See for example H. R. Kruyt, *Colloids,* translated by H. S. Van Klooster, 2nd ed. (New York, 1930), pp. 110 ff.

coagulation may be either slow or rapid, depending upon whether or not any charges are present on the particles.

The magnitude of the charge varies appreciably with slight changes in the electrolyte content of the suspension, and the effect of a given electrolyte appears to vary with the nature of the colloid. In clays, for example, the charges may be reduced by adding calcium chloride. Such electrolytes are called *coagulants*. Other electrolytes, as sodium carbonate, increase the charges on the clay particles and are called *peptizers*. Beyond certain limits of concentration the peptizing electrolytes also cause coagulation, so that part of the distinction between the two types may be due to the relative concentrations necessary to produce coagulation.

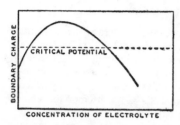

FIG. 12.—Diagram of boundary charge as a function of electrolyte content in colloidal suspensions, after Kruyt. When the curve is above the critical potential, the suspension is stable.

Two types of coagulation are recognized, *perikinetic* and *orthokinetic*. Perikinetic coagulation occurs in systems where essentially no sedimentation is taking place and where the probability of collision is equally likely in any direction, due only to the chaotic Brownian movement. Orthokinetic coagulation occurs in sedimenting systems where the probability of contact is greater in some directions than in others, due to the downward motion of the settling particles.

Perikinetic coagulation. Inasmuch as perikinetic coagulation involves no sedimentation of the individual particles, it is not as important from the point of view of mechanical analysis as orthokinetic coagulation. However, it is not out of place to consider the subject briefly. Von Smoluchowski[1] was the first to develop a mathematical theory of coagulation. He considered the factors that must be present before coagulation takes place. It is clear that particles can only adhere if they collide, and hence the probability of collision is of primary importance. Once they have collided it is necessary to consider the conditions under which they adhere, so that the probability of adherence is the second important factor. The probability of collision is controlled by the Brownian movement, and the probability of adherence is controlled by the electric charges on the particles. Von Smoluchowski first considered the progress of coagulation in a monodisperse system where no electrical charges were present on the particles, so that every collision resulted in adherence. Here the probability of adherence is 1.

In setting up his theory, he considered that each particle of radius r has about it a sphere of attraction of radius R, such that any other particle whose center enters this sphere of attraction is united to the first. Now the probability w that another particle will move into the sphere of attraction of a given particle, the latter being considered motionless, is $w = 4\pi DR$, where D is the displacement due to Brownian movement. From this starting point, von Smoluchowski developed an equation showing the number of primary particles

---

[1] M. von Smoluchowski, Versuch einer mathematischen Theorie der Koagulationskinetik kolloider Lösungen: *Zeits. Phys. Chem.*, vol. 92, pp. 129-168, 1916-1918.

remaining after a given time t. Similarly, he considered the formation of aggregates having two, three, and more primary particles and the change in their number with time. Figure 13, taken from his paper, shows the variation in the number of all particles, (Σn), of primary particles (n_1), of dyads (n_2), and of triads (n_3). The ratio n/n_0 is plotted as ordinate, and t/T as abcissa. The symbol n_0 represents the original number of particles present, and T is a measure of the rate of coagulation. It is clear that n and n_1 start at the point $n/n_0 = 1$ and decrease continually. The number of dyads, however, is zero at the start, but it rapidly rises to a maximum and then decreases as the number of triads becomes prominent. Similarly, the more complex aggregates all show a maximum at points successively farther to the right along the x-axis. This simple case of von Smoluchowski's theory was experimentally verified by several workers.

Von Smoluchowski next considered the case in which the charges on the particles were below the critical potential but not equal to zero. In this case every collision does not result in adherence, so that the rate of coagulation is slower than in the first instance. The net effect of the slower rate of coagulation on the mathematical theory was the insertion of the probability ξ, less than 1, for the certainty which distinguished the rapid coagulation.

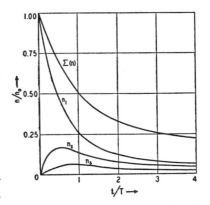

FIG. 13.—Progress of coagulation in a monodisperse system, after von Smoluchowski, 1916.

In 1926 Müller[1] developed a theory for the rapid perikinetic coagulation of bidisperse systems in which two sizes of particles were present. Müller's theory was experimentally verified by Wiegner and Tuorila.[2] The conclusions drawn were that the coagulation of bidisperse systems is more rapid than that of monodisperse systems, and that when the number of one size of particles is great compared to the number of the other, the rate of coagulation approaches that of a monodisperse system composed of the dominant-sized particle.

Orthokinetic coagulation. In mechanical analysis larger particles are continually overtaking smaller ones by virtue of their greater settling velocities. Hence the probability of collision is large, and it may be expected that coagulation would proceed at a rapid pace. The theory underlying this type of coagulation was developed by Tuorila.[3] Wiegner[4] published an excellent summary of the work done in this field, including also the earlier work of Von

---

[1] H. Müller, Die Theorie der Koagulation polydispersen Systeme: *Kolloid Zeits.,* vol. 38, pp. 1-2, 1926.
[2] G. Wiegner and P. Tuorila, Ueber die rasche Koagulation polydisperser Systeme: *Kolloid Zeits.,* vol. 38, pp. 3-22, 1926.
[3] P. Tuorila, Ueber orthokinetische und perikinetische Koagulation: *Kolloidchem. Beihefte,* vol. 24, pp. 1-122, 1927.
[4] G. Wiegner, Ueber Koagulationen: *Kolloid Zeits.,* vol. 58, pp. 157-168, 1932.

Smoluchowski and Müller. The present discussion, including parts of the preceding material, is based on these papers.

In developing his theory, Tuorila considered suspensions in which there were less than 10^8 particles per c.c., so that perikinetic coagulation did not take place during the time of observation. He set up the assumptions that large particles attract smaller ones in an attraction volume having a cross-sectional area of $\pi(A^2-R^2)$, where $A=(R+r)$, the sum of the radii of large and small particles. In an element of time, the large particles settle a distance L, so that the volume occupied by the attraction zone is $\pi(A^2-R^2)L$. This volume was designated by Tuorila as the *Hautraumvolumen* of the particles, and was represented by b. Since each large particle has this attraction volume, N particles would have $Nb = \pi(A^2-R^2)LN = B$ for their total attraction volume. Into this formula Tuorila inserted Stokes's Law (Chapter 5) of the radius, and obtained as his final result the expression

$$B = KMr(2 + S - 2S^2 - S^3)$$

where $S = r/R$, K is a constant, and M is the mass of large particles per c.c. It is clear from this expression that the total attraction volume, B, is proportional to the weight of large particles in the suspension, and dependent on the function $r(2 + S - 2S^2 - S^3)$. Since $S = r/R$, B is dependent on a function of the ratio of the radii present. When $r = 0$, $B = 0$, and the system is monodisperse. Likewise when $r/R = 1$, $B = 0$. The value of B as a whole may thus vary between $r/R = 0$, and $r/R = 1$. By plotting the function $(2 + S - 2S^2 - S^3)$ against r/R, Tuorila found the values were nearly constant between values of r/R from 0.0 to 0.4.

Tuorila reasoned that of n small particles per c.c. of suspension, the number nB would be in the attraction volume B, and hence swept along. Thus, the decrease in small particles, dn, per unit of time dt would be $dn/dt = -nB$, where the minus sign indicates a decreasing function. This differential equation yields the negative exponential.

$$n/n_0 = e^{-Bt}$$

where n is the number of small particles in suspension at any instant, and n_0 is the original number. This function requires that B be a constant; Tuorila's analysis of the function $(2 + S - 2S^2 - S^3)$ indicates that this condition is satisfied over the range indicated, i.e., $r/R < 0.4$. As sedimentation proceeds, the larger particles settle to successively deeper zones in the suspension, so that their effects on the small particles in a given zone are limited to the time required for the large particles to settle through the zone.

Tuorila confirmed his theories in a series of experiments, and much of the material that came to light is pertinent to the subject of mechanical analysis. The experiments showed that polydisperse systems in which the charges on the particles were above a given critical potential showed no coagulation effects, even in concentrations as high as 150 g. of solid per liter. When the charge was below the critical potential, orthokinetic coagulation took place, during which the larger particles swept the smaller ones along with them. The effects of orthokinetic coagulation were also found to increase rapidly with an increase in the concentration of the suspension.

In quartz suspensions, particles larger than 20 microns in radius (0.04 mm.

PREPARATION OF SAMPLES

diameter) did not enter into the coagulation. Thus an upper limit of coagulation was established, which appears to vary with the nature of the material being studied, inasmuch as in a clay the upper limit was higher. Quartz particles between 10 and 20 microns acted only slightly on the smaller particles; between 6 and 10 microns the effect was much stronger, and it reached a maximum at 5-6 microns. Particles under 4 microns in radius were completely swept out of suspension during the process.

Tuorila pointed out that in many instances coagulation is very slight at the beginning and hence hardly noticeable. As time goes on the effect increases rapidly, and toward the end of the process it slows down again. Thus an S-shaped curve results; Figure 14, taken from Tuorila's paper, illustrates the case. This effect may also occur in monodisperse systems, where perikinetic coagulation produces large aggregates which begin to settle and thus exert an orthokinetic effect on the remaining smaller particles. This orthokinetic effect naturally increases as the number of larger aggregates increases, until most of the particles are coagulated, when the process slows down again.

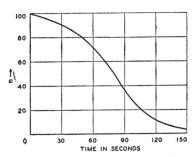

FIG. 14.—Progress of coagulation as a function of time, after Tuorila.

Summary of coagulation. The work of Tuorila and others has an important bearing on mechanical analysis. Inasmuch as sediments are polydisperse systems, the phenomena of orthokinetic coagulation may manifest themselves. In this type of coagulation the larger particles drag along the smaller ones and thus hasten coagulation effects. Likewise, orthokinetic coagulation proceeds at an accelerated pace in concentrated suspensions. This suggests that dilute suspensions may be preferable to concentrated suspensions for analysis.

General colloidal theory also indicates that coagulation does not take place if the particles are charged above a critical potential, and this state appears to be associated either with certain peptizing electrolytes or with suspensions entirely free from electrolytes. As a result, much thought has been devoted to the elimination of all electrolytes from the suspension or to the discovery of a peptizing electrolyte that may be applied to the widest possible range of sedimentary types.

A number of tests have been devised for determining whether coagulation has occurred in a suspension. These tests are discussed in detail in the generalized dispersion routine given at the end of this chapter.

Peptization procedures. Among chemical agents which have long been used for dispersing soils and sediments are ammonium hydroxide and sodium carbonate. The former was adopted by Briggs, Martin, and Pearce[1] for the United States Bureau of Soils in 1904, and sodium carbonate appears to have been introduced by Beam[2] in 1911. These pep-

---

[1] L. J. Briggs, F. O. Martin and J. R. Pearce, *loc. cit.*, 1904.
[2] W. Beam, The mechanical analysis of arid soils: Abst. in *Exp. Sta. Record*, vol. 25, p. 513, 1911.

tizers found wide favor among analysts until comparatively recent years, and they are still used at present, despite the competition offered by numerous other agents.

The use of ammonium hydroxide for dispersion was systematized by Odén[1] in 1919, when he developed his "normal method," which involves rubbing the soil or sediment with a stiff brush, adding ammonium hydroxide to a concentration of N/100, and shaking the suspension for 24 hr.

Odén's "normal method" was used by numerous workers. Correns and Schott,[2] in 1932, found it preferable to other methods for general work. For recent marine sediments, however, they recommended dialysis (see page 66).

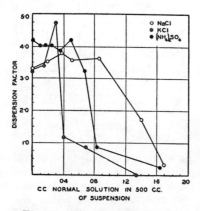

Fig. 15.—The dispersing effect of several electrolytes, after Puri and Keen, 1925. Relatively small concentrations of these electrolytes cause coagulation.

A detailed study of the effect of peptizers on soil suspensions was made by Puri and Keen[3] in 1925. This study marked a new epoch in the investigation of peptizers by comparing the effects of a number of electrolytes on dispersion, and it paved the way for further detailed studies which are being continued to-day.

Puri and Keen studied the effects of several electrolytes on soil suspensions which had previously been washed free of soluble salts. Varying amounts of the electrolyte were added and the degree of dispersion measured and plotted. It was found that the sodium carbonate curve displayed a prominent plateau, which indicated that quite a range of concentration caused approximately the same degree of dispersion. The plateau-effect is a distinguishing feature of good peptizers because it allows some flexibility in the concentration that may be used. Other electrolytes, as KCl, had a much smaller dispersive effect and displayed a sharp peak at the optimum concentration. The effects of several electrolytes are shown in Figure 15, adapted from Puri and Keen. Puri and Keen concluded that the effects of electrolytes on suspensions are not abrupt but

[1] S. Odén, loc. cit., 1919.
[2] C. W. Correns and W. Schott, Vergleichende Untersuchungen über Schlämm- und Aufbereitungsverfahren von Tonen: Kolloid Zeits., vol. 61, pp. 68-80, 1932.
[3] A. N. Puri and B. A. Keen, loc. cit., 1925.

PREPARATION OF SAMPLES 63

cause a continuous change in the degree of dispersion as the electrolyte concentration is varied. They related their results to the phenomenon of base exchange, but no detailed explanation was attempted at the time.

Winters and Harland [1] also studied the effects of sodium carbonate on dispersion. Their results agreed with Puri and Keen and showed that the dispersion effects vary somewhat with the soil horizon.

Olmstead, Alexander, and Middleton [2] compared several peptizers and decided that sodium oxalate was the most satisfactory. They pointed out that sodium and ammonium hydroxides yield good results when the calcium and magnesium carbonates have been removed by acid treatment and thorough washing. If calcium carbonate is present, sodium carbonate is better than either hydroxide because the carbonate decreases the solubility of the calcium carbonate, while the hydroxides produce coagulating calcium ions. Sodium oxalate, they found, was even better than sodium carbonate, because the calcium ions are completely removed by the oxalate. A comparison of the four peptizers on four soils showed that the oxalate had the greatest dispersive effect in every case.

Loebe and Köhler [3] also studied the dispersive effects of sodium oxalate and found it best suited for general work.

Krumbein [4] performed a series of experiments with sodium oxalate to determine whether it had a plateau-effect like sodium carbonate. Water suspensions containing 2.5 per cent of an unconsolidated calcareous Pleistocene lake clay were prepared by brush rubbing, and varying amounts of N/5 sodium oxalate or N/5 sodium carbonate were added. The percentage of material under 1 micron in the suspension was then measured by pipetting. Figure 16 shows the resulting curves. It is clear that sodium oxalate has both a greater dispersive effect and a wider range of safety. In both cases a concentration of about N/100 is optimum, as deduced from these curves.

Ungerer [5] conducted a detailed investigation of several methods of preparing soils for mechanical analysis. Tests were made with lithium chloride and lithium carbonate, by comparing the amounts of material smaller than 2 microns in the suspensions. Lithium chloride was found

---

[1] E. Winters, Jr., and M. B. Harland, Preparation of soil samples for pipette analysis: *Jour. Am. Soc. Agron.*, vol. 22, pp. 771-780, 1930.
[2] L. B. Olmstead, L. T. Alexander, and H. E. Middleton, *loc. cit.*, 1930.
[3] R. Loebe and R. Köhler, Beiträge zur Praxis der Schlämmanalyse: *Mitt. a. d. Lab. Preuss. Geol. Landesanst.*, vol. 11, Berlin, 1932.
[4] W. C. Krumbein, *loc. cit.*, 1933.
[5] E. Ungerer, Korngrössenbestimmungen nach dem Dekantier- und Pipettverfahren unter dem Einfluss verschiedener Vorbehandlungsmethoden: *Zeits. f. Pflanzenernährung., Düng, u. Bodenk.*, vol. 26A, pp. 330-336, 1932.

well suited to soils, whereas lithium carbonate was not recommended.

Vinther and Lasson [1] studied the effects of several electrolytes on the dispersion of kaolin, including sodium carbonate, ammonium hydroxide, lithium carbonate, calcium citrate, potassium silicowolframate, and sodium pyrophosphate. The sodium pyrophosphate in 0.002 $m.$ concentration was found most suitable. Lithium carbonate was next best after the sodium pyrophosphate. In all cases a 17-hr. period of shaking was used to effect dispersion.

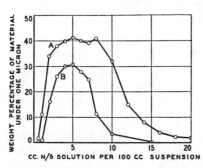

Fig. 16.—Effects of sodium oxalate (A) and sodium carbonate (B) on the dispersion of a lake clay.

Among recent detailed studies of dispersion, the work of Földvári [2] is of considerable importance. He compared the effects of ammonium hydroxide, sodium oxalate, and sodium metasilicate on a wide variety of soils and sediments. The choice of these three dispersing agents respectively was determined by Földvári's classification of peptizers into three broad groups: (1) those which supplied (OH)— ions to the suspension; (2) those which depended for their effects in part at least upon the removal of coagulating ions (notably Ca++) from the suspensions; and (3) those which supplied the particles with a "protective cover" and thus prevented the coagulating ions from reaching the particles. As a result of his comparative analysis, Földvári concluded that 0.005 N sodium oxalate gave the best results in most routine cases but, for sediments with a high content of gypsum or calcite, sodium metasilicate afforded the best dispersion. The concentration of the latter dispersing agent was 1 c.c. of waterglass (36°-38° Be) per liter of suspension.

The influence of base exchange phenomena on dispersion was investigated by Thomas.[3] He found that soils in which the exchangeable bases were replaced by sodium were most readily dispersed by sodium carbonate, while a magnesium soil flocculated immediately on the addition of sodium carbonate, due probably to the release of magnesium ions. Because of the difficulty of preparing sodium soils for routine purposes,

[1] E. H. Vinther and M. L. Lasson, Über Korngrössemmessungen von Kaolin- und Tonarten: *Ber. Deutsch. Keram. Ges.,* vol. 14, pp. 259-279, 1933.

[2] A. Földvári, Über die Wirkung einiger Tonstabilisitoren: *Kolloid-Beihefte,* vol. 44, pp. 125-170, 1936.

[3] M. D. Thomas, Replaceable bases and the dispersion of soil in mechanical analysis: *Soil Science,* vol. 25, pp. 419-427, 1928.

PREPARATION OF SAMPLES 65

Thomas recommended freeing the sample of exchangeable bases by acid treatment, followed by dispersion with sodium carbonate.

Soil scientists have devoted considerable attention to the subject of base exchange during the last decade, and numerous dispersion procedures make use of chemical agents which effect an elimination of coagulating ions by removing them by base exchange reactions.

Robinson [1] investigated the subject in a comprehensive treatise in 1933. He compared several methods of dispersion which involved base exchange phenomena, notably the International-A method, which relied on the use of HCl to remove the exchangeable bases and subsequent dispersion with ammonium hydroxide; the Sudan method,[2] which involves the direct use of 0.05 per cent sodium carbonate to effect dispersion; and the Puri method,[3] which consists essentially of removing the exchangeable bases with sodium chloride to yield sodium clay. Sodium hydroxide is added if necessary, to obtain an alkaline suspension. Robinson modified the International-A method by treating the soil with 4 c.c. of N NaOH per 10 g. of soil after acid treatment, instead of with ammonium hydroxide. The modified technique was called the International-soda method, and Robinson's work showed it to yield the best results for most types of soil. Sodium oxalate, on the other hand, while satisfactory in many cases, effected only incomplete dispersion with lateritic and ferruginous soils.

A problem of considerable importance in sedimentary petrology, and one which is not involved in most cases of soil analysis, is the content of primary carbonate particles in the size distribution of sediments. Sedimentary analysis is usually performed on unaltered or unweathered samples, in which the carbonate particles may represent an appreciable part of the size distribution. Acid treatment in such cases will seriously affect the analytical results by removing acid-soluble material and distorting the resulting statistical data. Methods involving base exchange, and relying on the formation of hydrogen clay, thus appear to be unsuitable to sediments in general, although they are applicable to non-calcareous sediments.

A more recent study by Puri [4] involved a modification of his original

[1] G. W. Robinson, The dispersion of soils in mechanical analysis: *Imp. Bur. Soil Sci., Techn. Comm. 26*, 1933.
[2] A. F. Joseph and F. J. Martin, The determination of clay in heavy soils: *Jour. Agric. Sci.*, vol. 11, pp. 293 ff., 1921. A. F. Joseph and O. W. Snow, *loc. cit.*, 1929.
[3] A. N. Puri, A new method of dispersing soils for mechanical analysis: *India Dept. Agric. Mem., Chem. Series*, vol. 10, pp. 209-220, 1929.
[4] A. N. Puri, The ammonium carbonate method of dispersing soils for mechanical analysis: *Soil Science*, vol. 39, pp. 263-270, 1935.

dispersion technique by the use of ammonium carbonate to effect base exchange. In his newer technique, the exchangeable bases are replaced with ammonia by boiling the soil with N ammonium carbonate solution. Boiling is continued until the volume of the solution is reduced by half, whereupon 4 to 8 c.c. of N sodium hydroxide or N lithium hydroxide is added per 10 g. of soil, to effect dispersion. Puri favored the lithium hydroxide in preference to sodium hydroxide because of its greater dispersive effect.

Puri's new method is somewhat more drastic than others in terms of the boiling involved, and for sediments it may be slightly modified as will be discussed later (page 75). The great advantage of Puri's approach is that it eliminates the need for acid treatment and makes available for sediments a dispersion technique which applies the advantages of base exchange phenomena.

Removal of water-soluble salts. Most sediments and soils contain water-soluble salts in varying amounts, and it is to be expected that the foreign electrolytes thus introduced into the suspension may have a marked effect on dispersion. Cations that appear to be commonly present in sediments are varying amounts of Ca^{++}, Na^+, Fe^{+++}, and Mg^{++}. Common anions are $SO_4^=$, $CO_3^=$, and Cl^-. Gypsum is an important salt in non-calcareous sediments, whereas calcium carbonate or bicarbonate is more common in the calcareous types.

Perhaps the most important study of the effects of foreign electrolytes was made by Wiegner[1] in 1927. Wiegner pointed out that if very small amounts of soluble salts are present, the charges on the particles are above the critical potential, whereas if appreciable amounts are present, the charges are below the critical potential, and dispersion may be seriously hindered. Dispersion procedures, such as shaking or boiling, increase the agitation of the particles and thus increase the number of collisions among them. If the charges are above the critical potential this added movement increases dispersion, whereas if they are below the critical potential the added collisions increase the rate of coagulation and thus slow down or prevent dispersion.

Wiegner compared the effects on soils of (1) shaking for 6 hr., (2) rubbing for an hour with a brush, and (3) boiling for an hour with a reflux condenser. The analyses were conducted in a Wiegner tube (Chapter 6), in which the same sample could be used after various treatments, because none of it is removed during the analysis. It was

[1] G. Wiegner, *loc. cit.,* 1927.

found that boiling was more effective than rubbing or shaking on washed soils, while shaking was the most effective on unwashed soils. The effect of N/10 ammonia was tested on washed and unwashed soils, and it was found that the washing-out of the foreign electrolytes was more effective than the use of the peptizer on unwashed soils.

Numerous other workers have discussed the removal of water-soluble salts either as a standard procedure in the routine analysis of soils or sediments or in connection with samples which do not respond to direct dispersion with peptizers. Among writers who have included the procedure in their methods are Olmstead, Alexander, and Middleton;[1] Correns and Schott,[2] who recommended it for recent marine sediments; Gessner,[3] Gallay,[4] Robinson,[5] and others.

Several methods are available for washing the sediments, and one of the simplest is by means of Pasteur-Chamberland filters. (See Figure 17.) Olmstead, Alexander, and Middleton's technique in this connection is effective:

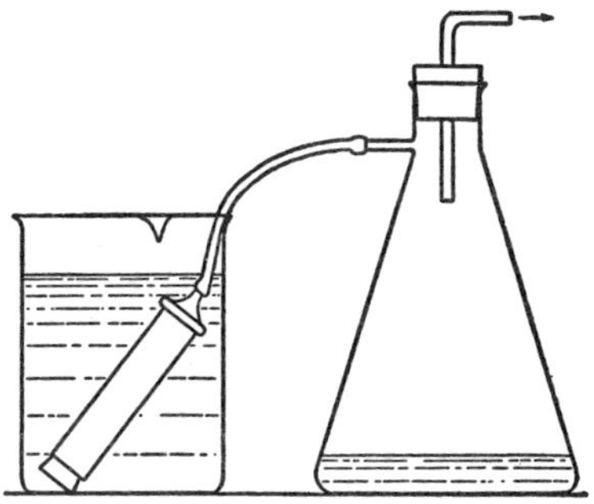

FIG. 17.—Apparatus for removing water-soluble salts by suction filtration.

The lower 12 cm. of the filter is sawed off, and fitted with a removable stopper. The suspension of sediment is placed in a beaker, and the filter, attached to a suction pump, is immersed within it. The suction is continued until as much as possible of the liquid is removed. The liquid within the filter is then removed by extracting the stopper, and the filter core is filled with distilled water. The stopper is re-inserted and back pressure is applied by means of a rubber bulb, to remove the material adhering to the outside of the filter. Additional distilled water is added to the beaker, and the process of washing is repeated. Usually six washings are sufficient to remove the soluble salts.

Robinson used a Buchner funnel fitted with 9-cm. diameter hardened filter paper (Whatman 50). The paper was fixed to the funnel with cellulose cement, after etching the funnel with hydrofluoric acid to insure adhesion. The funnels were fitted to filter flasks arranged in a battery of four attached to a single pump. Three washings with 20-30 c.c. of water each were generally used on soils.

[1] L. B. Olmstead, L. T. Alexander and H. E. Middleton, *loc. cit.*, 1930.
[2] C. W. Correns and W. Schott, *loc. cit.*, 1932.
[3] H. Gessner, *Die Schlämmanalyse* (Leipzig, 1931), pp. 164 ff.
[4] R. Gallay, *Kolloid-Beihefte*, vol. 21, pp. 431 ff., 1925.
[5] G. W. Robinson, *loc. cit.*, 1933.

The washing-out of foreign electrolytes is at best a tedious process, and some workers have taken a pragmatic view of the problem. That is, for average sediments without a large content of water-soluble salts, dispersion is effected by the use of a peptizer such as sodium carbonate or oxalate, or lithium chloride or hydroxide. Only in those cases where coagulation occurs despite this treatment is washing resorted to.

GENERAL CRITIQUE OF DISPERSION

The extensive literature on dispersion demonstrates that the problem is no simple one. Among the variables that enter the situation with respect to the dispersive effect of a given electrolyte are its composition, its concentration, associated base exchange phenomena, and the presence of foreign electrolytes. Moreover, the general problem is further complicated by the fact that all sizes of particles are not equally sensitive to dispersion or coagulation. Tuorila showed that in quartz suspensions the effects of coagulation begin to manifest themselves at diameters of about 0.04 mm., and become very pronounced between 5 and 10 microns, while particles under 4 microns are completely removed from suspension. Thus sediments made up predominantly of particles in the most sensitive range may be expected to be strongly affected by slight changes in the dispersion technique.

The interplay of numerous variables, some of which are independent of the others, strongly suggests that there can be no single dispersion technique for all types of materials. This statement has been repeated by numerous writers, and current researches appear increasingly to verify it. Soil scientists have made considerable progress toward standardized routines which apply to a wide range of soils, but among sediments there are problems of degrees of alteration and of induration which greatly complicate the problem.

Dispersion has been found to be a continuous process, rapid at first and slower later on. Puri and Keen showed this in their experiments on shaking, and Olmstead noted the same effect in his work on vibration. If this is universally true, it appears that complete dispersion can never be effected, or that there is a continuous increase of fine material due to the disruption or attrition of individual grains. Clark[1] raised the question whether there exists in soils any unique size frequency dis-

---

[1] C. L. Clark, The dispersion of soil-forming aggregates: *Soil Science*, vol. 35, pp. 291-294, 1933.

tribution or whether the distribution is not a function of the dispersion process. The difficulty of settling the question lies in the fact, as Clark pointed out, that the dispersion of aggregates cannot readily be distinguished from the disruption of crystal fragments by their end-products. This appears to be particularly true of the finer particles in the sediments.

Among fine-grained sediments in which few authigenic changes have taken place there should theoretically be little difficulty in effecting dispersion into the individual particles, but even here there is an increase of fine material with an increase of vigor or time in the dispersion process. If the sediment is indurated or altered by weathering, the original size distribution may have suffered considerable change due to secondary growth, dehydration, the introduction of secondary minerals, or the leaching-out of certain constituents. The reconstruction of the original distribution may accordingly be nearly impossible. Clark's point thus applies in part to sediments as well, and the problem raised is not one that can be readily solved.

Soil scientists, working in coöperation on several dispersion procedures, discovered that results obtained in different laboratories on the same soils were not consistent. As a result, considerable effort has been expended to develop standard methods which are free of subjective errors, and which may yield comparable results. In a recent paper Novak[1] compared several dispersion procedures and commented on the lack of uniformity among various laboratories. Sedimentary petrologists have not yet united in an attempt to adjust such difficulties in connection with sediments. Unquestionably the complexity of the general problem and the unknown influence of some of the variables account in part for the difficulties encountered.

The relative merits of chemical and physical methods of dispersing sediments have recently been investigated by Neumaier,[2] who reached the conclusion that chemical methods should not be used, but that reliance should be placed entirely upon physical means. Neumaier, however, included mainly earlier papers in his critique.

The relative merit of analyzing samples in their natural moist condition and analyzing air-dried samples has also been the subject of controversy. Some writers maintain that dried sediments undergo changes which seriously affect the subsequent analysis of the sample. Correns and

[1] W. Novak, Vorbehandlung der Bodenproben zur mechanischen Bodenanalyse: *Proc. 2nd Int. Congr. Soil Sci.*, vol. 1, pp. 14-39, 1932.
[2] F. Neumaier, *loc. cit.*, 1935.

Schott[1] investigated the problem in 1933 and reached the conclusion that the samples should not be dried. Földvári[2] contended, however, that in the case of ancient sediments the vicissitudes through which the material has passed render relatively meaningless the accidental moisture state in which the sample may have been found at the time of sampling. Among other writers who have expressed themselves on the question of damp vs. dry samples are von Sigmond,[3] Richter,[4] Hissink,[5] and Neumaier.[6]

GENERALIZED DISPERSION ROUTINE

The entire subject of dispersion revolves about a point earlier mentioned: the aggregates must be destroyed without affecting the sizes of the individual particles. There is a further condition implied in this process: the dispersed particles should not form aggregates again during the course of the analysis. To satisfy these conditions, shearing stresses and abrasive action on the individual particles should be kept at a minimum, while disaggregation and dispersion should be at a maximum.

Gessner[7] devoted considerable space to the dispersion of samples and presented a general routine which involved several procedures, each followed by tests for dispersion and coagulation, so that the treatment given depends on the difficulty of dispersing the material. In his routine the sample is shaken and then tested for coagulation. The test shows either complete dispersion, incomplete dispersion without coagulation, or coagulation. The first case is analyzed, the second is boiled, the third is washed. The boiled sample is tested and either analyzed or washed. The washed samples are either analyzed or rewashed.

As a result of a series of comparative tests made on certain of the dispersion procedures received earlier, Krumbein[8] developed a routine for dispersing fine-grained sediments in which the procedures become successively more vigorous, so that only the more resistant sediments receive the most vigorous treatment. Figure 18 shows the routine graph-

[1] C. W. Correns and W. Schott, Über den Einfluss des Trockens auf die Korngrossenverteilungen von Tonen: *Kolloid Zeits.,* vol. 65, pp. 196-203, 1933.
[2] A. Földvári, *loc. cit.,* 1936.
[3] A. A. J. von 'Sigmond, Bericht über den Int. Kom., u. s. w: *Int. Mitt. f. Bodenk.,* vol. 4, pp. 25-27, 1914.
[4] G. Richter, *loc. cit.,* 1916.
[5] D. J. Hissink, *loc. cit.,* 1921.
[6] F. Neumaier, *loc. cit.,* 1935.
[7] H. Gessner, *op. cit.,* p. 167, 1931.
[8] W. C. Krumbein, *loc. cit.,* 1933.

ically. The foreign electrolytes are washed out only when the need is indicated. In every case the sample is soaked in dilute peptizer for a preliminary period. This is followed by one of two sequences, depending on the amount of material above 0.06 mm. diameter in the sediment. By grinding a fragment between the teeth or rubbing between finger and

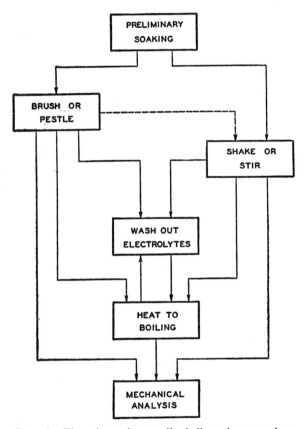

FIG. 18.—Flow-sheet of generalized dispersion procedure.

thumb, the amount of sand present may be estimated. No fixed proportion is involved, but the authors have found that in most cases, if only a trace of coarse material is present, brushing or pestling is more effective than shaking; while, if there is an appreciable amount of sand, shaking or stirring is preferable. This is particularly true of unconsolidated sediments, but in partially indurated cases it is desirable to brush the sample before shaking it. The dashed line in the figure indicates this possibility. The several arrows illustrate the paths that may be followed. An attempt was made to allow some flexibility and yet to pass from gentle to more vigorous steps as the need was felt.

Among the many dispersing agents that have been used for soils and sediments, sodium oxalate appears to be most favorable. In comparative tests it usually ranks among the best for routine purposes, and it has been adopted by the authors as the standard agent. The following routine accordingly describes dispersion in terms of sodium oxalate, but the reader may substitute other dispersing agents if he wishes. Among the older dispersers sodium carbonate (in concentrations from N/25 to N/100) is very effective in some cases. The writers have found it preferable to use sodium carbonate for weathered phases of glacial till. Among the newer dispersing agents lithium hydroxide has proved successful in many cases, and may be used in concentrations of about N/50 or N/100. In connection with dispersion procedures involving base exchange, the authors favor sodium hydroxide, but lithium hydroxide has also found considerable favor among some workers.

The following description of the several procedures is arranged in the order of the two main sequences, but any steps common to both are described only once.

The air-dried [1] sediment is crushed with a rubber-tipped pestle, a rolling pin, or a wooden mallet until the fragments are reasonably small. A test sample is then quartered and weighed. The weight of the sample depends on the range of sizes present. It is desirable to have a suspension of 2 to 3 per cent concentration for analysis by modern methods, so that about 25 g. are optimum for a liter of suspension. If the sample contains say 25 per cent of sand, the test sample should weigh about 30 g. When the sand is later sieved off, the remaining fine material will yield a suspension of about 2.3 per cent concentration.

The quartered sample is placed in a 250-c.c. Erlenmeyer flask with 100 c.c. of N/100 sodium oxalate solution [2] and allowed to soak for a period depending on the rate at which the lumps disaggregate. A minimum of 24 hr. and a maximum of eight weeks are possible limits to the time. Some Pennslyvanian shales required ten days of soaking before they yielded to the brush, while others needed only 24 hr. By occasional shaking the process of disintegration may be observed, and if the sample is taken out too early it may be returned to the flask after brushing or pestling is found ineffective on the lumps.

[1] If it is preferred to work with sediments in their natural moist condition, the moisture content may be determined from a separate sample.

[2] It is convenient to have an N/5 solution of sodium oxalate on hand, prepared by dissolving 13.4 g. of the salt in a liter of water. The N/100 solution is made as needed by adding 5 c.c. of this solution to 95 c.c. of water.

PREPARATION OF SAMPLES

After the preliminary soaking the sample is poured into an evaporating dish and rubbed with a stiff brush or rubber pestle. The writers favor a brush for the purpose. As the lumps disaggregate, water is added and the dispersed material poured into a beaker. The water used during brushing should have a concentration of N/100 sodium oxalate. The brushing process often requires the better part of an hour, but the final results justify the use of ample time. Whittles[1] prepared a rubber pestle by filing down a stopper and attaching it to a glass rod. Such pestles are advantageous as an aid to brushing, because the more resistant lumps may be gently crushed before brushing.

The volume of suspension after brushing may be about 400 c.c. It is tested for complete dispersion by placing a drop on a slide under a coverglass, allowing it to rest for a few minutes, and examining it under the microscope. If each grain stands out as an individual, and the smaller ones display Brownian movement, the dispersion is probably complete. If bead-like strings and clusters of individual particles extend through the field, the suspension is coagulated. If the field shows a mixture of individual grains and aggregates, not clustered together, coagulation is probably absent, but dispersion is incomplete. In practice there is considerable gradation among these three situations, and it is often difficult to distinguish among them. Further, if slow coagulation is present, the suspension may remain apparently dispersed for several hours. As a final test the writers allow the apparently fully dispersed suspension to stand overnight; if visible coagulation has not set in within that time, it will not occur during the analysis. In a very few cases suspensions have remained apparently dispersed for two or three days and finally coagulated.

If coagulation is rapid, a flocculent precipitate settles out, leaving essentially clear liquid behind. This precipitate differs from the sediment normally accumulating in that it behaves in a quasi-liquid fashion and "flows" as the container is tilted, whereas the normal sediment adheres rigidly to the bottom of the vessel. In cases of slow coagulation this effect may require several hours to manifest itself. In extremely slow coagulation the effect may be delayed for several days. Tuorila[2] has suggested a critical test for slow coagulation, which follows as a corollary from his researches. Inasmuch as coagulation increases rapidly with concentration, the sediment may be analyzed twice, in dilute and concentrated suspensions. If the results check within reasonable limits, no coagulation is present.

If the test shows complete dispersion, the suspension should be diluted to a liter with N/100 sodium oxalate and analyzed. If dispersion is incomplete, the suspension is diluted to about 800 c.c. with N/100 sodium oxalate and heated to the boiling point. As soon as the liquid boils it is withdrawn from the flame. After cooling, the suspension is tested, and in every case studied by the writers the samples were either fully dispersed or coagulation had set in.

[1] C. L. Whittles, *loc. cit.*, 1924.
[2] P. Tuorila, *loc. cit.*, 1927.

If coagulation is present at any stage of the routine, the suspension, diluted to about a liter, is poured into a tall beaker and the liquid filtered through a Pasteur-Chamberland filter with suction. The set-up involves the filter connected to a filter bottle, with the outlet of the bottle connected to the filter pump. The suspension passes into the filter (which is a cylindrical tube) and collects in the bottle. When the residue in the beaker is like paste, the filter is cleaned and the sediment again made up to a liter with N/100 sodium oxalate, since most of the original peptizer has been removed by the filtration. The suspension is again heated to boiling, and usually it will be dispersed. In some cases more than one washing is necessary. It is conceivable that some samples may be quite obstinate in their resistance to dispersion.

When the sediment has an appreciable amount of sand in it, the preliminary soaking is followed by shaking or stirring, unless the sediment is indurated, in which case brushing should precede the shaking or stirring. The sample is shaken for an hour, preferably in a reciprocating shaker, or it is stirred in an electric drink mixer for 5 or 10 min. It is important to use wire baffles in the cup. After the shaking or stirring, the suspension is tested for dispersion as described above.

If dispersion is incomplete, the suspension is diluted to about 800 c.c. and heated to boiling, as described above. If coagulation is present, the foreign electrolytes are washed out as described above.

The material above 1/16 mm. diameter is sieved from the sample before analysis. The suspension is poured through the sieve and the liquid collected in a beaker. The residue is washed with a gentle stream of water. The total volume of suspension should be less than a liter, including the wash water. The suspension may be sieved after brushing or shaking, but any convenient point in the routine may be used. The material above 1/16 mm. is sieved dry into grades if it is appreciable in amount. It should also be examined to determine whether it consists of individual grains or undisintegrated aggregates.

In many instances small undisintegrated fragments are found in the sieve, and in some sediments without coarse grains there are numerous small ironstone pellets. It is not easy in every case to decide what disposition should be made of such materials. Undisaggregated lumps may be treated further, unless they are so firmly cemented that the fragments tend to break rather than disintegrate. Ironstone pellets commonly are secondary materials, and a correction may be applied to the sample weight to allow for them. At best a pragmatic attitude suggests that small amounts of such materials be discounted by correcting for them; if appreciable parts of the sediment remain as undisaggregated lumps, the particular dispersion routine was not successful.

The experience of the authors and students in the laboratories of the University of Chicago has shown that the generalized dispersion routine described above is successful in a large majority of cases. Obstinate samples are encountered, however, in which induration prevents even an approximate disintegration or dispersion; in such instances thin-

PREPARATION OF SAMPLES

section mechanical analysis may be resorted to (Chapter 6). If the difficulty arises from a high content of soluble salts, especially calcium ions, washing usually proves adequate. In this connection, however, it seems desirable that the base exchange method of Puri[1] should be investigated for its general applicability to difficult sediments. Puri's method involves boiling the sample with ammonium carbonate for protracted periods of time, but this may be modified to a period of soaking in the electrolyte or to merely heating the suspension to the boiling point. The addition of sodium hydroxide to a concentration of 0.004 N to replace the ammonia results in a sodium clay with its high stability.

The authors have adopted the practice of experimenting on small samples in "test-tube" dispersion to determine the relative effects of various procedures and various dispersing agents. Likewise, qualitative tests on the nature of the foreign electrolytes are of value in determining the advisability of departing from the generalized routine. For "test-tube" dispersion, various amounts of the several peptizers are added to small volumes of suspension and the effects noted qualitatively by allowing the tubes to stand for several hours or over night.

[1] A. N. Puri, *loc. cit.*, 1935.

CHAPTER 4

THE CONCEPT OF A GRADE SCALE

INTRODUCTION

In most types of sedimentary analysis the data are arranged on some kind of size scale (which may be diameter, area, or volume) for convenience both in conducting the analysis and in tabulating the analytical data. This is especially true of mechanical analysis, but the general topic of grade scales is in itself so important that a separate chapter is devoted to it here.

A *grade scale* may be defined as an arbitrary division of a continuous scale of sizes, such that each scale unit or *grade* may serve as a convenient class interval for conducting the analysis or for expressing the results of an analysis. Against this grade scale may be plotted the amount of material in each grade (a size frequency diagram), or the amount of some particular mineral in the sediment (a mineral frequency diagram), or the average sphericity or roundness of material in each grade (a shape frequency diagram), and so on. In these cases size is usually chosen as the independent variable, and the grade scale is therefore arranged along the horizontal axis of the diagram, whereas the frequency is plotted along the vertical axis.

One of the commonest types of frequency diagram is the histogram, which is drawn very simply by setting a vertical block above each grade, proportional in height to the value of the other variable (amount of material, average sphericity, etc.) in each grade. Convention, in America at least, has been to draw each grade equal in width, whether it is so in fact or not. The subject of histograms receives more detailed consideration in Chapter 7; for the present only their relation to grade scales need be discussed.

Unfortunately, the shape of a particular histogram will vary according to the grade scale used in the analysis. That is, the same identical sediment, if analyzed on the basis of two different grade scales, will yield figures which may be quite unlike each other. Recognition of this fact has led to a considerable discussion of grade scales and methods of

GRADE SCALES

presenting data, in which one or another scale was proposed for all analyses to avoid the unfortunate variation of the histograms. An even greater volume of literature has discussed the relative merits of the several grade scales, and in some of these papers it was shown that one or another grade scale was more logical, more convenient, or more "natural" for sedimentary purposes.[1]

It seems that even to-day it is not universally recognized that the choice of grade scale is perfectly arbitrary. Except for minor differences in the statistical values obtained, the unique frequency curve is independent of any particular grade scale, whether equal or unequal in class interval.

MODERN GRADE SCALES

Udden grade scale. The first true geometric scale for soils or sediments, as far as the authors are aware, was introduced by Udden[2] in 1898. In choosing his grade limits, Udden mentioned his indebtedness to soil scientists but departed from their choice of grade scale because of the absence of a fixed geometric interval. In order to achieve a fixed ratio, he changed from the values 1, ½, ¼, 1/10 mm. to the values 1, ½, ¼, ⅛. By applying the same ratio of ½ (or 2, depending upon the sense of direction) Udden developed his original grade scale of twelve grades, extending from 16 mm. diameter to 1/256 mm., with the following limiting diameters: 16, 8, 4, 2, 1, ½, ¼, ⅛, 1/16, 1/32, 1/64, 1/128, 1/256 mm. Later, in 1914, Udden extended his scale in both directions, to include coarser and finer materials.[3] The introduction of the Udden grade scale marked the beginning of the modern period of grade scale development, although even to-day there are grade scales in wide use which do not follow the principle of strict geometric intervals.

The concept of a geometrical grade scale should be made explicit for the non-mathematical reader. A geometric series is defined as a progression of numbers such that there is a fixed ratio between successive elements in the

[1] Among papers on the subject may be listed the following: A. Atterberg, Die mechanische Bodenanalyse und die Klassifikation der Mineralböden Schwedens: *Int. Mitt. für Bodenkunde*, vol. 2, pp. 312-342, 1912. C. W. Correns, Grundsätzliches zur Darstellung der Korngrössenverteilung: *Centralbl. f. Min., Geol., u. Paläon.*, Abt. A., pp. 321-331, 1934. G. Fischer, Gedanken zur Gesteinssystematik: *Jahrb. d. Preuss. Geol. Landesanst.*, vol. 54, pp. 553-584, 1933. C. K. Wentworth, Fundamental limits to the sizes of clastic grains: *Science*, vol. 77, pp. 633-634, 1933.

[2] J. A. Udden, Mechanical composition of wind deposits: *Augustana Library Publications*, no. 1, 1898.

[3] J. A. Udden, Mechanical composition of clastic sediments: *Geol. Soc. America, Bulletin*, vol. 25, pp. 655-744, 1914.

series. Thus the series 1, ½, ¼, ⅛ is such a series because each number is one-half as large as the preceding one, so that any successive number may be found by multiplying its predecessor by ½. Any series of numbers may readily be tested for a fixed ratio by dividing any term by its successor. If the quotient is the same for all pairs, the series is geometric. In the series 1, ½, ¼, 1/10, the ratio $1 : \frac{1}{2} = 2$, but the ratio $\frac{1}{4} : 1/10 = 2.5$, so the ratio is not fixed. In some cases the geometric nature of the series is not immediately apparent by inspection, as in the series 1.000, 0.707, 0.500, 0.354, 0.250. A test indicates that the ratio is fixed, however: $1 : 0.707 = 0.707 : 0.500 = 0.500 : 0.354 = 0.354 : 0.250 = 1.414$. The number 1.414 is the square root of 2, and these numbers are on the grade scale based on $\sqrt{2}$. Another simple test for a geometric series is that the logarithms of the numbers to any base form an arithmetic series, i.e., a series of numbers *differing* by a fixed amount.

Hopkins grade scale. In 1899, the year following Udden's work, Hopkins[1] pleaded for a scientific basis for the division of particles into grades, suggesting a true geometric grade scale based on $\sqrt{10}$. Hopkins's grade scale was not adopted by the United States Bureau of Soils;[2] nevertheless, the scale forms an excellent basis for mechanical analysis.

Bureau of Soils grade scale. The grade scale used by the United States Bureau of Soils is shown in Table 4.

TABLE 4
SIZE CLASSIFICATION OF UNITED STATES BUREAU OF SOILS

Grade Limits (Diameters)	Name
2-1 mm.	Gravel
1-½ mm.	Coarse sand
½-¼ mm.	Medium sand
¼-1/10 mm.	Fine sand
1/10-1/20 mm.	Very fine sand
1/20-1/200 mm.	Silt
Below 1/200 mm.	Clay

Atterberg grade scale. In 1905 Atterberg[2] advanced the subject by seeking for fundamental physical properties as a basis for erecting a grade scale. His class intervals were based on the unit value 2 mm. and

---

[1] C. G. Hopkins, A plea for a scientific basis for the division of soil particles in mechanical analysis: *U. S. Dept. Agric., Dept. Chem., Bull.* 56, pp. 64-66, 1899.

[2] Briggs, Martin, and Pearce, *loc. cit.*

[3] A. Atterberg, Die rationelle Klassifikation der Sande und Kiese: *Chem. Zeitung*, vol. 29, pp. 195-198, 1905.

GRADE SCALES

involved a fixed ratio of 10 for each successive grade, yielding the limiting diameter 200, 20, 2.0, 0.2, etc. Each of these major grades was divided into two subgrades, chosen at the geometric mean of the grade limits. Thus, the division between the 20 and 2 mm. limits was found by taking the square root of the product of the grade limits: $20 \times 2 = 40$; and $\sqrt{40} = 6.32$. The value 6.32 was rounded off to 6.00 for convenience. This rounding-off process destroyed the geometric simplicity of the subgrades but did not affect the fundamental geometric nature of the main classes. Correns,[1] in discussing Atterberg's scale, pointed out that the subgrades, as well as the main classes, should be kept as geometric intervals. Atterberg's scale, with the descriptive names applying to each principal grade, is shown in Table 5.

TABLE 5
ATTERBERG'S SIZE CLASSIFICATION

Grade Limits (Diameters)	Name
2,000-200 mm.	Blocks
200-20 mm.	Cobbles
20-2 mm.	Pebbles
2-0.2 mm.	Coarse sand
0.2-0.02 mm.	Fine sand
0.02-0.002 mm.	Silt
Below 0.002 mm.	Clay

In choosing his grade limits, Atterberg observed that sand coarser than 2 mm. diameter does not hold water, whereas sand with smaller grains does to some extent, depending upon capillarity. The next significant boundary was found to be 0.2 mm., where a distinction was drawn between truly "wet sand" and relatively dry sand. Another change was noted at 0.02 mm., below which the individual grains could not be seen with the unaided eye, and in material finer than which root-hairs were not able to penetrate the pores. At the next grade limit, 0.002 mm., Atterberg pointed out that Brownian movement began. This correlation of physical properties with critical grain diameters is the outstanding characteristic of Atterberg's work, and his grade scale has been widely adopted by European workers with both soils and sediments. In 1927 the grade scale was adopted by the International Commission on Soil Science as the standard for all soil analyses. The United States Bureau of Soils, however, did not adopt the scale.

[1] C. W. Correns, Grundsätzliches zur Darstellung der Korngössenverteilung: *Centr. f. Min.*, Abt. A, pp. 321-331, 1934.

Wentworth grade scale. In America, sedimentary petrologists favor the Udden grade scale, as it was modified in 1922 by Wentworth.[1] Wentworth compared the usage of such terms as *cobble, coarse sand,* and the like, and on the basis of the usages he modified and extended Udden's scale, retaining, however, the geometric interval introduced in 1898. This grade scale, justly called the Wentworth scale, has been adopted by practically all American workers. The full grade scale is

TABLE 6
WENTWORTH'S SIZE CLASSIFICATION

Grade Limits (*Diameters*)	Name
Above 256 mm.	Boulder
256-64 mm.	Cobble
64-4 mm.	Pebble
4-2 mm.	Granule
2-1 mm.	Very coarse sand
1-½ mm.	Coarse sand
½-¼ mm.	Medium sand
¼-⅛ mm.	Fine sand
⅛-1/16 mm.	Very fine sand
1/16-1/256 mm.	Silt
Below 1/256 mm.	Clay

given in Table 6. In 1933 Wentworth [2] examined the limits of his grades in terms of the physical properties involved in grain transportation. He showed that given class limits in the grade scale, far from being arbitrary, agreed well with certain distinctions between suspension and traction loads.

Emphasis has been placed here on the Atterberg and Wentworth scales, largely because there is an increasing tendency for workers in sediments to use one or the other of them. It will be shown later that the rivalry between the grade scales is more apparent than real and that from the point of view of statistical analysis either scale is equally convenient. It is not to be assumed, however, that these two scales are the only ones that have been entertained by analysts. On the contrary, a great many grade scales have been proposed and used for soils and

---

[1] C. K. Wentworth, A scale of grade and class terms for clastic sediments: *Jour. Geology,* vol. 30, pp. 377-392, 1922.
[2] C. K. Kentworth, *loc. cit.,* 1933.

sediments, and some of these are discussed and compared by Fischer [1] and Zingg.[2] To a large extent these grade scales are similar to that of the Bureau of Soils or to the Atterberg scale, with more grades or with slightly different limits and class names. The Wentworth scale, also, has been modified, largely from the point of view of decreasing the class interval by using the ratio $\sqrt{2}$ or $\sqrt[4]{2}$, instead of 2.

TABLE 7

A.S.T.M. SIEVE SCALE

Mesh	Opening (mm.)
5	4.00
6	3.36
7	2.83
8	2.38
10	2.00
12	1.68
14	1.41
16	1.19
18	1.00
20	0.84
25	0.71
30	0.59
35	0.50
40	0.42
45	0.35
50	0.297
60	0.250
70	0.210
80	0.177
100	0.149
120	0.125
140	0.105
170	0.088
200	0.074
230	0.062
270	0.053
325	0.044

Engineering grade scales. In addition to the types of grade scales used by sedimentary petrologists and soil scientists, there is a wide variety of grade

[1] G. Fischer, *loc. cit.*, 1933.
[2] Th. Zingg, Beitrag zur Schotteranalyse: *Schweiz. Min. u. Pet. Mitt.*, vol. 15, pp. 39-140, 1935.

scales based on the mesh system, which are extensively used in engineering and commercial testing. Among the best known of these are the scale adopted by the American Society for Testing Materials [1] The scale was based on the fixed ratio $\sqrt[4]{2}$. Table 7 shows the relation between the sieve openings in millimeters and the corresponding mesh number. It may be noticed that every fourth value in this table agrees with the Wentworth class limits, starting with 4.00 mm.

Another well-known system based on mesh was that adopted in 1907 by the Institute of Mining and Metallurgy of England.[2] There is no fixed ratio between the successive sieve openings in the I.M.M. series, and hence it is not a true geometric scale. Table 8 lists the mesh numbers and corresponding openings in millimeters for this series. The I.M.M. series has been used widely in England for mechanical analysis of sediments.

TABLE 8

I.M.M. SIEVE SCALE

Mesh	Opening (mm.)
5	2.540
8	1.574
10	1.270
12	1.056
16	0.792
20	0.635
25	0.508
30	0.421
35	0.416
40	0.317
50	0.254
60	0.211
70	0.180
80	0.157
90	0.139
100	0.127
120	0.107
150	0.084
200	0.063

A criticism which may in general be directed against the mesh system of nomenclature is that, unless the openings in millimeters or some other unit are also given, it is not possible to convert the values to their metrical equivalents. A comparison of the openings corresponding to the various meshes in

[1] American Society for Testing materials, *A.S.T.M. Standards* (1930), part 2, p. 1120.
[2] Original reference not available. Data from W. S. Tyler Company, *Catalog 53,* p. 14.

GRADE SCALES

Tables 7 and 8 will indicate that they differ widely enough to be significant. Furthermore, when sieves are purchased merely in terms of so many meshes to the inch, without specifying a particular standard set, there may be no definite relation between opening and mesh, inasmuch as the number of meshes to the inch may be fixed, but the openings will vary according to the diameter of the wire or cloth used in weaving the sieve.

Robinson grade scale. In the preceding discussion, grade scales have been based on the diameters of the particles being classified. There are other types of grade scale, however. Robinson,[1] for example, considers expressions of size of irregular particles to be unsatisfactory in reporting mechanical analyses, and he recommends the direct use of settling velocities or their logarithms. By thus expressing size in terms of settling velocities, and the latter in terms of their logarithms, Robinson introduced the first logarithmic transformation scale. The great advantage of such scales is that they convert unequal geometrical intervals into equal arithmetical intervals and, with a suitable choice of logarithms, introduce integers instead of fractions as the grade limits. Such transformations are, of course, more appropriate for true geometric scales than for irregular unequal-interval scales, because the latter will not yield an arithmetic series of integers.

TABLE 9

RUBEY'S SIZE CLASSIFICATION BASED ON SETTLING VELOCITIES

Grade	Settling Velocity (in microns/sec.)
Very fine sand	> 3,840
Coarse silt	960-3,840
Medium silt	240-960
Fine silt	60-240
Very fine silt	15-60
Coarse clay	3.75-15
Medium clay	0.9375-3.75
Fine clay	< 0.9375

Rubey grade scale. Rubey [2] followed Robinson in the use of settling velocities instead of diameters directly, but carried the work to the

[1] G. W. Robinson, The forms of mechanical composition curves of soils, clays, and other granular substances: *Jour. Agric. Sci.,* vol. 14, pp. 626-633, 1924.
[2] W. W. Rubey, Lithologic studies of fine-grained Upper Cretaceous sedimentary rocks of the Black Hills region: *U. S. Geol. Survey, Prof. Paper 165A,* pp. 1-54, 1930.

development of an actual grade scale based on velocities. Rubey plotted the settling velocities and diameters of particles on double log paper and then drew in the size limits according to Atterberg, Wentworth, and Udden, on the same scale. A straight line, based on average settling velocity and size limit, was drawn through the graph, yielding Rubey's grade scale, which is shown in Table 9.

It will be noticed that the limiting velocities between the successive size fractions in Rubey's scale decrease by the constant ratio 1 to 4. In terms of diameters this means that the ratios are 1 to 2, because by Stokes's law (Chapter 5) it may be shown that the settling velocity varies as the square of the diameter. Thus Rubey's grade scale conforms to the principles of fixed geometric intervals.

TABLE 10
PHI AND ZETA GRADE SCALES

Wentworth Grades	ϕ	Atterberg Grades	ζ
32 mm.	−5		
16 mm.	−4	2000 mm.	−3
8 mm.	−3	200 mm.	−2
4 mm.	−2		
2 mm.	−1	20 mm.	−1
1 mm.	0		
½ mm.	+1	2 mm.	0
¼ mm.	+2		
⅛ mm.	+3	0.2 mm.	+1
1/16 mm.	+4		
1/32 mm.	+5	0.02 mm.	+2
1/64 mm.	+6		
1/128 mm.	+7	0.002 mm.	+3
1/256 mm.	+8		
1/512 mm.	+9	0.0002 mm.	+4
1/1024 mm.	+10		

Phi and zeta scales. In 1934 Krumbein [1] applied a logarithmic transformation equation to the Wentworth grade scale and obtained a "phi scale" which had integers for the class limits and increased with decreasing grain size. This grade scale was developed specifically as a statistical device to permit the direct application of conventional statistical

[1] W. C. Krumbein, Size frequency distributions of sediments: *Jour. Sed. Petrology*, vol. 4, pp. 65-77, 1934. See also The application of logarithmic moments to size frequency distributions of sediments: *Jour. Sed. Petrology*, vol. 6, pp. 35-47, 1936.

GRADE SCALES

practices to sedimentary data. More recently, Krumbein [1] also applied negative logarithms to the Atterberg scale and obtained a "zeta scale" with properties similar to the phi scale. The theory on which these transformed scales is based is that any true geometric scale may be converted to an equivalent scale with arithmetic intervals if logarithms of the scale limits are substituted for the diameter values. Krumbein chose the transformation equation $\phi = -\log_2 \xi$, (where ξ is the diameter in millimeters) for the Wentworth scale, and $\zeta = 0.301^{\cdot} - \log_{10} \xi$ for the Atterberg scale. The resulting phi and zeta scales are shown with their equivalent Wentworth and Atterberg scales in Table 10.

PROBLEMS OF UNEQUAL CLASS INTERVALS

It may be noted that without exception the grade scales proposed for soils and sediments have been based on unequal class intervals. To some extent this is due to necessity, inasmuch as the range of sizes in sedimentary particles, even within the same sediment in some cases, is so great that equal intervals are a practical impossibility. Thus, sandy shales may range in particle size from 1 mm. to less than 0.001 mm. in diameter. If an interval such as 0.1 mm. were used for the classes, the result would be, perhaps, that more than half the distribution would be in the smallest grade. To give full significance to the smaller sizes, it would be necessary to use a class interval of 0.001 mm. Two practical difficulties arise; one is that some thousand classes would be necessary, and the other is that it is virtually impossible to distinguish between grains of 0.999 and 1.000 mm. diameter, especially when the particles are not regular geometrical solids. Finally, differences of 0.001 mm. in the diameters of large particles are negligible, whereas the difference between particles of 0.001 and 0.002 mm. diameter may be significant.

The obvious conclusion to draw from these observations seems to be that a grade scale should be devised such that each grade bears a fixed size ratio to preceding and succeeding grades. This is the principle introduced by Udden and exemplified by the Atterberg and Wentworth scales. This is not to imply that unequal interval scales are not satisfactory unless they have a fixed ratio; on the contrary, any grade scale is satisfactory for descriptive purposes if it is mutually agreed upon by a sufficient number of workers. From an analytical point of view, on the other hand, an irregular interval in the grade scale may interfere with

[1] W. C. Krumbein, Korngrösseneinteilungen und statistische Analyse: *Neues Jahrb. f. Min., etc.*, Beil.-Bd. 73, Abt. A, pp. 137-150, 1937.

the convenient application of statistical analysis to the data. The recognition that two separate and distinct functions are associated with any grade scale [1] has not been sufficiently emphasized by soil scientists and sedimentary petrologists, and the topic deserves detailed discussion.

FUNCTIONS OF GRADE SCALES

Descriptive function. The first and perhaps the most important function of a grade scale is a descriptive function, which serves to place nomenclature and terminology on a uniform basis. If one reads the term *coarse sand* in a report, he would prefer to understand by the term exactly what the writer intended to convey. As long as there is no uniform terminology, each writer coins his own meanings, which may or may not be precisely defined. If, however, the reader knows that the writer is using the Atterberg classification, he may understand that material having a range of sizes from 2.0 mm. to 0.6 mm. diameter is meant. Likewise, if the Wentworth scale is being used, the term refers to material from 1 mm. to ½ mm. in diameter.

Obviously, no "justification" whatever is required for the descriptive function of a grade scale. The particular choice of such terms as *coarse sand, fine sand, clay,* and the like need be based on no other criterion than mutual agreement. If the limits chosen for each grade are also related to the physical properties of sediments, that fact may be taken as an added advantage.

Analytic function. In addition to the use of grade scales to establish uniformity of terminology, the classes or grades are used as units in performing various kinds of analyses on the sediment. The classes are used, for example, in determining the mechanical composition of the sediment, and in addition they may be used as units during statistical analysis. It is in connection with these analytic functions of grade scales that most of the confusion arose regarding the merits of one or another of the proposed scales. The recognition of the fact that histograms vary in form according to the grade scale used has led various writers [2] to the conclusion that some single scale should be used for all analyses, so that the unfortunate variation of the histogram could be avoided.

Unfortunately, there can be no single "correct" grade scale for all mechanical analyses, because the concept of discrete grades is absent

---

[1] W. C. Krumbein, *loc. cit.,* 1934.
[2] See, for example, L. Dryden, Cumulative curves and histograms: *Am. Jour. Sci.,* vol. 27, pp. 146-147, 1934.

from any continuous size range. That is, where the sizes change by infinitesimals along the entire range of diameters, one has a continuous function, in which any class interval whatever is purely artificial. From the nature of the particles comprising sediments it is clear that, with few exceptions, the diameters vary by infinitesimals along the entire range of sizes present, rather than by abrupt steps from one size to the next. In studying these distributions, however, convenient units are desirable; but the units themselves are quite arbitrary, and for practical purposes they need not be related to the descriptive aspects of the grade scale. If the grade scale is flexible enough to permit its use both for descriptive and for analytical purposes, a strong advantage of convenience is gained. Both the Atterberg and the Wentworth grade scales have such flexibility.

The recognition that a sediment is really a continuous size frequency distribution of particles, without any implication of a "natural" grouping of the material into classes, frees mechanical analysis from the confines of any single grade scale. From this point of view, the class intervals used in the actual analysis may be so chosen that they bring out most clearly the characteristics of the distribution itself. This continuous size frequency distribution may then be described in conventional statistical terms, which themselves may be related to any of a number of descriptive grade scales. Two common statistical methods of analysis are available, based either on the moments of the distribution or on the quartiles and median (see Chapters 8 and 9). In the moment method the class intervals are preferably chosen on a fixed geometric ratio, to facilitate computation. When quartile measures are used, however, it is immaterial what class intervals are used, or even whether or not they are based on a fixed geometric ratio. This high degree of independence with quartile measures depends on the fact that purely graphic methods are used, so that the choice of class interval may be determined by the preciseness with which the analyst wishes to construct his curves.

Although there is a growing recognition of the continuous nature of most sedimentary data, many current analyses are still conducted in terms of the descriptive units of which the sediment is composed. That is, instead of considering the sediment as a whole, one may be interested in the percentages of specific grades present. Soil scientists, for example, are more interested in the amounts of sand or colloid present in a soil than in the nature of the distribution as a whole. Likewise in commercial testing, specified amounts of sand, silt, or clay are desired, and the analyses are directed toward testing the material with this in view. In such cases the obvious technique is to analyze the material with the class units of the descriptive grade scale, so that results are obtained directly in terms of the grade scale being used.

It is to be explicitly pointed out, however, that when the mechanical analysis data are secured in terms of the continuous size frequency distribution, it is always possible to express the analysis in terms of specific grades on any grade scale, whereas when a fixed grade scale is used indiscriminately on all analyses, it may not be convenient to determine the nature of the continuous distribution, especially if the fixed scale has too few points along it, or if the intervals are not based on a fixed ratio.

CHOICE OF A GRADE SCALE

The wide choice of viewpoint possible in the analysis of sediments and other particulate substances suggests that for general purposes it may be preferable to choose some analytical class interval which would furnish data adequate for all the purposes outlined. The authors believe that a grade scale based on a fixed geometric interval and flexible enough to afford a number of relatively small subgrades is to be preferred. Either the Atterberg or the Wentworth scale is suitable as a base, and it is immaterial which is used, because the results obtained from the one scale may readily be expressed in terms of the other, if desired. In this connection, Hopkins's grade scale, mentioned earlier, deserves consideration because it also affords a convenient basis for analysis. The possibility of converting values from one descriptive scale to another depends simply on the fact that any continuous distribution is independent of the class intervals used in analyzing it, and, within the relatively small range of experimental error due to particular grade limits, the characteristics of the sample are constant.

It is in connection with the statistical manipulation of sedimentary data and with the conversion of the data from one descriptive scale to another that the logarithmic type of grade scale is most useful. In the Wentworth grade scale, the use of a logarithmic notation, such as the phi scale (Table 10), yields integers which mark the limits of each grade. If an analysis based on the Wentworth limits directly does not yield adequate data for the complete study of the sediment, the scale limits may be changed to half-phi units, which yields double the number of experimentally determined points. The analytical scale, however, still remains arithmetic, except that the phi intervals change from 1, 2, 3,... to 1, 1.5, 2.0, 2.5, 3.0,... In short, the substitution of the $\sqrt{2}$ grade scale does not affect the convenience of the phi notation. Likewise, even the use of the $\sqrt[4]{2}$ scale merely results in the units along the phi scale be-

GRADE SCALES

coming $\phi/4$, so that the series, still arithmetic, is 1.0, 1.25, 1.50, 1.75, 2.0,... In a similar manner the limits of the Atterberg scale may be too large for convenient analysis, and by choosing additional points at half- or quarter-zeta values (Table 11) any detail whatever may be brought out, limited only by imperfections of technique.

TABLE 11
LOGARITHMIC GRADE SCALES BASED ON $\zeta/4$ AND $\phi/4$ CLASS INTERVALS

ATTERBERG SCALE		WENTWORTH SCALE		
Grades (mm.)	ζ	Grades (mm.)	$\sqrt[4]{2}$-Scale (mm.)	ϕ
20.00	−1.00	2.	2.00	−1.00
			1.69	−0.75
11.25	−0.75		1.41	−0.50
			1.19	−0.25
6.32	−0.50	1.	1.00	0.00
			0.84	+0.25
3.56	−0.25		0.71	+0.50
			0.59	+0.75
2.000	0.00	½.	0.50	+1.00
			0.420	+1.25
1.125	+0.25		0.351	+1.50
			0.297	+1.75
0.632	+0.50	¼.	0.250	+2.00
			0.210	+2.25
0.356	+0.75		0.177	+2.50
			0.149	+2.75
0.200	+1.00	⅛.	0.125	+3.00
			0.105	+3.25
0.112	+1.25		0.088	+3.50
			0.074	+3.75
0.063	+1.50	1/16.	0.062	+4.00
0.035	+1.75			
0.020	+2.00			

The conversion of statistical values from one descriptive scale to another is a relatively simple matter as long as the several scales are based on geometric intervals and expressed as logarithms. The general approach may be illustrated in terms of the Wentworth and Atterberg scales, expressed in the phi and zeta notations respectively. Suppose a statistical measure of average size is computed for a sediment and expressed in phi terms. It is desired to convert this measure to its equivalent value on the Atterberg scale, as expressed in zeta terms. The phi scale is based on the fact that any diameter value may be expressed in terms of the Wentworth scale as $\xi = 2^{-\phi}$, where ξ is the diameter in millimeters and ϕ is a value along the phi scale. By taking logs of this expression, one obtains $\phi = - \log_2 \xi$. In like manner, any value on the Atterberg scale may be expressed as $\xi = 2 \times 10^{-\zeta}$, where ξ is the diameter in millimeters, as before, and ζ is a value on the zeta scale. When logs are taken of this last expression, there results $\zeta = 0.301 - \log_{10} \xi$. To convert values from one scale to the other, use is made of the general logarithmic equation for change of base: $\log_{10} \xi = \log_{10} 2 \log_2 \xi$. For the factor $\log_2 \xi$ is substituted the expression $-\phi$, and for the term $\log_{10} \xi$ is substituted the expression $0.301 - \zeta$. These substitutions yield the final transformation equation, $\zeta = 0.301 (\phi + 1)$. By means of this last equation any scale value in the zeta notation may be converted to its equivalent in the phi notation, and vice versa. The net effect is that analyses may be conducted on any convenient scale with true geometric intervals, and, by the choice of appropriate transformation equations, the statistical values may be expressed in terms of any other geometric scale. The relation between diameter, (ξ) and the ϕ and ζ scales is shown graphically in Figure 19.

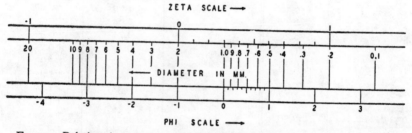

FIG. 19.—Relations between logarithmic grade scales and diameters in millimeters. The "zeta scale" is adapted to Atterberg grades, and the "phi scale" to Wentworth grades.

A full discussion of statistical methods available for sedimentary data is given in Chapters 8 and 9. These methods are illustrated by examples of mechanical analyses, and statistical computations based on the use of conventional and logarithmic grade scales are described. The principal purpose of the present chapter has been to lay the foundation for mechanical analysis by indicating that some kind of grade scale is important in the analysis of sediments, but that the particular choice may depend upon the convenience with which it may be used.

CHAPTER 5

PRINCIPLES OF MECHANICAL ANALYSIS

INTRODUCTION

MECHANICAL analysis is the quantitative expression of the size frequency distribution of particles in granular, fragmental, or powdered material. It does not necessarily involve the actual separation of the substance into grade sizes, nor does it require unconsolidated material.

Methods of mechanical analysis may be divided into two broad groups, the *modern precision methods* and the *older routines*. The fundamental differences between these groups are first, that the newer methods are underlain by a single mathematical theory of sedimenting systems which unifies the field and coördinates the methods; second, the older methods seek to separate the material into grade sizes, whereas the new techniques do not.

All methods of analysis are supported by several underlying principles, such as the settling velocities of particles, the dispersion and coagulation of suspensions, and theories of their operation. The factors involved in dispersion and coagulation were discussed in Chapter 3; the present chapter will concern itself with the remaining principles of mechanical analysis.

The historical development of methods of mechanical analysis is a topic which deserves consideration by workers in the field, but limitations of space prevent its treatment here. The interested reader is referred to Krumbein's paper [1] for a short history of the subject.

CLASSIFICATION OF DISPERSE SYSTEMS

Suspensions of solids in liquids are called *disperse systems* when the solid is so thoroughly distributed that the individual particles may no longer be of primary importance and interest is focused on the totality of the particles as a system. The solid constitutes the *dispersed phase*,

[1] W. C. Krumbein, A history of the principles and methods of mechanical analysis: *Jour. Sed. Petrology*, vol. 2, pp. 89-124, 1932. An error in the original paper is corrected in *Jour. Sed. Petrology*, vol. 3, p. 95, 1933.

and the liquid is the *dispersion medium*. Such systems are divided into *monodisperse* and *polydisperse* systems, depending upon whether the particles are all of the same size or of various sizes. Both types of systems may be classified according to the size of particles present; polydisperse systems may of course belong to more than a single size classification.

Coarse disperse systems. Particles larger than 0.1 micron (0.0001 mm.) in diameter constitute coarse disperse systems. This lower limit is generally accepted as the upper limit of the colloidal state, but it is recognized that there is a transition zone in which particles may have some of the attributes both of coarse systems and of colloids. This transition zone may extend above 1 micron.

Coarse disperse systems may be divided into three groups: (a) *macroscopic systems,* in which the individual grains may be resolved by the unaided eye; this group includes particles larger than about 10 microns (0.01 mm.) diameter; (b) *microscopic systems,* in which the individual grains may be resolved with a compound microscope; this group extends down to about 0.2 micron (0.0002 mm.) diameter; and (c) the *ultramicroscopic system,* in which particles are no longer seen as individuals under the microscope.

Colloidal disperse systems. Colloidal particles range in size from about 0.1 micron to 1 mumu (0.000,001 or 10^{-6} mm.) in diameter. Only the larger sizes within this category belong even to the ultramicroscopic group. The colloidal state may be defined as that state in which the dispersed phase is so finely divided that properties depending upon the surface area control its behavior. Such phenomena as dispersion and coagulation are outstanding phenomena among colloids.

Molecular disperse systems. Disperse systems containing only particles smaller than about 1 mumu are true solutions.

It should be understood that this classification of disperse systems is arbitrary, because the attributes of particles show complete transitions from one state to another. Especially is this true of polydisperse systems, in which it is not uncommon to find a range of sizes from very coarse particles through colloids and into soluble material. Mechanical analysis is concerned mainly with the first two types of disperse systems, the coarse systems and colloids. Methods of analysis depend very largely upon the predominant sizes, or the range of sizes in the material. Present methods of analysis, in fact, are most effective for diameters larger than 1 micron (0.001 mm.), so that essentially it is in connection with coarse disperse systems only that most mechanical analyses are conducted. Material smaller than 1 micron is frequently grouped into a single size class.

In order to illustrate the influence of size on the methods of analysis

PRINCIPLES OF SIZE ANALYSIS

commonly used, the following outline indicates the subdivisions of disperse systems and the analytical methods commonly used for each:

Coarse disperse systems. Particles larger than 10 mm.: direct measurement by macroscopic methods, sieving methods.
Particles between 10 and 0.05 mm. diameter: sieving methods, direct measurement by microscopic methods in part.
Particles between 0.05 and 0.001 mm. diameter: indirect sedimentation methods (pipette, Odén balance, Wiegner tube, etc.), but in some cases this group is subdivided as follows:

0.05 to 0.01 mm. diameter: rising current elutriation, decantation methods
0.01 to 0.001 mm. diameter: indirect sedimentation methods

Particles between 0.001 and 0.0001 mm. diameter: centrifugal methods.
Colloidal Disperse systems. Centrifuge, ultramicroscope, turbidity, various optical methods.

CONCEPT OF SIZE IN IRREGULAR SOLIDS

If all soils or sediments were composed of perfect spheres, a definition of size would be simple. The fact that natural materials are seldom regular in shape, combined with the fact that the particles composing a given mixture may range widely in their shapes, gives rise to a problem which has engaged the attention of numerous workers. In some cases the definition of size has depended upon the magnitude of the particle: large particles that could be conveniently handled were defined in terms of one set of criteria, and smaller particles were defined on entirely other bases. To a large extent definitions of size have been based on the most convenient and immediately applicable manner of obtaining a number which could be used for the purpose at hand.

One of the most thorough investigations of the concept of size of irregular particles has been made by Wadell,[1] and the following discussion is largely based on his work. Wadell's thesis is that "size" of a particle is best expressed by its simple volume value, because the volume is independent of its shape. The use of long, intermediate, and short diameters, or of the arithmetic or geometric mean of these, is, according to Wadell, relatively meaningless as a definition of size.[2] It is quite possible, for example, that the mean of three diameters of an irregular solid may be numerically equal to the diameter of a given sphere, and yet the volumes will be entirely different. The term *diameter* has a

---
[1] H. Wadell, Volume, shape, and roundness of rock particles: *Jour. Geology*, vol. 40, pp. 443-451, 1932.
[2] See, however, the discussion of definitions on page 127 of this chapter, under microscopic methods of analysis.

definite significance only in connection with a sphere; in that case diameter and size are synonymous, and calculations of surface area or volume may readily be made from the value of the diameter. For any other shape, however, the term *diameter* will not serve these purposes and hence cannot be used for any fundamental investigations of physical properties.

This line of reasoning impelled Wadell to define the size of irregular solids in terms of a *true nominal diameter,* which is equal to the diameter of a sphere of the same volume as the particle. The true nominal diameter has become a concept of great significance in sedimentary work, not only because of its adaptability to theoretical investigations, but also because of its immediate use in the laboratory study of sediments. Further details of its application to shape studies of particles will be given in Chapter 11.

In mechanical analysis various terms have been developed to express the size of irregular particles in terms of their settling velocities. Schöne [1] in 1868 introduced the term *hydraulischer Werth* (hydraulic value) to define the diameter of a quartz sphere having the same settling velocity as a given particle in water. The hydraulic value has no bearing on the actual size of the particle in terms of its volume, but it was used to express "size" in numerical terms. Odén [2] in 1915 improved the concept by introducing the term *equivalent radius* as the radius of a sphere of the same material as the particle and having the same settling velocity. More recently Wadell [3] sharpened the definition by introducing the term *sedimentation radius* as "the radius of a sphere of the same specific gravity and of the same terminal uniform settling velocity as a given particle in the same sedimentation fluid." Robinson,[4] previous to Wadell's work, recognized the apparently meaningless use of the term *radius* in connection with irregular particles studied by sedimentation and chose to ignore any expression of size in his mechanical analyses. Instead, he expressed his values directly as the logarithms of settling velocity.

Wadell's sedimentation radius will be accepted as standard in this text, and in any context referring to the size of irregular particles as determined by sedimentation methods, the sedimentation radius will be either explicit or implied. It will be developed shortly that two laws

[1] E. Schöne, Neber einen neuen Apparat für die Schlämmanalyse: *Zeits. f. anal. Chemie,* vol. 7, pp. 29-47, 1868.
[2] S. Odén, Eine neue methode zur mechanischen Bodenanalyse: *Int. Mitt. f. Bodenkunde,* vol. 5, pp. 257-311, 1915.
[3] H. Wadell, Some new sedimentation formulas: *Physics,* vol. 5, pp. 281-291, 1934.
[4] G. W. Robinson, The form of mechanical composition curves of soils, clays, and other granular substances: *Jour. Agric. Sci.,* vol. 14, pp. 626-633, 1924.

PRINCIPLES OF SIZE ANALYSIS

of settling velocities are generally applicable to sedimentary studies, either Stokes' law directly or a modification introduced by Wadell, designed to correct for the non-spherical particles in sediments. When Stokes' law is used directly, the computed size values may be called Stokes' sedimentation radius (or diameter), and when Wadell's practical formula is used, they may be called the practical sedimentation radius (or diameter).

THE SETTLING VELOCITIES OF SMALL PARTICLES

One of the fundamental principles on which mechanical analysis is based is that small particles will settle with a constant velocity in water or other fluids. It is universally true that small particles reach this constant velocity in a fluid medium as soon as the resistance of the fluid exactly equals the downward constant force (gravity) which acts on the particle. In general the settling velocity of the particle depends on its radius, its shape, its density, its surface texture, and the density and viscosity of the fluid. A number of mathematical expressions have been developed to show the relations among these factors, some based on empirical grounds and others on theoretical grounds. Several of these laws will be discussed in varying detail, depending upon their applicability to mechanical analysis.

STOKES' LAW OF SETTLING VELOCITIES

The classic formula for settling velocities, and the best known, is that of Stokes,[1] which confines itself to spheres. Since this equation is of such fundamental importance and of such widespread application, it will be considered in detail.

Theory of Stokes' law. Stokes first considered the resistance which a fluid offers to the movement of a sphere suspended in it, and arrived at the equation

$$R = 6\,\pi r \eta v \quad \ldots \quad (1)$$

where R = resistance in g. cm./sec.^{2}
r = radius of the sphere in cm.
η = viscosity of the fluid
v = velocity of the sphere in cm. sec.$^{-1}$

When a small sphere settles in a fluid, it is acted on by the force of gravity, $\frac{4}{3}\pi r^3 d_1 g$, acting downward; and by the buoyant force of the

[1] G. G. Stokes, On the effect of the internal friction of fluids on the motion of pendulums: *Trans. Cambridge Philos. Soc.*, vol. 9, part 2, p. 8-106, 1851.

liquid, $\frac{4}{3}\pi r^3 d_2 g$, given by Archimedes' principle, and acting upward, which results in a net force $\frac{4}{3}\pi r^3 (d_1 - d_2)g$ acting downward. At the instant when the resistance R exactly equals this net force, the velocity becomes constant and remains so. When this uniform state is reached there results:

$$6\pi r \eta v = \frac{4}{3}\pi r^3 (d_1 - d_2)g \quad \ldots \quad (2)$$

where the additional symbols are d_1 = density of the sphere
d_2 = density of the fluid
g = acceleration due to gravity (980 cm. sec.$^{-2}$)

By solving equation (2) for v, one obtains

$$v = \frac{2}{9}\frac{(d_1 - d_2)gr^2}{\eta} \quad \ldots \quad (3)$$

the equation of Stokes' law.

In general, if standard conditions are assumed, that is, a constant temperature, a given fluid, and a known specific gravity of the sphere, equation (3) may be expressed as

$$v = Cr^2 \quad \ldots \quad (4)$$

where C is a constant, and equal to $\frac{2}{9}\frac{(d_1 - d_2)g}{\eta}$. Tables for the value of this constant under various conditions have been computed and will be referred to later. It may be mentioned, however, that for water at 20° C., and particles with a specific gravity of quartz, 2.65, the value of the constant is $C = 3.57 \times 10^4$.

The assumptions of Stokes' law. Several assumptions underlie Stokes' law, and it is important to consider them in the light of mechanical analysis. The following assumptions are generally recognized:

(1) The particle must be spherical, it must be smooth and rigid, and there should be no slipping between it and the medium.
(2) The medium may be considered homogeneous in comparison to the size of the particle.
(3) The particle should fall as it would in a medium of unlimited extent.
(4) A constant settling velocity must have been reached.
(5) The settling velocity should not be too great.

Each of these assumptions deserves some consideration from the point of view of mechanical analysis to determine whether it is in fact satisfied in practice. Assumption 1 is satisfied to the extent that there is no slip between the particles and the fluid, inasmuch as they are wetted by the liquids commonly used in mechanical analysis. Similarly, since the particles are solid, the assumption of rigidity is satisfied, but it is seldom true that the grains are perfectly smooth. Arnold [1] has shown that pitted surfaces do not appreciably affect the settling velocities of small spheres, and consequently this factor may not be of paramount importance. The condition that the particle be a sphere is perhaps least satisfied, and this introduces several difficulties. The same sediment may have grains varying in shape from almost true spheres to irregular grains, plates, and laths. Experiments have been per-

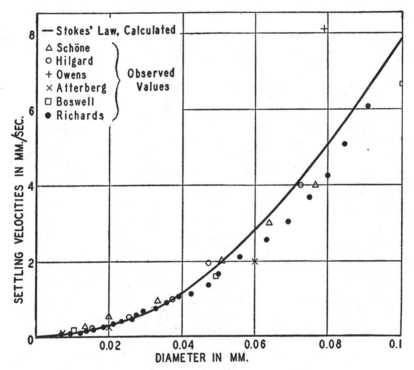

FIG. 20.—Comparison of some observed settling velocities with Stokes' law, in the size range 0 to 0.1 mm. diameter.

formed, however, to determine the agreement between the settling velocities of powders, soils, and sediments in terms of observation on the one hand and the expected theoretical values of Stokes' law on the other. The degree of agreement is quite remarkable within the range significant in most methods of mechanical analysis. Andreason and Lundberg,[2] for example, found a very satisfactory agreement between grades separated by an elutriator (Schöne's)

[1] H. D. Arnold, Limitations imposed by slip and inertia terms upon Stokes' law for the motion of spheres through liquids: *Phil. Mag.*, vol. 22, pp. 755-775, 1911.
[2] A. H. M. Andreason and J. J. V. Lundberg, Ueber Schlämmgeschwindigkeit und Korngrösse: *Kolloid Zeits.*, vol. 49, pp. 48-51, 1929.

and Stokes' law for grades ranging upward to 0.088 mm. diameter, although they mentioned that the larger grades showed some deviation. The general extent of the agreement may be shown by a comparison of the observed values of several experimenters and the theoretical values as given by Stokes' law. Figure 20 is such a curve based on values from Schöne,[1] Hilgard,[2] Owens,[3] Atterberg,[4] Boswell,[5] and Richards.[6] Stokes' law is shown as a solid line computed for quartz settling in water at a temperature of 15° C. These conditions are chosen as an average because of the lack of definite data regarding the exact conditions under which the experiments were carried on.

It will be noted that there is a fairly close agreement between observed and computed velocities until a diameter of about 0.05 mm. is reached. Richard's values begin to depart at about 0.04 mm., Schöne's at about 0.06 mm., and Hilgard's at about 0.07 mm. These deviations may of course depend upon varying experimental conditions as well as upon shape differences. It seems fairly safe, however, to consider the agreement satisfactory up to a diameter of at least 0.05 mm., but probably not beyond 0.07 mm. It will be seen later that these upper limits agree quite well with the theoretical upper limits of Stokes' law for true spheres.

Because of the difficulty of defining the size of the irregular particles commonly found in soils or sediments, the sizes are usually defined in terms of their settling velocities according to Stokes' law, by such terms as *hydraulic radius, equivalent radius,* or *sedimentation radius,* as described on page 94.

Assumption 2 merely states that the distances between the molecules of the fluid must be small compared with the sizes of the particles. This condition is fully satisfied down to the borders at least of the colloidal state, and perhaps well within it. For most practical purposes it may be ignored.

Assumption 3 is involved to a considerable extent in mechanical analysis, inasmuch as all methods of analysis involve vessels of finite size. Since the assumption states that the medium should be of unlimited extent, it is necessary to consider the error introduced by sedimentation cylinders and tubes of particular diameters. Several equations have been developed to express the influence of wall nearness on settling velocities. Lorentz[7] set up an equation for the resistance met by a particle settling parallel to a plane wall, and by using his terms instead of Stokes' R one finds that the ratio of the "true" velocity to Stokes' velocity is as follows:

$$v_t/v_s = 1 - 9r/16L \quad \ldots \quad \ldots \quad (5)$$

[1] E. Schöne, *loc. cit.*, 1868.
[2] E. W. Hilgard, On the silt analysis of soils and clays: *Am. Jour. Sci.*, vol. 6, pp. 288-296, 333-339, 1873.
[3] J. S. Owens, Experiments on the settlement of solids in water: *Geog. Journal*, vol. 37, pp. 59-79, 1911.
[4] A. Atterberg, Die mechanische Bodenanalyse und die Klassifikation der Mineralböden Schwedens: *Int. Mitt. f. Bodenkunae*, vol. 2, pp. 312-342, 1912.
[5] P. G. H. Boswell, *A Memoir on British Resources of Refractory Sands for Furnace and Foundry Purposes,* Part I (London, 1918).
[6] R. H. Richards and C. E. Locke, *A Textbook of Ore Dressing,* 2nd ed. (New York, 1925).
[7] H. A. Lorentz, *Abhandlungen über theoret. Physik* (Leipzig, 1911), vol. 1, p. 40.

PRINCIPLES OF SIZE ANALYSIS

where L is the distance between the sphere and the wall. Figure 21 shows the value of this ratio, expressed as a percentage, for spheres of radius 0.001 cm. and 0.0025 cm., at various distances from the wall. The values of the ratio are all smaller than 1, indicating that the effect of the factor is to reduce the velocity. The curves are hyperbolic, and the effects of wall nearness decrease very rapidly. The two curves also show that the effects vary with the size of the particles, being larger for larger particles. The suggestion offered by these data is that vessels of some appreciable radius (a minimum of about 2 cm. radius) should be used in mechanical analysis, to render wall nearness effects essentially negligible.

Ladenburg[1] attacked the problem from the point of view of a sphere of radius r settling in a cylinder of length L and radius R. Experiments by Arnold[2] in this case showed that the velocity according to Stokes' law is not appreciably affected until the radius of the particle equals 1/10 the radius of the cylinder. It would seem from this that tubes of very small radius may be used in mechanical analysis, but another factor may enter when a system of particles is present. In such cases each particle is influenced by its neighbors, so that an extremely complicated situation develops, which has not been fully elucidated mathematically. In dilute suspensions these effects are apparently not serious. The authors know of no quantitative data on the subject, but it is perhaps best to use suspensions containing not more than about 25 g. of solid to the liter. To be conservative, also, vessels of reasonably large diameter should be used, say 5 cm. or larger.

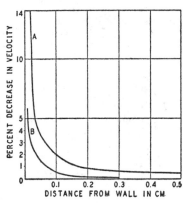

FIG. 21.—Effect of wall-nearness on settling velocities of spheres. Curve A, spheres of radius 0.0025 mm.; curve B, spheres of radius 0.001 mm.

Assumption 4 states that the constant velocity of fall must have been reached. It is clear that at time $t = 0$, the velocity is zero, so that the particle increases its velocity until the resistance of the fluid exactly counterbalances the downward force on the particle. There is thus an interval of time before the constant velocity is reached, and it is necessary to consider the order of magnitude of this interval. Weyssenhoff[3] has developed an equation which permits a computation of this interval. The equation is rather complex and need not be considered in detail; computations for a sphere of diameter 0.05 mm. (which is near the upper limit of applicability of the law) indicate that it requires about 0.003 sec. to achieve constant velocity. Hence assumption 4 need not concern practical mechanical analysis.

[1] R. Ladenburg, Über den Einfluss von Wänden auf die Bewegung einer Kugel in einer reibenden Flussigkeit: *Ann. der Physik*, vol. 23, pp. 447-458, 1907.
[2] H. D. Arnold, *loc. cit.*, 1911.
[3] J. Weyssenhoff, Betrachtungen über den Gültigkeitsbereich der Stokesschen und der Stokes-Cunninghamschen Formel: *Ann. der Physik*. vol. 62, pp. 1-45, 1920.

Assumption 5 provides that the motion should be slow. This condition imposes certain limits on the range of sizes that may be studied by Stokes' law, and it is important to consider it in some detail. The assumption is based on the fact that the viscosity of the medium should furnish all of the resistance which the sphere meets in its descent. When the sphere is so large that this no longer holds, the particle drags some of the liquid with it, and the radius no longer holds the same simple relation to the velocity as before. The limiting size in any given case will depend on a number of factors. If the liquid is particularly viscous, the particle may be larger than in less viscous liquids. Similarly, when the difference between the density of the sphere and that of the liquid is slight, the particle may be larger than in the reverse case. Allen,[1] discussing the upper limit of Stokes' law, pointed out that the law was valid as long as the inertia terms are neglected in comparison with viscosity. This requires that the velocity times the radius times the density of the fluid must be small compared to the viscosity:

$$vd_2r < \eta \qquad (6)$$

In seeking an upper limit, Allen defined as the critical radius that value of r which established equality of the two sides of the expression. By setting this equation up in the form $v = \eta/d_2r$, and substituting $v = Cr^2$ from Stokes' law, one obtains

$$Cr^2 = \frac{\eta}{d_2r}$$

from which it is clear that

$$r = \sqrt[3]{\frac{\eta}{d_2C}} \qquad (7)$$

Arnold[2] subsequently showed that the inertia terms begin to manifest themselves when a radius of 0.6 the value of the critical radius is reached.

By considering a sphere of quartz (specific gravity = 2.65) settling in water at a temperature of 20° C. (the value of C in Stokes' law in this case is 3.57×10^4), it may be shown that the uncorrected critical radius has a value of about 0.006 cm., which is a diameter of 0.12 mm. Six-tenths of this value is about 0.08 mm. diameter, which sets an upper limit to the application of Stokes' law in ordinary mechanical analysis. This corresponds to a grain slightly larger than 1/16 mm., and involves a settling velocity of about 5 mm. per second. It will be noted that this value is of the same order of magnitude as the experimental data shown in Figure 20.

The problem of the lower limit of Stokes' law has also received the attention of several workers. Perrin's work[3] in this connection is particularly noteworthy. He prepared essentially monodisperse systems of very small particles by centrifuging gamboge suspensions. The radii of the particles were determined by three methods, one of which was their settling velocities. The values found by Stokes' law agreed strikingly with the values obtained by

---

[1] H. S. Allen, The motion of a sphere in a viscous fluid: *Phil. Mag.*, vol. 50, pp. 323-338, 519-534, 1900.

[2] H. D. Arnold, *loc. cit.*, 1911.

[3] J. Perrin, *Atoms*, translated by D. L. Hammick (London, 1920).

PRINCIPLES OF SIZE ANALYSIS

the other two methods. Inasmuch as the particles ranged in radius from about 0.15 micron to 0.5 micron, Perrin concluded that Stokes' law held despite Brownian movement and that it was valid within the borders of the colloidal state. Van Hahn,[1] after considering the results of several observers, concurred with Perrin in this conclusion. That there must be a lower limit seems obvious, or colloidal suspensions would not be essentially permanent, as they are.

For the purposes of mechanical analysis we may consider it sufficient that the law holds to a diameter of 0.1 micron, which is at the lower limit of coarse disperse systems as earlier defined. Thus both the upper and lower limits of Stokes' law fall at convenient points, as far as mechanical analysis is concerned, because the upper limit of the law occurs just about at the lower limit of sieving, and the lower limit occurs within the colloidal state, at sizes smaller than ordinary methods of analysis can separate. This permits a composite analysis to be made by sieving down to about 0.06 mm. diameter, and below that by settling velocities computed from Stokes' law.

Summary of Stokes' law. In summary of the discussion of Stokes' law it may be said that the highly practical results that follow from its application to mechanical analysis warrant its use as a fundamental equation in the development of any method of analysis. In applying the law there are certain precautions to be observed as to the sizes of particles to be separated and the variables in the equation that may affect the results.

The limits of the sizes which may be studied by Stokes' law have already been discussed; among the other variables are the density of the particles and the viscosity of the liquid. In a pigment, where all the particles have the same specific gravity, the first problem does not enter, but among sediments and soils there is a mixture of particles ranging rather widely in specific gravity. The results of heavy mineral separations show, however, that in the average case more than 95 per cent of the material is quartz or feldspar. The specific gravity of quartz is 2.65, and that of feldspar about 2.6. Hence by far the greatest amount of material is under 2.7, so that the two most common values to be adopted are either 2.65 or 2.7, depending upon the percentage of heavy minerals in the sediment. In special cases, as where magnetite concentrates occur, the value assumed will have to be adjusted to the special merits of the case. The effect of specific gravity on the settling velocity may be seen from a simple example. For a sphere of diameter 0.05 mm. in water at 20° C., an increase in specific gravity from 2.6 to 2.7 causes an increase of nearly 6 per cent in the rate of settling.

The viscosity of water varies rather considerably in small ranges of

[1] F.-V. von Hahn, *Dispersoidanalyse* (Leipzig, 1928), pp. 270 ff.

temperature. Figure 22 shows the curve of viscosity of water from 0° to 30° C. The value drops about 50 per cent in this range. As in the case of varying specific gravities, changes in the viscosity of the water cause considerable changes in the settling velocity of the particle. A quartz sphere of diameter 0.05 mm. has a settling velocity of 0.196 cm./sec., in water at 15°, and 0.223 cm./sec. at 20° C. This is an increase of 11.4 per cent for five degrees, or an average of 2.3 per cent per degree. From this it is clear that some temperature control should be exercised, so that no considerable fluctuations occur during an analysis.

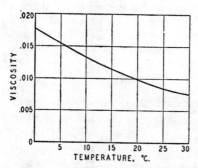

Fig. 22.—Viscosity of water as a function of temperature.

In addition to the changes in viscosity which accompany changes of temperature, convection currents may be set up, which materially affect the normal settling of the particles.

Reference was made earlier to a form of Stokes' law (equation 4), in which all variables except the velocity and the radius were combined into a single constant. By choosing standard conditions of temperature and average specific gravity of the material being analyzed, the velocities for spheres of any given radii may easily be computed. The most common temperatures at which analyses are made are 15° and 20° C., and the average specific gravities usually chosen for sediments are 2.65 and 2.70. Table 12 (page 110) shows the values of the constant in Stokes' law under these conditions, and the velocities of particles ranging in diameter from 0.06 mm. to 0.5 micron. This table furnishes basic information used in several methods of analysis and will be referred to later. It should also be mentioned that the velocities in cm./sec. of any intermediate-sized particles may be found by multiplying the value of the constant by the square of the radius in centimeters.

Rubey's Formula

In 1933, Rubey[1] developed a general formula for settling velocities which agrees with observed values over a wider range than Stokes' law. In extending the law of settling velocities beyond the critical value of

[1] W. W. Rubey, Settling velocities of gravel, sand, and silt particles: *Am. Jour. Sci.*, vol. 25, pp. 325-338, 1933.

PRINCIPLES OF SIZE ANALYSIS

Stokes' law, Rubey conceived that the total force acting on a large particle was the sum of the forces due to viscous resistance and the impact of the fluid. By equating these forces to the effective weight of the particle, Rubey obtained the expression

$$\frac{4}{3}\pi r^3(d_1 - d_2)g = 6\pi r\eta v + \pi r^2 v^2 d_2 \quad . \quad . \quad . \quad . \quad (8)$$

from which he obtained his formula by solving for v:

$$v = \left[\frac{4}{3}gd_2(d_1 - d_2)r^3 + 9\eta^2 + 3\eta\right]^{\frac{1}{2}} \Big/ d_2 r \quad . \quad (9)$$

Figure 23, adapted from Rubey's paper, shows the gradual transition between the ranges of viscous resistance and fluid impact. The heavy

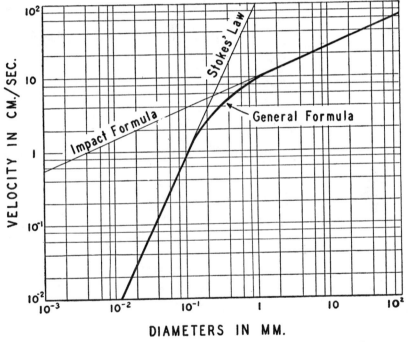

FIG. 23.—Rubey's general formula for settling velocities. The heavy line agrees well with observed data.

line agrees well with observed settling velocities of quartz or galena, as Rubey's original figure indicates.

Analysis of Rubey's formula, as expressed in equation (8), shows that, when the velocity of the particle is so small that the inertia terms

$(\pi r^2 v^2 d_2)$ may be neglected, the expression simplifies to Stokes' law. It is thus a generalization of the latter.

Wadell's Sedimentation Formula

The most recent work on settling velocities has been done by Wadell,[1] who opened a new approach to the problem by examining the functional relationship between the coefficient of resistance, C_r, and Reynolds number, R_e.

The coefficient of resistance is defined by equating the force producing motion of a sphere to the force resisting motion, expressed as the coefficient of resistance times the dynamic pressure acting on the cross-sectional area of the sphere:[2]

$$\frac{4}{3}\pi r^3(d_1-d_2)g = C_r \frac{\pi r^2 v^2 d_2}{2} \quad \ldots \quad (10)$$

from which $C_r = \frac{8}{3}\frac{g(d_1-d_2)r}{d_2 v^2}$. The symbols have the same significance as in the case of Stokes' law, equation 3.

Reynolds number is defined in terms of the radius of the sphere, its velocity, and the density and viscosity of the fluid, as follows:[3]

$$R_e = \frac{2rvd_2}{\eta} \quad \ldots \quad (11)$$

Reynolds number is dimensionless, i.e., it is a pure number.

Wadell chose as his starting point an equation relating the coefficient of resistance to Reynolds number in the following manner:

$$C_r \frac{\pi r^2 v^2 d_2}{2} = f(R_e) \frac{\pi r^2 v^2 d_2}{2} \quad \ldots \quad (12)$$

from which $C_r = f(R_e)$. By plotting a number of observed settling velocities and radii in terms of R_e and C_r, with R_e as abscissa and C_r as ordinate, on double log paper, Wadell developed an empirical formula for settling velocities which not only extends the range of practical settling velocities to much larger diameters than those afforded by Stokes' law, but in addition enabled him to elucidate the influence of shapes of

---

[1] H. Wadell, The coefficient of resistance as a function of Reynolds number for solids of various shapes: *Jour. Franklin Inst.*, vol. 217, pp. 459-490, 1934.
H. Wadell, Some new sedimentation formulas: *Physics*, vol. 5, pp. 281-291, 1934.
H. Wadell, Some practical sedimentation formulas: *Geol. Fören. Forhändl.*, vol. 58, pp. 397-408, 1936.
[2] J. E. Christiansen, Distribution of silt in open channels: *Trans. Am. Geophysical Union*, part II, pp. 478-485, 1935.
[3] H. Wadell, *loc. cit.*, 1934.

PRINCIPLES OF SIZE ANALYSIS

particles on their settling velocities. Moreover, the opening of this new approach to the problem of settling velocities affords important means of studying actual sedimentary problems in terms, for example, of deposition in water as against air. The theoretical aspects of these applications do not belong in the present volume, which considers only petrographic applications. In the latter connection, Wadell's formula extends the use of pure sedimentation methods to particles which at present are generally sieved, and it may be expected that such sedimentation methods of analysis may be developed for practical work. Likewise shape factors may be included, an important item when it is recalled that truly spherical particles are practically non-existent among sediments.

Wadell's formula, expressed in terms of a correction to be applied to Stokes' radius, is

$$r_a = r_s \left[1 + 0.08(2r_s v_a d_2/\eta)^{0.69897} \right] \quad \ldots \quad (13)$$

where r_a is the actual radius, r_s is the radius according to Stokes' law, and v_a is the actual settling velocity. The exponent $0.69897 = \log_{10} 5$ was determined by statistical methods from the observational data.

Figure 24, adapted from Wadell, shows a portion of the range covered by the original graph. The heavy line is Wadell's curve, which was shown to agree quite closely with observed settling velocities of spheres up to values of Reynolds number of about 3,000. In the same figure are shown Stokes', Oseen's, and Goldstein's laws (see below), to indicate their departure from Wadell's curve in the higher ranges. Stokes' law is valid to values of about $R_e = 0.2$, Oseen's and Goldstein's laws apply to about $R_e = 0.5$. Rubey's formula, according to Christiansen[1] agrees with observed values on quartz and galena to values of $R_e = 1000$.

Rouse[2] recently extended the curve developed by Wadell to values of R_e up to one million. Beyond 10^5 the curve for spheres shows an abrupt change in slope, due to the onset of turbulence in the boundary layer at the front of the sphere.

Wadell[3] also derived a modified form of Stokes' law which has immediate application to mechanical analysis. Arguing on the basis that sedimentary particles are not true spheres, but a mixture of particles of varying shape, he developed a resistance formula for a hypothetical particle intermediate in shape between a disc and a sphere. By applying his reasoning to Stokes' resistance equation, $R = 6\pi r \eta v$, Wadell obtained

---

[1] J. E. Christiansen, *loc. cit.*, 1935.
[2] H. Rouse, Nomogram for the settling velocity of spheres: *Report of the Committee on Sedimentation 1936-37*, pp. 57-64, Nat. Research Council, 1937.
[3] H. Wadell, *loc. cit.*, 1936.

the value $R_w = 9.44\pi r \eta v$. By equating this value to the effective weight of the particle, $\frac{4}{3}\pi r^3(d_1 - d_2)g$, Wadell obtained

$$v_p = \frac{1}{7}\frac{(d_1 - d_2)gr_p^2}{\eta} \quad \ldots \ldots \quad (14)$$

where v_p is the "practical settling velocity," and r_p is the "practical sedimentation radius." It may be noted from a comparison of this expression

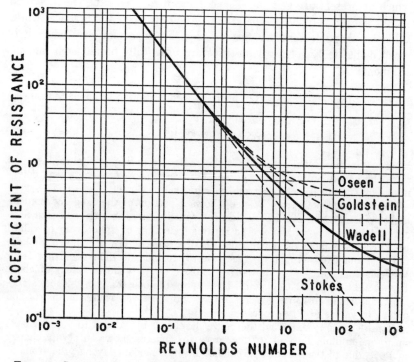

FIG. 24.—Comparison of several laws of settling velocities of spheres, expressed in terms of Reynolds number and the co-efficient of resistance. The heavy line (Wadell's formula) agrees closely with observed data. (Adapted from Wadell, 1934.)

with Stokes' law, (equation 3), that the only difference between the two equations is the numerical constants. Stokes' law has the fraction $\frac{2}{9}$, whereas Wadell's value is $\frac{1}{7}$.

This simple relation offers an immediate method of correcting Stokes' law to allow for shape variations in the sediment, namely, by finding the value of the ratio v_p/v_s, where v_s refers to Stokes' velocity. This ratio may be found by dividing equation (14) by equation (3):

$$v_p/v_s = \frac{1}{7}\left[\frac{(d_1-d_2)gr^2}{\eta}\right] \Big/ \frac{2}{9}\left[\frac{(d_1-d_2)gr^2}{\eta}\right] = \frac{1}{7}\Big/\frac{2}{9} = \frac{9}{14} = 0.64$$

Thus the practical settling velocity of a given sedimentary particle is 64 per cent of the theoretical settling velocity of the corresponding sphere. Likewise the ratio r_p/r_s may be computed from the same equations and is found to be $r_p/r_s = \sqrt{7}/\sqrt{9/2} = \sqrt{14/9} = 1.25$, and hence for a given settling velocity the practical radius is 1.25 as large as the radius of the corresponding sphere. These results are in strict accord with theory, because an irregular particle of the same volume as a sphere will have a greater surface area and hence a smaller settling velocity; likewise for a given settling velocity an irregular particle will be larger than the corresponding sphere.

As in the case of Stokes' law, the practical sedimentation formula may be expressed as $v_p = Kr_p^2$, where $K = \frac{1}{7}\frac{(d_1-d_2)g}{\eta}$. For quartz particles (sp.g. = 2.65) at 20° C., K has the value 2.28×10^4. From the relation $v_p/v_s = Kr^2/Cr^2$, it follows that K/C should also equal 0.64, so that it becomes a simple matter to compute K from published tables for the value of C.

Oseen's Law of Settling Velocities

Oseen[1] developed a resistance formula which differs from Stokes' resistance (equation 1) in that the latter is the first term of Oseen's series

$$R = 6\pi r\eta v \left(1 + \frac{3}{8}\frac{d_2 r}{\eta}|v|\right) \quad \ldots \ldots \quad (15)$$

where the symbols have the same meaning as in Stokes' law, and $|v|$ is the absolute value of the velocity. By equating this expression to the force acting downward on the particle, $\frac{4}{3}\pi r^3(d_1-d_2)g$, and solving for v, Oseen obtained his settling velocity equation

$$v = \frac{-\frac{3\eta}{r} + \sqrt{\frac{9\eta^2}{r^2} + 3d_2(d_1-d_2)gr}}{\frac{9}{4}d_2} \quad \ldots \ldots \quad (16)$$

From the nature of equation (16), it follows that Oseen's equation becomes identical with Stokes' law when all but the first term of equation

[1] C. W. Oseen, Ueber den Gultigkeitsbereich der Stokes'schen Widerstandformel: *Ark. Mat., Astron. Fys.*, vol. 6, 1910; vol. 7, 1911; vol. 9, 1913.

(15) is neglected. In this case v becomes so much smaller than unity that v^2 may be neglected in comparison with v. This relation may readily be seen by solving equation (16) for r and dropping powers of v higher than 1.[1]

Oseen's law is directly applicable to mechanical analysis by sedimentation for diameters above the upper limit of Stokes' law. Figure 25 shows the curves of Stokes' and Oseen's laws, and indicates how the latter departs from Stokes' curve beyond diameters of about 0.01 mm. For the smaller sizes of particle, in the silt and clay ranges, the greater ease of computing diameters according to Stokes' law or Wadell's practical formula renders these laws more convenient for general use.

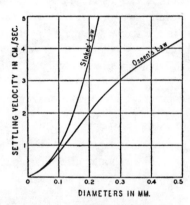

Fig. 25.—Departure of Stokes' law from Oseen's law. Below diameters of about 0.05 mm., Stokes' law is a special case of Oseen's more general equation.

Goldstein's Law of Settling Velocities

Goldstein[2] began his consideration of the problem from the standpoint of Oseen's resistance formula (equation 15), in which terms depending upon the square of the velocity were neglected, and solved the equation for the complete series introduced by Oseen, restricting himself, however, to small values of Reynolds number. Goldstein's solution was expressed entirely in terms of Reynolds number and a "drag coefficient" k_D, which he defined as $k_D = \dfrac{D}{\pi d_2 v^2 r^2}$, where D is the "drag" and the other symbols have their previous meaning. In his final solution Goldstein obtained the expression

$$k_D = \frac{12}{R_e}\left(1 + \frac{3}{16}R_e - \frac{19}{1{,}280}R_e^2 + \frac{71}{20{,}480}R_e^3 - \ldots\right) \quad . \quad . \quad (17)$$

for the complete law. In this series Stokes' law is $k_D = 12/R_e$ and Oseen's law is $k_D = (12/R_e) + 2.25$. Goldstein pointed out that for

---

[1] H. Gessner, *Die Schlämmanalyse* (Leipzig, 1931), p. 20, shows the steps involved.

[2] S. Goldstein, The steady flow of viscous fluid past a fixed spherical obstacle at small Reynolds numbers: *Proc. Roy. Soc. London*, pp. 225-235, 1929.

PRINCIPLES OF SIZE ANALYSIS

values of R_e less than 1.6 it was unnecessary to consider the influence of corrections beyond Oseen's expression.

It is interesting in connection with the abbreviated forms of Stokes' and Oseen's laws just given, that Stokes' law and Rubey's equation may also be expressed very briefly in terms of the coefficient of resistance, C_r, and Reynolds number as $C_r = 24/R_e$ and $C_r = (24/R_e) + 2$, respectively.[1]

To a large extent mechanical analysis by sedimentation methods is performed on silt and clay, but as methods are developed for sedimentation studies of sand and gravel, Oseen's law and perhaps Goldstein's further modifications may be more extensively used. These same considerations affect the more extensive use of Wadell's theoretical law, which agrees with observed settling velocities up to values of Reynolds numbers of about 3,000. Moreover, the allowance for shape factors which may be included in Wadell's formulas becomes increasingly important as large particles are studied by their settling velocities.

SUMMARY OF LAWS OF SETTLING VELOCITIES

The preceding discussion of settling velocities indicates that for practical purposes Stokes' law (equation 3) and Wadell's practical sedimentation formula (equation 14) are most applicable to mechanical analysis. Stokes' law and Wadell's formula both extend to values of Reynolds number of at least 0.2 (diameter about 0.06 mm. for quartz spheres at 20° C.).

Stokes' law is widely used in mechanical analysis and affords good values even for sedimentary particles, although it is generally recognized that perfect agreement is not theoretically possible because of the nonspherical shapes involved. Wadell's practical formula is a modification of Stokes' law, designed to adjust this limitation, and it undoubtedly is the most accurate expression for general use. It may be mentioned that the consideration of the assumptions of Stokes' law is pertinent also to the practical sedimentation formula, with the exception of the assumption of sphericity.

For practical purposes, the most convenient manner of using Stokes' equation or Wadell's equation is from tabulated values of the settling velocities. Table 12 furnishes these values for Stokes' law, and Table 13 makes available the corresponding values from Wadell's formula for the same conditions of sedimentation. It may be noted that any two corresponding values in Tables 12 and 13 satisfy the ratio $K/C = 0.64$.

[1] J. E. Christiansen, loc. cit., 1935.

TABLE 12

Settling Velocities of Spheres of Specific Gravity 2.65 and 2.70 at Temperatures of 15° and 20° C., Computed from Stokes' Law*

DIAMETERS IN MILLIMETERS	SETTLING VELOCITIES IN CM./SEC.			
	Specific Gravity of Particles = 2.65		Specific Gravity of Particles = 2.70	
	15° C. $C=3.14 \times 10^4$	20° C. $C=3.57 \times 10^4$	15° C. $C=3.24 \times 10^4$	20° C. $C=3.67 \times 10^4$
1/16 .... 0.0625	0.305	0.347	0.315	0.357
.0442	.153	.174	.158	.179
1/32 .... .0312	.0764	.0869	.079	.0893
.0221	.0382	.0435	.0394	.0447
1/64 .... .0156	.0191	.0217	.0197	.0223
.0110	.0096	.0109	.0099	.0112
1/128 ...... .0078	.0048	.00543	.00493	.00558
.0055	.00239	.00272	.00247	.00280
1/256 ...... .0039	.00120	.00136	.00123	.00140
.00276	.00060	.00068	.00062	.00070
1/512 ...... .00195	.00030	.00034	.00031	.00035
.00138	.000148	.000168	.000153	.000173
1/1024 ..... .00098	.000075	.000085	.000077	.000087
.00069	.000038	.000043	.000039	.000044
1/2048 ..... .00049	.000019	.000021	.000019	.000022

* Computations by slide rule.

THEORY OF SEDIMENTING SYSTEMS

Laws of settling velocities confine themselves to the settling rates of individual particles falling through a fluid of infinite extent. In mechanical analysis, however, one deals with a system of particles, and it is important to consider how the system behaves as a whole during sedimentation. Such systems of particles may be all of one size (monodisperse) or of various sizes (polydisperse).

In the following discussion it will be assumed that the concentration of the system is so dilute that the particles do not interfere with one another during descent, that the particles are small spheres (to permit the direct application of Stokes' law), and that no coagulation phenomena are present. In actual mechanical analysis, of course, these simplifications may not hold rigorously, but it will be shown that for all practical purposes the theory affords a sound basis for a number of new techniques of mechanical analysis.

PRINCIPLES OF SIZE ANALYSIS

TABLE 13

Settling Velocities of Spheres of Specific Gravity 2.65 and 2.70 at Temperatures of 15° and 20° C., Computed from Wadell's Practical Sedimentation Formula*

DIAMETERS IN MILLIMETERS	SETTLING VELOCITIES IN CM./SEC.			
	Specific Gravity of Particles = 2.65		Specific Gravity of Particles = 2.70	
	15° C. $K=2.01\times10^4$	20° C. $K=2.28\times10^4$	15° C. $K=2.07\times10^4$	20° C. $K=2.35\times10^4$
1/16 ....0.0625	0.195	0.221	0.200	0.228
.0442	.098	.111	.101	.114
1/32 .... .0312	.0487	.0554	.050	.057
.0221	.0244	.0278	.0251	.0286
1/64 .... .0156	.0122	.0139	.0126	.0143
.0110	.0061	.0070	.0063	.0072
1/128 ...... .0078	.00306	.00348	.00316	.00358
.0055	.00153	.00174	.00158	.00179
1/256 ...... .0039	.00077	.00087	.00079	.00089
.00276	.000384	.000435	.000395	.000448
1/512 ...... .00195	.000192	.000217	.000197	.000224
.00138	.000096	.000108	.000098	.000112
1/1024 ..... .00098	.000048	.000054	.000049	.000056
.00069	.000024	.000027	.000024	.000028
1/2048 ..... .00049	.000012	.000013	.000012	.000014

* Computations by slide rule.

Sedimentation of monodisperse systems. The laws governing the sedimentation of monodisperse systems are quite simple, inasmuch as all the particles are of one size. Assuming a dilute suspension of spheres settling in a fluid at constant temperature, it follows that each particle will settle with the same velocity v, and in time t will have reached a point h units lower in the column of fluid, on the basis of the relation $v = h/t$. If one considers a cylinder of height h, which has the particles uniformly distributed through the liquid at time t_0, it is clear that the entire suspension will settle with a uniform speed and collect on the bottom of the vessel. Now the amount of material p which has settled out at any given time t is dependent on the total amount of dispersed material P, on the time, on the velocity, and inversely on the depth h. This permits the setting-up of the relation

$$p = \frac{kPvt}{h} \quad \ldots \quad (18)$$

in which k is the constant of proportionality.

At the time when all the material has settled to the bottom, $p = P$, and

$$1 = \frac{kvt}{h}$$

but $h = vt$ from the relation $v = h/t$, so that $k = 1$, and the equation (18) may be written simply as

$$p = \frac{Pvt}{h} \quad \ldots \quad (19)$$

Sedimentation of polydisperse systems. In polydisperse systems the problem may be considered from the point of view of a series of monodisperse systems, with the radii of the successive groups differing by infinitesimals from each other. This was the approach made by Odén[1] in 1915, and it forms the first clear expression of a mathematical theory of sedimenting systems. Odén later generalized his theory in collaboration with Fisher,[2] and this latter form of the theory furnishes the foundation of all the modern precision methods of mechanical analysis. Odén's original theory is an excellent example of the application of mathematical analysis to the solution of a complicated problem, and an abbreviated form of the theory is presented here.

Odén's original theory. Consider a polydisperse suspension with its particles uniformly distributed through the liquid. As sedimentation proceeds, each fraction having a given radius settles as a unit, and at any given time the amount of material which settles to the bottom consists of fractions which have completely settled from the suspension, plus some part of the fractions which have not completely settled out, because their velocities are not great enough to carry them from the top to the bottom of the cylinder in the time involved. The total amount settled on the bottom may be indicated by $P(t)$, and this is to be divided into the two parts mentioned. The fraction that has completely settled from the suspension has a velocity greater than h/t, and the portion of the partially sedimented fractions that have settled out has a velocity less than h/t. The value of this portion is given by equation (19).

The letter P represents a mixture of particles in which the successive radii differ from each other by infinitesimals. It may therefore be written as $P = F(r)dr$. In order to express p as a function of the radii, we may substitute this value for P in equation (19):

$$p = \frac{F(r)dr \cdot vt}{h}$$

and since by equation (4), $v = Cr^2$,

$$p = \frac{F(r)r^2 dr \cdot Ct}{h} \quad \ldots \quad (20)$$

where C is the constant of Stokes' law.

[1] S. Odén, *loc. cit.*, 1915.
[2] R. A. Fisher and S. Odén, The theory of the mechanical analysis of sediments by means of the automatic balance: *Proc. Roy. Soc. Edinburgh*, vol. 44, pp. 98-115, 1923-24.

PRINCIPLES OF SIZE ANALYSIS

When a fraction is completely sedimented, $p = P$, so that $1 = \dfrac{r^2 Ct}{h}$, and

$$r = \sqrt{\dfrac{h}{Ct}} \qquad \ldots \ldots (21)$$

The value for r in equation (21) is the critical radius for any time t, which determines the fractions that completely settle to the bottom and those which only settle in part. We may now use these values in setting up an expression for $P(t)$:

$$P(t) = \int_0^{\sqrt{h/Ct}} \dfrac{F(r) r^2 dr \cdot Ct}{h} + \int_{\sqrt{h/Ct}}^{\infty} F(r) dr \qquad \ldots \ldots (22)$$

In this equation the first integral is the sum of the portions of the partially sedimented fractions, whose radii are smaller than $\sqrt{h/Ct}$; and the second integral is the sum of the completely sedimented fractions, whose radii are greater than this critical radius.

By differentiating equation (22) there results:

$$\dfrac{dP(t)}{dt} = \int_0^{\sqrt{h/Ct}} \dfrac{F(r) r^2 dr \cdot C}{h} \qquad \ldots \ldots (23)$$

It will be noticed that equation (23) is exactly like the first term on the right-hand side of equation (22), except that t is absent. If equation (23) is multiplied through by t, we obtain:

$$t \cdot \dfrac{dP(t)}{dt} = \int_0^{\sqrt{h/Ct}} \dfrac{F(r) r^2 dr \cdot Ct}{h} \qquad \ldots \ldots (24)$$

Thus the important point has been established that the sum of the partially sedimented fractions is equal to the first derivative of $P(t)$, multiplied by t.

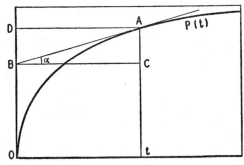

FIG. 26.—Principles of graphic analysis of Odén curves. See text for details.

This result may now be used in interpreting the experimental curve and obtaining the frequency distribution of the analyzed material. Figure 26 shows the $P(t)$ curve as a function of the time, obtained by weighing the amount of material that has accumulated at the bottom of a cylinder of suspension. At the point t an ordinate has been erected, and at A, where this intersects the $P(t)$ curve, a tangent has been drawn, intersecting the Y-axis at B. Two horizontal lines, AD and BC, are also drawn as indicated. We know that the distance OD is the total weight of sediment at time t, and it is required to prove that the distance OB is the amount of material having radii greater than that radius which is obtained by substituting the particular value of t in equation (21).

The derivative of the $P(t)$ curve at any point is its slope at that point. This is represented by $\tan \alpha$. But $\tan \alpha = $ AC/BC, so that $\dfrac{d P(t)}{dt} = \dfrac{AC}{BC}$, or

$$BC \cdot \frac{d P(t)}{dt} = AC$$

Now the distance BC represents the time of sedimentation, and hence by substituting t for BC we obtain

$$t \cdot \frac{d P(t)}{dt} = AC = BD \quad \cdots \quad (25)$$

We have already seen from equation (24) that $t \cdot \dfrac{d P(t)}{dt}$ represents the partially sedimented fractions, and by equation (25) it was shown that this equals the line segment BD. This means graphically that of the total amount OD sedimented in a particular time t, the distance OB represents the portion

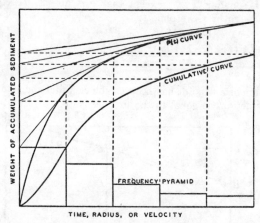

Fig. 27.—Relation between Odén's $P(t)$ curve, the cumulative curve, and the frequency pyramid or histogram of a sediment.

which has completely settled out of suspension and the distance BD represents the portion only partially sedimented.

If we now choose a series of time intervals corresponding to the settling velocities or radii of particular grade sizes, it is necessary only to draw tangents to the $P(t)$ curve at those points and to read off the intercepts (the

PRINCIPLES OF SIZE ANALYSIS

values of OB) on the Y-axis. The data so obtained are cumulative, and by subtracting one value from the next the amount of material in any grade is directly obtained. Figure 27 illustrates an original P(t) curve, the cumulative curve obtained from it, and the histogram derived from the cumulative curve.

The modified Odén theory. It was pointed out that the theory developed by Odén in 1915 was generalized in 1923-1924 by Fisher and Odén to include all possible methods of determining the frequency distributions of soils and sediments by indirect methods. In 1925 Odén[1] summarized the general theory, and his last paper is followed in the accompanying discussion.

It is assumed that a suspension of particles which has complete dispersion and a uniform distribution of the particles at time $t = 0$ is at a constant temperature and has a concentration so dilute that the particles do not interfere with each other during their descent. If G is the total weight of the particles suspended in V c.c. of water and s is the specific gravity of the particles, then at time t_o every cubic centimeter of the suspension contains G/V g., or G/Vs c.c., of solid particles, and therefore (1 — G/Vs) c.c. of water. Hence the uniform specific gravity of the suspension at the start is

$$\phi_0 = 1 + \frac{G}{Vs}(s-1) \quad \ldots \quad (26)$$

Let z represent the fraction or percentage by weight of particles having a velocity less than $v = x/t$. Then at time t and depth x there will remain zG g. of particles per V c.c., because at that time all particles with a velocity greater than x/t will have settled below this depth, and those particles whose velocity is less than x/t will continue in the same concentration as at the start. Hence at time t and depth x the specific gravity of the suspension will be $\phi = 1 + \frac{zG}{Vs}(s-1)$, or

$$\phi = 1 + kz \quad \ldots \quad (27)$$

where $k = \frac{G(s-1)}{Vs}$, a constant under the given experimental conditions. Equation (27) is fundamental to the derivation of the several methods which may be used to determine the frequency curve of the sediment. By definition z is a function of v, and it is an ordinate of the cumulative curve of the sediment. If we let Y $= dz/dv$, or $dz =$ Ydv, it is clear that Y must be an ordinate of the frequency curve, since dz is the proportion of particles between v and $v + dv$ and is equal to Y

[1] S. Odén, The size distribution of particles in soils and the experimental methods of obtaining them: *Soil Science*, vol. 19, pp. 1-35, 1925.

times dv. Here Y is the frequency function $f(v)$ of the sediment. It is usually more convenient to represent settling velocities as log v or radii as log r for purposes of graphing them. In the former case we may let $dz = Y dv/v = Y \cdot d(\log_e v)$. Since by Stokes' law $v = Cr^2$, and hence $dv = 2Cr dr$, we may substitute these values for v and dv in the preceding equation and obtain $dz = 2Y dr/r = 2Y \cdot d(\log_e r)$, where Y is now the frequency function $f(\log_e r)$. Either of these values of Y may be used in the following treatment, depending on the method of presentation desired.

For the sake of completeness a summary of the important equations and their significance is included here, but the interested reader is referred to the writers mentioned for a full discussion of the mathematical details. In developing the equations which follow, equation (27) is fundamental, and in addition the two partial derivatives of $v = x/t$, $\delta v/\delta x = 1/t$ and $\delta v/\delta t = -x/t^2$, are of importance.

There are four general methods by which the frequency function $f(\log_e v)$ or $f(\log_e r)$ may be obtained from equation (27). Changes in the density of the suspension may be considered as a function of the time at a constant depth, or as a function of the depth at a constant time. Similarly, changes in the hydrostatic pressure may be measured as a function of the time or of the depth. In addition to these four general methods, Odén discussed two others. One concerns the change in weight of an immersed body, and the other considers the weight of sediment accumulating at the base of a suspension.

If we consider first changes in the density, equation (27) may be differentiated with respect to x. Multiplication by $\delta v/\delta v$ yields an expression in which several substitutions may be made to obtain

$$Y = \frac{x}{k} \frac{\delta \phi}{\delta x} \quad \ldots \quad (28)$$

By differentiating equation (27) with respect to t, and multiplying by $\delta v/\delta v$, an expression is obtained in which substitution yields

$$Y = -\frac{t}{k} \frac{\delta \phi}{\delta t} \quad \ldots \quad (29)$$

In order to consider changes in the hydrostatic pressure of the suspension, it should be recalled that if p denotes the hydrostatic pressure at depth x in the suspension, then at depth $x + dx$, the increment to p, δp, is equal to $\phi \delta x$, because it is the density times the added depth which equals the added weight, and hence measures the added pressure. Consequently, $\phi = \delta p/\delta x$, and $\delta \phi/\delta x = \delta^2 p/\delta x^2$. We may substitute the value of Y from equation (28) for $\delta \phi/\delta x$ in the last equation, and obtain

$$Y = \frac{x}{k} \frac{\delta^2 p}{\delta x^2} \quad \ldots \quad (30)$$

Equation (30) measures the hydrostatic pressure as a function of the depth at a constant time. In order to consider the decrease as a function of the time

PRINCIPLES OF SIZE ANALYSIS

at a constant depth, a more lengthy mathematical process is necessary, which yields as its final result the equation

$$Y = \frac{t^2}{kx} \frac{\delta^2 p}{\delta t^2} \qquad \cdots \cdots (31)$$

We may next consider the change in weight of an immersed body. A cylinder of weight W in air and cross-sectional area f is counterpoised in the suspension by a weight A, which just balances when the upper end of the cylinder is at the surface of the liquid. As the density of the suspension decreases, A will increase. The relation that holds for any time then is $A = W - fp$, or $\delta A/\delta t = -f\delta p/\delta t$. We may take the second derivative of the same expression, and by substitution from equation (31) gain the result

$$Y = -\frac{t^2}{kfx} \frac{\delta^2 A}{\delta t^2} \qquad \cdots \cdots (32)$$

Finally, the rate of accumulation of sediment on a plate of area a suspended at a depth x below the surface of the suspension may be considered. To obtain the value of Y in this case it is necessary to compute the pressure exerted on the pan by the column of water of height x, and the downward force exerted by the particles remaining in suspension. The relation obtained is differentiated with respect to t, and substitution in the resulting derivative yields, after some simplification, the equation

$$Y = -\frac{t^2}{A_1} \frac{\delta^2 A}{\delta t^2} \qquad \cdots \cdots (33)$$

where A_1 represents the total weight of all the particles in the suspension.

MODERN METHODS OF MECHANICAL ANALYSIS

Any of the four general equations, (28), (29), (30), or (31), may be used to determine the ordinate of the frequency curve. The possible methods fall into two groups. Either the density or the hydrostatic pressure is measured as a function of the time at a constant depth, or they are measured as a function of the depth at a constant time. It is clear that the apparatus needed for the latter group would be more unwieldy than for the former, because it would be necessary to take measurements at a number of points in the suspension simultaneously. Specifically, equation (28) requires the simultaneous observation of the density at various depths in the suspension at a given time t. From these data $\delta\phi/\delta x$ for the different values of x may be computed, and their substitution in the equation furnishes a series of values for Y. Each value of Y corresponds to a certain value of v or of r, and the frequency curve may be plotted with the Y's as ordinates and $\log_e v$ or $\log_e r$ as abscissae In the case of equation (30) it is necessary to measure the hydrostatic pressure simultaneously at various depths. From the data obtained, the second derivative of the pressure with respect to the depth may be computed, and by substitution the corresponding values of Y may be found.

Odén[1] developed an apparatus for the application of equation (30). He used a long sedimentation tube, to which ten capillary tubes were attached. The capillary tubes were filled with pentane, and at any instant the values measured by them presented a curve across the tubes in the rack. The method was discarded by Odén because of a number of sources of error which could not be corrected by any simple device.

By far the greatest number of methods of analysis utilize equations (29) and (31). Equation (29) is the foundation of the pipette method,[2] but in practice the equation itself is seldom used, because sufficient data for cumulative curves or histograms can be obtained directly from the successive weights of the pipette residues. The hydrometer method is also based on this equation. Equation (31) forms the basis of such methods as Wiegner's tube. Equation (32) affords a method of determining the frequency curve from the apparent change in weight of a plummet counterpoised in the suspension. Equation (33) was developed especially for the Odén sedimentation balance. For the application of centrifugal force to modern methods, see page 123.

Summary of modern methods: the inherent error. The discussion of methods of analysis based on Odén's theory would not be complete without a discussion of the sources of error involved in their application. The theory itself raised mechanical analysis to new high levels of development, but in the practical application of the theory various complicating factors arise. Several writers [3] have discussed the errors involved. Coutts and Crowther showed that an "inherent error" is involved in Odén's method due to currents set up in the suspension, owing to differences below and at the edges of the suspended balance pan. These currents interfere with normal settling of the particles. Shaw and Winterer pointed out that another source of error is due to electrical charges on the wall of the vessel which tend to draw the smaller particles to the outer portions of the suspension. Gessner pointed out that in Wiegner's tube clear water enters the suspension from the manometer and interferes with sedimentation. The pipette method is also subject

---

[1] S. Odén, *loc. cit.*, 1925.
[2] Details of this and other methods of analysis are given in Chapter 6.
[3] J. R. H. Coutts and M. Crowther, A source of error in the mechanical analysis of sediments by continuous weighing: *Trans. Faraday Soc.*, vol. 21, pp. 374-380, 1925-1926. C. F. Shaw and E. V. Winterer, A fundamental error in mechanical analysis of soils by the sedimentation method: *Proc. 1st Int. Congr. Soil Sci.*, vol. 1, pp. 385-391, 1928. H. Gessner, *op. cit.*, 1931, p. 96.

PRINCIPLES OF SIZE ANALYSIS

to error, inasmuch as the withdrawal of the sample affects a spherical rather than a thin horizontal zone.

Despite these practical defects, modern methods may still be referred to as precision methods in the sense that they are based on sound principles, and the errors may in many cases be evaluated sufficiently closely so that their limitations may be known. Köhn,[1] for example, showed that the error in the pipette method is only a fraction of 1 per cent in a 10-c.c. sample taken from a depth of 10 cm. Likewise, various other workers[2] have argued that the "inherent errors" may in many cases be small enough to neglect. Correns and Schott questioned their importance, and Vendl and Szádeczky-Kardoss found that errors of the type mentioned by Coutts and Crowther are generally quite small.

OLDER METHODS OF ANALYSIS

SEDIMENTATION AND ELUTRIATION METHODS

The preceding section discussed the theories underlying modern methods of analysis, but the use of older routine techniques is still current, and it may be well briefly to review the principles on which they are based.

Theory of decantation methods. Decantation methods of mechanical analysis are among the oldest techniques. With them a separation of the several grades is effected by allowing the suspension to stand until particles larger than a given radius have settled to the bottom of the vessel. At that instant the supernatant liquid is decanted or siphoned off, and clear water poured in. The sediment is resuspended, and an equal interval of time is allowed to elapse, so that all the larger particles may completely settle again, and the supernatant liquid is drawn off as before. By a repetition of this process a practically complete separation of the material into grades may be made. The decanted liquid may itself be put through the same process to separate still smaller grades.

No complete mathematical theory has been developed for decantation methods, but it is possible to set up a relation which will indicate the course of the process. We may consider a bidisperse system, such that the settling

---

[1] M. Köhn, Beiträge zur Theorie und Praxis der mechanischen Bodenanalyse: *Landwirts. Jahrb.*, vol. 67, pp. 485-546, 1928.

[2] C. W. Correns and W. Scott, Vergleichende Untersuchungen über Schlämmund Aufbereitungsverfahren von Tonen: *Kolloid Zeits.*, vol. 61, pp. 68-80, 1932. M. Vendl and E. V. Szàdeczky-Kardoss, Über den sogennanten grundsätzlichen Fehler der mechanischen Analyse nach dem Odén'schen Prinzip: *Kolloid Zeits.*, vol. 67, pp. 229-233, 1934.

velocity of the larger particles is twice that of the smaller. In successive decantations the amount of the smaller grade remaining will obviously be ½, ¼, ⅛.... This series may be written as ½, ½², ½³, ... ½ⁿ, where the exponent indicates the number of decantations involved. In the general case, if the larger particles have a velocity p times as great as the smaller, the series is $1/p$, $1/p^2$, $1/p^3$, ... $1/p^n$. This relationship enables us to express the amount of fine material remaining after any number of decantations as a proportion of the original amount of fine material:

$$w/w_0 = p^{-n} \quad \ldots \quad (34)$$

where w is the amount of fine material left after n decantations, and w_0 is the original amount of fine material. Since by Stokes' law the velocity varies as the square of the radius, equation (34) may be expressed in terms of radii as

$$w/w_0 = p^{-2n} \quad \ldots \quad (35)$$

where p now represents the relation between the radii, instead of between the velocities. Figure 28 illustrates two curves of the type developed by equation (35). The amount of fine material remaining is plotted as ordinate and the number of decantations as abcissa. The upper curve represents the case in which the radius of the larger particles is 10/9 that of the smaller ($p = 10/9$), and in the lower curve twice that of the smaller ($p = 2$). The steepness of the curve is thus determined by the value of p.

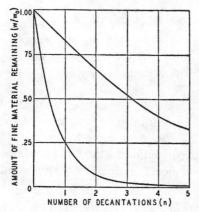

FIG. 28.—Progress of decantation methods in separation of bidisperse systems. Curve A represents the system $R/r = 10/9$; curve B represents the system $R/r = 2$, where R and r are the large and small radii, respectively.

By means of equation (35) it is possible to compute the number of decantations necessary to effect a separation of the two grades in a bidisperse system to any desired degree of accuracy. From the nature of the equation it is clear that a complete separation can never be made, but we may consider the case in which not more than 1 per cent of finer material remains. If $p = 2$, and $w/w_0 = 0.01$, these values may be substituted in the equation. After taking logs, we obtain $n = -\dfrac{\log 0.01}{2 \log 2}$ from which the value of n is found to lie between 3 and 4. In actual practice the suspensions are polydisperse, which introduces complexities because there is a continuous decrease in size of radius from one grade to the next and the nature of the frequency distribution determines to

PRINCIPLES OF SIZE ANALYSIS

some extent the successive weights of material left. To determine the exact error involved in any given number of decantations it would be necessary to know the frequency distribution of the sediment. The approximate number of decantations necessary to effect a practically complete separation of the fine material may be determined, however. If it is desired to have not more than 1 per cent of material finer than 9/10 of the critical radius at which the separation is to be made, the value of n from equation (35) is found to be 22.

It would seem from these points that decantation methods are subject to very definite errors in practice, and in addition they are affected by errors of the same nature as those discussed under modern methods. That is, when the liquid is decanted or siphoned off, a noticeable amount of the sedimented material is often carried over. This of course acts as a compensating error, but it renders more difficult the exact evaluation of the total error.

Rising current elutriation. Rising current elutriation was more extensively used in the past than it is at present, although the elutriator devised by Schöne[1] is still used to a considerable extent. In the general method a current of water is sent up through a vertical tube, and all the particles whose settling velocities in quiet water are less than the upward velocity of the water are carried away. By varying the strength of the current, or by introducing several tubes of varying dimensions, a separation into several grades may be effected.

The theory underlying elutriation by rising currents involves first the relation between the settling velocity of the particle and the upward velocity of the water. It is clear that the downward velocity v of the particle at any instant is the settling velocity v_0 of the same particle in quiet water, decreased by the upward velocity v_w of the current:

$$v = v_0 - v_w \quad . \quad . \quad . \quad . \quad (36)$$

If the upward velocity of the water is greater than the settling velocity of the particle, the latter is carried upward with a velocity equal to that of the current, diminished by the settling velocity of the particle. In this case the v of equation (36) has a negative value.

The velocity of the particles carried upward by the current depends on the difference between v_w and v_0. In separating the smaller particles of a bidisperse system, the velocity of the water is made equal to the settling velocity of the larger particles, so that they remain suspended, while the smaller ones are borne away. In polydisperse systems the separation between two grades is made in an analogous manner. The current is so adjusted that it is the same as the settling velocity of the particles having the critical radius, and all the smaller material is carried off by the rising current.

[1] E. Schöne, *loc. cit.*, 1868.

The time required to separate the smaller particles from a sediment depends on the length of the vessel and on the velocity with which they rise through the tube. Gessner,[1] computing the time necessary to effect complete separations between grades of various sizes, found that it increased very rapidly for the finer sizes. The time element, therefore, is one of the factors which limits the usefulness of rising current elutriators.

Another of the difficulties met in rising current elutriators is the lack of uniformity of the current throughout the cross section of the tube. This is due to wall friction, and in many instances particles are seen to be carried upward in the central part of the tube, only to settle down again along the sides. It is necessary, therefore, to assume an average velocity of the current and to base separations on that average value.

Rising current elutriation has been applied to a wide range of sizes, but the large volumes of water required and the time necessary to separate the smaller grades impose practical limits. Another factor which should be kept in mind is that the dispersing electrolyte becomes greatly diluted as fresh water enters the tube; when the concentration of the dispersing agent approaches zero, a coagulation of the smaller particles may follow. Thus the method may be strictly applicable only to sizes above the limit of coagulation; in general it is best suited to particles above 0.01 mm. in diameter. When used within the range of sizes to which it is suited, rising current elutriation appears to afford a convenient and practical method of sorting sediments into grades.

Air-current elutriation. The theory of rising current elutriation discussed in the preceding section applies equally well to fluids other than water. Rising currents of air have been used extensively for the separation of fine powders, but the method has not been used by sedimentary petrologists. In general, the same theoretical considerations apply in the separation of the grades by air, except that the much lower density of air permits a more thorough separation of very fine grades, lessens the effects due to wall friction, and renders less important variations in temperature. Instead of coagulation phenomena, which commonly accompany water elutriation, air elutriation often develops electrical charges on the particles, which must be eliminated. Roller [2] has recently shown that air elutriators may be designed to eliminate most of the difficulties commonly encountered. Details of his apparatus and technique are given in Chapter 6.

[1] H. Gessner, *op. cit.*, 1931, p. 118.
[2] P. S. Roller, Separation and size distribution of microscopic particles: *U. S. Dept. Commerce, Bur. of Mines, Tech. Paper 490*, 1931.

PRINCIPLES OF SIZE ANALYSIS

Application of centrifugal force to mechanical analysis. A number of workers have applied centrifugal force to sedimenting systems in order to hasten the settling of small particles. Among the earliest users of centrifugal force for this purpose was the United States Bureau of Soils.[1]

The general theory of centrifugal force as applied to the sedimentation of small particles was developed by Svedberg and Nichols,[2] who used the method to determine the size of colloidal particles. Other workers applied the theory to modern methods of analysis; among these are Trask,[3] who developed a centrifugal modification of Odén's method, and Steele and Bradfield,[4] who developed a centrifugal modification of the pipette method.

Full details of the theory of centrifuging may be had from Svedberg and Nichols's paper; the treatment here will briefly indicate the method of approach. Under centrifugal acceleration the force applied to cause movement of a particle in a fluid is $\frac{4}{3}\pi r^3 (d_1 - d_2) \omega^2 (x + a)$. The symbols here have the same meaning as the right-hand term of equation (2), except that for g has been substituted the value $\omega^2(x+a)$, where ω is the angular velocity, a the distance of the particle from the axis of rotation before fall, and x the distance of fall. By setting the above equation equal to Stokes' resistance, one obtains $6\pi r \eta v = \frac{4}{3}\pi r^3 (d_1 - d_2) \omega^2 (x+a)$, and solving the equation for v, there results

$$v = \frac{2(d_1 - d_2)\omega^2(x+a)r^2}{9 \quad \eta}. \quad \ldots \quad (37)$$

Inasmuch as the velocity of fall through the distance x may be expressed as dx/dt, equation (37) can be written as a differential equation in which dx/dt is substituted for the v on the left. By rearranging the terms for integration one obtains

$$r^2 \int_0^t dt = \frac{9\eta}{2\omega^2(d_1 - d_2)} \int_0^x \frac{dx}{x+a}$$

This expression yields, upon integration,

$$r^2 t = \frac{9\eta}{2\omega^2(d_1 - d_2)} \cdot \log_e \frac{x+a}{a}. \quad \ldots \quad (38)$$

Equation (38) may be solved for r or for t. Trask applied the theory to Odén's method, using the equation to solve for r; Steele and Brad-

---

[1] L. J. Briggs, F. O. Martin and J. R. Pearce, loc. cit., 1904.
[2] T. Svedberg and J. B. Nichols, Determination of size and distribution of size of particle by centrifugal methods: *Jour. Am. Chem. Soc.*, vol. 45, pp. 2910-2917, 1923.
[3] P. D. Trask, Mechanical analysis of sediments by centrifuge: *Econ. Geology*, vol. 25, pp. 581-599, 1930.
[4] J. G. Steele and R. Bradfield, The significance of size distribution in the clay fraction: *Rep. Am. Soil Survey Assn.*, Bull. 15, pp. 88-93, 1934.

field solved for t, to determine the time of centrifuging for the pipette method.[1]

THEORY OF SIEVING

The use of sieves to separate the coarser portions of soils or sediments dates back to the early days of mechanical analysis, and the simplicity and convenience of the method have been the greatest factors in its continued use. In practice, sieving is exceedingly simple. The material to be sieved is placed in a sieve and shaken until the particles smaller than the mesh openings fall through. By repeating the process with successively smaller meshes, the material may be separated into any given number of grades.

The theory of sieving is not so simple as the practice, and if due consideration is given to all the factors involved, a number of complexities are found to enter which limit the accuracy of the usual operations of sieving. A number of opponents to the use of sieves as instruments of mechanical analysis have written critiques of the method; one of the most antagonistic was Mitscherlich,[2] who pointed out a number of years ago that sieves sort grains not only according to size, but also according to shape. This may be illustrated by considering spherical and lath-shaped grains. The largest sphere that can pass through a given sieve has a diameter equal to the mesh, whereas a lath of any length, theoretically, can pass through the sieve, providing only that its two smaller dimensions are less than the maximum dimensions of the mesh, including its diagonals. A long lath may have a much larger volume than a sphere of the same cross section, and hence if size is defined in terms of the nominal diameter (i.e., based on volume), the sieving process does not sort according to size.

If size is defined in terms of some average diameter, sieves again fail to make a sharp distinction, because for non-spherical shapes the maximum length has no direct bearing on passage through the sieve. Instead, the intermediate and shortest diameters are the deciding factors, and hence it may be seen that sieves sort grains on the basis of the least cross-sectional area, which may or may not have any fixed relation to the volume of the particle.

Despite the general validity of these criticisms, sieving is a well established procedure in mechanical analysis, and it is possible to use sieve data for a number of purposes in sedimentary studies. The fact

---

[1] Steele and Bradfield used a different notation, but the results are the same.
[2] E. A. Mitscherlich, *Bodenkunde für Land- und Forstwirte* (Berlin, 1905), pp. 37 ff.

PRINCIPLES OF SIZE ANALYSIS

that many sands, separated into size frequency distributions by sieves, plot as straight lines on logarithmic probability paper (page 189), indicates that in general all the significant data are not obscured by sieving. Of course, for precise shape studies and their influence on the properties of sediments sieving may be merely a preliminary procedure of separating the sample into convenient size units.

In developing a theory of sieving, the simplest case may be considered: that of a mixture of two sizes of spheres, one of which is slightly larger and the other slightly smaller than the sieve openings. If a mixture of such spheres be placed in a sieve to an appreciable depth, so that a number of layers of spheres are involved, it may be seen that as the sieve is shaken, the separation of the smaller spheres depends upon the number of such spheres which come into contact with the screen at a given moment. During early stages of the sieving the number of smaller spheres falling through is relatively large because of the high proportion of small spheres in the mixture, but as the remaining small spheres decrease in proportion in the mixture, the number falling through in any given instant is reduced.

As a first approximation one may assume that the number of small spheres, dy, falling through the sieve in any small interval of time, dt, is proportional to the number of small spheres present at that moment. This assumption leads to the differential equation

$$dy/dt = -ay$$

where y is the number of small spheres in the mixture at time t and a is a constant of proportionality. The negative sign indicates that the function decreases with time.

Integration of the differential equation yields $\log y = -at + \log C$, and by evaluating $\log C$ at time $t = 0$, it is found that $\log C = \log y$, so that $C = y_0$, the original number of small spheres in the mixture. Substituting y_0 for C and rearranging yields the equation

$$y/y_0 = e^{-at} \quad \ldots \ldots \ldots \ldots (39)$$

For spheres this function would assume various values of the constant a; in general, the finer the sieve, the smaller the value.

In practice it may be doubtful whether so simple a relation holds, because of the complexity introduced by the shape factor. Not only must the grains come into contact with the screen, but in order to fall through they must assume the proper position so that the short and intermediate diameters are approximately normal to the plane of the screen. In practice it is also common to have a finite difference between successive sieve meshes, so that on any sieve there is a relatively wide range of particle sizes. The smallest particles will readily fall through the meshes, but the larger ones, closer to the critical radius, will suffer delays due to the need for proper orientation. Thus it is likely that the function may show a more marked decrease at first than the exponential function, and a slower decrease later.

Wentworth[1] investigated the question of sieving quantitatively and found that an empirical equation of the type $y = at^{-m} + b$ fitted the data fairly well. Figure 29, adapted from Wentworth, indicates the progress of sieving on a ½-mm. sieve. Time is shown along the x-axis, and the

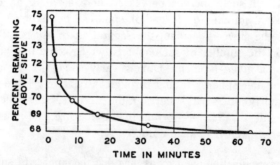

FIG. 29.—Progress of sieving on sieve with meshes of ½ mm. (Adapted from Wentworth, 1927.)

percentages remaining above the sieve are plotted along the y-axis. Wentworth's study also showed that the separation of the grains is probably never quite complete, especially among the smaller sizes. In general, however, he concluded that a five- or ten-minute period of shaking in an automatic shaker is usually sufficient.

THEORY OF MICROSCOPIC METHODS OF ANALYSIS

The microscope has been used by numerous workers for determining particle size and particle-size distribution. A number of techniques have been developed, ranging from direct measurement of diameters by means of micrometer oculars to microprojections of the particles onto screens or photographic plates, followed by measurement of the enlarged images. The wide variety of techniques is in part a response to the large number of materials which have been studied microscopically. Pigments, dust, ceramic products, sensitivity of photographic emulsions, pulverized coal,

---

[1] C. K. Wentworth, The accuracy of mechanical analysis: *Am. Jour. Sci.*, vol. 13, pp. 399-408, 1927.

PRINCIPLES OF SIZE ANALYSIS 127

and many other materials [1] have been investigated. Sedimentary petrologists have not utilized microscopic methods as extensively as sieving and sedimentation methods, but there appears to be an increasing tendency to adapt the microscope to measurements of size attributes.

The microscope is a convenient instrument for measuring grains from diameters of about 0.5 mm. down to the limit of resolving power of the microscope. The lower limit varies with different instruments and with the wave length of the light being used. By a combination of oil immersion and blue light, particles as small as 0.0002 mm. diameter (0.2 micron) have been resolved.[2] For most general purposes, however, microscopic methods may be used down to diameters of about 0.001 mm. (1 micron).

Definition of "size" of microscopic particles. Inasmuch as most natural particles are irregular in shape, the influence of shape factors on definitions of size should be considered. In addition, it is necessary to consider the orientation of the particles on the microscopic slide. If the grains are sprinkled on the slide in a dry state, the tendency will be for them to assume positions of rest such that the shortest diameter will be approximately vertical.[3] The section exposed to view will accordingly represent approximately the long and intermediate dimensions of the grain. If the grains are mounted in balsam or another medium, the orientation of the three axes may be random, with the result that it is difficult to determine whether the grain is viewed along an edge or broadside. If the grains are nearly equidimensional the influence of the random orientation is usually negligible, but with flat or lath-shaped grains it may influence the values obtained.

Regardless of the manner of mounting the grains, several definitions of size are possible. One may define size as the arithmetic mean of the diameters exposed to view, or one may add in the estimated thickness of the grain as a third diameter. For rapid work one may express size in terms of the intermediate diameter only, or if the grains are oriented at random, the maximum horizontal intercept through the grain may be used.[4] For more detailed work Wadell [5] has recommended that the area of the grain image (in a camera lucida drawing, a photograph, or a

[1] A complete bibliography of non-sedimentary studies of particle size is given by E. M. Chamot and C. W. Mason, *Handbook of Chemical Microscopy* (New York, 1931), vol. 1, Chap. 12.
[2] F.-V. von Hahn, *op. cit.* (1928), p. 38.
[3] H. Wadell, Volume, shape, and roundness of quartz particles: *Jour. Geology*, vol. 43, pp. 250-279, 1935.
[4] W. C. Krumbein, Thin section mechanical analysis of indurated sediments: *Jour. Geology*, vol. 43, pp. 482-496, 1935.
[5] H. Wadell, *loc. cit.*, 1935.

screen projection of the images) be measured with a planimeter, and expressed as the diameter of a circle having the same area. This diameter is called by Wadell the "nominal sectional diameter." For grains of marked elongation or flattening, the harmonic mean of the diameters may be used.[1] If the three diameters of the grain are a, b, and c, the harmonic mean is defined as $d_h = (3abc)/(ab + bc + ac)$. Perrot and Kinney[2] considered this diameter the most logical to be used for pigments and ceramics because of its relation to the specific surface of the materials.[3] Roller[4] pointed out that, in addition, diameters defined in this manner are closely related to diameters calculated from Stokes' law. For practical purposes, however, Roller also showed that in the average case the harmonic mean lies within about 6 per cent of the value of the arithmetic mean, and that the arithmetic mean may therefore be used directly for computing mean surface diameters. The arithmetic mean of the three diameters is simply $d_a = (a + b + c)/3$.

In addition to the arithmetic and harmonic mean diameters, the geometric mean diameter of the particle may be used. This is computed by finding the cube root of the product of the three diameters, $d_g = \sqrt[3]{abc}$.

Determination of size distribution from microscopic counts. Whatever definition of particle size is chosen, the succeeding step is to measure a number of grains and arrange them into grades or classes to determine the distribution of sizes within the sample. Practice varies considerably; the grades may be chosen as equal arithmetic intervals or on a geometric basis, and the number of grains counted ranges from a few hundred to more than a thousand. For most routine analyses, involving grains more or less equidimensional and having a restricted range of sizes (quartz sand, for example), a few hundred grains probably suffice.

It should be recognized that the frequency data obtained by microscopic measurement are expressed in terms of numbers of grains rather than by weights, as in sieving and sedimentation. If the material is homogeneous, the weight frequency may be computed from the number frequency, but as a general rule microscopic size determinations by

---

[1] H. Green, A photomicrographic method for the determination of particle size of paint and rubber pigments: *Jour. Franklin Inst.*, vol. 192, pp. 637-666, 1921.
[2] G. St. J. Perrot and S. P. Kinney, The meaning and microscopic measurement of average particle size: *Jour. Am. Ceram. Soc.*, vol. 6, pp. 417-439, 1923.
[3] For a sphere the specific surface S is defined as follows in terms of the diameter d and the density ρ: $1/d = (1/6) \rho S$.
[4] P. S. Roller, Separation and size distribution of microscopic particles: *U. S. Dept. Commerce, Bur. of Mines, Tech. Paper 490*, 1931.

PRINCIPLES OF SIZE ANALYSIS

number should not be directly compared with sieve determinations by weight.[1]

The usual statistical methods may be applied to microscopic data for expressing the average size of sediments, as well as other statistical values. These methods are discussed in Chapters 8 and 9; practical examples of microscopic size analyses are given in Chapter 6.

Mechanical analysis of thin sections of indurated sediments. For the study of grain size distributions of indurated sediments, in which the particles are too firmly cemented to be disaggregated, microscopic methods afford practically the only method of attack. There are several precautions which must be taken, however, when size is estimated from thin sections, because in detail the problem is quite complex.

The problem of thin-section mechanical analysis has been attacked from several angles by sedimentary petrologists, but the mathematical theory of random sections through groups of spheres has also been treated by astronomers and biologists. Krumbein,[2] working independently, approached the problem from the moments of the grain distribution, but after publication learned through correspondence that Hagerman[3] had previously attacked the problem in a similar manner, but from a different mathematical approach. The essential features of Krumbein's mathematical analysis had, however, been used by astronomers in the study of globular star clusters.[4] To increase the complexity, it was further learned that similar but more rigorous mathematical treatment had been applied by Wicksell to the study of spherical corpuscles embedded in tissues.[5] To cap the situation, Fisher[6] had also approached the problem of indurated sediments, but from a point of view different from that of Hagerman or Krumbein. As a result there are several methods of approach to the study of thin-section mechanical analysis, and all are complex in terms of ordinary methods (sieving or sedimentation) because of the restrictions placed upon analysis by the sectioning of the grains. The discussion here will follow essentially the method used by Krum-

---

[1] The relation between number frequency curves and weight frequency curves has been studied by Hatch for distributions which are symmetrical on a logarithmic size scale. See T. Hatch, Determination of "average particle size" from the screen analysis of non-uniform particulate substances: *Jour. Franklin Inst.*, vol. 215, pp. 27-38, 1933. The question of number vs. weight frequencies is discussed further in Chapter 8.
[2] W. C. Krumbein, *loc. cit.*, 1935.
[3] T. H. Hagerman, Ein metod för bedömning av kornstorleken och sorteringsgraden inom finkorniga mekanist sedimentära bergarter: *Geol. Förening, Förk.*, vol. 46, pp. 325-353, 1924.
[4] S. D. Wicksell, A study of the properties of globular distributions: *Arkiv f. Matematik, Astron., och Fysik*, vol. 18, 1924.
[5] S. D. Wicksell, The corpuscle problem: *Biometrika*, vol. 17, pp. 84-99, 1925. S. D. Wicksell, The corpuscle problem (ellipsoidal case): *ibid.*, vol. 18, pp. 152-172, 1926.
[6] G. Fisher, Die Petrographie der Grauwacken: *Jahrb. d. preuss. geol. Landesanst*, vol. 54, pp. 320-343, 1933.

bein, inasmuch as it is most familiar to the authors, and because it is based on the mathematical foundation developed by Wicksell.

It is common knowledge that if random grain diameters are measured from thin sections, the distribution of observed diameters will not be an accurate indication of the grain diameters themselves. This is because in only a very small number of cases will the random sections be exactly through the center of the grain. Generally the average size of the grain sections will be less than the average size of the grains. However, with spherical grains there is a definite relation between the random sectioning and the true size distribution, so that mathematical analysis may determine what corrections must be applied to the random sections.

A complete solution of the problem is not simple, but fortunately the most important statistical values of the sediment, including the average grain size and the standard deviation, may be determined satisfactorily even though the grains are not true spheres.

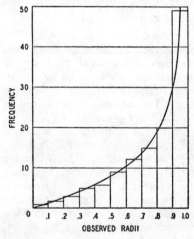

Fig. 30.—Frequency distribution of observed sectional radii of uniform lead shot.

If a number of lead shot, all of the same radius, are embedded at random in sealing wax and ground down to a polished section, the section will disclose a number of lead circles of varying radius. The radii of the observed circular sections may be measured, and the data arranged in classes equal to tenths of the true radius. The observed radii range from zero to unity, and when the observations are plotted, as in Figure 30, the result is striking. A smoothed curve passed through the histogram yields a frequency curve in which the line rises to the right but does not descend again. The average radius, computed from the observed data, is 0.763 of the actual radius of the shot.

If one confined himself to the observed data he would assume that he had spheres of various sizes, and he would find that the observed average size is some 24 per cent smaller than the actual. This serves to illustrate the general effect of sectioning through a rock, and emphasizes that one cannot argue from observation alone that he has either the true value of the average size, or even an approach to the true frequency curve. In practice the picture is more complex than in the example given. The grains of sediments are not all of one size, nor are they true spheres.

Krumbein considered the mathematical relations between the moments [1] of the grain distribution and the moments of the observed sectional distribution. For spheres of one size, as in the example given, the treatment is fairly simple, and the same approach was made by both Hagerman and Krumbein:

Consider a random section through a sphere of radius r; the problem is to determine the frequency function of observed radii which occurs when a number of spheres of this same radius are sectioned at random. Fortunately a two-dimensional analysis suffices. In Figure 31 is a sphere to be cut by a random section anywhere along the y-axis from $-r$ to $+r$, with an equal likelihood for any point of intersection. One may, in fact, restrict himself to the range from 0 to r, since the circle is symmetrical. The equal likelihood of cutting the section between y and $y + dy$ may then be stated as $P_1 = 1/r$. Call the observed radius of the random section x, a variable measured along the x-axis. To find $P(x)$, the distribution of observed radii, let $y = H(x)$ and $dy = H'(x)dx$. From the circle, $x^2 + y^2 = r^2$, so that

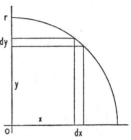

Fig. 31.—Part of a sphere to be cut by a random section along the y axis from $-r$ to $+r$.

$$y = H(x) = \sqrt{r^2 - x^2} \quad \ldots \ldots \quad (40)$$

$$dy = H'(x)dx = -\frac{xdx}{\sqrt{r^2 - x^2}} \quad \ldots \quad (41)$$

For transforming the probabilities the usual relation $P(x)dx = P_1[H(x)]H'(x)dx$ may be used; substitution yields

$$P(x)dx = \frac{1}{r} \frac{x}{\sqrt{r^2 - x^2}} dx \quad \ldots \quad (42)$$

It will be noted that the sign of dx has been changed, to keep the probabilities positive.

Fig. 32 shows the form of the $P(x)$ curve. It is the same as the curve obtained experimentally in Figure 30. The curve is asymptotic to the line $x = r$, but the area is finite.

In the mathematical treatment of the general problem Hagerman and Krumbein followed different lines of reasoning. The latter considered a frequency distribution of radii $F(r)$, where $F(r)dr$ is the probability that r lies between r and $r + dr$. From the

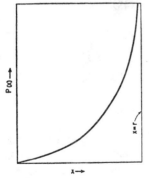

Fig. 32.—Form of the $P(x)$ curve.

[1] The moments of a frequency distribution are parameters which describe the properties of the distribution. The theory of moments is discussed in Chapter 8.

random sections is observed a distribution of x's or apparent radii, which may be called $Q(x)$. Here $Q(x)dx$ represents the probability that x lies between x and $x + dx$. Now the probability that r is between r and $r + dr$, and the probability that x is between x and $x + dx$, is $P(x)F(r)drdx$. Since r may have any value greater than x, $Q(x)$ extends over all the possible values of r, and hence is the integral of the expression from x to infinity:

$$Q(x) = \int_x^\infty P(x)F(r)dr = x\int_x^\infty \frac{F(r)}{r\sqrt{r^2-x^2}}dr \quad . \quad . \quad . \quad (43)$$

where dx has been dropped from both sides of the equation.

Equation (43) is an integral equation, and in practice $Q(x)$ will be a set of empirical data, $F(r)$ will be entirely unknown, and $P(x)$ will apply strictly only when perfect spheres are involved. For these reasons the solution is restricted to a consideration of the moments of the distributions, because from the observed moments of $Q(x)$ may be computed and converted into the corresponding moments of $F(r)$.

To solve equation (43) several steps are necessary.[1] The solution is obtained in such manner that the integrals are in the form of the nth moments of the distributions Q and F about the origins of x and r respectively:

$$\int_0^\infty x^n Q(x)dx = C\int_0^\infty r^n F(r)dr \quad . \quad . \quad . \quad . \quad . \quad (44)$$

where C is a constant which depends upon n. By letting $n = 1, 2, 3, \ldots$, the moments of $F(r)$ may be obtained in terms of $Q(x)$. The final solutions for the first four moments are:

$$n_{x1} = \frac{\pi}{4} n_{r1} \quad . \quad . \quad . \quad . \quad . \quad (45)$$

$$n_{x2} = \frac{2}{3} n_{r2} \quad . \quad . \quad . \quad . \quad . \quad (46)$$

$$n_{x3} = \frac{3\pi}{16} n_{r3} \quad . \quad . \quad . \quad . \quad . \quad (47)$$

$$n_{x4} = \frac{8}{15} n_{r4} \quad . \quad . \quad . \quad . \quad . \quad (48)$$

In practice the analysis is performed by determining the moments of the observed distribution $Q(x)$ and correcting them by means of the preceding equations for at least the first two moments, which afford the average size and "degree of sorting" of the sediment. Examples of sediments studied from this approach are given in Chapter 6.

The above mathematical approach is the same as that made by Wicksell in the study of globular star clusters. In his later paper on corpuscles

---

[1] The steps are given in Krumbein's paper, *loc. cit.*, 1935.

PRINCIPLES OF SIZE ANALYSIS

Wicksell attacked the more general case of sectioning spheres and included the probability that a sphere of radius r to $r + dr$ would be cut by the sectioning plane. Thus Krumbein's solution is a special case of the general problem, based on the assumption that the thin section is a typical sample of the grain distribution. This assumption has been accepted in most thin-section work in geology, but it may require closer scrutiny. If the probability of sectioning is included in the problem, it is found that the probability of a sphere being sliced is the ratio of its radius to the mean radius of the spheres, r/r_m.

Thus in order to generalize Krumbein's solution it is necessary to introduce this probability of slicing, $P_0 = r/r_m$ into equation (42), with the result that the r in the denominator is canceled and replaced by r_m. With this change, the mathematical operations proceed essentially as before, resulting finally in the expression

$$\int_0^\infty x^n Q(x) dx = \frac{c}{r_m} \int_0^\infty r^{n+1} F(r) dr \quad \cdots \quad (49)$$

for equation (44). This general solution differs from the particular case in that the nth moment of the x-distribution is equated to the $(n+1)$th moment of the r's. Hence the relations between the moments differ in the final results. The general equations corresponding to (45), (46), (47), and (48) become, in the general case:

$$\frac{\pi}{2} x_{hm} = n_{r1} \quad \cdots \quad (50)$$

$$n_{x1} = \frac{\pi}{4} \frac{n_{r2}}{n_{r1}} \quad \cdots \quad (51)$$

$$n_{x2} = \frac{2}{3} \frac{n_{r3}}{n_{r1}} \quad \cdots \quad (52)$$

$$n_{x3} = \frac{3\pi}{16} \frac{n_{r4}}{n_{r1}} \quad \cdots \quad (53)$$

where x_{hm} of equation (50) is the harmonic mean of the x-distribution.

The equations of the general solution were applied by Krumbein to the data used for his particular solution, and it was found that the agreement between observed and expected values was better by a considerable percentage with the original solution than with the more general theory. The general theory implies that the spheres are suspended in a medium, and thus presumably are not in contact. In sediments, on the other hand, the grains are in actual contact, and it may be that when the spheres are packed closely together the chances of slicing are nearly equal for all sizes. From the evidence thus far in hand it appears that the original solution (equations 45 to 48) has a more direct application.

An example of an analysis by thin section is given in Chapter 6, using the original simplified theory for the calculations.

SUMMARY OF PRINCIPLES OF MECHANICAL ANALYSIS

The increasing emphasis on the quantitative aspects of sedimentary studies requires that workers in the field be informed regarding the fundamental principles upon which so much of their technique depends. The reaching of sound conclusions about analytical data and the development of theories to account for sedimentary phenomena require that the influence of technique on the resultant data be known. It is for these reasons that the principles underlying mechanical analysis have received detailed discussion in the present chapter. In practice many of the finer points are ignored by common consent, but when a particularly precise and exhaustive study is to be made, a knowledge of the underlying principles may be of considerable aid.

The complete shift of emphasis in analytical techniques which was introduced by Odén's theory of sedimenting systems illustrates the advantages that accrue from an investigation of underlying principles, as well as the greater precision possible because methods can be developed which are in accord with the demands of the theory. The result of Odén's work has been that older routine methods of analysis, such as decantation and elutriation, are gradually being displaced by more modern precision methods. One cannot, however, ignore the older methods, which are capable of yielding good results in the range of sizes directly suited to them, and it seems likely that by a more careful consideration of their underlying principles significant improvements may be made in their usefulness.

CHAPTER 6

METHODS OF MECHANICAL ANALYSIS

INTRODUCTION

THE number of methods at present available for mechanical analysis is so great that an entire volume could be devoted to their enumeration and description.[1] Many of the methods represent minor variations of fundamental techniques, some depend upon slight changes in established apparatus, and some were developed for the analysis of special materials. It is virtually impossible for any single worker to have personal experience with every device known, and it becomes necessary to choose from among the wide variety a few methods which may be adapted to sediments primarily on the basis of their soundness and secondarily on the basis of convenience or cost.

In the present chapter both old and new methods will be described, but special emphasis will be given to methods based on Odén's theory of sedimenting systems. Among these latter, the pipette method will be stressed as one of practically universal application to fine sediments. This emphasis on the pipette method is a natural consequence of the authors' greater familiarity with the method, which they have adopted for all laboratory work at the University of Chicago.

In contrast to the wide variety of methods available for the finer sediments, the process of sieving has remained by far the most popular method for material in the sand ranges and above. To some extent microscopic methods may take the place of sieving in the future, but certainly at present sieve techniques are practically universal in America at least. Mention was made in Chapter 5 of a few of the theoretical objections to sieving. Some workers prefer to use elutriation methods exclusively for analysis, to avoid composite data based partly on sieving and partly on sedimentation. With this end in view, elutriators and

---

[1] Perhaps the best known and most complete volumes on the subject are F. V. von Hahn, *Dispersoidanalyse* (Leipzig, 1928), and H. Gessner, *Die Schlämmanalyse* (Leipzig, 1931). The former is written primarily from the point of view of the colloid chemist; the latter treats the subject from the soil scientist's point of view.

settling tubes for sand have been developed as substitutes for sieving (see page 157).

Numerous sediments have a range of sizes from coarse to fine. Among these are sandy silt, loess, glacial till, sandy shale, and the like. For such sediments a composite method of analysis is necessary, involving, usually, a splitting of the sample at some convenient size, so that the coarser material may be sieved and the finer material analyzed by a sedimentation method. Purely for convenience, the line between coarse and fine sediments will be chosen at $\frac{1}{16}$ mm. This is the lower limit of the sand size in Wentworth's classification, and sieves may readily be obtained with meshes fine enough to separate material at about this dimension. Further, $\frac{1}{16}$ mm. is near the upper limit of applicability of Stokes' law or Wadell's practical sedimentation formula and so furnishes a convenient line of demarcation.[1]

With composite types of sediments the coarse and fine portions are analyzed separately, and the analytical data combined into a single size frequency distribution. Such composite analyses often show an abrupt "break" in the distribution in the vicinity of $\frac{1}{16}$ mm., because the principle on which sieves separate the particles is not the same as the principles operative in sedimentation analysis. In many cases this hiatus between the two methods is not serious, but in general any unusual features of the data in the vicinity of $\frac{1}{16}$ mm. should be examined from the point of view of possible experimental errors.

COARSE SEDIMENTS

In Chapter 5 sedimentary particles were classified in terms of the disperse systems to which they belong. The outline on page 92 includes all particles larger than 0.1 micron diameter as coarse disperse systems. For mechanical analysis the limit $\frac{1}{16}$ mm. (0.0625 mm.) has been chosen as a convenient point to distinguish between techniques for coarse and fine sediments. Coarse particles may be further subdivided; the smaller group includes sand and pebbles conveniently analyzed by sieving, and the larger group comprises pebbles and cobbles large enough to be handled as individual particles.

The line between the two groups of coarse particles may be chosen, merely for convenience of discussion, at 16 mm. diameter. Pebbles of

[1] See C. K. Wentworth, Methods of mechanical analysis of sediments: *Univ. Iowa Studies in Nat. History*, vol. 11, no. 11, 1926, for a more detailed discussion of the advantages of the lower limit of sand as a limit for sieving. With other grade scales, such as the Atterberg scale, the limiting value will not be precisely 1/16 mm., but it will be of the same order of magnitude.

METHODS OF SIZE ANALYSIS

this diameter have a sufficiently large volume to render that measure convenient, and they are large enough to be handled individually.

Analysis by Sieving

Sieve analysis is rather well standardized, and in the following discussion the instructions will involve sieves arranged on the Wentworth scale or related scales, such as $\sqrt{2}$ or $\sqrt[4]{2}$ scales. The reader may understand from the discussion in Chapter 4 on grade scales, however, that any convenient set of sieves may equally well be used.

TABLE 14

Wentworth Grade Scale, $\sqrt{2}$ Scale, $\sqrt[4]{2}$ Scale, and Corresponding Tyler Sieve Openings

Wentworth Grade Scale (mm.)	$\sqrt{2}$ Scale (mm.)	$\sqrt[4]{2}$ Scale (mm.)	Tyler Screens (mm.)
4	4.00	4.00	3.96
		3.36	3.33
	2.83	2.83	2.79
		2.38	2.36
2	2.00	2.00	1.98
		1.68	1.65
	1.41	1.41	1.40
		1.19	1.17
1	1.00	1.00	0.991
		0.840	.833
	0.707	.707	.701
		.595	.589
½	0.500	.500	.495
		.420	.417
	.354	.354	.351
		.297	.295
¼	.250	.250	.246
		.210	.208
	.177	.177	.175
		.149	.147
⅛	.125	.125	.124
		.105	.104
	.088	.088	.088
		.074	.074
1/16	.062	.062	.062

There are many kinds of sieves on the market, including bolting cloth sieves, plates with round holes punched through them, and woven wire sieves with square meshes. The preferred type for general analyses are those with woven wire meshes, double crimped to prevent distortion. Sieves are available in brass rings of several diameters, and either 6- or 8-in. diameters are commonly used.

Unfortunately most commercial types of sieves have been developed in connection with engineering uses, so that they may not agree precisely with the Wentworth grade scale, which is most commonly used by sedimentary petrologists in America. Among the better known products are Tyler Standard Screen Scale Sieves,[1] which are based on a 200-mesh sieve having openings of 0.0029 in., and increasing uniformly of the $\sqrt[4]{2}$ scale. The millimeter equivalents of these sieves do not agree precisely with the Wentworth grade limits, but they lie so close that the difference is well within the United States Bureau of Standards limit of tolerance for sieves.

Table 14 lists the grade limits of the Wentworth, $\sqrt{2}$, and $\sqrt[4]{2}$ grade scales and indicates the corresponding sizes of the Tyler screen meshes. In the first column the Wentworth scale is shown alone; the second column lists the $\sqrt{2}$ scale, and the third column has the complete $\sqrt[4]{2}$ scale from the range of 4 mm. to $\frac{1}{16}$ mm.

In actual practice it is largely immaterial what particular meshes of sieves are used as long as a sufficiently small interval is involved between sieves to bring out the continuous nature of the frequency distribution. The discussion of grade scales in Chapter 4 covers this topic. For most routine analyses, however, it is customary to use either the Wentworth intervals directly, or $\sqrt{2}$ intervals. When the data are to be plotted as cumulative curves, it is usually convenient to sieve the material first with $\sqrt{2}$ sieves and, by inspection of the sieve residues, to resieve the heaviest loaded sieves through the intermediate $\sqrt[4]{2}$ sieves. An example will indicate the procedure: A beach sand is sieved with the $\sqrt{2}$ scale, yielding the following weights of material on each sieve

0.701 mm.	0.01 g.
0.495	0.35
0.351	2.90
0.246	14.75
0.175	5.75
0.124	0.45
0.088	0.06
	24.36 g.

[1] Manufactured by the W. S. Tyler Company, Cleveland, Ohio.

METHODS OF SIZE ANALYSIS

In comparison with the other sieve residues, the quantity on the 0.246-mm. sieve is so large that it would be convenient to divide the amount into two grades, inasmuch as it represents more than 60 per cent of the sample. By using the 0.289 sieve, on the $\sqrt[4]{2}$ scale, this grade was separated into the sub-grades 0.351-0.289 = 4.91 g., and 0.289-0.246 = 9.84 g. Thus an additional point on the cumulative curve results in a more accurate smoothing of the data. If the data are to be

FIG. 33.—Ro-Tap Automatic Shaking Machine. (Courtesy of W. S. Tyler Co., Cleveland, O.)

used in computing the moments of the distribution (Chapter 9), either the Wentworth or the $\sqrt{2}$ grades may be used directly, without the necessity of resieving.

Most commonly sieves are furnished with flanges so that one sieve may be fitted above another. In this manner an entire column of sieves may be used simultaneously. In hand sieving the column of sieves may be set up in decreasing mesh downward from the top, and with a pan at the base of the column. The material is poured into the top sieve, and the entire column rocked and tapped with the flat of the hand until sieving is completed in the top, coarsest sieve. This may then be removed from the column, and the process repeated until sieving is complete. In this manner the finer sieves are worked for a longer period than the coarser, and the critical sieve is always open to view.

A simple test for the completeness of sieving is to shake the sieve over a large sheet of glazed paper. As long as any appreciable number of grains pass through, the sieving should be continued.

When many sieve analyses are to be made, it is convenient to have an automatic shaking machine. Such devices are marketed and take an entire set of sieves. They are operated by electricity, and may combine a rotary motion with a tapping effect. The Tyler automatic "Ro-Tap" shaker is such a machine, and it may be equipped with an automatic clock which times the sieving interval. The shaker and clock are shown in Figures 33 and 34.

FIG. 34. — Automatic timer. (Courtesy of W. S. Tyler Company.)

Other shaking devices are on the market, and simple but efficient hand-operated machines can be constructed at low cost. Andreason [1] described a shaking machine constructed by placing a nest of sieves in a support which itself was mounted to a baseboard by means of flexible bands, so that during operation the nest of sieves is agitated to and fro by an eccentric shaft. Figure 35 is a diagram of this apparatus, from Andreason's paper.

Wentworth's study of sieving, described in Chapter 5, indicated that an interval of about 10 min. in an automatic shaker is usually sufficient for approximately complete separations, and that interval has been fairly widely adopted by sedimentary petrologists in America. The weight of sample to be sieved depends upon the sizes of material present, but in general a sample of 25 g.[2] is suffi-

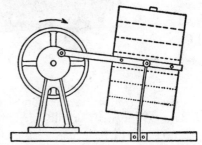

FIG. 35.—Andreason's shaking machine.

cient for material between $\frac{1}{2}$ and $\frac{1}{16}$ mm. diameter. The test sample is split from the field sample by one of the methods described in Chapter 3, weighed, and placed in the top of a column of sieves. After shaking, each sieve may be emptied in turn onto a large sheet of glossy paper

[1] A. H. M. Andreason, Zur Kenntnis des Mahlgutes: *Kolloidchem. Beihefte*, vol. 27, pp. 349-458, 1928.
[2] C. K. Wentworth, *loc. cit.*, 1926.

METHODS OF SIZE ANALYSIS

(12 x 12 or 16 x 16 in.) and the separate transferred to the balance pan for weighing. When a number of samples are to be worked, time is saved by having as many sheets as there are sieves, each labeled with a corresponding sieve opening. The material from each sieve is removed to its corresponding sheet, and in this manner the sieves are freed for the next sample. The weighing of one set of separates may then be performed while the next sample is being sieved.

In many cases, when the grains are angular, it may be found that a number of grains remain lodged in the sieve, and cannot be removed readily. The use of a fairly stiff brush, rubbed over the bottom of the sieve, is often useful in loosening the grains, and a similar effect can be had by tapping the rim of the sieve with a wooden mallet, taking care to strike the rim along the general diagonals of the wire mesh, to prevent distorting the sieve.

The scale used in weighing sieve separates need not be an expensive analytical balance. Any good beam scale, with sliding weights, and sensitive to 0.01 g., may be used. It is seldom necessary to weigh the sieve separates to more than two decimal places. Each sample should be recorded on a separate sheet as the weighing progresses. The accompanying report of an analysis indicates a convenient method of setting up the form.

Report of Sieve Analysis

Sample Number 33   *Analyzed by* WCK   *Date* 1/5/38
Description of Sample **Beach sand**
Weight of Test Sample 28.54 g.   *Method of Analysis* 10 min. shaking in Ro–Tap

Screen Opening	Grade Size	Weight Retained	Weight Per Cent	Cumulative Per Cent
0.701	1–0.707	0.03 g.	0.1	0.1
0.495	0.707–.500	0.06	0.2	0.3
.351	0.500–.354	1.02	3.6	3.9
.246	0.354–.250	16.37	57.5	61.4
.175	0.250–.177	10.22	35.8	97.2
.124	0.177–.125	0.74	2.6	99.8
.088	0.125–.088	0.03	0.1	99.9
		28.47 g.	99.9	
Sieve Loss		0.07	0.1	
		28.54 g.	100.0	

When it is desired to sieve coarse pebbles or cobbles of a size larger than is commonly handled with sieves, several devices are available. Metal squares or rings may be used, made of heavy wire, to extend the sieve sizes as far as necessary. Squares are probably preferable inasmuch as common sieves have square meshes, and the same shaped rings insure uniformity of the basis on which separations are effected. A simple and convenient device may be made from a sheet of zinc measuring about 12 x 12 in. Square holes may be cut into it, ranging from 8 to 64 mm. on the $\sqrt[4]{2}$ scale. The sieving is accomplished by dropping the pebbles through corresponding openings one by one, and either placing them in separate piles or tallying them. The method is slow, but is effective when sieving is to be done directly in the field. The pebbles in each group may be weighed or counted, depending upon the manner in which frequency is to be expressed.

Wet sieving of coarse particles. Wet sieving is sometimes resorted to for sieving coarse material, but for general purposes it is not as satisfactory as dry sieving, because the separation is not complete. It appears that the film of water in the wet sieve prevents some small particles from passing through. However, when the sediment is dirty or partially aggregated, wet sieving often aids in separating the aggregates and obtaining a cleaner product. In all cases, however, the wet sieve separates should be resieved through the same sieves when dry, to remove the finer particles.[1]

For wet sieving two general procedures may be used. The loaded sieve may be agitated in a pan of water, taking care that none of the material is washed over the edge of the sieve, or a spray of water may be directed into the sieve or a nest of sieves.

Preliminary splitting of composite sediments. Thus far the assumption has been that the grains of the sediment were all within the sieving range, namely larger than $\frac{1}{16}$ mm. diameter. If there is any material smaller than $\frac{1}{16}$ mm., it collects in the pan and may be grouped into a single class, smaller than $\frac{1}{16}$ mm., or it may be analyzed further by a sedimentation method. Many sediments have particles both coarser and finer than $\frac{1}{16}$ mm., so that composite analyses are necessary. The usual procedure is to split the sample at the $\frac{1}{16}$-mm. point. The following routine is suggested, to prevent aggregates of fine material from remaining on the sieves:

A weighed test sample of the sediment is disaggregated by means of the routine suggested in Chapter 3, either by shaking in a dilute solution of a peptizer or by rubbing with a brush. Disaggregation should be continued until all the grains are clean and no more aggregates can be de-

[1] H. Gessner, *op. cit.* (1931), p. 145.

METHODS OF SIZE ANALYSIS 143

tected. The entire suspension is then poured through a sieve with $\frac{1}{16}$-mm. mesh, and the fine material caught in a beaker. A convenient device is to have a tin funnel large enough to accommodate the sieve, so that the wash water is not lost. A fine stream of water from a rubber tube is directed over the sieve residue until it remains clear. The sieve material is then washed into filter-paper and dried. The residue is sieved, care being taken to use a pan below the smallest sieve to catch any fines that were not separated by the washing. These fines are added to the coarsest grade separated by sedimentation. Methods for the analysis of the fine material are given in later portions of this chapter.

Direct Measurement of Large Particles

Sedimentary particles larger than about 16 mm. diameter may be measured individually, and from a compilation of the size data the frequency may be determined. The measurements to be performed on the particles depend upon the definition of size adopted. The concept of size of irregular particles was discussed in Chapter 5 (pages 93, 127); in the present discussion three common measurements of size will be included: the nominal diameter, the mean diameter, and the intermediate diameter.

Measurement of the nominal diameter. The nominal diameter of a particle is found by determining the diameter of a sphere having the same volume as the particle. The equipment required consists of a cylinder graduated with metric divisions. The diameter of the cylinder should be large enough to accommodate the pebbles, and the scale divisions should be sufficiently small so that the volume of the pebble may be read with a reasonable degree of accuracy. The graduate is partially filled with water, and a rubber stopper is dropped in to prevent breakage as the pebbles are introduced. The initial volume of water is recorded, a pebble is dropped in, and the new reading made. The difference is the volume of the pebble in cubic centimeters. The process is repeated with successive pebbles until the graduate is filled. Care should be exercised to avoid having air bubbles on the pebbles, especially the smaller ones. By wetting the pebbles in a beaker of water before introducing them into the graduate the air bubbles may be eliminated.

For pebbles appreciably smaller than 16 mm. diameter an ordinary burette may be used, with scale divisions of 0.1 c.c. In such a tube pebbles as small as about 4 mm. diameter may be measured.

A number of special devices have been used for measuring volumes. Fancher, Lewis, and Barnes [1] describe several of them.

After the volumes of the pebbles have been found, the corresponding

---

[1] G. H. Fancher, J. A. Lewis, and K. B. Barnes, Some physical characteristics of oil sands: *Penn. State College, Bull. 12,* p. 72, 1933.

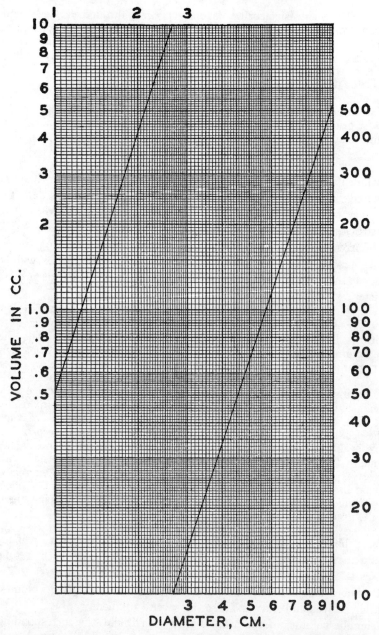

Fig. 36.—Graph for converting volume to diameter. Use upper and left scales for volumes from 0.5 to 10 c.c., and lower and right scales for volumes from 10 to 500 c.c.

METHODS OF SIZE ANALYSIS

diameter of a sphere of equal volume may be calculated by the equation $d_n = \sqrt[3]{\frac{6V}{\pi}} = \sqrt[3]{1.92V}$, where V is the volume in cubic centimeters and d_n is the nominal diameter in centimeters. It may be converted to millimeters by multiplying d_n by 10. The computations may be performed on a slide rule, but a graph showing volume along one axis and diameter along the other is convenient. The graph is constructed by plotting several corresponding values for d and V on double logarithmic paper and drawing a straight line through the points. Figure 36 is such a graph for the range 1.0 to 10.0 cm. diameter.

A convenient method of recording the results of nominal diameter measurements is to write the observed volume in the first column and the calculated nominal diameter in the second column. If the number of pebbles to be measured is not too large, it is often desirable to number each pebble with a pencil, for later reference and comparison, inasmuch as pebbles of apparently different "sizes" to the eye may have very similar nominal diameters, depending upon their shape. The recorded size data may be arranged into grades if frequency data are desired. In some studies involving shapes (Chapter 11) the volumes may be used directly for the assembling of frequency classes. In such cases the nominal diameter need not be calculated, the volumes being arranged into grade sizes at once.

Measurement of long, intermediate, and short diameters. For some purposes it may be desirable to know the longest, intermediate, or shortest diameters of particles, or their arithmetic mean, the mean diameter. These lengths may be measured with a caliper, but a convenient device is available for rapid work. This device, developed by Appel in the laboratories of the University of Chicago, consists of a board about 12 in. long, 3 in. broad, and 1 in. deep, on which a centimeter scale is embedded by countersinking. One end of the base is fastened to a block of wood, flush with the end of the centimeter rule. Another block, mounted on side runners, slips over the base as shown in Figure 37. The pebble is laid against the end block, in proper orientation, and the sliding block is moved against it. The scale reading yields the appropriate diameter directly. Three measurements are made on each pebble, the longest, intermediate, and shortest diameters, each of which can be found by simple inspection or trial. The arithmetic mean of the three values is the mean diameter. It is computed according to the equation

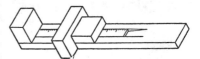

FIG. 37.—Device for measuring diameters of pebbles.

$$d_m = (a + b + c)/3$$

where a, b, and c are the three measured diameters. If the intermediate diameter itself is to be used as a measure of size, the other diameters need not be measured.

Theoretically, much criticism may be directed against the mean and intermediate diameters of irregular particles, but in actual practice it is often found that these values closely approach the value of the nominal diameter. They may, accordingly, be used as approximations of the latter if the departure of the particle from true sphericity is not too great.[1] As long as the particles are nearly spherical, the mean diameter will be very nearly equal to the nominal diameter, but with increasing departure from sphericity the mean diameter becomes larger than the nominal diameter. For very flat disc-like pebbles there may be no direct relationship between the two values, but within limits the agreement may be close enough for practical purposes. To test this, a random sample of 100 pebbles was taken from a beach and all three-types of diameter measured. The pebbles were approximately disc-shaped, had an average sphericity of 0.705, and an average roundness of 0.678. The ratio of the shortest to the longest diameter was 1/2.1, and the ratio of the shortest to the intermediate diameter was 1/1.5.

TABLE 15

COMPARISON OF AVERAGE MEAN AND INTERMEDIATE DIAMETERS WITH AVERAGE NOMINAL DIAMETER OF 100 BEACH PEBBLES FROM LITTLE SISTER BAY, WISCONSIN

Data Compared	Nominal Diameter, d_n (mm.)	Mean Diameter, d_m (mm.)	Intermediate Diameter, d_i (mm.)
Largest pebble	46	45	49
Smallest pebble	23	23	22
Arithmetic mean size of 100 pebbles	30.0	30.9	30.6
Ratio	$d_n/d_n = 1.00$	$d_m/d_n = 1.03$	$d_i/d_n = 1.02$
Geometric mean size of 100 pebbles	22.4	21.2	21.6
Ratio	$d_n/d_n = 1.00$	$d_m/d_n = 0.95$	$d_i/d_n = 0.97$

Table 15 shows that the range of sizes is roughly the same in all three cases, and that the departure of the arithmetic average of the mean and

[1] It was pointed out in Chapter 5 that Roller demonstrated the usefulness of the mean diameter as an approximation of the harmonic mean diameter. Thus the mean diameter may have practical, if not rigorous theoretical, significance.

METHODS OF SIZE ANALYSIS

intermediate diameters of the pebbles from the nominal diameter is within 3 per cent. With the geometric means the agreement is somewhat less satisfactory, although here the values agree within 5 per cent. Roller[1] considered an agreement of 6 per cent between the harmonic and arithmetic means to be satisfactory for most purposes.

Although the agreement between average values of the pebbles appears to be satisfactory, comparisons of the frequency distributions as histograms may yield figures which differ widely in appearance.

FINE SEDIMENTS

Direct Separation into Grades

Sediments with particles smaller than $\frac{1}{16}$ mm. may be separated directly into grades by either of two general methods, *decantation,* or *rising current elutriation;* in the former the centrifuge may be used to hasten sedimentation, giving rise to one type of centrifugal separation. Among the disadvantages of the direct methods are that complete separation is practically never accomplished and that, except for centrifugal methods, the techniques are not practical for diameters smaller than about 0.01 mm. Indirect methods are generally more precise and apply without difficulty to particles at least as small as 0.001 mm. diameter.

Decantation methods. In Chapter 5 it was pointed out that decantation methods include all methods of mechanical analysis in which the grades are separated by starting with a thoroughly mixed suspension, allowing sufficient time to elapse for particles above a given diameter to settle to the bottom and at that moment drawing off the supernatant liquid, including the smaller particles still in suspension.

The theory of decantation methods indicates that complete separation of the grades is seldom effected and that for precise work the number of decantations is high. Decantation methods are among the simplest techniques, however, in terms of apparatus and ease of operation. The apparatus required may be merely a set of beakers, a liter graduate, or a specially constructed settling tube with a side outlet. The following simple procedure requires only a liter graduate and a large beaker; it is based essentially on the technique recommended by Wentworth in 1926.[2]

A weighed quantity of the sediment, disaggregated and dispersed in accordance with the techniques given in Chapter 3, and with all particles

[1] P. S. Roller, Separation and size distribution of microscopic particles: *U. S. Dept. Commerce, Bur. of Mines, Tech. Paper 490,* 1931.
[2] C. K. Wentworth, *loc. cit.,* 1926.

above $\frac{1}{16}$ mm. sieved off, is diluted to a sufficient volume so that it fills the liter graduate to a depth of 30 cm. To separate the $\frac{1}{16}$-$\frac{1}{32}$ mm. grade, sufficient time is allowed after thorough shaking for particles larger than $\frac{1}{32}$ mm. to settle to the bottom. From Stokes' law [1] it is found that particles with a diameter of $\frac{1}{32}$ mm. settle 0.0869 cm./sec., or require 11.5 sec. to settle 1 cm. This is equivalent to 345 sec., or $5\frac{3}{4}$ min., for 30 cm. At the end of that time the supernatant suspension is withdrawn with a rubber tube as a siphon, taking care to avoid drawing up any of the bottom sediment. The siphoned liquid is drawn into a beaker. The original graduate is filled with clear water, shaken thoroughly, and the settling process is repeated for the same length of time, after which the drawn-off liquid is combined with the previously decanted suspension. The process is repeated three or four times, until the supernatant liquid is clear after the settling period. The sediment in the graduate is then collected in filter-paper, dried, and weighed as the $\frac{1}{16}$-$\frac{1}{32}$ mm. grade.

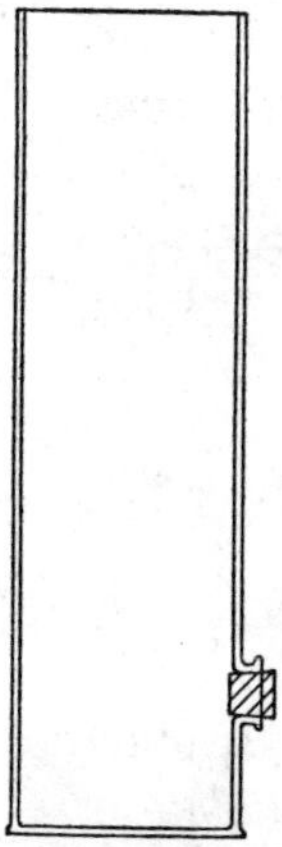

FIG. 38.—Kühn's settling tube.

For the separation of the $\frac{1}{32}$-$\frac{1}{64}$ mm. grade, the decanted water is combined and poured into liter beakers to a depth of 10 cm. The settling period for particles $\frac{1}{64}$ mm. in diameter for this depth of suspension is 46 sec. per centimeter or 7 min. 40 sec. for 10 cm. The separation below $\frac{1}{64}$ mm. is seldom carried on, so that the siphoned suspension may be run directly into the drain and material below $\frac{1}{64}$ mm. computed by difference. The decantations are continued until the water is clear, the residue is filtered as before, and weighed. The result of the analysis is three grades: $\frac{1}{16}$-$\frac{1}{32}$ mm.; $\frac{1}{32}$-$\frac{1}{64}$ mm.; and all material below $\frac{1}{64}$ mm. by difference. If the $\frac{1}{64}$-$\frac{1}{128}$ mm. grade is to be recovered, the time required to settle 10 cm. is 31 min. The time required thus increases by a factor of four for each smaller grade, and the volume of water to be handled also increases by a factor of three or four, depending upon the number of decantations for each grade.

Numerous workers have introduced special apparatus for decantation. The tube and siphon were originally used by Wagner in 1891;[2] at a still earlier date Kühn[3] used a cylinder with a side opening to facilitate withdrawing the

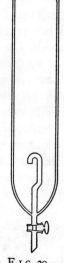

FIG. 39.—Appiani's settling tube.

[1] Wentworth used a series of empirical settling velocities based on experiment. It is preferable to use Stokes' law or Wadell's practical formula to compute settling velocities. See Table 16 (page 166) for numerical data.

[2] E. Wolff, Die Bodenuntersuchung: *Landwirts. Versuchs-Stat.*, vol. 38, pp. 290-292, 1891.

[3] H. W. Wiley, *Principles and Practice of Agricultural Analysis*, 2nd ed. (Easton, Pa., 1906), vol. I, p. 203.

METHODS OF SIZE ANALYSIS 149

supernatant liquid. Appiani [1] introduced a siphon with stopcock near the base of the tube to facilitate withdrawal of the liquid. Atterberg [2] in 1914 raised the method to the peak of its development by introducing an efficient side outlet. Trask [3] subsequently introduced a removable bottom cup to facilitate removal of the collected sediment. In all of these devices the underlying method of operation is similar, although various authors use different settling times, depending upon the hydraulic values which seemed most logical to them. The types of apparatus mentioned above are shown in Figures 38-41.

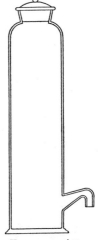

FIG. 40.—Atterberg's sedimentation cylinder.

Atterberg's apparatus has been extensively used by European workers, and several writers [4] have discussed it at some length. Köhn [5] studied the streamlining and turbulent effects in the withdrawal of the suspension from the outlet. Figure 42, adapted from his photograph, shows the eddy set up during the flow.

Centrifugal decantation methods. Owing to the time required for small particles to settle, especially in decantation methods where a number of settling periods must be allowed, various workers have hastened the process by the use of centrifugal force. Among the earliest devices for this purpose was a centrifugal cream separator of the Babcock type used by Whitney to hasten sedimentation.[6] Perhaps the most extensive use of the centrifuges was by

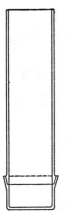

FIG. 41.—Trask's decantation tube.

[1] G. Appiani, Ueber einen Schlämmapparat für die Analyse der Boden und Thonarten: *Forsch. Geb. Agrik. Physik,* vol. 17, pp. 291-297, 1894.
[2] A. Atterberg, Die mechanische Bodenanalyse und die Klassifikation der Mineralböden Schwedens: *Int. Mitt. für Bodenkunde,* vol. 2, pp. 312-342, 1912.
[3] P. D. Trask, Sedimentation tube for mechanical analysis: *Science,* vol. 71, pp. 441-442, 1930.
[4] G. Richter, Die Ausführung mechanischer und physikalischer Bodenanalysen: *Int. Mitt. für Bodenkunde,* vol. 6, pp. 198-208, 318-346, 1916. J. P. Van Zyl, Der Atterbergsche Schlämmzylinder: *Int Mitt. für Bodenkunde,* vol. 8, pp. 1-32, 1918. A. A. J. von 'Sigmond, Ueber die Methoden der mechanischen und physikalischen Bodenanalyse: *Publik. der k. ungar. Geol. Reichsanstalt,* Budapest, 1916.
[5] M. Köhn, Beiträge zur Theorie und Praxis der mechanischen Bodenanalyse: *Landwirts. Jahrb.,* vol. 67, pp. 485-546, 1928.
[6] L. J. Briggs, F. O. Martin and J. R. Pearce, The centrifugal method of mechanical soil analysis: *U. S. Dept. Agric., Bur. of Soils, Bull. 24,* 1904.

Briggs, Martin, and Pearce in the United States Bureau of Soils.[1] Their method involved the removal of the coarsest particles with a sieve, after which the finer material was dispersed in a sterilizer bottle. Sedimentation was effected in the same bottle by shaking it and allowing the suspension to rest until all the sand settled out, as determined by microscopic examination. The silt and clay were decanted off into a centrifuge tube, and the material was centrifuged until all the silt settled out, leaving the clays and colloids still in suspension. In each case the decantations were repeated until satisfactory separations were made, using the microscope to check results. In this manner three grades were obtained: the sand remained in the sterilizer bottles, the silt in the centrifuge, and the clays and colloid in the decantations from the latter.

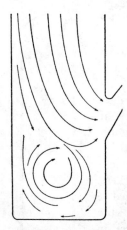

FIG. 42.—Streamlines in Atterberg's tube. Sketched from a photograph by Köhn, 1928.

More recently Truog and others[2] made a detailed study of centrifugal decantation methods, developing procedures for the more complete separation of the finer clay fractions.

Rising current elutriation. The separation of particles into grades by rising currents may be accomplished either with water or with air, and either in a single vessel or in a series of vessels of different sizes. The simplest type of apparatus is the single-tube water elutriator, the best known of which is that of Schöne,[3] introduced in 1867. Schöne's elutriator is a conical vessel about 20 in. tall, with an inlet tube at the base and an opening at the top for the passage of the water. As indicated in Figure 43, a glass tube is inserted at the top, which acts simultaneously as an outlet and a piezometer. The sediment, properly dispersed, is placed in the vessel, and a current of water is

FIG. 43.—Schöne's rising current elutriator.

[1] *Loc. cit.*, 1904.
[2] E. Truog, J. R. Taylor, R. W. Pearson, M. E. Weeks and R. W. Simonson, Procedure for special type of mechanical and mineralogical soil analysis: *Proc. Soil Sci. Soc. America*, vol. 1, pp. 101-112, 1936.
[3] E. Schöne, Ueber einen neuen Apparat für die Schlämmanalyse: *Zeits. f. anal. Chemie*, vol. 7, pp. 29-47, 1867.

sent through with a velocity just greater than the settling velocity of the smallest grade to be removed. The flow is continued until the water flows out clear. The finest grade is then separated from the collected water by filtration, dried, and weighed. The next grade is obtained by increasing the velocity to a value just above the settling velocity of the next larger grade size, and the process repeated until the desired number of grades is removed. Various modifications have been made of Schöne's apparatus, principally in connection with the piezometer and the tube connections, but in principle the apparatus still follows Schöne's original.

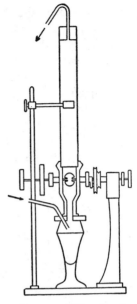

FIG. 44.—Hilgard's elutriator.

There is a strong tendency for aggregates of particles to form in rising current elutriators, and these interfere with the separation of the smaller particles. To overcome this difficulty, Hilgard [1] devised his churn elutriator, illustrated in Figure 44. In principle this operates like the Schöne elutriator, except that near the base of the tube is an automatic stirring device which aids in breaking up the floccules and preventing streaming of the fluid.

One of the most recent elutriators, designed to overcome the disadvantages of the older type, is Andrews's kinetic elutriator,[2] introduced in 1927. This elutriator consists of several upright tubes in a vertical column, as shown in Figure 45. The upper tube feeds the undispersed sediment to a restricted zone in the main vessel, where an upward current of water is directed against a stationary cone. The impact of the particles against the cone disag-

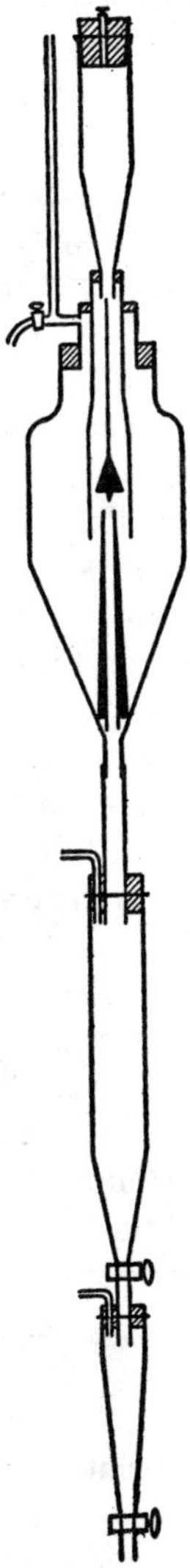

FIG. 45.—Andrews' kinetic elutriator.

[1] E. W. Hilgard, On the silt analysis of soils and clays: *Am. Jour. Sci.*, vol. 6, pp. 288-296, 333-339, 1873.

[2] L. Andrews, Elutriation as an aid to engineering inspection: *Int. Engineering Inspection, Separate,* 1927. Andrews's elutriator is obtainable from Internal Combustion, Ltd., Aldwych, London W. C. 2, England.

gregates them, and the fine material is carried off through a spout. The coarse particles settle to the bottom of the vessel where they are again picked up by the current and redirected against the cone. This bombardment is continued until the aggregates are destroyed, after which the remaining coarse particles are separated into grades in the lower vessels of the apparatus. Andrews's elutriator thus combines the efficiency of a single-tube elutriator with the separation effects of multiple-tubed devices and simultaneously eliminates the tendency toward flocculation which is so common to most devices.

Another device designed to overcome the disadvantages of early-type elutriators was introduced in 1929 by Gross, Zimmerley, and Probert.[1] A rotary rubber impeller provided a mobile bed for the material to be separated. Gum arabic was used to aid dispersion.

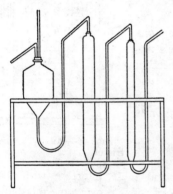

FIG. 46.—Kopecky's elutriator.

Among multiple-tubed elutriators the Kopecky[2] apparatus is also well known. This apparatus consists of three tubes arranged side by side, as shown in Figure 46. The sediment is placed in the smallest tube, and the water enters from the bottom of this vessel. The narrow diameter of the first tube results in a fairly high velocity of the water within it, so that all particles except the coarsest are removed. In the second vessel, with its larger diameter, an intermediate grade is removed, because the same volume of water passing through it results in a lower net velocity. Finally, in the third and largest vessel, all but the finest particles remain. The result of the analysis is to obtain four grades, the limits of which depend upon the original volume of water per second and on the respective diameters of the tubes. As in the case of the single-vessel elutriators, the water is allowed to flow until it becomes clear.

Studies of Kopecky's apparatus[3] showed that the largest vessel is rather poorly designed for the continuous flow of water at a uniform velocity. Tests with colored fluids showed distinct "streaming" at the center. Andreason[4] attacked the problem of multiple-tubed elutriators

---

[1] J. Gross, S. R. Zimmerley and A. Probert, A method for the sizing of ore by elutriation: *U. S. Bur. Mines. Reps. of Investigations, Serial 2951*, 1929.

[2] J. Kopecky, *Die Bodenuntersuchung zum Zwecke der Drainage-arbeiten* (Prag, 1901).

[3] H. Gessner, *op. cit.* (1931), p. 203.

[4] A. H. M. Andreason, *loc. cit.*, 1928.

METHODS OF SIZE ANALYSIS

by using three vessels shaped similarly to Schöne's, as shown in Figure 47. Andreason devised his three vessels so that the successive ratios of the velocities would be 1.00, 2.08, and 5.40. With a water flow of ¼ liter per minute, the velocities in the three vessels were 0.27, 0.13, and 0.05 cm./sec. Thus Andreason separated particles at the size limits 0.06, 0.04, and 0.02 mm. diameter. Among multiple-tubed elutriators of the Kopecky type, Andreason's apparatus appears to furnish the best general results because of the more efficient shape of the vessels.

A multiple-tube rising current elutriator which achieved considerable popularity in England, was developed by Crook.[1] It consists of a cylindrical tube surmounted by a larger vessel. In operation, the water velocity is so adjusted that the sand remains in the lower tube, the silt remains in the upper vessel, and the clay is carried through an outlet tube into a beaker. A constant head apparatus for water flow was made by using a funnel as an overflow in a large bottle. Full instructions for the operation of his elutriator are given by Crook in the reference cited.

Gollan[2] developed two models of rising current elutriators. The first consisted of a single conico-cylindrical vessel with auxiliary tubes which could be inserted within the main tube to effect the separation of additional grades. The second model consisted of three vertical tubes and is essentially a modification of Kopecky's apparatus.

FIG. 47.— Andreason's elutriator.

A slightly modified Kopecky apparatus was used by Rhoades[3] for the aggregate analysis of soils.[4] Other types of elutriators are discussed by Gessner;[5] more recent devices include Rauterberg's[6] single-vessel elutriator, constructed very simply from a cylindrical separatory funnel.

[1] T. Crook, The systematic examination of loose detrital sediments. Appendix to Hatch and Rastall's *Petrology of Sedimentary Rocks* (London, 1913), pp. 348 ff.

[2] J. Gollan, Nouvel Appareil de levigation pour l'analyse mecanique des sols: *Anales de la Sci. Agron.*, pp. 145 ff., 1930. J. Gollan, L. Hervot and V. Nicollier, Análisis mecánico de Suelos: *Rev. Fac. Quim. Ind. Agr.*, vol. 2, 1932.

[3] H. F. Rhoades, Aggregate analysis as an aid in soil structure studies: *Rep. Am. Soil Survey Assn.*, Bull. 13, pp. 165-174, 1932.

[4] *Aggregate analysis* is a term applied to a form of mechanical analysis in which an attempt is made to preserve the soil structure during analysis. A comparison of such an analysis with a mechanical analysis based on complete dispersion affords a means of determining the nature of soil aggregates. Aggregate analysis has not been used widely in connection with sediments, but it appears to afford a means of studying coagulation effects during and after deposition.

[5] H. Gessner, *op. cit.* (1931), pp. 115 ff.

[6] E. Rauterberg, Ein einfacher Schlämmapparat: *Zeits. J. Pflanzenernäh, Düng., u. Bodenkunde*, vol. 15 A, pp. 263-269, 1930.

Stokes' law may be used directly in connection with rising current elutriators if the apparatus is calibrated for the range of velocities desired. The volume of the tube, assumed to be cylindrical, is $V = qh$, where q is the cross section and h is the height. The velocity of a column of water may be expressed as $v = h/t$, where h is the height. From these two relations one may derive the following expression for the velocity of the water when a given volume flows through the tube: $v = V/qt$. To calibrate the tube, the piezometer is set in place and a liter beaker is set below the outlet. The water is turned on to a given extent, and the time is measured until the beaker is filled. Knowing V, q, and t, the velocity v can be computed. By trial and error the volume per second is controlled until the desired value of v is obtained. When this is accomplished, the level of water in the piezometer is marked on the tube. After the several velocities have been established (assuming a single-vessel elutriator), the water flow may thereafter be adjusted until the piezometer stands at the required height.

Air elutriation. The use of air currents to separate fine particles has received considerable attention, especially in its application to the analysis of pigments, cement, and ceramic materials. Air analyzers or elutriators have not been applied extensively for the study of sediments, however. Most generally the apparatus used consists of an upright cylindrical vessel through which an upward current of air passes, carrying with it the finer particles.

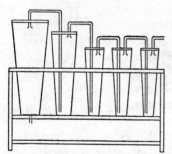

FIG. 48.—Cushman and Hubbard's air elutriator.

The use of air elutriation dates back to the early years of the present century. In 1906 Gary[1] described an air separator which consisted essentially of a conical container for the powder charge surmounted by a tall cylinder. The air blast impinged vertically on the powder and blew out the fine particles. Another early elutriator was introduced by Cushman and Hubbard in 1907.[2] Five percolating jars were arranged in series, as shown in Figure 48, the first of three-gallon, the second of two-gallon, and the last three of one-gallon capacity. The powder was placed in the largest jar, at the bottom of which an air blast entered the system. A series of tubes connected each jar with the next, and a suction device at the end of the

---

[1] M. Gary, Determination of a uniform method for the separation of the finest particles in Portland cement by liquid and air processes: *Int. Assn. Testing Materials, Brussels Congr.*, 1906.

[2] A. S. Cushman and P. Hubbard, Air elutriation of fine powders: *Jour. Am. Chem. Soc.*, vol. 29, pp. 589-596, 1907.

METHODS OF SIZE ANALYSIS 155

system was so adjusted that none of the particles was carried beyond the last jar. Pearson and Sligh [1] developed an analyzer in 1915 similar in principle to Gary's but with an automatic device for tapping the cylinder during separation. Among the more recent devices is that of Gonell,[2] introduced in 1929. Gonell's apparatus consists of three main parts, as sketched in Figure 49. The bottom vessel contains an air blast inlet which terminates just above the bottom, where the powder charge is placed. Above this is a conical vessel which supports a cylindrical tube. A glass plate near the top of the apparatus serves to support a bell jar with a cone inverted over the cylinder. As the analysis proceeds, the fine material is carried into the bell jar, some settling in the cone and part falling on the glass plate. Through an opening in the top of the bell jar, part of the fine material is withdrawn to a collecting vessel. Gonell used Stokes' law in determining the velocities required for separation.

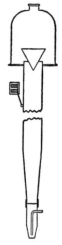

F I G. 49.—Gonell's air elutriator.

Perhaps the most intensive study of air elutriators was made by Roller in 1931.[3] He reviewed earlier work and developed his own apparatus, which consists of a cylindrical separator with a conical top and bottom. The sample is placed in a U-tube at the bottom of the main chamber, and as the air blast passes through the tube, an automatic hammer agitates the U-tube to expose fresh charges of powder to the current. Figure 50 shows the general set-up, without details of the hammer. The particles carried upward by the air current are collected as fractions in a paper thimble at the top of the apparatus, and by proper control of the current Roller was able to effect separations down to about 3 microns. Roller used Stokes' law for computing velocities and checked his results with microscopic measurements. Full details for the operation of the apparatus, sample analyses, and an excellent discussion of methods of graphic presentation and statistical analysis of the data are given in Roller's report.

F I G. 50.—Roller's air elutriator. The hammer is shown at A.

[1] J. C. Pearson and W. H. Sligh, An air analyzer for determining the fineness of cement: *U. S. Dept. Comm., Bur. Standards Tech. Paper 48,* 1915.

[2] H. W. Gonell, Determination of size distribution of powder, especially cement: *Zement,* vol. 17, pp. 1786 ff., 1929.

[3] P. S. Roller, Separation and size distribution of microscopic particles: *U. S. Dept. Comm., Bur. of Mines, Tech. Paper 490,* 1931.

Other direct separation methods. In the usual decantation methods the sediment is uniformly distributed through the suspension at the start of the separation, but in contrast to this is a technique in which the sediment is introduced into the settling tube as a unit. The principle here is that if a mixture of various-sized particles is introduced at the top of a column of water at the start of the separation, the differential settling velocities of the particles will result in the segregation of the several grades during their fall through the tube.

FIG. 51.—Bennigsen's silt flask.

Numerous devices have been developed to operate on this principle, which was first described by Rham[1] in 1840. Perhaps the best known of the devices is Bennigsen's silt flask,[2] introduced *circa* 1860 and shown in Figure 51. The sample is introduced into the flask and shaken. A cork is inserted in the neck, and the flask inverted. The material settles out according to its size, and the respective amounts present are read in cubic centimeters from graduations on the neck of the flask.

Clausen[3] modified Bennigsen's flask by separating the bulb of the flask from the volumetric settling tube. A rubber tube connects the two parts of the apparatus. Tube and flask are filled with water, and the sediment is placed in the flask. The flask itself is agitated at the side and then superimposed over the tube. This permits the material to enter the tube at a given instant and results in a more effective separation of sizes. A more recent modification of Clausen's tube was introduced by Löber[4] in 1932. Instead of relying on a volumetric reading of the grades, Löber's tube was closed with the finger at the bottom. As the successive grades settled (based on computed settling velocities), the tube was dipped into a dish of water, the finger removed, and the grade collected. A separate dish was used for each grade, and by collecting the separates the weight composition of the material was readily obtained.

Other types of apparatus, utilizing the principle of the silt flasks, are described by Gessner.[5] Most recent of the separatory techniques in this

---

[1] W. L. Rham, An essay on the simplest and easiest mode of analyzing soils: *Jour. Roy. Agric. Soc. England*, vol. 1, pp. 46-59, 1840.

[2] F. Wahnschaffe and F. Schucht, *Anleitung zur wissenschaftlichen Bodenuntersuchung*, 4th ed. (Berlin, 1924), p. 24.

[3] F. Wahnschaffe and F. Schucht, *op. cit.* (1924), p. 25.

[4] H. Löber, Ein besonders einfaches Verfahren der Schlämmanalyse: *Centralblatt für Mineralogie*, Abt. B, pp. 364-368, 1932.

[5] H. Gessner, *op. cit.* (1931), pp. 73-78.

METHODS OF SIZE ANALYSIS

category is that of Emery,[1] who used a glass tube 5 ft. long, with one end tapered and connected with a stopcock. The tube is filled with water and the sand introduced at the top. Emery proposed his tube as an alternate method to the sieving of sands.

Indirect Determination of Sizes

Odén's sedimentation balance. In 1915 the principles underlying the mechanical analysis of fine-grained sediments were abruptly placed on an entirely new foundation by the publication of Odén's theory of sedimenting systems.[2] As a practical application of his theory, Odén developed his continuous sedimentation balance, the essential principles of which are shown in Figure 52. The apparatus consists of a balance pan, suspended near the bottom of a cylinder of soil suspension, upon which the falling particles accumulate. The pan is counterpoised with another in such a manner that when the sediment lowers the pan below a critical level, the scale beam operates an electrical contact which releases shot into the counterpoise. By recording the number of shot and the times of adjustment, a curve is constructed showing the weight of accumulated sediment as a function of the time.

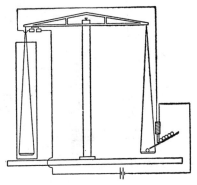

Fig. 52.—Diagram of Odén's continuous sedimentation balance.

More than any other device, perhaps, Odén's balance has been studied and modified by numerous workers. Odén himself developed elaborate controls for his method, and in 1924 Coutts, Crowther, Keen, and Odén [3] developed an automatic recording balance, which represents the ultimate yet developed in such apparatus.

Figure 53 is a diagram of the automatic balance. The activating mechanism is the rod M in the solenoid S, which controls the recording pen on the drum H. The weights which drop into the pan R serve to

---

[1] K. O. Emery, Rapid method of sand analysis: *Geol. Soc. America, 50th Ann. Meeting, Abstracts,* p. 15, 1937.

[2] S. Odén, Eine neue Methode zur mechanischen Bodenanalyse: *Int. Mitt. für Bodenkunde,* vol. 6, pp. 257-311, 1915. Odén's original theory is given in Chapter 5, which also describes the graphic method devised by Odén for evaluating his analytical data.

[3] J. R. H. Coutts, E. M. Crowther, B. A. Keen and S. Odén, An automatic and continuous recording balance: *Proc. Roy. Soc.,* vol. 106A, pp. 33-51, 1924.

lower M into the solenoid at intervals during the procedure, and result in a record as shown on the drum in the diagram. Full details and additional data on the theory of the apparatus are given by Keen.[1]

Other modifications of Odén's balance were made by various workers. Johnson[2] substituted an ingenious recording device for Odén's weight-dropping attachment. The recording device punctures holes in a record on a revolving drum by sending electric sparks through at timed intervals. After

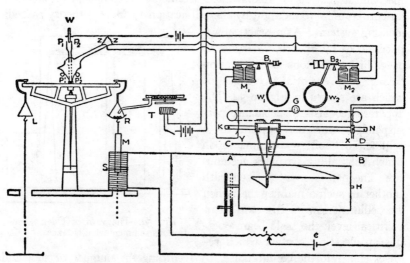

FIG. 53.—Diagram of Odén-Keen self-recording apparatus. (After Keen.)

the analysis, the paper record shows a series of holes which can be connected with a continuous line to furnish an Odén sedimentation curve. Werner[3] developed a simplified apparatus in which a volumetric method replaces the continuous weighing technique of Odén. The sedimented material is collected in a graduated tube, and a record kept of the volume of material deposited as a function of time. Werner's device has the advantage that it is easily constructed and requires no expensive equipment. The essential features are sketched in Figure 54. Vendl and Szádeczky-Kardoss[4] described another modification of Odén's method in which a delicate spring balance is used for determining the weight of sedimented material.

[1] B. A. Keen, *The Physical Properties of the Soil* (London, 1931), pp. 82 ff.
[2] W. H. Johnson, A new apparatus for mechanical analysis of soils: *Soil Science*, vol. 16, pp. 363-366, 1923.
[3] D. Werner, A simple method of obtaining the size distribution of particles in soils and precipitates: *Trans. Faraday Soc.*, vol. 21, pp. 381-394, 1925-26.
[4] M. Vendl and E. v. Szádeczky-Kardoss, Über den sogennanten grundsätzlichen Fehler der mechanischen Analyse nach dem Odén'schen Prinzip: *Kolloid Zeits.*, vol. 67, pp. 229-233, 1934.

METHODS OF SIZE ANALYSIS

Schramm and Scripture[1] used a series of test-tubes instead of a balance to obtain the Odén sedimentation curve. Given volumes of suspension are poured into a series of tubes and the liquid above a mark on the side is drained from successive tubes at stated intervals. The sedimented material in each tube is dried and weighed, to determine how much had accumulated during the intervals. The data so obtained furnish points along the sedimentation curve. Tickell[2] describes the Schramm and Scripture method in full detail.

In 1930 Trask[3] applied centrifugal force to aliquot portions of the sample to hasten sedimentation of the smaller particles. The complete method includes decanting the sands, separating the suspension of fine materials into a number of aliquots, and determining the weight of material that separates from each aliquot after centrifuging for definite times at specified speeds. The sedimentation curve is constructed from the data obtained from the aliquots. The method is rapid and requires less expensive apparatus than Odén's original method.

FIG. 54.—Werner's modification of Odén's method.

In the discussion of Odén's theory of sedimenting systems (Chapter 5) it was pointed out that equation (33) (page 117) was developed for the Odén balance. In practice the equation is seldom used, inasmuch as simple graphic methods (page 114) are available for analyzing the Odén curve. All of the techniques described above yield Odén curves, which can be treated in the standard graphic manner. The principal difference among the methods is, perhaps, the preciseness with which the curve is determined.

Continuous sedimentation cylinders. In 1918 Wiegner[4] introduced a manometric sedimentation cylinder which rested upon the principle that two columns of liquor of different specific gravities will rise to levels inversely proportional to their densities, when confined in separate tubes which are joined at some point.

Wiegner's apparatus consists of a long glass cylinder to which is attached a parallel manometric tube of about the same length but of a smaller diameter, as shown in Figure 55. A stopcock controls the point of

---

[1] E. Schramm and E. W. Scripture, Jr., The particle analysis of clays by sedimentation: *Jour. Am. Ceram. Soc.*, vol. 8, pp. 243-258, 1925.

[2] F. G. Tickell, *The Examination of Fragmental Rocks* (Standard University Press, 1931), pp. 10-16.

[3] P. D. Trask, Mechanical analysis of sediments by centrifuge: *Econ. Geology*, pp. 581-599, 1930.

[4] G. Wiegner, Ueber eine neue Methode der Schlämmanalyse: *Landwirts. Versuchs-Stat.*, vol. 91, pp. 41-79, 1918.

juncture. With this closed, water is poured into the manometer, and the soil suspension into the cylinder. When the stopcock is opened, the manometer registers the hydrostatic pressure at the point of juncture, and as particles in the suspension settle below this level, the pressure decreases. By observing the decrease in the height of water in the manometer tube as a function of the time, a continuous curve is obtained, from which the size frequency distribution may be determined graphically.

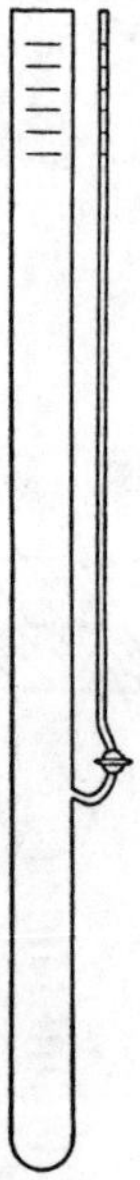

In terms of Odén's theory, Wiegner's apparatus is based on equation (31) of Chapter 5 (page 117). That is, the manometer measures the hydrostatic pressure at a fixed depth as a function of the time. The hydrostatic pressure diminishes with time as the particles settle below the juncture of the manometer, and hence the sedimentation curve obtained is concave upward, instead of convex as in the case of an Odén curve. However, the Wiegner curve is a reflection of an Odén curve with respect to the x-axis,[1] as Figure 56 shows, and hence the frequency distribution may be determined in the same graphic manner that is used with an Odén curve.

Gessner[2] added a major improvement to Wiegner's tube in 1922, when he added an automatic photographic recording device. A sheet of photographic paper is placed on a revolving drum in a light-tight box. A beam of light is directed to a mirror behind the manometer tube and reflected back to a lens leading to the sensitive paper on the drum. As the column of water falls, a continuous photographic record is obtained. Figure 57 is a vertical diagram of the essential features of the apparatus.[3] With Gessner's apparatus it is not difficult to analyze the range of sizes from 0.1 to .002 mm. diameter.

FIG 55. —Wiegner's continuous sedimentation tube.

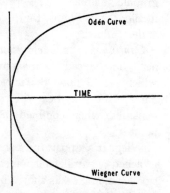

FIG. 56. — Relation between Odén and Wiegner curves. (After Odén, 1925.)

---

[1] S. Odén, The size distribution of particles in soils and the experimental methods of obtaining them: *Soil Science*, vol. 19, pp. 1-35, 1925.
[2] H. Gessner, *op. cit.*, p. 98.
[3] Gessner has a detailed discussion of the apparatus in his book (*op. cit.*, pp. 191 ff.), including full directions for its operation.

Wiegner's tube, like Odén's balance, has been the subject of considerable modification by other workers. Zunker[1] added an auxiliary manometric tube near the top of the main cylinder so that the difference in heights could more accurately be measured by eye from a parallel scale of the two manometers. One of the difficulties of Wiegner's original apparatus was that small differences of height in the two tubes could not be accurately read by eye. Zunker's tube was designed to overcome this difficulty. Another modification, also directed toward this end, was introduced by Kelley,[2] who bent the manometer tube through an angle. Kelley's modification is shown in Figure 58. By bending the tube through an angle θ from the vertical, a given vertical difference in height, Δh, becomes $\Delta h/\cos \theta$ along the inclined tube, so that the reading may be made to $1/\cos \theta$ of the original vertical scale, thus permitting a much closer reading. Other modifications of Wiegner's tube were made by Odén,[3] who introduced an internal manometric tube having a liquid with a lighter specific gravity than water, and Von Hahn,[4] who developed two modifications, one of which involves an inverted U-tube, one arm of which extends downward into a vessel of the suspension, and the other into a vessel of clear water as shown in Figure 59. A stopcock at the juncture of the tubes leads to a suction pump, and by opening the stopcock columns of liquid are drawn into the two tubes to heights depending upon their specific gravities. The stopcock is closed during an analysis, and the uniform gas pressure within the tubes results in varying relations between the heights of the two columns of liquid. The net effect is the same as in Wiegner's tube.

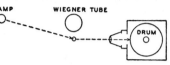

Fig. 57.—Vertical view of Gessner's continuous recording device for Wiegner's tube.

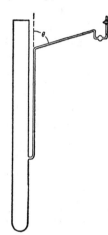

Fig. 58.—Kelley's modification of Wiegner's tube.

A more recent modification of the Wiegner apparatus was made by Barnes.[5] A needle was geared to a dial on the manometer to obtain precise readings of the water level. The instrument detected changes of level of the order of magnitude of 10^{-4} cm. A galvanometer was used to detect the point of contact, and the corresponding value was read from the dial.

In 1934 Knapp[6] described a patented automatic sedimentation unit, called

[1] F. Zunker, Die Bestimmung der specifischen Oberfläche des Bodens: *Landwirts. Jahrb.*, vol. 58, pp. 159-203, 1922.
[2] W. J. Kelley, Determination of distribution of particle size: *Jour. Ind. Eng. Chem.*, vol. 16, pp. 928-930, 1924.
[3] S. Odén, *loc. cit.*, 1925.
[4] F.-V. von Hahn, *op. cit.*, p. 310.
[5] A method for the determination of size distributions in soils: *Rep. Am. Soil Survey Assn.*, Bull. 11, pp. 169-173, 1930.
[6] R. T. Knapp, New apparatus for determination of size distribution of particles in fine powders: *Ind. and Eng. Chem.* (Analytical edition), vol. 6, pp. 66-71, 1934.

the "Microneter," which utilizes the principle of Wiegner's tube. The sedimentation cell is mounted in a brass casting which contains a small pressure-measuring orifice, connected to a pressure cell. The pressure cell is a metal bellows so arranged that a change of pressure moves a small mirror, which reflects a beam of light to a photographic plate. The curve obtained is analyzed graphically in the manner of Odén or Wiegner curves.

In 1927 Crowther [1] developed a continuous sedimentation tube with a manometer which measured the hydrostatic pressure at two points in the suspension. The essential details are shown in Figure 60, which illustrates the manometric attachment. Aniline (specific gravity = 1.02) was used for the manometer liquid. The three-way stopcocks are added for convenience in washing the side tubes without losing the aniline.

Although the differential manometer of Crowther's tube actually measures the excess of the hydrostatic pressure between the two points in the suspension over that of an equal column of water, this pressure difference is equal to the density of the suspension half way between the entry tubes. Keen [2] thus points out that Crowther's tube may be used in connection with Odén's theory of sedimenting systems, specifically equation (29) of Chapter 5 (page 116). It will be noted that although both Crowther's tube and Wiegner's tube are continuous sedimentation devices, they do not operate on identical principles, inasmuch as Wiegner's apparatus is related to equation 31 of Chapter 5.

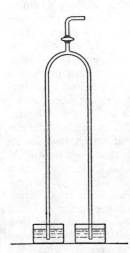

FIG. 59.—Von Hahn's d o u b l e sedimentation tube.

The method of Crowther is unique among sedimentation tube methods because of the fact that it furnishes a cumulative curve instead of an Odén-type curve. When the manometer readings are plotted directly against diameters of particles (as determined from settling velocities), an S-shaped curve is obtained, which can be interpreted as the cumulative curve, because the manometer readings are directly proportional to the ordinates of the cumulative curve.

The pipette method. Among all the methods of mechanical analysis related to Odén's theory, the pipette method has received by far the

---

[1] E. M. Crowther, The direct determination of distribution curves of particle size in suspension: *Jour. Soc. Chem. Ind.*, vol. 46, pp. 105T-107T, 1927.
[2] B. A. Keen, *op. cit.* (1931), p. 60.

METHODS OF SIZE ANALYSIS

widest official recognition, both because of its convenience and because of the simplicity of the required apparatus. Actually no more equipment is required than a 10- or 20-c.c. pipette, a liter graduate, several 50-c.c. beakers, a hot-plate, and an analytical balance—equipment to be found in almost any laboratory.

The pipette method was developed independently in 1922-1923 by Robinson[1] in England, Jennings, Thomas, and Gardner[2] in the United States, and Krauss[3] in Germany. In theory the pipette method is based on equation (29) of Chapter 5 (page 116), inasmuch as the method actually determines the density of the suspension at a fixed depth as a function of the time. For practical purposes of analysis, however, the principles on which the pipette method is based may be considered as follows. If a suspension is thoroughly shaken so that the particles are uniformly distributed and is then set at rest, all particles having a settling velocity greater than h/t will have settled below a plane of depth h below the surface, at the end of an interval of time t. All particles having a velocity

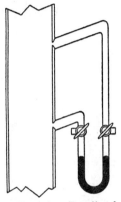

Fig. 60.—Detail of Crowther's continuous sedimentation tube.

less than h/t, however, will remain in their original concentration at depth h, because they will have settled only a fraction of this distance in time t. A small sample is taken from depth h at time t and evaporated to dryness. The weight of the residue, multiplied by a proportionality factor based on the ratio of the pipette volume to the total suspension volume, will represent the total amount of material having settling velocities less than h/t.

After the first pipette sample has been withdrawn, the suspension is again shaken and a greater period of time is allowed to elapse, so that particles of a next smaller size may settle below depth h. The second pipette sample will then contain a residue smaller than that of the first sample by an amount equal to the weight of material lying between the two chosen sizes or settling velocities. The process may obviously be repeated, and by simply subtracting the weights of successive residues

---

[1] G. W. Robinson, A new method for the mechanical analysis of soils and other dispersions: *Jour. Agric. Science*, vol. 12, pp. 306-321, 1922.

[2] D. S. Jennings, M. D. Thomas and W. Gardner, A new method of mechanical analysis of soils: *Soil Science*, vol. 14, pp. 485-499, 1922.

[3] G. Krauss, Ueber eine ... neue Methode der mechanischen Bodenanalyse: *Int. Mitt. für Bodenkunde*, vol. 13, pp. 147-160, 1923.

(each multiplied by the proportionality factor) the amount of material in any grade may be determined directly.

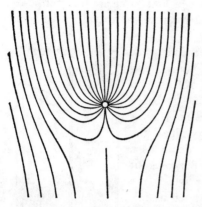

FIG. 61.—Streamlines of flow in pipette method. (After Köhn, 1928.)

Theoretically the pipette sample should represent a horizontal stratum of depth h and essentially infinitesimal thickness. Practically, the pipette taps a spherical zone, and Köhn[1] studied the influence of this fact on the accuracy of the analysis. He photographed the streamlines of liquid entering the pipette and concluded that inasmuch as part of the sphere is above the theoretical stratum and part below, the error is essentially compensatory. Figure 61 is drawn from Köhn's photograph. Keen[2] and Gessner[3] concur with Köhn in his conclusion that the net error is practically negligible.

Many devices have been developed for pipetting the suspension. Robinson used an ordinary pipette, Jennings, Thomas, and Gardner used a multiple-intake pipette fixed at a constant depth, as shown in Figure 62, and Krauss used a series of three pipettes with side openings, which were introduced to the desired depth by a rack and pinion. A diagram of this device is shown in Figure 63. Other special pipettes, including devices for raising and lowering them, as well as constant-temperature jackets for the cylinder of suspension, were developed by various workers.[4]

Steele and Bradfield[5] applied centrifugal force to the pipette method to obtain detailed analyses of material smaller than 5 microns in diameter. Ordinary gravity

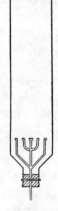

FIG. 62.—Jennings, Thomas, and Gardner's pipette.

---

[1] M. Köhn, loc. cit., 1928.
[2] B. A. Keen, op. cit. (1931), p. 73.
[3] H. Gessner, op. cit. (1931), p. 79 ff.
[4] M. Köhn, loc. cit., 1928. A. H. M. Andreason, loc. cit., 1928. L. B. Olmstead, L. T. Alexander and H. E. Middleton, A pipette method of mechanical analysis of soils based on improved dispersion procedure: U S. Dept. Agric., Tech Bull. 170, 1930. T. M. Shaw, New aliquot and filter devices for analytical laboratories: Ind. and Eng. Chem., vol. 4, pp. 409-413, 1932.
[5] J. G. Steele and R. Bradfield, The significance of size distribution in the clay fraction: Rep. Am. Soil Survey Assn., Bull. 15, pp. 88-93, 1934.

METHODS OF SIZE ANALYSIS

settling was used to diameters of 0.5 micron, after which 25-c.c. portions of the suspension were placed in tubes and centrifuged at 2,200 r.p.m. to hasten sedimentation of the finest particles. Analyses down to 0.0000625 mm. diameter were successfully carried on.

The authors have found that the use of an ordinary pipette, with a rubber tube for suction,[1] and supported by hand, yields amply satisfactory results after a preliminary period of practice. A 20-c.c. pipette is used, on the stem of which have been engraved marks at 5, 10, and 20 cm. from the tip. The pipette is held by both hands, one resting on the edge of the cylinder of suspension. The pipette is lowered to the proper mark, and an even suction is applied with the mouth. When the pipette is filled, the end of

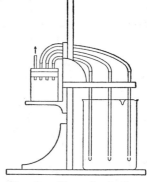

FIG. 63.—Krauss's pipette.

the rubber tube is clamped with the teeth, and the pipette transferred above a 50-c.c. beaker. By releasing the tube the contents are transferred without loss. A single rinse of the pipette with clear water suffices. If one prefers, a simple suction device suggested by Whittles[2] may be used (Figure 64). The aspirator bottle is attached to a suction pump,

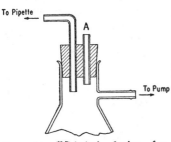

FIG. 64.—Whittles' device for withdrawing pipette sample.

and a rubber tube on the left leads to the pipette. A short glass tube (A in Figure 64) extends through the stopper. In taking a sample, the rubber tube from the pipette is pinched, the pipette inserted in the suspension, a finger is placed on A, and the tube is released until the pipette is filled. At that instant the tube is again pinched and the finger removed from A. The suction pump is operated at a slow uniform rate during the entire process.

Considerable experimentation has been performed on the pipette method in the laboratories of the University of Chicago, and because of the authors' wide experience with the method, the following detailed procedures are given here. The general remarks about the computation of settling velocities, the preliminary preparation of the samples, and similar items apply to any method of analysis, but all are included here for the sake of completeness.

[1] W. C. Krumbein, The mechanical analysis of fine-grained sediments: *Jour. Sed. Petrology*, vol. 2, pp. 140-149, 1932.
[2] C. L. Whittles, Methods for the disaggregation of soil aggregates and the preparation of soil suspensions: *Jour. Agric. Sci.*, vol. 14, pp. 346, 369, 1924.

PROCEDURE FOR PIPETTE ANALYSIS

Choice of grade sizes. For most routine analyses workers in America probably use Wentworth grades directly. If the pipette method is restricted to material finer than $1/16$ mm. in size, Stokes' law or Wadell's practical sedimentation formula may be used for computing the settling velocities of the limiting sizes. Tables 12 and 13 of Chapter 5 include the settling velocities of particles on the $\sqrt{2}$ grade scale. From these tables one may prepare the time schedule for analysis. For example, on the basis of Stokes' law, a quartz particle $1/32$ mm. in diameter has a settling velocity of 0.0869 cm./sec. at 20° C. For sampling depths of 10 cm., the time of settling may readily be found. From the relation $v = h/t$, where $v = 0.0869$ and $h = 10$, one obtains $t = h/v = 10/0.0869 = 115$ sec. In a similar manner the time for any other limiting diameters may be computed. Table 16 shows the depths and times for grades on the $\sqrt{2}$ scale from $1/32$ mm. to $1/2048$ mm., based on Stokes' law. In

TABLE 16

TIMES OF SETTLING COMPUTED ACCORDING TO STOKES' LAW *

Diameters in Millimeters	Velocity (cm./sec.)	h (cm.)	Hr.	Min.	Sec.
1/16............0.0625	0.347	20	0	0	58
.0442	.174	20	0	1	56
1/32............ .0312	.0869	10	0	1	56
.0221	.0435	10	0	3	52
1/64............ .0156	.0217	10	0	7	44
.0110	.0109	10	0	15	..
1/128............ .0078	.00543	10	0	31	..
.0055	.00272	10	1	1	..
1/256............ .0039	.00136	10	2	3	..
.00276	.00068	10	4	5	..
1/512............ .00195	.00034	10	8	10	..
.00138	.000168	10	16	21	..
1/1024............ .00098	.000085	5	16	21	..
.00069	.000043	5	32	42	..
1/2048............ .00049	.000021	5	65	25	..

* The values in this table are based on temperature of 20° C. and an average specific gravity of the sediment equal to 2.65. Seconds are neglected in lower part of table.

the table the values assume a temperature of 20° C. and a specific gravity of the sediment equal to 2.65. A similar table of time values based on Wadell's formula may be prepared from the data in Table 13 of Chapter 5.

METHODS OF SIZE ANALYSIS

Experience has shown that a standard depth of 10 cm. for sampling is inconvenient for some sizes of material. During the first few moments after shaking a suspension one may observe irregular currents in the cylinder; it seems desirable to allow a sufficient length of time for these to become quiet, and hence the first few values of the time schedule have been computed for sampling depths of 20 cm. Likewise for the finest sizes the time required for settling 10 cm. is quite long; to eliminate the time factor, the last several values have been computed for depths of 5 cm.

Preparation of samples and technique of analysis. The sediment is dispersed in accordance with the techniques of Chapter 3 (page 72), and if there is any material coarser than $1/16$ mm. present, it is removed by wet sieving as described on page 142. The sieve residue is dried, weighed, and sieved into grades. The suspension passing the sieve is poured into a liter graduate and water added to bring the volume to exactly 1,000 c.c. The suspension is well shaken by holding the palm of one hand over the mouth of the graduate and inverting the vessel, or a simple stirring device may be made.[1] This device, illustrated in Figure 65, consists of a narrow brass rod about 16 in. long, at the base of which a perforated disc is fastened. The device is inserted into the graduate and moved rapidly up and down. Agitation is continued until the material collected at the bottom of the vessel has been distributed through the suspension. As soon as the agitation has been completed, the time is noted, or a stopwatch is started. Exactly 1 min. 56 sec. later the pipette is inserted to a depth of 20 cm., and a 20-c.c. sample withdrawn with a uniform suction.[2] The sample is transferred to a 50-c.c. beaker and set on a hot-plate to evaporate. The hot-plate should have a temperature of about 100° C., to prevent boiling or spattering.

FIG. 65.—Stirring rod for suspensions.

After the first pipette sample has been withdrawn, the suspension is again agitated, and at the expiration of the next time interval another pipette sample is withdrawn. Each pipette sample is taken with respect to the new level of the suspension—no water should be added to the suspension during the analysis.

Computation of results. After the several pipette samples have been taken, and the beakers evaporated to dryness, the weight of residue in each beaker is determined with an analytical balance to 3 or 4 decimal places. For each beaker the weighing notation may be as follows:

---

[1] This device was called to the authors' attention by G. Rittenhouse of the United States Soil Conservation Service, Washington, D. C.

[2] If there is no material coarser than 1/16 mm. in the suspension, it is not necessary to take a sample for material coarser than 1/32 mm. Some analysts prefer, however, to withdraw a sample immediately after the first shaking, as a check on the total amount of material in the suspension.

```
Weight of beaker and residue .......  17.938 g.
Less weight of beaker ..............  17.406
Weight of residue ..................   0.532 g.
```

The following example of the first several separations will indicate the computational routine: 1 liter of suspension contains 27.44 g. of sediment finer than 1/16 mm. and was dispersed with N/100 sodium oxalate. N/100 sodium oxalate is equivalent to 0.67 g. sodium oxalate per liter of suspension, or 0.013 g. per 20- c.c. of suspension. This value must be subtracted from the weight of residue in each beaker to correct for the dispersing agent.

The weight of the residue in beaker #1, representing material finer than 1/32 mm., is 0.532 g., and that in beaker #2, representing material finer than 1/64 mm., is 0.446 g. Subtracting 0.013 g. from each of these yields 0.519 g. and 0.433 g. The volume of the pipette, 20 c.c., is 1/50 the volume of the suspension, so that each of the weights found are to be multiplied by 50, to convert the results into terms of the original volume. After this multiplication, a table is set up as follows, showing the amount of material in the successive grades:

```
Weight of material finer than 1/16 mm. ...........  27.44 g.
Weight of material finer than 1/32 mm. ...........  25.95
    Difference: amount in 1/16–1/32 mm. grade......   1.49 g.

Weight of material finer than 1/32 mm. ...........  25.95 g.
Weight of material finer than 1/64 mm. ...........  21.65
                                                     4.30 g.
Etc.
```

These weights may be converted into percentages of the total sample weight for histograms or cumulative curves. If material coarser than 1/16 mm. was present, that material is sieved into grades and the combined results expressed as the size distribution of the sample.

Various time-saving procedures have been developed for the pipette routine. Rittenhouse[1] found by experiment that for the finer sizes no serious error is introduced if the successive pipette samples are withdrawn without shaking the suspension between pipettings. The time saved by this procedure is considerable for such small sizes as 1/512 and 1/1024 mm. Rittenhouse has also found that a battery of thirty or forty analyses may be conducted simultaneously by setting up a time schedule which allows intervals of about one minute between the sampling times of successive suspensions. In this manner he has run as many as 100 analyses in two or three days. Rittenhouse has also developed a short method for computing the percentages in each grade.[2]

[1] G. Rittenhouse, A suggested modification of the pipette method: *Jour. Sed. Petrology*, vol. 3, pp. 44-45, 1933.

[2] G. Rittenhouse, a laboratory study of an unusual series of varved clays from northern Ontario: *Am. Jour. Sci.*, vol. 28, pp. 110-120, 1934.

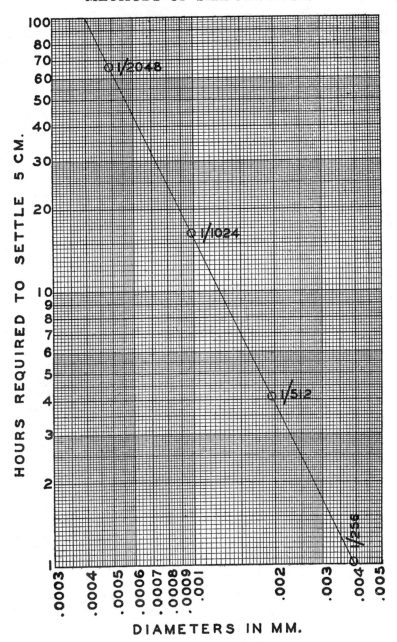

Fig. 66.—Time chart for pipette method.

When cumulative curves are to be constructed from the analytical data, instead of histograms, it is not necessary to sample the suspension for precise grade limits. Instead, the number of determinations to be made may depend upon the detail with which the curve is to be drawn.

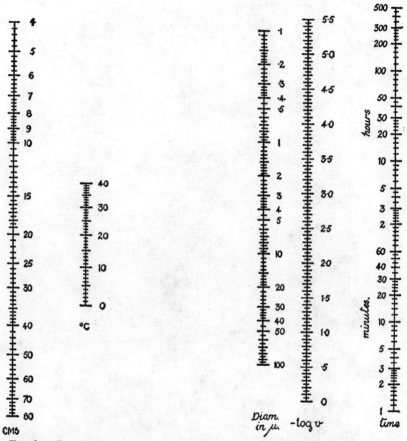

Fig. 67.—Crowther's nomogram for time of settling, computed for particles with a specific gravity of 2.70. Example: Compute the time for a particle of diameter 2 microns to settle 10 cm. at a temperature of 20° C. A straight edge is laid between points 20 and 2 of scales II and III, and the line extended to scale IV, which it intersects at 3.45. This last point is connected with 10 on scale I and the line extended to scale V, which it intersects at 8 hours.

A time chart showing diameters against settling time [1] may be prepared, as shown in Figure 66. By means of this chart any convenient values may be used, especially along the steeper parts of the cumulative curve, to bring the slope of the curve out in greater detail.

[1] W. C. Krumbein, A time chart for mechanical analyses by the pipette method: *Jour. Sed. Petrology*, vol. 5, pp. 93-95, 1935.

METHODS OF SIZE ANALYSIS

Several writers[1] have prepared nomograms for computing velocities, radii, or times of settling for various-sized particles, according to Stokes' law. Figure 67 is a reproduction of Crowther's chart.

One of the variables which may affect the accuracy of pipette analyses, but which applies equally well to any method, is the temperature at

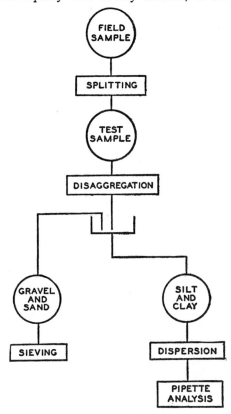

FIG. 68.—Flow-sheet for mechanical analysis. The process indicated below "disaggregation" refers to wet sieving through 0.061 mm. sieve.

which the analysis is conducted. The settling velocities given above assume a temperature of 20° C., and the discussion in Chapter 5 showed that the viscosity of water varies with the temperature. Inasmuch as the settling velocity depends in part upon the viscosity, it is clear that temperature fluctuations should be avoided during analysis. Various methods are available for maintaining uniform temperatures. Andreason[2]

[1] E. M. Crowther, Nomographs for use in mechanical analysis calculations: *Proc. 1st Int. Congr. Soil Sci.*, Part II (1927), pp. 399-404, 1928. H. Rouse, Nomogram for the settling velocity of spheres: *Rep. Com. Sed., 1936-37*, Nat. Research Council, 1937, pp. 57-64.

[2] A. H. M. Andreason, *loc. cit.*, 1928.

used an insulated cell for his analyses, but a simple and effective device is to have a water-tight box or compartment deep enough to submerge liter graduates nearly to the top. The entire suspension is thus surrounded by water, and even though the room temperature may fluctuate several degrees, the effect on the water-bath will be negligible.

The steps involved in routine mechanical analyses by the pipette method are shown in the accompanying flow-sheet (Figure 68). The sheet shows all the procedures from the splitting operation to the final analysis. Processes are shown in rectangles in the flow-sheet, and materials are shown in circles.

The hydrometer method. The hydrometer method of mechanical analysis was introduced by Buoyoucos [1] in 1927. A hydrometer, calibrated to read grams of soils per liter, is introduced into the suspension at intervals, and readings are taken. From the data obtained a cumulative curve may be drawn directly. Theoretically the hydrometer measures the density of the suspension at a given depth as a function of time, and consequently is based on equation (29) (Chapter 5) of Odén's theory. As in most other devices, however, the equation itself is not used in practice.

Bouyoucos's hydrometer is shown in Figure 69. It consists of a cylindro-conical base, weighted with lead, and a narrow stem with a scale calibrated directly in grams.[2] The rapidity of the method, compared with most other techniques, has led to an extensive study of it in terms of its accuracy and theoretical soundness, as well as in terms of the most effective shape of the hydrometer bulb.

FIG. 69.—Bouyoucos's hydrometer.

Numerous comments have been made for and against the hydrometer as an accurate device. Keen,[3] Joseph,[4] Gessner,[5] and Olmstead, Alexander, and Lakin[6] have criticized the method from the point of view of accuracy and

[1] G. J. Bouyoucos, The hydrometer as a new method for the mechanical analysis of soils: *Soil Science,* vol. 23, pp. 343-353, 1927.
[2] The instrument and glass cylinder are obtainable from the Taylor Instrument Company, Rochester, N. Y.
[3] B. A. Keen, Some comments on the hydrometer method for studying soils: *Soil Science,* vol. 26, pp. 261-263, 1928.
[4] A. F. Joseph, The determination of soil colloids: *Soil Science,* vol. 24, pp. 271-274, 1927.
[5] H. Gessner, *op. cit.* (1931), p. 114.
[6] L. B. Olmstead, L. T. Alexander and H. W. Lakin, The determination of clay and colloid in soils by means of a specific gravity balance: *Rep. Am. Soil Survey Assn.,* Bull. 12, pp. 161-166, 1931.

METHODS OF SIZE ANALYSIS

theoretical soundness. Bouyoucos replied to his critics,[1] undertook a series of experiments [2] to demonstrate that the hydrometer method agrees well with other standard methods of analysis, and argued that the method conformed to the principles of Stokes' law.[3]

More recently Bouyoucos developed a more sensitive hydrometer for soils.[4] The instrument has a range from 0-10 g. per liter. It has a large stream-lined bulb and a short stem. Numerous other workers have developed special types of hydrometers in the decade since 1927. Puri,[5] for example, introduced a hydrometer with a short bulb and a long stem. To increase the accuracy of the readings, a pin was mounted on the top of the stem, and its level read with reference to a burette scale mounted above the cylinder.

One of the most thorough studies of the hydrometer method was made by Casagrande,[6] who recognized that a method affording the basic simplicity and convenience of hydrometer readings as compared to other methods justified an attempt to place it upon a firm foundation. Casagrande developed the theory of the hydrometer method in considerable detail, including the influence of hydrometer shape on the results. The several sources of error of the technique were evaluated, including such items as temperature corrections, effects of concentration of the suspension, accuracy of the hydrometer readings, and the like. As a result of his investigations Casagrande developed a hydrometer having the form shown in Figure 70. For routine purposes a glass instrument was used; for precise studies he used a hydrometer having a rust-proof metal stem fitted at the top with a thin horizontal disk. The level of the hydrometer was read by referring the edge of the disk to a scale mounted on an adjoining stand. Casagrande's work showed that the hydrometer method, when used with proper precautions, yields results comparable to those obtained with other precision methods. For such exact work the time element is, however, of the same order of magnitude as with other techniques.

Another detailed study, largely in terms of Bouyoucos's hydrometer, was

---

[1] G. J. Bouyoucos, The hydrometer method for studying soils: *Soil Science*, vol. 25, pp. 365-369, 1928.

[2] G. J. Bouyoucos, The hydrometer method for making a very detailed mechanical analysis of soils: *Soil Science*, vol. 26, pp. 233, 238, 1928. G. J. Bouyoucos, A comparison between pipette and hydrometer methods for making mechanical analyses of soil: *Soil Science*, vol. 38, pp. 335 ff., 1934. G. J. Bouyoucos, Further studies on the hydrometer method for making mechanical analyses of soils and its present status: *Rep. Am. Soil Survey Assn.*, Bull. 13, pp. 126-131, 1932.

[3] G. J. Bouyoucos, Making mechanical analyses of soils in fifteen minutes: *Soil Science*, vol. 25, pp. 473-480, 1928.

[4] G. J. Bouyoucos, A sensitive hydrometer for determining small amounts of clay or colloids in soils: *Soil Science*, vol. 44, pp. 245-246, 1937.

[5] A. N. Puri, A new type of hydrometer for the mechanical analysis of soils: *Soil Science*, vol. 33, pp. 241-248, 1932.

[6] A. Casagrande, *Die Aräometer-Methode zur Bestimmung der Kornverteilung von Böden* (Berlin: Verlag von Julius Springer, 1934).

made by Wintermyer, Willis, and Thoreen[1] of the United States Bureau of Public Roads. A technique was developed in which readings were made with Bouyoucos's hydrometer at intervals up to 1,440 min. (24 hr.). Settling velocities were computed according to Stokes' law, and the percentage of soil remaining in suspension, P, was determined by the equation P = 100 (R/W), where R is the hydrometer reading and W is the weight of material originally dispersed per liter of suspension. Various correction coefficients were involved in the final evaluation of the results. Subsequently Thoreen[2] showed that an ordinary specific gravity hydrometer could be used in place of Bouyoucos's special instrument. Graphic methods of evaluating the correction coefficients were developed by Willis, Robeson, and Johnston,[3] also of the United States Bureau of Public Roads.

The widespread interest in the hydrometer method reflects the need for a simple, rapid, and yet accurate method of mechanical analysis. The authors have not had enough experience with the hydrometer method to evaluate it in terms of general sedimentary analysis, but it appears to be useful for the preliminary study of large groups of samples. Although minor irregularities of the cumulative curve may not be brought out, significant differences between samples may be evaluated; if more complete analyses are desired, the pipette method may be used.

FIG. 70.—Casagrande's hydrometer.

Bouyoucos furnishes a simplified instruction sheet with his instruments.[4] Essentially the technique involves preliminary dispersion, followed by hydrometer readings at 40 sec., 1 hr., and 2 hr. These three readings furnish the data for the following four classes of material: sand (coarser than 0.05 mm. diameter), silt (0.05 to 0.005 mm.), clay (0.005-0.002 mm.), and "fine clay" (finer than 0.002 mm.). Although these limits do not agree with the Wentworth grade limits, they nevertheless furnish several points along an approximate cumulative curve.

The plummet method. A method closely related to the hydrometer method involves the use of a plummet suspended within the suspension, but near the surface. The change in its apparent weight is observed as a

---

[1] A. M. Wintermyer, E. A. Willis and R. C. Thoreen, Procedures for testing soils for the determination of the subgrade soil constants: *Public Roads,* vol. 12, no. 8, 1931.
[2] R. C. Thoreen, Comments on the hydrometer method of mechanical analysis: Mimeographed report, *U. S. Bureau Public Roads,* 1932.
[3] E. A. Willis, F. A. Robeson and C. M. Johnston, Graphical solution of the data furnished by the hydrometer method of analysis: *Public Roads,* vol. 12, no. 8, 1931.
[4] Supplied by the Taylor Instrument Company, Rochester, N. Y.

METHODS OF SIZE ANALYSIS 175

function of the time. The data so obtained yield the change of density at a constant depth as a function of the time, and hence the method may be related to equation (29) of Chapter 5. Odén, however, developed a special equation for the plummet method (equation (32) of Chapter 5). Schurecht [1] used the method in 1921; he suspended a small glass tube, partially filled with mercury, in the suspension. The plummet is attached to an analytical balance, and weighings are made at intervals ranging from a few minutes to a number of days. The general principle of the apparatus is shown in Figure 71. Ries [2] describes the method in some detail and furnishes an outline of the computations to be made.

Other workers who used or investigated the plummet method include Van Niewenberg and Schoutens,[3] von Hahn,[4] and Olmstead, Alexander, and Lakin.[5] Von Hahn used a Mohr specific gravity balance (see Figure 150, Chapter 14, for illustration) and concluded that the method could not be recommended for general application. Olmstead and his associates used a small pear-shaped plummet suspended from a chainomatic balance. Their study was made primarily to find a rapid method, having a convenience equal to the hydrometer, but with the accuracy of the pipette method. The results of the study showed sources of error which prevented the method's recommendation. The paper contains an excellent discussion of the problem involved.

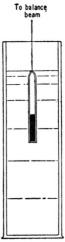

FIG. 71.— Principle of the plummet method, after Schurecht.

Photocell method. In 1934 Richardson [6] applied a photoelectric cell to the problem of determining the size distribution of soils and clays. The method consists essentially of directing a beam of light through a sedimenting system and against a photocell. The photocell is connected to a galvanometer, which indicates the intensity of the beam in terms of current. To obtain a continuous record of the change in light intensity, Richardson used a string galvanometer, the motion of the string being

---

[1] H. C. Schurecht, Sedimentation as a means of classifying the extremely fine clay particles: *Jour. Am. Ceram. Soc.,* vol. 4, pp. 812-821, 1921.
[2] H. Ries, *Clays, Their Occurrence Properties, and Uses* (New York, 1927), pp. 204 ff.
[3] C. J. Van Niewenberg and Wa. Schoutens, A new apparatus for a rapid sedimentation analysis: *Jour. Am. Ceram. Soc.,* vol. 11, pp. 696-705, 1928.
[4] H.-V. von Hahn, *op. cit.,* pp. 296 ff.
[5] L. B. Olmstead, L. T. Alexander and H. W. Lakin, *loc. cit.,* 1931.
[6] E. G. Richardson, An optical method for mechanical analysis of soils, etc.: *Jour. Agric. Sci.,* vol. 24, pp. 457-468, 1934.

recorded on a strip of bromide paper fastened to a revolving drum. The curve obtained decreased rapidly at first and more slowly later on.

Richardson showed that the light extinction as measured by the cell is proportional to Σnd^2, where n is the number of particles of diameter d and Σ is a summation sign. Experiments showed that this relation holds down to diameters of 12 microns (0.012 mm.) at least. In theory, the settling velocity of a particle is, by Stokes' law, $V = Cr^2$. The proportionality found above in terms of light intensity is $I = k\Sigma nd^2$. For a given depth y_1, the light cut off at time t_1 will be due to all particles having velocities less than y_1/t_1, or d^2 less than y_1/Ct_1. That is,

$$I = k \sum_{d^2=0}^{y_1/Ct} nd^2.$$

By constructing a curve of I against t, the slope at any point t will be proportional to the number n of particles for which $d^2 = y_1/Ct_1$. Thus plotting the slope against y (or $1/t$) yields the corresponding frequency curve as n against d^2. In similar manner Richardson showed that a second approach was possible, involving the simultaneous measurement of light extinction over the entire cell at a fixed instant. Of the two methods, the simpler involves measurements at a given depth as a function of the time.

Richardson subsequently [2] improved the apparatus used and developed an ingenious spiral drum which records the log of time, so that the record yields the light intensity as a function of log t directly.

Other indirect methods. In addition to the techniques described in the foregoing section, many other methods of analysis have been described in the literature. Among these methods are several based on the absorption of X-rays, the use of ultracentrifuges, the turbidity of suspensions (Tyndall effect), and the like. Von Hahn[1] discusses these and other techniques in some detail.

Mechanical Analysis under the Microscope

Numerous techniques have been used for the microscopic measurement of particles. Perhaps the most extensively used method with loose grains has been direct measurement with a micrometer eyepiece. In their simplest form such eyepieces are merely a scale engraved on glass at the focal plane of the eyepiece, so that object and scale are simultaneously visible. Figure 72 illustrates the scale; other types of micrometer eye-

---

[1] E. G. Richardson, A photo-electric apparatus for delineating the size frequency curve of clays or dusts: *Jour. Scientific Instruments,* vol. 13, pp. 229-233, 1936. The instrument in its improved form is offered for sale by A. Gallencamp and Company, Finsbury Square, London E. C. 2.

[2] F.-V. von Hahn, *op. cit.,* 1927.

METHODS OF SIZE ANALYSIS 177

pieces are described by Johannsen.[1] In order to use a micrometer eyepiece it is necessary to calibrate it for the microscopic combination being used. This is accomplished by placing an accurately ruled scale on the microscope stage and focusing the micrometer eyepiece on it so that a line of the eyepiece scale coincides with a line on the stage micrometer. The number of divisions of the eyepiece scale which correspond to a given number of divisions of the other furnishes the data for calibration. If, for example, fifty divisions of the eyepiece scale correspond to eleven divisions of the stage micrometer, the relation is $50x = 1.1$ mm., or $x = 0.022$ mm. Thus, one eyepiece division equals 22 microns.

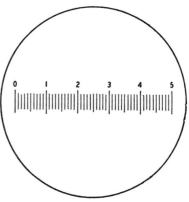

FIG. 72.—Micrometer eyepiece.

To measure individual grains with the micrometer eyepiece, the grain is brought to proper orientation along the scale (a mechanical stage is highly desirable for this), and the chosen lengths are measured. If areas are to be measured, a grid micrometer may be used, which has squares ruled on it. The area of the grain may be estimated by referring it to the size of the smallest enclosing set of scale lines. For the measurement of a number of particles a tally sheet is convenient. The range of sizes present (as represented, say, by long or intermediate diameters) is roughly determined, and a series of size classes is set up covering this range. The number of grains in each class interval is then indicated by tally marks.

Measurement of loose grains. In conducting mechanical analyses by means of a microscope, it is important that the sample used be representative of the material being studied. This is true in all methods of analysis, of course, but inasmuch as relatively small samples are usually mounted on the slide for microscopic measurements, particular care should be used in splitting the sample. The field sample may be split to a smaller sample of 20 or 25 g. by means of a Jones sample splitter (page 45), and this smaller sample may be quartered down by hand or preferably with a microsplit (page 358), to obtain a sample small enough to mount on the slide. No fixed rules can be given for the size of the final sample, inasmuch as individual practice varies from counting

---

[1] A. Johannsen, *Manual of Petrographic Methods,* 2nd ed. (New York, 1918), pp. 287 ff.

a few hundred grains to counting several thousand. One "rule of thumb" that may be used is to count about 300 grains and convert the numbers to percentages by classes; this is followed by a count of an additional hundred or so grains, and percentages are recalculated on the entire number of grains counted. If the percentages remain fairly fixed, the sample is probably adequate; if large differences occur, an additional number is counted until only minor fluctuations remain.

If the grains are to be mounted in the dry state, the final sample may be transferred to a clean microscope slide, and the edges of the slide tapped with a pencil, so that the grains assume their most stable position of rest.[1] This is especially important for measurements of the nominal sectional diameter (page 296), which are defined in terms of the area of the grain section in the plane of the long and intermediate diameters.

The actual operation of measuring and tallying the grains may be based on a count of all the grains in the sample; or of all the grains in several random fields over the sample; or along certain horizontal or vertical lines through the sample as laid off by traverses with the mechanical stage.

The reader is referred to Table 33, and Figure 143 of Chapter 11, for an example of a microscopic size and shape analysis.

The measurement of numerous grains under the microscope is usually a tedious process, and various other techniques have been developed. A field of grains, with eyepiece micrometer in place, may be photographed, and the individual grains measured from the print or an enlargement. Direct drawing of the field may be made with a camera lucida, or, much more conveniently, with a microscopic projection device,[2] which projects an enlarged image of the field. Figure 73 illustrates such an instrument. By focusing the image on drawing paper at some convenient magnification, and drawing outlines of the grains, an entire field may be covered in a short time. The resulting images may be measured directly with a centigrade scale and translated into correct dimensions in terms of the magnification used. For measurements of the nominal sectional diameter a planimeter may be used to determine the area of each grain.

Thin-section analysis. The methods of measurement described here apply equally well to loose grains or thin sections. In the former case the measurements are used directly in tabulating the frequency distribution; if thin sections are used, the observed radii must be corrected in accordance with the theory of thin-section analysis described in Chapter 5. The procedure to be followed involves setting up the frequency table

[1] H. Wadell, Volume, shape, and roundness of quartz particles: *Jour. Geology*, vol. 43, pp. 250-280, 1935.

[2] A very satisfactory device for this purpose is the "Promar Microscopic drawing and projection apparatus" offered by the Clay-Adams Co., New York, N. Y. A less expensive device, called the "Microprojector," is offered by Bausch and Lomb, Rochester, N. Y.

METHODS OF SIZE ANALYSIS

of observed radii (or diameters) and computing the moments [1] of the observed distribution. The observed moments are then corrected, which yields the characteristics of the grain distribution but does not yield the entire distribution.

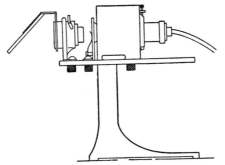

FIG. 73.—Bausch and Lomb's "Microprojector."

Table 17 shows the data of a thin-section mechanical analysis of Palms Mine quartzite, including the computation of the first two moments.

TABLE 17

DISTRIBUTION OF INTERCEPT DIAMETERS IN THIN SECTION OF QUARTZITE FROM PALMS MINE, BESSEMER, MICHIGAN *

Classes (mm.)	Number Frequency (f)	m	fm	m^2	fm^2
0.08–0.16	16	0.12	1.92	0.014	0.224
.16– .24	87	.20	17.40	.040	3.480
.24– .32	155	.28	43.40	.078	12.100
.32– .40	150	.36	54.00	.130	19.500
.40– .48	65	.44	28.60	.192	12.500
.48– .56	32	.52	16.64	.271	8.670
.56– .64	8	.60	4.80	.360	2.880
.64– .72	4	.68	2.72	.463	1.850
0.72–0.80	1	0.76	0.76	0.579	0.579
Total	518	....	170.24	....	61.783

* Computations by slide rule.

[1] The moments of the frequency distribution are described in Chapter 9. In the present type of analysis, arithmetic rather than logarithmic moments should be used, inasmuch as the theory of thin-section analysis is based on the arithmetic moments directly.

In determining the moments of the distribution, the mid-point of each class is entered in the column headed m. The grain frequency f in each class is multiplied by m, and the products entered in the fm column. This column is totaled and divided by the total frequency, here 518. The resulting quotient is the observed first moment, $n_{x1} = 0.329$. For the second moment the mid-point of each class is squared and entered in the m^2 column. The grain frequencies are now multiplied by these squared values, as shown in the fm^2 column. The total is divided by 518 to yield the observed second moment, $n_{x2} = 0.119$. Higher moments may be computed by following a similar process with successively higher powers of m.

Details of computation are summarized below the table; a fuller discussion of arithmetic moments is given in Chapter 9. The diameters used in the study were defined as the maximum horizontal intercepts through the variously oriented grains.[1]

The characteristics of the grain distribution of the quartzite may be found by means of equations (45) and (46) of Chapter 5 (page 132). The observed first moment of the grain sections, n_{x1}, is 0.329 mm. From equation (45), $n_{r1} = 1.27 n_{x1} = 0.418$ mm., the arithmetic mean size of the quartzite grain distribution. The observed second moment, n_{x2}, is 0.119. From equation (46), $n_{r2} = 1.50\, n_{x2} = 0.179$. If the standard deviation (page 219) of the grain distribution is to be used as a measure of spread, it may be obtained from the relation $\sigma^2 = n_2 - (n_1)^2$. In the present case this is $\sigma = \sqrt{0.418 - (0.179)^2} = 0.071$ mm.

COMPARISONS OF METHODS OF MECHANICAL ANALYSIS

The wide choice of methods available for mechanical analysis, especially for fine-grained sediments, has resulted in numerous workers' comparing the relative accuracy and convenience of two or more techniques. Earlier papers [2] compared various decantation methods with

---

[1] W. C. Krumbein, Thin-section mechanical analysis of indurated sediments: *Jour. Geology*, vol. 43, pp. 482-496, 1935.

[2] Among the large number of such papers may be mentioned the following: G. M. Darby, Determination of grit in clays: *Chem. and Met. Engineering*, vol. 32, pp. 688-690, 1925. A. F. Joseph and F. J. Martin, The determination of clay in heavy soil: *Jour. Agric. Sci.*, vol. 11, pp. 293-303, 1921. W. Novak, Zur Methodik des mechanischen Bodenanalyse: *Int. Mitt. für Bodenkunde*, vol. 6, pp. 110-141, 1916. C. W. Parmelee and H. W. Moore, Some notes on the mechanical analysis of clays: *Trans. Am. Ceram. Soc.*, vol. 11, pp. 467-493, 1909. N. Pellegrini, Ueber die physikalisch-chemische Bodenanalyse: *Landwirts. Versuchs-Stat.*, vol. 25, pp. 48-52, 1880. H. Puchner, Ein Versuch zum Vergleich der Resultate verschiedener mechanischer Bodenanalyse: *Landwirts. Versuchs-Stat.*, vol. 56, pp. 141-148, 1902.

rising current elutriation. More recent studies [1] compared the newer methods based on Odén's theory among themselves or compared them with the older routine methods. Many of the studies compared individual grade sizes, others compared the cumulative curves obtained by the several methods, and some included the effects of dispersing agents on the results. It is difficult in all cases to decide upon the most favorable method, owing in part to conflicting results. In addition to differences in methods of dispersion, the personal element enters the study to some extent, inasmuch as familiarity with a method often results in a degree of success not obtained with limited experience. The most complete comparisons were made by committees in connection with soil analysis,[2] who decided on the pipette method as the most suitable for general work.

[1] Among such papers may be mentioned the following: A. H. M. Andreason, *loc. cit.*, 1928. G. J. Bouyoucos, A comparison of the hydrometer method and the pipette method for making mechanical analysis of soil, with new directions: *Jour. Am. Soc. Agron.*, vol. 23, pp. 747-751, 1930. A. Carnes and H. D. Sexton, A comparison of methods of mechanical analysis of soils: *Agric. Engineering*, vol. 13, pp. 15 ff., 1933. C. W. Correns and W. Schott, Vergleichende Untersuchungen über Schlämm- und Aufbereitungsverfahren von Tonen: *Kolloid Zeits.*, vol. 61, pp. 68-80, 1932. M. Köhn, *loc. cit.*, 1928. A. Kuhn, Die Methoden zur Bestimmung der Teilchengrösse: *Kolloid Zeits.*, vol. 37, pp. 365-377, 1925. L. B. Olmstead, L. T. Alexander and H. W. Lakin, *loc. cit.*, 1931. O. Pratje, Die Sedimente des Südatlantischen Ozeans: *Wiss. Ergeb. der Deutsch. Atlantischen Expedition auf dem..."Meteor,"* vol. 3, part 2, Lief. 1, 1935.

[2] See for example, Subcommittee of A. E. A., The mechanical analysis of soils; a report on the present position, and recommendations for a new official method: *Jour. Agric. Sci.*, vol. 16, pp. 123-144, 1926.

CHAPTER 7

GRAPHIC PRESENTATION OF ANALYTICAL DATA

INTRODUCTION

GRAPHIC presentation is one of the first steps in an analysis of the results of any sedimentary study. Not only does a graph present the results visually, but it serves an additional purpose in suggesting new lines of attack.

There are certain principles of graphic presentation, which depend upon conventions of analytic geometry, and which should be followed in order to introduce uniformity into methods of presentation. At the risk of discussing familiar material, some of these elementary principles will be presented here.

GENERAL PRINCIPLES OF GRAPHS

Most graphs involve the plotting of one set of data against another, usually by drawing points on a plane for each pair of observations. It is conventional to choose two coördinate axes at right angles to each other, which are used as axes of reference. The vertical axis is called the y-axis, and the horizontal axis is called the x-axis.

Choice of dependent and independent variables. It is a well established mathematical convention that the *independent variable* shall be plotted along the x-axis, and the *dependent variable* along the y-axis. The independent variable increases or decreases by arbitrarily chosen amounts, and the dependent variable is measured at each of these given values. For example, if the change in heavy mineral content is studied in a linear series of samples, the distance between the samples is arbitrarily chosen, and the mineral content of these samples is then investigated. This procedure defines distance as the independent variable and percentage of heavies as the dependent variable. Similarly, when frequency is plotted against diameter of grain, diameters are arbitrarily fixed (by a choice of sieve meshes, for example), and the frequency of grain on each sieve is then determined. Here diameter is the independent variable, and frequency the dependent variable.

In some instances it is not obvious which is the independent variable, and in such circumstances an arbitrary choice is made, depending upon the emphasis which is to be given to the results. One may compare the organic content of a series of samples with their skewness. These two apparently unrelated sets of data may justifiably be examined either in terms of how the organic content varies as the skewness changes, or how the skewness varies as the organic content changes. In the first case skewness is the independent variable, and in the second organic content is the independent variable.

Choice of scale units. Most graphs are drawn on ordinary arithmetic graph paper, which is divided into squares, with a given number of rulings per inch or per centimeter. In such paper the successive rulings are equally spaced; that is, the actual measured distance between the values 1 and 2 is the same as the distance between 2 and 3, and so on.

A second type of ruling, which is extensively used with sedimentary data, is the logarithmic scale. Here the measured distances between successive units are not equal, but decrease in geometric intervals through

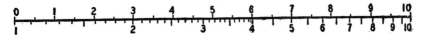

FIG. 74.—Relation between arithmetic scale (top) and logarithmic scale (bottom).

cycles of 10. Each cycle is of equal length, so that the measured distance from 1 to 10 is the same as that from 10 to 100, and so on. The use of logarithmic scales achieves the same result as plotting the logarithms of the original data on ordinary arithmetic graph paper. The logarithmic ruling is used when rates of change are to be compared, or when wide fluctuations in the values of experimental data are to be smoothed. Graph paper may be obtained either with logarithmic scales along both axes, or with a logarithmic scale along the y-axis and arithmetic ruling along the x-axis. The former type of paper is called *double log paper*, and the latter is *semi-log paper*.

The relations between arithmetic and logarithmic scales are shown in Figure 74. In using arithmetic scales the zero point is indicated on the scale, whereas with a logarithmic scale there is no zero point; instead, the cycles extend simply from the highest to the lowest values. When a range of values greater than 10 is involved in logarithmic plotting, multiple cycle paper is used, in which more than a single cycle is included.

Even though original data are plotted directly on a logarithmic scale, it is not correct to consider the result as a graph of the original data. The use of a logarithmic scale changes the variable to its logarithm. If one plots mineral frequencies on a logarithmic scale against distance on an arithmetic scale, the graph shows the relation of log mineral frequency against distance, and not mineral frequency directly. The subject of changing the variables in connection with frequency curves is discussed more fully in Chapter 9.

GRAPHS INVOLVING TWO VARIABLES

It is convenient to distinguish between graphs based on two variables and those based on three or more variables because of the increasing complexity of higher-dimensional figures. Two-dimensional graphs include a wide variety of forms—the frequency of grain diameters, changes of average size with distance, comparisons of size and degree of sorting, comparisons of shapes of grains, mineral compositions, and so on.

HISTOGRAMS

The simplest manner of depicting the results of mechanical analyses is to prepare a histogram[1] of the data. For this purpose the results of the analysis are compiled into a frequency table, which shows the class intervals in millimeters or any other convenient units, and the frequencies of each class or grade, usually as a percentage of the total weight. Diameters in millimeters, their logs, or whatever size-equivalent is used is chosen as the independent variable, and frequency is the dependent variable. In general, the class intervals are laid off along the horizontal x-axis, and above each of the classes a vertical rectangle is drawn, with a width equal to the class interval and a height proportionate to the frequency in the class.

Conventions among sedimentary petrologists have varied widely in plotting histograms. Some writers plot diameters on the vertical axis and frequency on the horizontal axis, but as long as the size is more conveniently chosen as the independent variable, it is preferable to standardize the procedure by plotting size always on the horizontal axis. Another common convention for sediments is to plot the size scale such that values of x (the diameter) increase to the left. This results in a reversed scale of values in terms of conventional mathematical practice,

---

[1] Some writers in sedimentary petrology have used the term *frequency pyramid* instead of the word *histogram*. The latter word is, however, a statistical term of common usage, accepted universally by statisticians, and its use will be retained here.

but despite that there are numerous arguments in favor of such a reversed scale for sedimentary data.[1] Merely for convenience of comparison, it is perhaps desirable that the reversed-scale convention be adopted as a standard practice.

Two general types of histograms are used. In one, the diameters are laid off directly on an arithmetic scale, with the result that the successive vertical rectangles decrease in width as the class intervals become smaller. In the other case, the class limits are drawn on a logarithmic scale, either directly or by implication, so that each vertical bar is equal in width, regardless of the original difference in absolute class interval.[2] In preparing the first type of histogram certain precautions must be followed, because it is necessary to preserve the area under the curve as a constant, equal to 100 per cent of the frequency.

In order to preserve the area under the curve when a histogram is drawn on the basis of class intervals in millimeters, each histogram block must be

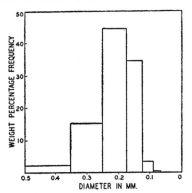
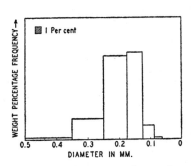

FIG. 75.—Incorrect method of drawing histogram with *diameter in mm.* as independent variable.

FIG. 76.—Histogram of same data as in Figure 75, showing representation by areas.

drawn in terms of area rather than height. Areas involve both width and height, and hence for a given frequency percentage the height of the block will depend on the class interval. Figures 75 and 76 illustrate the wrong and right way to draw a histogram on this basis. It will be noted that the use

[1] The data of most mechanical analyses are obtained in order from coarsest to finest, and certain special sedimentation curves, as Odén curves (page 113), are obtained automatically in the reversed sense. Choice of direction of a scale is perfectly arbitrary; in fact, in astronomy stellar magnitudes are expressed on such a reversed scale, as is hydrogen-ion concentration in chemistry. The practice of reversing scales for convenience is therefore no radical departure from accepted practices.

[2] In the latter case it is assumed that the Wentworth or Atterberg scale, or some other true geometric grade scale, is used in preparing the logarithmic graph.

of a single scale along the vertical axis results in a histogram with apparently much of the material in the larger classes. The correct diagram on the right is drawn on an areal basis, as all histograms should be, with the small square representing 1 per cent. This histogram is quite noticeably different from its neighbor.

It was noted years ago that if each grade is indicated as of equal width on the horizontal axis, the histogram becomes much more symmetrical. Because of this, presumably, it has become customary for sedimentary petrologists to draw their histograms with each rectangle equal in width, so that the sediment shown in Figure 76 appears like that in Figure 77. Actually this procedure does not show diameters directly, because the scale has been transformed to a logarithmic scale, whether that fact is so indicated on the figure or not. It is important that the practical worker be aware of the differences involved in plotting data on arithmetic and logarithmic scales, because any statistical devices that may be used to describe the sediment are strikingly affected by the change of independent variable from diameter to log diameter.

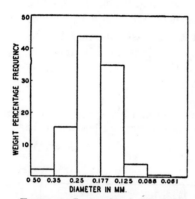

FIG. 77.—Conventional histogram of same data as in Figure 76, plotted on an implied logarithmic scale, with class intervals shown as equal. (Data from Pettijohn and Ridge, 1932.)

That the intervals become equal when logs are used is easily shown. Thus, $\log_{10} 2 = 0.301$; $\log_{10} 1 = 0.000$; $\log_{10}(\frac{1}{2}) = -0.301$. Here each class is equal in width (interval = 0.301), with the origin at 1 mm. The logs of numbers smaller than 1 are negative, and, if the base 10 is used, the class limits are not marked by integers. This suggests that logs be taken to such a base that the class limits become integers and, for the sake of convention, so that negative numbers may extend to the left instead of to the right. These ends are accomplished by taking negative logs to the base 2 of the diameter values, whereupon the above class limits become $-\log_2 2 = -1$; $-\log_2 1 = 0$; $-\log_2(\frac{1}{2}) = +1$, and the scale is transformed to an arithmetic scale with equal units.[1]

Frequency polygons. In addition to histograms as frequency diagrams, a common statistical device is to indicate variations in frequency by

[1] This type of transformation substitutes a new variable for diameters in millimeters. This concept is discussed in Chapter 4, as the ϕ scale, and is treated more fully in Chapter 9. In an analogous manner, a ζ scale may be derived for the Atterberg grades.

means of a line diagram instead of with rectangular blocks. Such frequency diagrams are called *frequency polygons,* and they are prepared by plotting the frequency corresponding to a given grade size on a line midway between the grade limits. The resulting points are then connected with a continuous line, made of straight line segments, as shown in Figure 78. The continuous line is brought down to the zero point at the centers of the grades just larger and smaller than the limiting grades in the analysis. Frequency polygons are recommended by some statisticians [1] as a device to be used when the data vary continuously, to avoid the implication that each grade is an individual entity. Frequency polygons have not been used widely by sedimentary petrologists,[2] but they may occasionally express the frequency more clearly than histograms, and hence may be considered for use. Either an ordinary arithmetic scale or a logarithmic scale may be used, as in the case of histograms.

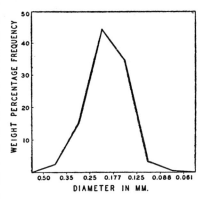

FIG. 78.—Frequency polygon of same data as in Fig. 77.

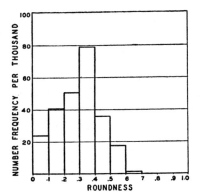

FIG. 79.—Histogram of roundness variation of quartz particles. (After Wadell, 1935.)

The use of histograms for attributes other than size. Although histograms have been discussed from the point of view of size characteristics, it is possible to plot many other attributes of sediments in that manner. A histogram is essentially a statistical device used to represent frequency. Hence any frequency attribute may be so expressed. Mineral frequencies, shape frequencies, surface texture frequencies, and others are included here. In general, the same principles of construction apply, and care

---

[1] F. C. Mills, *Statistical Methods* (New York, 1924), pp. 79-81.
[2] Miss Gripenberg has used a combination histogram and frequency polygon in expressing the composition of sediments. See Stina Gripenberg, A study of the sediments of the North Baltic and adjoining seas: *Fennia,* vol. 60, no. 3, pp. 191 ff.

must be exercised to represent areas correctly, as previously pointed out. Figure 79 is a histogram of the distribution of roundness of quartz grains in the 0.001-0.01 cu. mm. volume class, after Wadell,[1] to illustrate the type of diagram involved.

Cumulative Curves

The cumulative frequency curve is a curve based on the original histogram data, and is made by plotting ordinates which represent the total amount of material larger or smaller than a given diameter. Two types of cumulative curves are possible, the "more than" curve and the "less than" curve. It is immaterial which is used, inasmuch as either furnishes the same type of information. The commoner type in sedimentary data is perhaps the "more than" type. It is made by choosing a size scale along the horizontal axis, and a frequency scale from 0 to 100 per cent along the vertical axis. The horizontal scale may be either arithmetic or logarithmic, as in the case of the histograms. In either case the procedure is the same, and it is not necessary to consider the areas in drawing the curve.

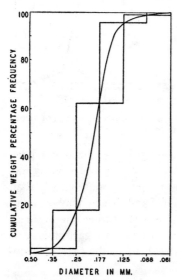

Fig. 80.—Diagram showing relation between histogram and cumulative curve. Data as in Fig. 77.

At the upper limit of the first class interval an ordinate is erected equal in height to the frequency in that class. At the end of the next class another ordinate is drawn, equal in height to the sum of the frequencies in the first two classes, and so on. In short, the cumulative curve is equivalent to setting one histogram block above and to the right of its predecessor, so that the base of each block is the total height of all preceding blocks. Strictly speaking, this would yield a step diagram, as shown in Figure 80. However, it is common practice to draw only points at the upper limit of each ordinate, and to connect the points with short line segments. It is also customary to draw a smoothed curve through the points, to obtain a continuous curve representing the continuous distribution of sizes. Cumulative curves have come into wide use in sedimentary work

[1] H. Wadell, Volume, shape, and roundness of quartz particles: *Jour. Geology*, vol. 43, pp. 250-279, 1935.

GRAPHIC PRESENTATION 189

because of the convenience with which statistical values are drawn from them (see Chapter 9).

A type of graph paper of considerable value in analyzing cumulative curves is logarithmic probability paper,[1] which has a logarithmic scale along one axis and a probability scale along the other. The probability scale is so designed that a symmetrical cumulative curve will plot as a straight line on the graph. Many sands show straight lines on this paper, and it affords an excellent method of comparing sedimentary data. A further use of the paper is to study the effect of using combined sieving and sedimentation methods on

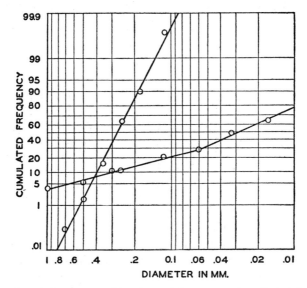

FIG. 81.—Cumulative curves of beach sand (steep curve) and glacial till (gentle curve) drawn on logarithmic probability paper. The till shows a "break" at 1/16 mm., due to change from sieving to sedimentation method of analysis.

the same sample. It was mentioned on page 136 that there often is a hiatus between the portions sieved and sedimented, and such an hiatus will appear on the probability paper as a change in the slope of the line. The paper is also of much use in the statistical study of samples, inasmuch as from it the average size and degree of spread may be directly read for sediments which plot as straight lines. This latter point is further discussed in Chapter 9. Figure 81 shows two samples plotted on probability paper, one of which is a straight line and the other shows an abrupt change at 1/16 mm. When the line is curved, the sediment does not have a symmetrical cumulative curve.

[1] Probability paper, designed by Hazen, Whipple, and Fuller, may be obtained from the Codex Book Co., New York, N. Y.

FREQUENCY CURVES

Frequency curves are smooth curves which show the variation of the dependent variable as a continuous function of the independent variable. Histograms, cumulative curves, and frequency curves are related mathematically, and any one may be obtained from any other. The relation between histograms and frequency curves is more direct than between these two and cumulative curves, but in the construction of frequency curves from histogram data the cumulative curve may be most conveniently used. The relation between the histogram and the

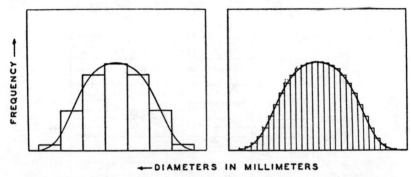

FIG. 82.—Diagram showing transition from histogram to frequency curve.

frequency curve is that the latter represents the limit of a histogram as the class intervals become smaller and smaller and finally reach zero, while the frequency increases without bound. Figure 82 illustrates the transition from one to the other, and in fact a common manner of drawing frequency curves is simply to superimpose a smoothed curve over the histogram bars. This procedure is not always accurate, however, owing to the relatively large classes used in sedimentary work, and because of an unfortunate variation in the histograms of the same sediment when different grade sizes are used in the analysis.

There is a unique frequency curve which may be obtained fairly satisfactorily from the cumulative curve by a graphic method. The smoothed cumulative curve is a continuous curve, and from it by graphic differentiation may be obtained the smooth frequency curve, independent of the particular grade sizes used in the analysis.

GRAPHIC DIFFERENTIATION OF CUMULATIVE CURVES

There are several methods by which an approximation to the unique frequency curve may be obtained from sieve data. One such is obtained

GRAPHIC PRESENTATION

by drawing a smooth curve over the histogram itself, if care be exercised to preserve areas in the smoothing process. Methods for the numerical or semi-graphic computation of frequency curve ordinates have been described by several writers.[1] However, since the frequency curve is the derivative of the cumulative curve (see page 215), the usual method of graphic differentiation may be used,[2] but cognizance should be taken of the fact that the x-axis is logarithmic when the cumulative curve is drawn on the basis of equal intervals for each grade.

As an illustration, data based on Wentworth class intervals will be used, so that the final frequency curve will be obtained in terms of that descriptive grade scale. For this purpose it is convenient to use 3-cycle semi-logarithmic paper (Eugene Dietzgen #340-L 310), in which the length of a cycle is about 8.1 cm., so that when the Wentworth scale is laid off along the x-axis, the actual distance between successive points is about 2.4 cm. Also by convention, 10 per cent on the vertical scale is chosen so that it is about half the length of the horizontal scale unit, or about 1.2 cm. This assures a cumulative curve which is usually not too steep for convenient handling. By adopting these conventions the final curves are directly comparable, because they are all obtained in reference to fixed scale relations.

When the variables are plotted on arithmetic scales a pole p is usually drawn to the left of the vertical axis at a distance equal to a unit along the x-axis. By this means the ordinates of both the cumulative curve and frequency curve may be read from the same numerical units of the y-axis. In the present case a logarithmic x-scale is used, but fortunately the same relations hold as in the arithmetic case, and the pole p is drawn to the left of the vertical axis a distance equal to the arithmetic value (actual length) of the Wentworth units along the x-axis, here 2.4 cm. Since in such logarithmic scales the geometric ratio between successive points yields equal arithmetic intervals, the length of the interval determines the pole distance.

The cumulative curve is divided into any convenient number of units, as shown in part in Figure 83. At each of the points an ordinate is erected to the curve as at x_1A, x_2B. These ordinates need not necessarily coincide with the experimentally determined points of the cumulative

---

[1] S. Odén, On clays as disperse systems: *Trans. Faraday Soc.*, vol. 17, pp. 327-348, 1921-22. D. S. Jennings, M. D. Thomas and W. Gardner, A new method of mechanical analysis of soils: *Soil Science*, vol. 14, pp. 485-499, 1922. C. E. Van Orstrand, Note on the representation of the distribution of grains in sands: *Researches in Sedimentation in 1924*, Nat. Research Council, pp. 63-67, 1925.

[2] H. Von Sanden, *Practical Mathematical Analysis*, translated by H. Levy (New York, 1924), Chap. VII.

curve. A tangent is now drawn to the cumulative curve at A. The angle α, made by this tangent with the horizontal, is then laid off at P, and the line PQ is drawn. From Q, where the ray of the angle intersects the y-axis, a horizontal line is drawn to the ordinate x_1A or its projection. The intersection of the horizontal line from Q with this ordinate locates a point on the desired frequency curve. The process is repeated at x_2B, where the tangent at B makes an angle β with the horizontal. When this angle is laid off at P, it yields the point R on the y-axis, and finally

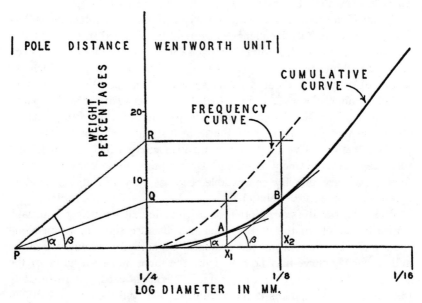

Fig. 83.—Graphic differentiation of cumulative curve. See text for explanation.

the second point along the frequency curve. Further repetition yields as many points on the frequency curve as the number of originally chosen points along the cumulative curve. In general, these arbitrarily chosen points of tangency may not include the inflection point [1] of the cumulative curve. Hence a separate determination should be made at that point because the mode of the frequency curve lies on the ordinate of the inflection point.

Proof that the method of graphic differentiation applies in the case of a logarithmic x-scale is not given in the references cited, but it may

---

[1] The inflection point of a curve is the point of maximum slope of the tangent. It is the point where the tangent to the curve changes its direction of rotation. It may accordingly be located by moving a ruler along the curve so that it is always tangent to it, until the point is reached where the direction of rotation of the ruler reverses itself.

be shown that the method satisfies the requirements of mathematical rigor:[1]

In the conventions adopted in the illustration, it is possible to plot directly the values of the experimentally determined points of the cumulative curve in terms of their logs to the base 10. At the same time, the Wentworth units determine the areal relations of the resulting frequency curve by fixing the position of the pole P. For plotting purposes we may lay off the logs of the Wentworth scale to the base 10, in which case the distance between successive points will be $k \log_{10} 2$, where k is the length of a logarithmic cycle on the graph paper. If the convention of decreasing the values of the diameters to the right is followed, the choice for plotting is $x = - k \log_{10} \xi$, where x is the actual distance along the x-axis from the arbitrary origin $\log_{10} 1 = 0$, k, is the length of a cycle, and ξ is the numerical value of the diameter.

The transformation $\log_{10} \xi = \log_{10} 2 \log_2 \xi$ may be used to locate the pole P in terms of the Wentworth scale. Substitution in the equation for x yields $x = - k \log_{10} 2 \log_2 \xi$. Call the unit along the x-axis λ_x, and set $k \log_{10} 2 = \lambda_w$. Now let $\phi = - \log_2 \xi$, and by substitution we have

$$x = \lambda_x \phi \quad \quad \quad \quad \quad \quad (1)$$

One may next choose y and y', the actual heights of the ordinates of the cumulative and frequency curves respectively, such that

$$y = \lambda_y \eta \quad \quad \quad \quad \quad \quad (2)$$
$$y' = \lambda_y \psi \quad \quad \quad \quad \quad \quad (3)$$

where η is a function of ϕ, and λ_y is the unit along the y-axis, chosen equal in both cases so that the same numerical unit applies to the ordinates of each curve.

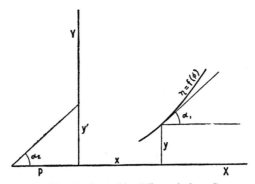

Fig. 84.—Proof of graphic differentiation. See text.

From these parametric equations, allow $\eta = f(\phi)$ to be the cumulative curve, so that $\psi = d\eta/d\phi$ will be the corresponding frequency curve. In Figure 84 a tangent is drawn to the cumulative curve at y, making an angle α_1

[1] W. C. Krumbein, Size frequency distributions of sediments: *Jour. Sed. Petrology*, vol. 4, pp. 65-77, 1934.

with the horizontal. By the calculus, $\tan a_1 = dy/dx$, and from equation (2),

$$\tan a_1 = dy/dx = \lambda_y \, d\eta/dx = \lambda_y \frac{d\eta}{d\phi} \frac{d\phi}{dx} \quad \ldots \quad (4)$$

But $d\eta/d\phi = \psi$, and from equation (1), $dx/d\phi = \lambda_x$, so that $d\phi/dx = 1/\lambda_x$. Substituting in equation (4),

$$\tan a_1 = \frac{\lambda_y \psi}{\lambda_x} \quad \ldots \ldots \ldots \ldots \ldots \quad (5)$$

Also from Figure 84, $\tan a_2 = y'/p$, where p is the pole distance. From equation (3), $y' = \lambda_y \psi$, and by substitution

$$\tan a_2 = \frac{\lambda_y \psi}{p} \quad \ldots \ldots \ldots \ldots \ldots \quad (6)$$

But by construction in the graphic method, $a_1 = a_2$. Hence (5) and (6) may be equated:

$$\frac{\lambda_y \psi}{p} = \frac{\lambda_y \psi}{\lambda_x}$$

from which there results

$$p = \lambda_x$$

and the method applies.

In choosing intervals along the cumulative curve, it is well to include at least ten. When the form of the frequency curve develops, additional points may be chosen to bring out needed details. Both curves represent continuous functions, and consequently one is not limited to specific points along either curve. The possible errors introduced by the smoothing process are discussed below.

Figure 85 is inserted to illustrate several types of frequency curves obtained by the method described, and it shows as well the corresponding cumulative curves. Certain points may be noted. In curve A, which represents a beach sand, the highest ordinate of the frequency curve extends above the 100 per cent line of the cumulative curve. This means that the steepness of the cumulative curve at that point is so great that the rate of change of percentage per unit of the grade scale here is 105 per cent. In other words, the grains are very highly concentrated about this modal value.

Curve B of Figure 85 represents a silt (loess) in which the cumulative curve is less steeply inclined, so that the mode of the frequency curve is much less accentuated than in the sand.

These examples focus interest on another point, which is the accuracy of the graphic method. It may confidently be stated that if the curvature of the cumulative curve is known at every point, and if the tangents are correctly drawn, the method yields rigorous results. In experimental work, however, interpolation of the cumulative curve is necessary, and

GRAPHIC PRESENTATION

to the extent that the smoothing introduces errors, the final results are inaccurate. Also the tangents to the curve cannot be drawn correctly by inspection, and another error is introduced by this fact. There are devices for reading the tangents to curves,[1] and hence this error may be made practically negligible. The correction of the error involved in smoothing the cumulative curve requires some discussion.

As stated earlier, cumulative curves are independent of the grade scale used in the analysis, but in smoothing such curves it is obvious that the precise locations of the known points will influence the smoothing process, so that the cumulative curves obtained from different grade scales may not be completely identical, especially if too few points are known. Since both cumulative curves and frequency curves of sediments are continuous functions of the diameters, the most logical way of avoiding these errors of smoothing is to determine by experiment as many points as possible along the cumulative curve. It should be borne in mind that uneven class intervals do not distort the cumulative curve, and consequently as many points may be experimentally determined as one wishes, regardless

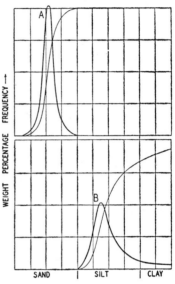

Fig. 85.—Examples of graphic differentiation of beach sand (A) and loess (B).

of the particular intervals between successive points. It is not convenient, however, to construct a histogram from such unequally spaced points, unless the points have a fixed ratio to each other. Thus in using the analytic data to construct a cumulative curve, one is independent of the need of a fixed scale for analysis.

SEDIMENTATION CURVES

A type of graphic presentation commonly used in connection with such methods of analysis as the Odén balance (Chapter 6) is included here to complete the classification of curves representing the size attributes of sediments, although the details of the construction of such

---

[1] The Richards-Roope tangent meter, offered by Bausch and Lomb, is such a device.

curves has been given in Chapter 5. The Odén curve is related mathematically to the cumulative curve, and in turn to the frequency curve, despite the fact that time rather than size is the original independent variable in the Odén curves. Other sedimentation curves include the Wiegner curve, also described elsewhere (Chapter 6).

Graphs with Distance or Time as Independent Variable

Distance. Among graphic devices much used in the study of sediments are graphs of the linear variation in the average size of sediments along a line of samples, variations in thickness with distance, and in short, the linear variation of any measurable attribute as a function of distance along a formation, stream, beach, or the like. In all these cases distance is used as the independent variable, and is plotted along the horizontal axis. With very few exceptions, the distance scale is chosen as arithmetic, and if a logarithmic function is sought, the logarithmic scale is generally used along the y-axis. An exception to this generalization is Baker's use of a logarithmic scale for distance in the study of sands in the London Basin.[1] If a general rule may be adduced regarding the choice of scale for the independent variable, it is perhaps that of convenience. With regard to mathematical convenience, it may be mentioned that when parabolic or hyperbolic functions are suspected, double log paper may be used to test the relation, as explained in a later portion of this chapter.

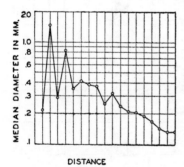

Fig. 86.—Size variation diagram. The samples are spaced at half-mile intervals. (Data from Pettijohn and Ridge, 1932.)

As illustrations of graphs using distance as the independent variable, Figures 86 and 87, adapted from originals[2] in the literature, indicate types of data commonly presented.

There is no general term applied to graphs of the nature just discussed, but various writers have coined terms to describe them. Pettijohn and Ridge[3] used the term "Size Variation Series" to describe the rela-

---

[1] H. A. Baker, On the investigation of loose arenaceous sediments by the method of elutriation, etc.: *Geol. Mag.*, vol. 57, pp. 321-332, 363-370, 411-420, 463-470, 1920.

[2] F. J. Pettijohn and J. D. Ridge, A mineral variation series of beach sands from Cedar Point, Ohio: *Jour. Sed. Petrology*, vol. 3, pp. 92-94, 1933. C. Burri, Sedimentpetrographische Untersuchungen an Alpinen Flussanden: *Schweiz. Min. u. Pet. Mitt.*, vol. 9, 205-240, 1929.

[3] F. J. Pettijohn and J. D. Ridge, *loc. cit.*, 1933.

tions of size to distance, and Krumbein[1] used the term "Median Procession Curve" to describe the variation of average grain size (the median) with distance.

Time. The direct use of time as an independent variable in sedimentary studies has not found wide application, perhaps because few studies have been made of the same phenomenon over any appreciable interval. Related studies by engineers, however, such as measurements of the flow of streams as a function of time, or measurements of the silt content of streams as a function of time, are not uncommon. Likewise, in the studies of cyclical sedimentary phenomena, such as varved clays, where the length of each cycle is uniform (a year, say), time

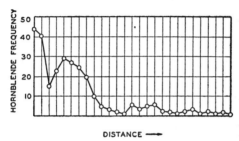

FIG. 87.—Mineral variation diagram. (Data from Burri, 1929.)

may be used instead of distance along the horizontal axis, because each unit along the x-axis is an equal unit of time. Likewise it is sometimes convenient to replace thickness by time if the rate of deposition of a sediment can be shown to have been constant.

For ordinary graphical purposes time is usually plotted on an arithmetic scale rather than on a logarithmic scale, with the exception perhaps of studies involving possible parabolic or hyperbolic functions, in which mathematical convenience is involved. The dependent variable may be plotted either on arithmetic or logarithmic intervals.

The investigation of time series and trends is itself a major field in the study of economic data, and most textbooks of statistics devote considerable space to the analysis of graphs with time as independent variable. Inasmuch as the methods used in conventional statistics are identical with those used in the analysis of such data in sedimentary studies, the reader is referred to standard texts[2] for details beyond the relatively few included here.

[1] W. C. Krumbein, Textural and lithological variations in glacial till: *Jour. Geology*, vol. 41, pp. 382-408, 1933.
[2] F. C. Mills, *op. cit.* (1924), Chap. 7.

The analysis of time series usually involves one or more of four elements: the long-time trend of the variate, such as the increase or decrease in average size over a long period of time; seasonal variations, such as the increase or decrease of sediment carried by a stream during times of high and low water; cyclical movements, such as recurrent increases or decreases of some variable (thickness, size) during regular time intervals; and finally accidental variations introduced by any of a number of random causes.

TABLE 18

COMPUTATION OF THREE-YEAR MOVING AVERAGE OF THICKNESS OF VARVED SLATE *

Arbitrary Year	Varve Thickness (mm.)	3-year Totals	3-year Averages
1	28	..	....
2	4	37	12.3
3	5	26	8.7
4	17	50	16.7
5	28	59	19.6
6	14	48	16.0
7	6	44	14.6
8	24	81	27.0
9	51	98	32.6
10	23	102	34.0
11	28	77	25.6
12	26	78	26.0
13	24	88	29.3
14	38	69	23.0
15	7	93	31.0
16	48	..	....

* Data from F. J. Pettijohn. The varves are from Lake Minnitaki, Ontario. See his paper, Early Pre-Cambrian varved slate in northwestern Ontario: *Geol. Soc. Am., Bull.*, vol. 47, pp. 621-628, 1936.

It is to be expected that variations will occur in the measured values through successive intervals of time, and the essential attack on time series problems involves the smoothing of these irregularities to disclose the underlying trend. A simple method commonly used for studying time trends is the "moving average" method, which involves simply the taking of averages over the data arranged chronologically, in such a manner that each average involves three, five, or more successive values, and the succeeding averages are computed by dropping the first item

GRAPHIC PRESENTATION

of each group, and adding the next item from the table. In this manner a series of average values is obtained, each average representing a group of observations, and thus disclosing the underlying trend.

The method of computing moving averages is illustrated in Table 18. The data represent the thickness of a series of Pre-Cambrian varved slates, referred chronologically to an arbitrary succession of yearly intervals. The second column shows the observed thicknesses of the varves, and the third column indicates the 3-yr. totals. The first figure is found by adding the thicknesses of years 1, 2, 3; the second figure is the sum of thicknesses for years 2, 3, 4, and so on. The averages for the 3-yr. periods are given in the last column. Figure 88 includes the individual varves and the moving average, to indicate how individual irregularities are smoothed by the average.

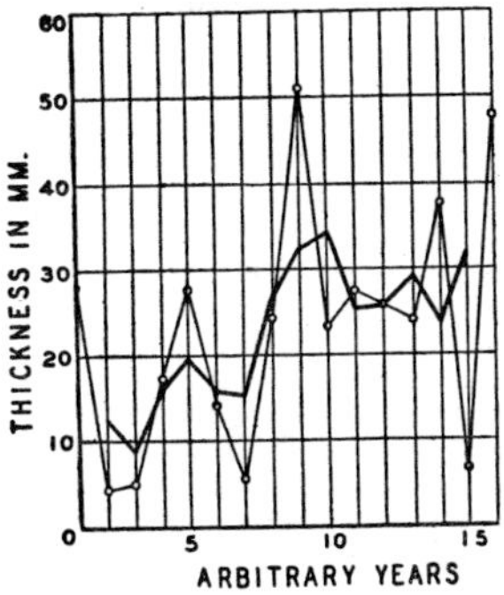

FIG. 88.—Three-year moving average of varve thickness. The average is indicated by the heavy line.

SCATTER DIAGRAMS

A scatter diagram is any graph in which the values of one variable are plotted against another. In the broadest sense, therefore, scatter diagrams include all graphs involving two variables. As the term is used here, scatter diagrams include those graphs in which any two independent sets of data taken from the sedimentary study are plotted against each other to learn whether there may be any relation between them. For example, a study of sediments may include an investigation of the average "degree of sorting" of the sediment, and an analysis of the carbon content. Each of these sets of data is obtained from an independent laboratory investigation: the problem is whether there is any relation between the two characteristics.

The choice of independent variable often presents a problem in the preparation of scatter diagrams. This is not true when size, distance, or time are used, which are usually chosen as independent variables by convention. In the example given, however, the choice of independent variable may be entirely arbitrary (page 183).

In scatter diagrams there is no fixed rule regarding the choice of scales for the two axes. Either or both may be arithmetic or logarithmic; the essential problem is to find the simplest relation between the variables, if any relation exists. Scatter diagrams are usually made as a preliminary to a statistical study of the correlation between the variables.

Methods of computing the correlation coefficient are given in Chapter 9. It is not necessary in every case to compute the correlation coefficient, because inspection of the graph often indicates whether some relation exists, although it may not indicate the exact nature of the relation. It is usually safe to conclude that there is no fixed relation between the variables if the points scatter widely over the field.

Figure 89 is included here as an example of a scatter diagram. It shows the relation between the geometric mean size and the average degree of roundness of samples of beach pebbles. The data in this scatter diagram are used in Chapter 9 as an example for computing the correlation coefficient.

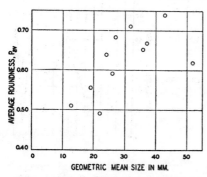

Fig. 89.—Scatter diagram of average roundness and geometric mean size of beach pebbles.

GRAPHS INVOLVING THREE OR MORE VARIABLES

Strictly speaking, each variable involved in a graph represents a dimension, and the problem involved in plotting three or more variables becomes one of indicating more than two dimensions on a sheet of paper. There are numerous devices for accomplishing this. The most familiar example, perhaps, is a contour map, in which three variables are involved, two of them (length and breadth) being considered as independent variables, and the third (elevation) being the dependent variable. The contour map is itself an illustration of a broader type of diagram, the three-variable surface, and the discussion may appropriately begin wtih it.

Three-variable surfaces. Any three variables may be plotted against each other along three axes, one of which is vertical, and the other two horizontal but at right angles to each other. One of the variables may be independent and the other two dependent, or two may be independent and the third dependent, depending upon the particular situation. In sedimentary studies perhaps the most common application of surfaces is to areal studies, in which length and breadth are the independent variables, and the other variate (size, mineral content, organic content, or any other measurable attribute) is the dependent variable. The usual procedure in such cases is to plot distances along the x- and y-axes,

GRAPHIC PRESENTATION

and the dependent variable along the vertical z-axis. This most general manner of plotting the points must be modified for use on graph paper, which has no vertical dimension. Isometric projection paper, which has one vertical axis and two diagonal axes at 120° to each other, serves the purpose. Such figures, illustrated in Figure 90, are not convenient for immediate visualization, especially if the points are to be connected by a series of lines of equal value. For the latter purpose the plane of the paper is taken as the x-y plane, and the z-axis is implied by contour lines which show the configuration of the surface at every point as a projection. The general term applied to such contoured surfaces is *isopleth maps,* which may include any variable as the surface form.

Isopleth maps. An isopleth may be defined as a line of equal abundance or magnitude. An isopleth map, therefore, shows the areal distribution of a variable in terms of lines of equal magnitude. A common example is an isopleth map of average size of sediment.[1] Figure 91 is such an illustration. The sampling points along a beach are indicated by dots, and the average size of each sample is indicated at the appropriate sampling site. Isopleths are drawn through points of equal magnitude, or are interpolated between the sampling points. The result is a surface which expresses simply and clearly the areal variation of average size.

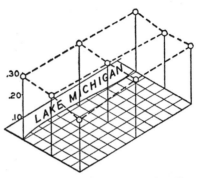

FIG. 90.—Isometric surface of median grain size of beach sand. The vertical lines represent the median diameters. The grid spaces are 10 ft. on edge.

Isopleth maps may be prepared with any variable which shows a continuous gradation of value. Practically all sedimentary data, as far as present researches extend, are continuous, and the areal variation of such items as average size, degree of sorting, shape of particle, heavy mineral content, organic content, porosity, and the like may be conveniently represented in this manner. The increasing use of areal sets of samples suggests that sedimentary petrologists may make more extensive use of such maps in the future. Not only are isopleth maps useful for depicting the areal variation of sedimentary characteristics,

[1] Trask (*Econ. Geology,* vol. 25, pp. 581-599, 1930) suggested the term *median map* for areal representations of the median grain size. Shepard and Cohee (*Geol. Soc. Am. Bull.,* vol. 47, pp. 441-458, 1936) substituted the term *iso-megathy map* for such areal representations of the median.

but they may be used for comparative purposes. By drawing the maps on translucent paper, one map may be superimposed on the other and areal relations sought. Likewise it is possible to prepare ratio isopleth maps, in which the ratio of size to sorting, say, for each sample is computed and the results are plotted as a map. In such cases irregularities in the surface may furnish clues to changed conditions, or to sampling errors, and thus suggest areas for more detailed study.

In the construction of isopleth maps it is conventional to use arithmetic scales throughout. This is not necessary, of course, when isometric paper is used to plot a perspective view of the data, but the use of such

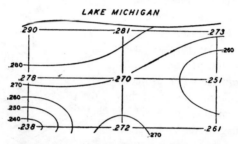

Fig. 91.—Portion of a median map of beach sand. Data are the same as in Figure 90. The contour interval is 0.01 mm.

involved graphs may likely find its widest use in future studies of the functional relations among the variables. It is possible to use a logarithmic scale for the dependent variable by means of an indirect device. That is, the logarithms of the variable may be plotted at the sampling points, and lines of equal logarithmic magnitude drawn through them. This device was used in a study of beach pebbles by Krumbein and Griffith,[1] who plotted the logarithms of average size on an isopleth map.

Triangle diagrams. An effective device for comparing three variables is the use of triangle graph paper. An illustration may serve as the simplest manner of indicating the use of such paper for mechanical analysis data. The analytical data are arranged into three groups for each sample, percentages respectively of sand, silt, and clay. The result is three numbers for each sample, and these numbers are used for plotting. The three vertices of the triangle are labeled, one for each of the three variables. Assume that a sample has the following composition: 26 per cent sand, 43 per cent silt, and 31 per cent clay, a total of 100 per cent. To plot this, one locates the point which lies 26 units upward

[1] W. C. Krumbein and J. S. Griffith, Beach environment at Little Sister Bay, Wisconsin: *Geol. Soc. Am., Bull.*, vol. 49, pp. 629-652, 1938.

along a vertical axis, 43 units along the axis joining the silt vertex and the opposite side of the triangle, and 31 units along the third axis. The result is a single point as shown in Figure 92.

Triangle diagrams have found wide use in sedimentary studies, not only for plotting size attributes, but also for mineral attributes. The heavy minerals in sediments may be classified according to their ultimate origin from igneous, sedimentary, and metamorphic rocks. A triangle plot of the result will indicate the relative contributions of each. In using triangle paper, the three values must be expressed as percentages totaling 100 per cent, because single points will only result when parts per hundred are plotted.

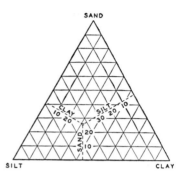

Fig. 92.—Triangle diagram, showing method of locating point.

Another common use of triangle diagrams is to subdivide the field into classes for descriptive purposes. Gessner[1] used such graphs to classify soils into groups, such as sand, silt, sandy clay, and the like. Figure 93 illustrates the method of subdivision used. Each vertex is chosen in terms of a primary constituent, and the field subdivided into groups. It may be seen that such devices offer logical methods for describing sediments in terms of fixed percentages of material, and in a manner such that the relation of any group to the others is immediately apparent.

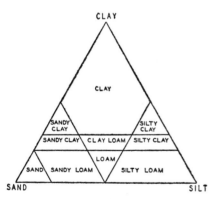

Fig. 93.—Triangular field showing classification of soils. After Lakin and Shaw, 1936.

It is also possible to use triangle graphs to plot four variables. For that purpose the four variables are recalculated to 100 per cent, and three of them are plotted as before. The result will be not a point in the field, but an area enclosed by three lines, each perpendicular to the respective axes. These three lines form a small triangle, the size of which indicates the amount of the fourth variable. Figure 94

[1] H. Gessner, *Die Schlämmanalyse* (Leipzig, 1931), p. 217.

illustrates the plotting of a sediment composed of 16 per cent sand, 43 per cent silt, 30 per cent clay, and 11 per cent gravel. The first three are plotted on the three axes; the fourth variable (gravel) is indicated by the small triangle.

Ratio charts. Various types of charts and graphs have been developed to indicate the ratios between variables. One may wish to compare the relative abundance of garnet and hornblende in a series of samples collected along a traverse. A simple device for presenting these data is to choose a horizontal distance scale and at a given distance above this line to draw a parallel line which represents the amount of garnet present, called unity for convenience. The ratio of hornblende to garnet is computed for each sample; if the garnet has the frequency value 24 (percentage, number of grains, or the like) and hornblende has the frequency value 10, the ratio $10/24 = 0.42$ furnishes a point to be drawn at the scale value 0.24, using the garnet line as unity. Values of the ratio larger than 1 are plotted above the unity line. The result is a curve which varies above or below the garnet line and indicates the variation of hornblende to garnet, on the assumption that the garnet frequency is fixed. Figure 95 illustrates such a chart, as used by Pettijohn[1] in comparing the ratio of hornblende to garnet. Here the garnet line was chosen as 100, and a logarithmic scale was used on the vertical axis. The logarithmic axis serves to decrease the absolute range of the values and is suitable when wide fluctuations occur.

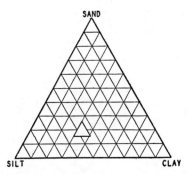

FIG. 94.—Method of plotting four variables on triangle diagram. The size of the small triangle indicates the amount of gravel in the sediment.

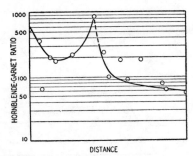

FIG. 95.—Graph of hornblende-garnet ratio (garnet = 100). After Pettijohn, 1931.

[1] F. J. Pettijohn, Petrography of the beach sands of southern Lake Michigan: *Jour. Geology*, vol. 39, pp. 432-455, 1931.

GRAPHIC PRESENTATION

Miscellaneous Graphic Devices

In many instances it is desirable to show a number of related phenomena on a single chart, without regard to the arrangement of the data according to variables or axes. Among the wide variety of such charts which are available, three will be mentioned.

Bar charts. These charts are simply constructed by choosing a vertical scale of frequency (percentage, number, amount) and representing the several variables to be compared as vertical bars, all of the same width (Figure 96). Bar charts are not suitable for detailed analysis, but they serve for rapid comparisons.

Pie diagrams. Pie diagrams are used for the same purposes as bar charts, except that segments of a circle indicate the relative magnitudes. Figure 97 shows the same data as the bar chart, arranged in a pie diagram.

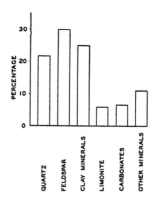

Fig. 96.—Bar chart of average composition of shale. (Data from Clarke, U. S. G. S. Bulletin 770.)

Such diagrams are useful for rough presentation and for popular illustrations.

Sailboat and star diagrams. Sailboat and star diagrams have been used to illustrate the chemical composition of igneous rocks,[1] and they are applicable to a number of situations in sedimentary data. In their construction a central point is chosen, from which radiate as many lines as there are variables to be compared. Along each of these a length is laid off proportionate to the magnitude, and the termini of the lines are joined, yielding a figure similar to a sailboat.

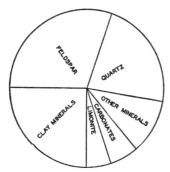

Fig. 97.—Pie diagram of same data as in Figure 96.

MATHEMATICAL ANALYSIS OF GRAPHIC DATA

Graphs and charts are seldom an end in themselves but are used to draw conclusions from data, to investigate the relations or lack of relations among variables, or merely to simplify text descriptions. Conclusions to be

---

[1] J. F. Kemp, *A Handbook of Rocks* (New York, 1911).

drawn from data are often clarified, however, if a definite mathematical relation can be found between the variables. When such a relation is found, it is possible to investigate the geological implications more precisely, because behind each mathematical function is a set of conditions which must be true if the relation holds.

The detailed examination of curves based on experimentally determined points is beyond the scope of this book, and there are available a number of excellent treatises on the subject.[1] However, there are several fairly simple relations which may be encountered in sedimentary studies.

Special attention will be given to three common types of functions of two variables, applicable to scatter diagrams, or graphs with time or distance as independent variables. These three functional relationships are (a) linear functions, (b) power functions, and (c) exponential functions; they are easily recognized by the fact that they plot as straight lines respectively on ordinary graph paper, double logarithmic graph paper, and semi-logarithmic graph paper.

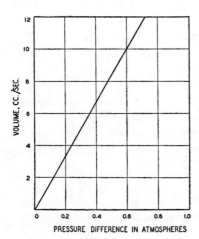

FIG. 98.—Flow of water through Cisco sandstone, after Plummer, Harris, and Pedigo, 1934.

Linear functions. A linear function of two variables may be defined as such a relation that a change in one variable induces an exactly proportional change in the other variable. This constant proportionality results in a straight line graph when one variable is plotted against the other on ordinary graph paper. An example of such a function is given by Plummer, Harris, and Pedigo,[2] in connection with the flow of water and other fluids through sandstone. By plotting the volume of water that flowed through a rock core per unit of time under given pressure differences, a straight line was obtained, as shown in Figure 98.

The fundamental characteristic of a linear function is that the rate of change of the dependent variable with respect to the independent variable is constant. In the calculus this is expressed by the relation

[1] See for example C. O. Mackey, *Graphical Solutions* (New York, 1936); also T. R. Running, *Empirical Formulas* (New York, 1917).

[2] F. B. Plummer, S. Harris and J. Pedigo, A new multiple permeability apparatus: *Am. Inst. Min. and Met. Eng. Tech. Pub. 578,* 1934.

$dy/dx =$ const. If, in the example cited, volume in cubic centimeters per second is chosen as the dependent variable, it must suffer a constant rate of change with respect to the pressure difference: at twice a given pressure difference the quantity must be doubled, etc., for the range in which the linear function holds. The fact that the line ascends to the right indicates that the constant in the differential equation $dy/dx =$ const. is positive.

Linear functions are the simplest mathematical relations which exist between variables, and it may be anticipated that among sedimentary data in general they will not be as common as other functions, because of the large number of factors which are present in most sedimentary situations. In the present case the occurrence of a linear function is due to the laminar motion of the fluid through the sandstone, in accordance with Darcy's Law.

If one is interested in determining the actual analytical expression involved in the linear function, he may read it directly from the graph. The equation of any straight line may be written as $y = ax + b$, where a is the slope of the line and b is the point where the line crosses the y-axis. Both of these values are constants. To determine a, especially when the scale units are not equal in length, it is perhaps best to choose two convenient points on the line rather far apart, and divide the change in y by the increase in x. This quotient, expressed as a decimal, will at once furnish the value of a. Likewise, b is determined simply by reading the value of the point where the line crosses the y-axis. In the example given, a change of 10 units in volume occurs in an increase of 0.6 units of pressure difference. Thus the value of a is $10/0.6 = 16.7$. The value is positive because the curve rises to the right. The value of b is zero because the curve passes through the origin, and hence the equation of the line is $y = 16.7x$.[1]

Power functions. A power function of two variables may be defined as such a relation that if the independent variable is changed by a fixed multiple, the dependent variable will also change by a fixed multiple. The general equation for such a function is $y = ax^n$ where a is a constant and n is either positive or negative, a whole number or a fraction.[2] Thus both parabolas and hyperbolas are included, the latter when n is negative. From the nature of the function it may be seen that if logs are taken of both sides of the equation, the expression $\log y =$

---

[1] The actual case has been simplified in this example. The y-intercept has a small value because some pressure difference is required in practice before liquid flows through the core. Likewise the value $a = 16.7$ cannot be used directly for the coefficient of permeability. The expression given is merely an empirical statement of the relation between y and x.

[2] When $n = 1$, the function is linear, which is thus a special case of the power function.

$n \log x + \log a$ results. This means that if the original data are plotted on double logarithmic paper the graph will be a straight line.

The presence of a power function requires that the rate of change of the dependent variable must itself depend on the value of the independent variable, or on some power of it. Thus in a simple parabola, the rate of change of y will be proportional to x, $dy/dx = mx$, where m is a constant.

An example of a power function, perhaps not strictly sedimentary in nature, occurs in connection with the probable error of collecting samples. This probable error, discussed earlier, was found to decrease as the number of samples in a composite was increased. By plotting the values of Table 3 (page 41) on double logarithmic paper, the straight line shown in Figure 99 results, demonstrating that the relation is a power function. Moreover, the fact that the line descends to the right indicates that the exponent n is negative, and hence the relation is a hyperbola rather than a parabola.

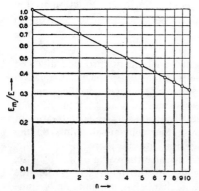

Fig. 99.—Double logarithmic graph of $E_m/E = 1/\sqrt{n}$. (Data from Table 3, page 41.)

The determination of the analytical expression for the curve is more complicated with power functions in general than with linear functions. However, in the present case the relation is fairly simple. Inasmuch as the logarithmic units are of the same length on both axes, the value of n may be determined very simply by finding the tangent of the angle of slope of the straight line. This angle is 26.5° measured in a clockwise direction. The tangent of 26.5° is 0.50, and since the line descends to the right, $n = -0.5 = -\frac{1}{2}$. Also, the line intersects the axes at the point (1, 1), and hence the value of a is unity. Thus the equation is $y = x^{-\frac{1}{2}}$, or $y = 1/\sqrt{x}$.

The implications of a hyperbolic function are that the value of the dependent variable must decrease in fixed ratio as the independent variable increases in a fixed ratio. In the present case it may be seen that, as a result of this property, the value of the probable error decreases rapidly at first and then more slowly, as the number of samples in a composite increases.

In physics parabolic and hyperbolic functions are numerous; as an illustration it may be mentioned that Stokes' law, $v = Cr^2$, is a parabolic function.

Exponential functions. An exponential function may be defined as a relation in which the dependent variable increases or decreases geometrically as the independent variable increases arithmetically. All exponential functions may be written as $y = mb^{ax}$, where m, b, and a are constants. In the nature of the case, m is always the value of y at the origin, and b is usually chosen as a certain constant $e = 2.7182\ldots$, the base of natural logarithms. Hence it is common convention to express exponential functions as $y = y_0 e^{ax}$, where a may be either positive or negative. If logs are taken of both sides of the expression, there results $\log y = \log y_0 + ax \log e$. By taking the logs to the base e, the expression simplifies to $\log_e y = ax + \log_e y_0$. If y and y_0 are combined into a single term, it may be seen that a log appears on the left, but none on the right: $\log(y/y_0) = ax$. Thus if an exponential function is plotted on semilogarithmic paper, the points will lie on a straight line. Hence it is only necessary to plot the observed values on such paper to determine whether an exponential function is involved.

Exponential functions may be found to occur rather commonly in sedimentary situations. Krumbein [1] discussed several negative exponential functions, and the following example is repeated here. The average size of beach pebbles was determined at several points along a beach, and the data were plotted on semi-logarithmic paper. A straight line, descending to the right, was found as shown in Figure 100. This demonstrates that the equation is of the type $y = y_0 e^{-ax}$, and several interesting properties follow from the nature of the function. In any exponential function of this type, it is necessary that the rate of change of the dependent variable be proportional to the value of the dependent variable at any given point, $dy/dx = -ay$, where a is constant and y is the dependent variable. This means, in a negative exponential, that if the pebbles are being worn down as they move along the beach, the rate of wear is proportional to the average size of the pebble. Complexities which may enter into the interpretation of exponential functions are discussed in the original paper.

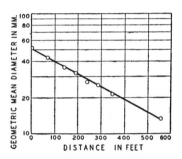

FIG. 100. — Semi-logarithmic graph of geometric mean size of pebbles as a function of distance along the beach.

In considering exponential functions analytically, the problem reduces to the determination of the constant a. In the example cited, the function is a

[1] W. C. Krumbein, Sediments and exponential curves: *Jour. Geology*, vol. 45, pp. 577-601, 1937.

negative exponential, and the following treatment indicates the steps involved.

There are several methods for determining a; one of the most convenient is a simple analytical method in which it is only necessary to find the value of x at the point where y is reduced to half its original value. In that case $y/y_0 = \frac{1}{2}$, and the original equation becomes $\frac{1}{2} = e^{-ax}$. By taking logs of this expression to the base e and changing the sign, there results $-\log_e(\frac{1}{2}) = ax$. But $-\log_e(\frac{1}{2}) = \log_e 2 = 0.693$, and hence $a = 0.693/x$, where the value of x is called the "half distance" and corresponds to the point where y is half its original value. In the present case y_0 is 52 mm., and the x-value at the point where $y = 26$ mm. (half its original value) is 260 ft., as determined from Figure 100. The unit of distance was chosen as 100 ft. for convenience, so that the half-distance value of x is 2.6. Placing this value in the equation just given, we obtain $a = 0.693/2.6 = 0.26$. Hence the required equation for pebble size is $y = 52\, e^{-0.26x}$.

It is customary to refer to a as the coefficient of the physical attribute being considered; in the present case the value $a = 0.26$ may be called the coefficient of pebble size.

TABLE 19

PROPERTIES OF LINEAR, POWER, AND EXPONENTIAL FUNCTIONS

Function	Rate of Change of Dependent Variable	Nature of Curve
Linear	$dy/dx =$ const., and hence slope of curve is constant	Straight line, with fixed slope. Plots as straight line on ordinary arithmetic graph paper
Power	$dy/dx = mx^p$, and hence slope of curve is dependent on x	Parabolas and hyperbolas. Plot as straight lines on double log paper. Parabolas rise to right, hyperbolas descend to right
Exponential	$dy/dx = ay$, and hence slope of curve is proportional to the value of y at any given point	Exponential curves. Plot as straight lines on semi-log paper. Positive exponentials rise to right; negative exponentials descend to right

Comparison of linear, power, and exponential functions. The three functions considered here do not by any means exhaust the possible mathematical relations which may be found in sedimentary situations. For example, periodic functions, such as rhythmic variations in grain size or the like, may be expected to occur, but their analysis is usually complex.

Perhaps the most significant comparison that may be made of the three functions is in terms of their rates of change. This topic was mentioned in connection with each function, but for comparison they are shown in Table 19, which summarizes the mathematical properties of the functions.[1]

In any given case, the physical significance of the constants in the equations depends partly on the geological set-up of the data. The mathematical implications are fixed, but the interpretation placed on the data depends on the individual case. In any event, the physical interpretations must not violate any of the mathematical principles involved.

Many workers in sedimentary petrology question the value of mathematical analysis, owing to the large number of variables and errors which are involved in the simplest situation. In general, this may be granted in the present stage of development of the science, but it is equally true that an approximate determination of the nature of the functional relationship may point the way for more rigorous studies of problems and eventually establish underlying principles of universal application. Work in sediments has already reached a fair state of rigorous analysis in some connections, and it seems that elementary types of mathematical analysis at least may be applied in cases where graphic presentation suggests simple functional relationships.

[1] An excellent table which includes numerous empirical functions and the conditions under which they plot as straight lines is given by C. O. Mackey, *op. cit.* (1936), p. 117.

CHAPTER 8

ELEMENTS OF STATISTICAL ANALYSIS

INTRODUCTION

A THOROUGH laboratory study of sediments includes quantitative data on the sizes, the shapes, the mineral composition, the surface textures, and perhaps the orientation of the grains. These fundamental data are related to the physical and chemical factors in the environment of deposition. To relate characteristics with environment one may investigate the areal variation of the sediment, which implies the comparison of one sample with the next. This comparison is most conveniently accomplished by means of statistical analysis.

The word *statistics* is defined [1] as "The science of the collection and classification of facts on the basis of relative number or occurrence as a ground for induction; systematic compilation of instances for the inference of general truths." This definition shows that the study of sediments is largely statistical in nature. Sedimentary petrologists are interested in the collection and classification of sedimentary data as a basis for inferences about sediments. Mechanical analysis is concerned with the ranges of diameters present, and the relative abundance of particles in each diameter range. This is clearly a statistical operation. The analysis of particle shape, of mineral content, and of particle orientation are all concerned with the collection of facts in terms of the number of occurrences of any particular attribute.

Statistical technique may be divided into several operations. The first operation is the collection and classification of data. In sedimentary terms this refers to mechanical analysis, mineral analysis, and so on. The second step may be the presentation of the data in the form of tables and graphs. Finally, the data themselves may be analyzed statistically, and from the values obtained, inferences may be drawn about the sediment.

Several approaches to statistical analysis are possible, depending on the nature of the data. A statistical series involving magnitude (as size

[1] *Webster's New International Dictionary* (Springfield, Mass., 1926).

of grains, or percentage of heavy minerals), is called a *frequency distribution*. If geographic location is involved (as in the comparison of samples over the areal extent of a formation), the statistical series is called a *spatial distribution*. If time is an important factor (as in the changes of sedimentary characteristics as time goes on), the statistical series is called a *time series*. Each of these cases is of importance in the study of sediments.

THE CONCEPT OF A FREQUENCY DISTRIBUTION

The discussion of frequency distributions will be confined to size frequency distributions (mechanical analysis data), although it should be borne in mind that the same principles may apply to the study of mineral distribution or the shape distribution of particles in a sediment. In all size frequency distributions there are two principal variables, "size" and frequency. The frequency distribution itself is simply the arrangement of the numerical data according to size. Size is considered to be the independent variable, and frequency the dependent variable. This choice means that frequency is a function of size, expressed as $y = f(x)$, where y is the frequency and $f(x)$ is some function of size. By convention any graph of the frequency distribution is drawn with size (diameters in millimeters or any numbers representing size, such as the logarithms of the diameters) plotted along the horizontal x-axis, and frequency (percentage by weight or by number or any other symbol representing frequency) plotted along the vertical y-axis.

Frequency distributions may be of two types. A *discrete series* is one in which the independent variable increases by finite increments, with no gradations between. A pile of coins, made of pennies, nickels, dimes, and quarters, if assembled into a frequency distribution by counting the number of each coin present, constitutes such a discrete series. In discrete series each value of the variable (in coins 1¢, 5¢, 10¢, etc.) is a separate group of items, so that drawing a smooth curve through the data is quite erroneous. The second type of frequency distribution is the *continuous distribution*, in which the independent variable increases by infinitesimals along its range of values. That is, if the individuals were arranged side by side, there would be complete gradation from one to the next. The heights of men form such a series. It is obvious also that, with few if any exceptions, sediments fall within the class of continuous distributions. Within a single sediment there is a continuous range of sizes from largest to smallest.

In continuous data there is no inherent grouping. Whatever classes of size are erected are perfectly arbitrary, a point which is emphasized in Chapter 4 in the discussion of grade scales. However, some sort of grouping is necessary in analyzing the data, so that frequency may be expressed as the amount of material within selected intervals along the size scale. Mechanical analysis is the operation of determining this abundance or frequency within chosen size classes or grades.

HISTOGRAMS AS STATISTICAL DEVICES

Perhaps the most common graphic device used in presenting frequency data of sediments is the histogram, described in Chapter 7. This simple frequency diagram is readily understood and has a universal appeal because of its clarity and simplicity. From the histogram itself much may be learned. In the first place, one may see that there is a particular class which has the greatest frequency of individuals within it, and that the frequency decreases on either side. The class of greatest frequency is called the *modal class*, and from the extent to which it towers above its neighbors one may note whether it is a conspicuous modal group or not. Likewise, from the rectangles stretching away on either side of the modal class one may see the range of size in the population. From the extent of the spread one may roughly note whether the tendency is for the individuals to cluster about the most prevalent size or to spread widely on either side. Finally, one may note whether or not the distribution of individuals on either side of the modal class is symmetrical or not.

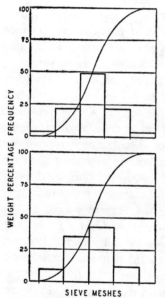

FIG. 101.—Two histograms from the same cumulative curve. The units on the *x*-axis represent the logarithms of the sieve sizes.

Among the earliest workers to use histograms in the study of sediments was Udden.[1] He observed that the histograms of sediments varied considerably, according to the type of sediment involved. That is, dune or beach sands have well-defined central groups, whereas histograms of such sediments as glacial till were wide-spread and irregular. Udden defined the modal class as the "maximum grade" and contrasted this maximum with the material on either side, which he designated as the "coarse and fine admixtures." In this manner Udden obtained a sorting factor which was used in his geologic reasoning about the sediments.

[1] J. A. Udden, The mechanical composition of wind deposits, *Augustana Library Publications*, no. 1, 1898.

ELEMENTS OF STATISTICS 215

Unfortunately, the histogram is influenced by the class intervals used in the analysis, and its shape varies according to the particular class limits chosen. Figure 101 shows the same continuous frequency distribution represented as a cumulative curve, analyzed according to two different grade scales, and it may be noted that the histograms are not at all alike. One of the diagrams is symmetrical and one is definitely unsymmetrical. It would appear from this that the particular form assumed by a histogram is accidental and depends on the nature of the classes used in the analysis. If one therefore bases conclusions on a given histogram, he may never be quite certain that his conclusions are correct. This applies especially to situations in which only a few classes are used. The difficulty with histograms is due to the fact that the diagrams attempt to illustrate a continuous frequency distribution as though it were made of discrete classes. Consequently, the diagram may not furnish much visual information about the frequency distribution considered as a continuous variation of size. It is because of this that statisticians recommend the use of smooth curves to represent continuous distributions.

If one wishes to generalize from a picture of the frequency distribution, it is much safer to use the unique frequency curve, because within small experimental limits the essential shape of the distribution will be brought out by the continuous curve. A graphic method of obtaining the frequency curve is given in Chapter 7.

CUMULATIVE CURVES AS STATISTICAL DEVICES

The difficulty attendant upon the variation of histograms has resulted in the wide adoption by sedimentary petrologists of the cumulative curve. Experience has shown that whereas histograms vary depending upon the class interval used, the cumulative curve remains fairly constant regardless of the particular class limits used. Within the limits of errors of smoothing, the cumulative curve is a more reliable index of the nature of the continuous distribution than the histogram.

It was shown in Chapter 7 that the frequency curve is the limit approached by the histogram as the class intervals decrease to zero and the number of individuals increases without bound. By the calculus it is possible to show that every continuous curve has associated with it an integral curve and a derivative curve. The integral curve is such that its ordinate at any point represents the area under the given curve up to that point, whereas the derivative curve is such that its ordinate at any point represents the slope of the given curve at that point. The relation is such that if one curve is the integral of a second curve, the second curve is itself the derivative of the first curve.

It is a widely recognized fact that the cumulative curve is the integral of its corresponding frequency curve, and consequently that the frequency curve is the derivative of its cumulative curve. This relationship may be demonstrated as follows: When the cumulative curve is prepared from the histogram, the percentage of material in each class is summed to obtain the successive ordinates of the cumulative curve. Thus, inasmuch as the histogram classes represent areas, the ordinates of the cumulative curve are linear representations of the area under the histogram up to that point. Further, the total area under the histogram is the sum of the areas in the successive blocks. This may be expressed in the shorthand of mathematics as follows. Let Σ be the summation symbol, let $f_1, f_2, \ldots f_n$ be the frequencies in each histogram block, and let Δx be the class interval, assumed constant. Then, if there are n classes or blocks, the total area under the histogram is

$$\sum_{1}^{n} f_i \Delta x = \text{total area under histogram, where } f_i \text{ represents the several}$$

frequencies. However, as the classes become smaller and smaller, or in other words as Δx approaches zero, the limit of this sum is the integral of the function taken over the range involved. In mathematical notation this is

$$\lim_{\Delta x \to 0} \sum_{1}^{n} f_i \Delta x = \int_{a}^{b} f \cdot dx$$

where a and b are the limits of the range of sizes in the distribution. This mathematical relation, often called the fundamental theorem of the integral calculus, is proved in all standard texts on the subject.

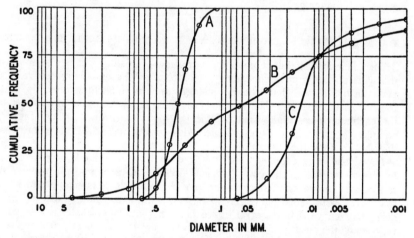

Fig. 102.—Examples of cumulative curves. A, beach sand; B, glacial till; C, loess.

The fact that the cumulative curve is the integral of the frequency curve explains why cumulative curves are less liable to fluctuations than histograms. In cumulating the original frequency data on which the histogram is based, a process of finite integration is performed which converts the finite class intervals into a continuous function when the curve is smoothed. The histograms, on the other hand, are plotted as "raw" data and so preserve the accidents of treating a continuous function as a series of discrete intervals.

The cumulative curve may be used in the same manner as the histogram in interpreting the nature of sediments, and the fact that it is less liable to

ELEMENTS OF STATISTICS

fluctuations due to accidents of the grade scale has led various writers to use it to the exclusion of histograms. The most abundant grains are associated with the inflection point (page 192) of the cumulative curve, and the degree to which the grains cluster about or spread away from the modal group may be seen from the steepness of the curve. Irregularities in the smooth rise of the curve indicate secondary modal groups. Likewise, the approximate degree of sorting or sizing of the sediment may be read from the general slope of the curve and the range of sizes included within it.

Figure 102 illustrates several cumulative curves of sediments. It may be seen at a glance that curve A (a beach sand) is symmetrical and well sorted, curve B (a loess) is well sorted but is not symmetrical, and curve C (a glacial till) is poorly sorted. In each case the modal class is associated with the steepest part of the curve.

INTRODUCTION TO STATISTICAL MEASURES

Although much can be done by purely graphic methods in the interpretation of frequency curves, it is more convenient to have the characteristics of the curve expressed as numbers. Statisticians have developed analytical devices so that the numbers themselves, instead of the pictures of the curves, may be used in referring to the distribution.

Figure 103 shows six frequency curves, all drawn to the same scale. The top row of three curves are all symmetrical, but curve A is less peaked than B, and both A and B are less widely spread out than C. Similarly, the curves in the lower row are all unsymmetrical, but curves E and F, while equally unsymmetrical, are inclined or skewed in opposite directions. Curve D shows an extreme degree of asymmetry. In order to describe and compare this wide range of curves, a number of statistical measures are necessary.[1]

Measures of the central tendency. Perhaps the most important measure is a measure of the central tendency, the value about which all other values cluster. In general this value corresponds to the size which is most frequent, although in asymmetrical curves this may not be so. Such measures of the central tendency are called *averages*. They include such diverse measures as the *arithmetic mean size*, the *median size*, the *modal size*, the *geometric mean size*, and others.

From a sedimentary point of view, the average size of a sediment is

---

[1] The reader is referred to any standard textbook of statistics for more detailed definitions of the terms used in this section. Among elementary, non-mathematical references may be mentioned F. C. Mills, *Statistical Measures* (New York, 1924). A more detailed discussion of theory may be found in B. H. Camp, *The Mathematical Part of Elementary Statistics* (New York, 1931). A general reference of much value is R. A. Fisher, *Statistical Methods for Research Workers* (Edinburgh and London, 1932).

of interest because it indicates the order of magnitude of the grains. Average size is also useful for comparing samples collected in the direction of transport as along a beach or stream. Curves of the average

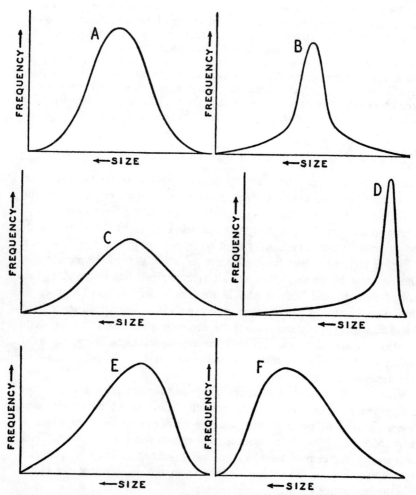

Fig. 103.—Frequency curves. The abscissae may be diameter in millimeters, logs of the diameters, or any other expression of "size."

size against distance may disclose some underlying law of variation. In a similar manner, maps may be prepared of the areal variation of size within a given environment, as a basis for reasoning geologically about the causes of the variation. The average grain size is thus an important value, and the choice of particular averages will receive detailed attention in Chapter 9.

Measures of the degree of scatter. Two frequency curves with the same average size may have entirely different degrees of spread, such as curves A and C of Figure 103, because the average value merely represents the central point and does not indicate the spread of the data on either side. Hence, a second measure needed to describe the curves is a measure of the *degree of spread* or *degree of dispersion* of the data about the central tendency. Such measures of spread may be the *mean deviation,* the *standard deviation,* the *quartile deviation,* the *range,* and so on. As with measures of average size there is a choice of devices available, and one of the problems of the practising statistician is to determine the appropriate measure to use.

From a geological point of view, the average spread of the curve, which means the tendency of the grains to cluster about the average value, is another important characteristic of sediments. Some geological agents are more selective in their action than others, and this may manifest itself in the extent to which the grains tend to be selected or "sorted" according to size. Theoretically, perhaps, one may expect a perfectly sorted sediment to consist of only one size of grain, but in any natural situation there are deviations about this size, due to fluctuations in velocity, shape and density of the grains, and the like. Consequently, the degree of spread may prove of importance as a clue to the nature of the deposit. For example, it is not known whether the selectivity of sediments increases or decreases in the direction of transport; what meager evidence there is on beaches suggests that the average spread may be fairly constant over a given stretch of beach. Profiles of average spread of the curves along a traverse line or maps of the average spread over a formation may furnish clues to variations in the depositing agent.

Measures of the degree of asymmetry. The average size and the degree of spread of two curves may be the same, but one may not be symmetrical. This situation is illustrated approximately by curves A and E of Figure 103. Hence it is necessary to have a measure of the tendency of the data to spread on one side or the other of the average. Such asymmetry is called *skewness,* and various skewness measures are available. Because skewness may be either to the left or to the right, a positive or negative sense is usually assigned to it. Thus, curves E and F are skewed in opposite directions; the choice of positive and negative directions may be conventional. In extreme types of skewness, such as shown by curve D of Figure 103, additional measures may be needed to describe the shape of the curve, or mathematical methods may be used to "symmetrize" the curve by changing the independent variable.

Skewness is an attribute of sediments about which relatively little is known. When sedimentary curves are plotted with diameter in millimeters as the independent variable, they almost invariably show extreme types of skewness, but the asymmetry is reduced when logarithms of the diameters are used as independent variable. For this reason it is often simpler mathematically to analyze sedimentary data on a logarithmic basis. The physical meaning of skewness is not easily interpreted, for several reasons. For example, sampling errors may manifest themselves in skewness, either if more than one size frequency distribution is included in the sample due to improper selection of samples, or if the sample of a single population is too small to reflect the attributes of the original distribution. Likewise, skewness may result if a symmetrical distribution is later acted upon by a transporting agent which removes only a portion of the material. A sandy gravel may have some of its finer material removed, leaving behind a skewed lag sediment. Miss Gripenberg[1] has suggested that skewness has a genetic significance in some instances and that a sediment deposited by a uniform current may increase in skewness as the material is followed along in the direction of transport.

Relatively few studies have been made of the areal variation of skewness within given deposits, and the data are perhaps too meager for generalizations. The almost universal presence of skewness in sediments, especially in terms of diameter as the independent variable, suggests that there is a genetic relation between agent and skewness, as Miss Gripenberg points out, and that the skewness may vary areally in accordance with definite laws.

Measures of the degree of peakedness. Frequency curves which are alike in their degree either of symmetry or asymmetry may nevertheless vary in the degree to which peakedness is present. Curves A and B of Figure 103 illustrate this difference. Curve B has a more pronounced peak than A. Statistical measures designed to express this attribute are measures of *kurtosis;* here also a choice is available.

Not much is known about the significance of kurtosis in sediments. It appears to be related to the selective action of the geological agent, but the sum total of the factors entering into the selective process are not known. The kurtosis of a curve, and especially of symmetrical

[1] Stina Gripenberg, A study of the sediments of the North Baltic and adjoining seas: *Fennia,* vol. 60, no. 3, 1934.

ELEMENTS OF STATISTICS

curves, has a definite geometrical significance, however, whether or not it may have a physical or geological significance. No complete investigation of the areal variation of kurtosis has been made, and virtually nothing is known of its magnitude or prevalence in sediments.

ARITHMETIC AND LOGARITHMIC FREQUENCY DISTRIBUTIONS

The extreme type of skewed curve shown as D in Figure 103 is commonly encountered in sedimentary practice, especially when the data are plotted with diameter in millimeters as the independent variable. When the same data are plotted with log diameter as independent variable, the curve becomes much more symmetrical. The symmetrizing influence of a logarithmic size scale was mentioned above; it will be discussed here in terms of the application of statistical measures to the data. From a mathematical point of view it is simpler to describe a symmetrical curve than one which is asymmetrical. A symmetrical frequency curve may be completely described by two measures, an average size and the degree of spread about the average. If the curve is moderately skewed, three measures usually suffice, but for extreme skewed curves the labor involved in computing the necessary number of measures becomes quite tedious. From the point of view of convenience alone, the symmetrizing effect of logarithmic plotting is ample justification for its use. There is another "justification," however; most workers in sediments prefer to plot their data on an implied log scale, by drawing the classes equal in width, to facilitate interpretation of the data.

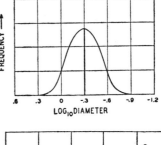
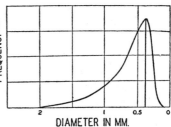

Fig. 104.—Logarithmic and arithmetic graphs of the same frequency distribution, showing shift in position of central ordinate.

From a strict statistical point of view, a sedimentary histogram based on a geometric grade scale, but drawn with its blocks of equal width, is no longer a picture of the frequency distribution of diameters. This may be more effectively demonstrated with frequency curves, as shown in Figure 104, which shows the same data plotted as an arithmetic and a logarithmic frequency curve. In the logarithmic curve an ordinate has been drawn in such a manner that the areas on both sides are equal. This same ordinate, converted to its diameter equivalent, is shown in the arithmetic curve: it no longer divides the area into halves.[1] Inasmuch as histograms or frequency curves are areal representations of the frequency, it must be obvious that the geometrical interpretations of the two diagrams must differ. One may say, for example, that the distribution of (logarithmic) individuals is symmetrical about the central ordinate, but it is not correct to say that the distribution of sedimentary particles, considered as grains of a given size, is equal and symmetrical about the average value.

The distinction between two types of frequency curves, one based directly on diameters as the independent variable, and the other based on logs of the diameters, is important in sedimentary data. Numerous workers [2] have discussed the merits of one or another manner of plotting the data, and various statistical measures have been proposed to take cognizance of the shift in geometrical significance introduced by the logarithmic plotting. Three broad types of measures have been developed as a result. The first includes arithmetic measures based directly on grain diameters in millimeters; these measures include a size factor. The second type of measure is also based on grain diameters, but involves ratios between sizes to eliminate the size factor and to emphasize the geometric nature of the frequency distribution. The third type foregoes the diameter distribution entirely and applies a series of logarithmic measures to the logarithmic frequency curve.

All three kinds of measures are used at present, but some confusion has arisen due to the use of one kind of measure coupled with inferences drawn from postulates underlying another kind of measure. An important precaution to be used in sedimentary statistical practice is that the identity of the independent variable must be known at all times. It

---

[1] A complexity enters this analysis, due apparently to a shift of the mode during the transformation.

[2] The following papers are among those which bear on the problem: C. W. Correns, Grundsätzliches zur Darstellung der Korngrössenverteilung: *Zentr. f. Min.*, Abt. A., pp. 321-331, 1934. T. Hatch and S. P. Choate, Statistical description of the size properties of non-uniform particulate substances: *Jour. Franklin Inst.*, vol. 207, pp. 369-387, 1929. W. C. Krumbein, Application of logarithmic moments to size frequency distributions of sediments: *Jour. Sed. Petrology*, vol. 6, pp. 35-47, 1936. P. D. Trask, *Origin and Environment of Source Sediments of Petroleum* (Houston, Texas, 1932), pp. 67 ff. C. E. Van Orstrand, Note on the representation of the distribution of grains in sands: *Researches in Sedimentation in 1924*, pp. 63-67, Nat. Research Council, 1925. C. K. Wentworth, Method of computing mechanical composition types of sediments: *Geol. Soc. America, Bulletin,* Vol. 40, pp. 771-790, 1929.

is not necessary for statistical purposes that the independent variable have any immediately comprehensible significance; it is only required that a curve be given. The identity of the independent variable may, however, be preserved by defining it in terms of diameters, so that at any point in the analysis one may convert his results back to diameter terms if he wishes.

It would appear at first glance that the application of logarithmic methods would complicate the essential simplicity of the sedimentary picture, but actually this is not so. In order to avoid difficulties, it is only necessary to set up a mathematical relationship such that a new logarithmic variable is substituted for the diameters of the grains. Methods of analysis and the grouping of the data into size classes are not changed; the new variable is used only in the computation of statistical measures, and the geometrical meaning of the measures is directly related to the logarithmic frequency diagram of the sediment. The phi and zeta scales mentioned in Chapter 4 and discussed more fully in Chapter 9 afford one method of attacking this statistical problem.

QUARTILE AND MOMENT MEASURES

Coupled with the problem of choosing suitable independent variables for sedimentary data is the choice of sets of measures in terms of their underlying mathematical theory. In conventional statistical practice two main types of measures have been used.

Quartile measures. If a size frequency distribution is arranged in order of magnitude, with the smallest particle at one end and a continuous gradation upward to the largest particle at the other end, it is possible to choose certain particles as representing significant values. The size of the middlemost particle represents an average of the group, and is called the *median*. To determine the spread of the distribution about the median, two other particles are measured. The first is just larger than one fourth of the distribution (the *first quartile*), and the second is just larger than three fourths of the distribution (the *third quartile*). Measures of spread are based on differences or ratios between the two quartiles, depending upon whether arithmetic or geometric measures are to be used. Similarly, logarithmic measures are based on logs of the quartiles. For measuring the asymmetry or skewness, a comparison is made of the median value with an average of the first and third quartiles, either arithmetically, geometrically, or logarithmically.

The outstanding feature of quartile measures is that they are confined to the central half of the frequency distribution and the values obtained are not influenced by extreme particles, either very large or very small. Furthermore, quartile measures are very readily computed, and most of the data may be obtained directly from the cumulative curve by graphic means. For these reasons quartile measures are extensively used in sedimentary data, and they apply even to incomplete sets of data. This is an advantage for fine-grained sediments, where part of the material is beyond the range of ordinary methods of mechanical analysis.

Moment measures. In contrast to quartile measures are moment measures, which are influenced by every individual in the distribution, from coarsest to finest. Moment measures are much more complex mathematically than quartile measures, and they involve rather tedious computations compared with the quartiles. Nevertheless, moment measures are more extensively used in conventional statistical practice because of their greater sensitivity to the influence of each member of the distribution and because of their more unified mathematical basis. A full understanding of the nature of moments cannot be had without recourse to the calculus, but fortunately the computations may be made, and the geometrical significance evaluated, without mathematical knowledge.

The first moment of a frequency distribution is its center of gravity and is called the *arithmetic mean.* It is a measure of the average size of the sediment. The second moment, or more properly its square root, measures the average spread of the curve and is expressed as the *standard deviation* of the distribution. It is analogous in physics to the radius of gyration of a system. The third moment, or its cube root, is a measure of the skewness of the data. The moment measures are thus a set of parallel measures to the quartile measures, but their geometric significance is different.

Mathematically, the rth moment of a distribution is defined as

$$n_r = \frac{1}{N} \int_{-\infty}^{+\infty} x^r f(x) \, dx,$$

where $f(x)$ is the frequency function and N is the total frequency. By setting $r = 1, 2, 3, \ldots$, the successive moments result. The moments are thus related as the successive powers of x into the integral. In practice, the first four moments are used; unfortunately, there is no physical analogue of moments higher than the second, so that the moments cannot conveniently be expressed in simple terms. In practice, where $f(x)$ is unknown, the frequency in the several classes of width Δx is multiplied by some power of their distance from the origin and the result is translated to a value in terms of the first moment, as described in Chapter 9.

ELEMENTS OF STATISTICS

The tediousness of computing moment measures, combined with their mathematical complexity, has militated against their extensive use in sedimentary petrology. A further difficulty has arisen from the fact that conventional statistics books afford only methods of computation based on equal intervals, whereas with sediments the data are usually expressed in unequal grade sizes. Finally, moment measures expressed in terms of diameters in millimeters often result in complexities owing to the extreme skewness of the data. Fortunately, however, methods are available for the direct use of logarithmic moments, which appear to have physical significance in sediments and which may be converted to their diameter equivalents.

The fact that moments are affected by the value of every grain in the distribution may limit their application to sedimentary data. When analyses are so expressed that all material finer (or coarser) than a given grade is grouped into one class, the values of the higher moments are distorted. Improved techniques of analysis, especially among the fine-grained sediments, may remedy this difficulty, however.

THE QUESTION OF FREQUENCY

There is one conspicuous manner in which the statistical data of sedimentary petrology differ from most conventional statistical data. Frequency in sedimentary data is usually expressed by weight instead of by number, and it is usually expressed as percentage frequency rather than absolute frequency. No complete investigations of this aspect of sedimentary usage have been made, and the problem of weight vs. number is still largely unsolved. In a given sample there may be only one or two large pebbles to a gram, whereas there may be literally millions of small particles to a gram. Hence if the grades are weighed, the resulting frequency distribution will give greater significance to the larger sizes, where a few pebbles will outweigh a tremendous number of fine grains. The curve, then, may be inclined toward the coarser sizes. If the grades are counted, on the other hand, the several large pebbles would be quite negligible in contrast to the millions of fine particles.

Conventional statistical measures are defined in terms of number frequency, represented by N, the total number of individuals. One may raise the question, however, whether it is not possible to redefine exactly the same types of measures in terms of weight frequency. It is possible to set up statistical measures on a weight basis (or weight percentage) which are directly applicable to conventional usage and may be related to probabilities, areas under curves, and the like equally as conveniently as number measures. It will not be true in general, however, that there is any necessary simple relation between the measures defined on a weight basis and the measures defined by number.

It would be convenient, however, to know whether number or weight is a more important concept in the interpretation of sediments. As far as the writers are aware, no mathematical statistician has attacked the problem, and the following discussion is to be taken as a tentative qualitative evaluation of the problem. Part of the discussion will be based on the only apparent research that has been done along these lines, and part on geologic reasoning. In 1933 Hatch[1] showed that, if a frequency distribution of grains is symmetrical when plotted on a logarithmic basis, there is a simple relation between the weight frequency curve and the number frequency curve. In such cases only two parameters are involved, which Hatch defined as the log *geometric mean* and the *log standard geometric deviation*. His demonstration showed that the log standard deviation remains constant when the frequency is changed from weight to number, but the log geometric mean diameter changes from one distribution to the other. Thus two frequency curves of the same sediment, symmetrical on logarithmic plotting, one based on number frequency and the other on weight frequency, have the relations[2] shown in Figure 105. Note that the number frequency curve lies to the right of the weight frequency curve, which means that the average value has shifted toward smaller sizes, an expected result due to the greater numerical significance of many smaller grains as opposed to a few larger grains.

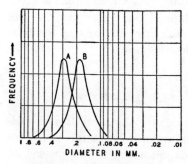

Fig. 105.—Weight-percentage frequency curve (A) and number-percentage frequency curve (B) of same sediment (a beach sand).

When the frequency curve is skewed on a logarithmic basis, the simple relation between weight and number frequency no longer holds, inasmuch as an additional parameter enters the situation. However, the fact that

[1] T. Hatch, Determination of "average particle size" from the screen-analysis of non-uniform particulate substances: *Jour. Franklin Inst.*, vol. 215, pp. 27-37, 1933.

[2] The relations between the curves are determined graphically by the use of logarithmic probability paper (page 189). The necessary condition is that the cumulative data plot as a straight line. The corresponding frequency on the alternate basis is then found by translating the curve parallel to itself in accordance with an equation given by Hatch (*loc. cit.*). The curves of Figure 105 were obtained by graphic differentiation of the cumulative curves.

there is a relation between the symmetrical curves suggests that there also is a relation between skewed curves, however complex that relation may be.

From a geological point of view, the problem of frequency may be considered from several angles. In terms of the kinetic energy of the transporting medium, the work performed in moving a pebble varies directly as its mass (weight), assuming the stream velocity constant. Thus there is a physical relation between weight and energy. With a given geometrical form, on the other hand, number may be converted into weight, and hence there is also a relation between number and energy. Whether number or weight may be taken as the more significant value, however, is not clear. If surface area of a given geometrical form is considered, it is possible to relate surface area either to weight or to number of particles. Unfortunately in none of these cases is the mathematical relation simple, especially in a distribution of diverse shapes, densities, and diameters.

If convenience be taken as a criterion, there is little doubt that in the average case weight is a variable more readily determined than number, especially among fine grains. With gravels it may be more convenient to count the pebbles; likewise microscopic methods usually involve counting the grains. As long as current methods of analysis remain in use, namely, sieving and sedimentation, then weight will be more convenient to use than number. However, if microscopic measurements of limited samples increase in use as a device, then number will be more convenient than weight. The number of factors entering the problem suggests that for immediate purposes it may be immaterial whether one or the other is used, as long as measures based on weight be not directly confused or compared with measures based on number. The safest procedure to follow is to indicate in the published results the type of frequency data involved and to define the measures specifically in connection with the variables used.

One important point may be stressed here: frequency is always the dependent variable, and although the physical interpretation of the data may vary with the manner of expressing frequency, the geometrical significance of the statistical measures is the same regardless of the particular choice of frequency used.

CHAPTER 9

APPLICATION OF STATISTICAL MEASURES TO SEDIMENTS

INTRODUCTION

AT least three points of view have been manifested by sedimentary petrologists in the application of statistical measures to sedimentary data. One group has developed measures designed to furnish a series of numbers for each sample, as an aid in describing and classifying sediments. This group has not concerned itself directly with statistical theory, on the ground that conventional devices furnish too few numbers for detailed work.

A second group has applied conventional statistical measures to sediments, so that the relation of the measures to the body of statistical theory could be known. The contention is that unless the measures can be related to the background of statistical theory, little more can be done than to classify sediments; that the relation of the measures to environmental factors which control the characteristics of sediments cannot be brought out with arbitrary measuring devices.

A third group includes those who maintain that statistical procedures are essentially meaningless as applied to sedimentary data. The errors of sampling and analysis are so large, the contention runs, that the data have little quantitative significance, and no matter how much statistical manipulation is involved, the final data are no better than the original.

It is true that statistical manipulation cannot create data, but by applying statistical reasoning it is possible to determine how large the errors are and to devise corrections so that more reliable data may be obtained. Furthermore, statistical operations furnish a means of summarizing large amounts of information in a convenient manner, such that comparisons and descriptions are greatly simplified. Workers as a whole are becoming increasingly aware of the advantages of a statistical approach, although there is as yet little tendency to standardize the techniques.

It is the opinion of the authors that a fuller understanding of the

STATISTICAL METHODS

significance of statistical devices will do much in clearing the present confusion about means and ends. Toward that end, the discussion in the present chapter is concerned as much with the geometrical meaning of the measures as it is with the mere mechanical process of arriving at the numbers.

QUARTILE MEASURES

Quartile measures are perhaps as widely used as any other device for describing and comparing sediments. The first use of quartile measures in sedimentary data was by Trask,[1] who introduced a set of geometric quartile measures in 1930, and discussed the theory of them more fully in 1932. In 1933 Krumbein[2] used conventional arithmetic quartile measures for describing glacial tills. In 1934 Miss Gripenberg[3] demonstrated that Trask's geometric measure of spread was related to the logarithmic probability curve, and in 1936 Krumbein[4] showed the relations among arithmetic, geometric, and logarithmic quartile measures in terms of conventional statistical theory.

The great advantage of quartile measures is the ease with which they are determined from the analytical data. Three values usually suffice for the computation of the measures. These are the median and the first and third quartiles, each read directly from the cumulative curve.

The median. The median diameter is defined[5] as the middlemost member of the distribution; it is that diameter which is larger than 50 per cent of the diameters in the distribution, and smaller than the other 50 per cent. For its graphic determination, therefore, it is only necessary to draw a cumulative curve of the sediment and to read the diameter value which corresponds to the point where the 50-per cent line crosses the cumulative curve, as shown in Figure 106.

The median has the advantage that it is not affected by the extreme grains on either end of the distribution, and it is not necessary to have the complete analysis to determine it. Among the disadvantages of the median are that it cannot be manipulated algebraically; that is, the

---

[1] P. D. Trask, Mechanical analysis of sediments by centrifuge: *Econ. Geology*, vol. 25, pp. 581-599, 1930. P. D. Trask, *Origin and Environment of Source Sediments of Petroleum* (Houston, Texas, 1932), pp. 67 ff.
[2] W. C. Krumbein, Lithological variations in glacial till: *Jour. Geology*, vol. 41, pp. 382-408, 1933.
[3] Stina Gripenberg, A study of the sediments of the North Baltic and adjoining seas: *Fennia*, vol. 60, no. 3, 1934, pp. 214 ff.
[4] W. C. Krumbein, The use of quartile measures in describing and comparing sediments: *Am. Jour. Sci.*, vol. 37, pp. 98-111, 1936.
[5] F. C. Mills, *Statistical Methods* (New York, 1924), p. 112.

medians of each grade cannot be averaged to give the median of the distribution. This disadvantage is not great, however, inasmuch as graphic methods are generally used in determining the median.

Quartile deviation. The measure of average spread which is commonly used with the median is the quartile deviation. The quartiles lie on either side of the median and are the diameters which correspond to frequencies of 25 and 75 per cent. By convention, the smaller diameter value is taken as the first quartile, Q_1. It is that diameter which has 25 per cent of the distribution smaller than itself and 75 per cent larger than itself. It is found from the cumulative curve by reading the diameter value which corresponds to the point where the 75-per cent line intersects the cumulative curve. The third quartile, Q_3, is that diameter which has 75 per cent of the distribution smaller than itself, and 25 per cent larger than itself. It is found by determining the diameter value corresponding to the intersection of the 25-per cent line and the cumulative curve. (See Figure 106.)

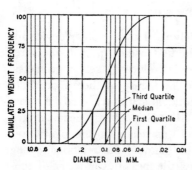

Fig. 106.—Method of reading median and quartiles from cumulative curve.

Three types of quartile deviation are used in sedimentary petrology, depending upon the particular features of the curve which are to be emphasized. They are the arithmetic, geometric, and logarithmic quartile deviations. For convenience these three types of measures will be discussed together, the better to bring out their relations to one another.

The simplest form of quartile deviation is the arithmetic quartile deviation,[1] QD_a which is a measure of half the spread between the two quartiles. The difference is so chosen that positive values always result:

$$QD_a = (Q_3 - Q_1)/2 \quad . \quad . \quad . \quad . \quad . \quad (1)$$

The second possibility is a geometric quartile deviation, QD_g, which is based on the ratio between the quartiles, instead of on their differences. Specifically it is the square root of the ratio of the two quartiles, so chosen that the value is always greater than unity:

$$QD_g = \sqrt{Q_3/Q_1} \quad . \quad . \quad . \quad . \quad . \quad (2)$$

Trask[2] introduced this measure as a "sorting coefficient," but owing to

---

[1] F. C. Mills, *op. cit.* (1924), p. 158.
[2] P. D. Trask, *op. cit.* (1932).

his reversal of the usual definition of the quartiles, his equation is $So = \sqrt{Q_1/Q_3}$. This is identical with equation (2), however, as long as the larger quartile is used as numerator. For convenience, then, the symbol So will be used for equation (2), *with the understanding that the larger quartile is always in the numerator.*

Finally, there is a log quartile deviation, which is equal to half the difference between the logs of the quartiles. This measure is simply the log (to any base) of equation (2):

$$\text{Log } QD_g = \log So = (\log Q_3 - \log Q_1)/2 \quad \ldots \quad (3)$$

Obviously equation (3) may be computed either with the logs directly, or the log of So may be found in logarithmic tables.

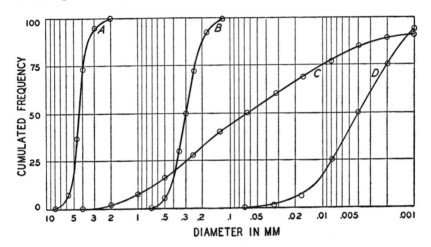

Fig. 107.—Examples of cumulative curves. A, beach gravel; B, beach sand; C, glacial till; D, a Pennsylvanian underclay. (After Krumbein, 1936.)

These three measures each have certain characteristics which may be examined by means of a few samples. Figure 107 shows the cumulative curves of four sediments; a beach gravel, a beach sand, an underclay, and a glacial till. Table 20 lists the median and quartiles of these sediments as well as the quartile measures defined by equations (1), (2), and (3), the latter measure to the base 10.

Inasmuch as the QD_a measures the difference between the quartiles, its value depends both on the size of particles involved and on the units of measurement used. The values for the beach sand and the gravel illustrate this size factor; similarly, if centimeters were used instead of millimeters, the QD_a for the beach sand would be 0.0060 instead of 0.060.

Thus the arithmetic quartile deviation does not directly compare the relative spread of the curves, because the size factor colors the result. This does not imply that the QD_a finds no use in sedimentary petrology; on the contrary, where the size factor is to be brought in, it shows up most clearly with this arithmetic form of the quartile deviation.

TABLE 20

COMPARISON OF QUARTILE MEASURES OF UNDERCLAY, GLACIAL TILL, BEACH SAND, AND BEACH GRAVEL *

Sediment	Median (mm.)	Q_1 (mm.)	Q_3 (mm.)	QD_a	So	Log_{10}So
Underclay	0.004	0.002	0.008	0.003	2.00	0.301
Glacial till	0.062	0.010	0.290	0.140	5.48	0.740
Beach sand	0.300	0.240	0.360	0.060	1.22	0.087
Beach gravel	4.420	3.900	4.970	0.535	1.13	0.054

* Data from Figure 107.

The geometric quartile deviation, So, being essentially a ratio between the quartiles, at once eliminates the size factor and the units of measurement. Thus in Table 20 the beach sand and the beach gravel have very similar values for So, showing that the relative spread of the curves is very much the same. This is borne out by inspection of the first two curves in Figure 107. Here the difference in coarseness has no bearing on the sorting coefficient, nor would it make any difference if centimeters were used instead of millimeters as the units of measurement. Thus in general, So is a convenient measure to use for describing the spread of the curve, uninfluenced by size factors.

On the basis of nearly 200 analyses, Trask found that a value of So less than 2.5 indicates a well sorted sediment, a value of about 3.0 a normally sorted sediment, and a value greater than 4.5 a poorly sorted sediment. These numbers do not, however, lend themselves directly to a visualization of what they signify in terms of the actual spread of the curve. That is, one cannot say that a sediment with $So = 3.0$ is twice as widely dispersed (i.e., half as well sorted) as another sediment with $So = 1.5$. This is because the values of So are geometric rather than arithmetic. However, it is a simple matter to transform the values of So into measures that may be directly compared with each other.

It is here that the significance of the log quartile deviation becomes apparent. Since So increases geometrically, the logs of So will form an arithmetic series, so that the values of log So may be directly compared with each other. The last column of Table 20 lists the logs of So to the base 10, as an illustration. By comparing the beach sand and the underclay, for example, we may see that the spread of the grains in the underclay is some 3.4 times as great as that in the sand, because 0.301/0.087 = 3.4. This is the same as saying that the sand is 3.4 times as well sorted as the underclay, but this information cannot be read by comparing the So values directly.

Although logs to the base 10 are most convenient to use in ordinary cases, it is possible to choose logs to such a base that they describe the sediments in terms of some easily visualized characteristic. For example, if one were able to say that some sediment A had two Wentworth grades between the first and third quartiles, as against sediment B, which had three Wentworth grades between the quartiles, one would have not only an easily visualized measure, but as well one that would satisfy the condition that the relative spread of the curves could be directly compared.

Application of the phi scale to the quartile deviation.[1] It is in connection with logarithmic measures that the phi scale (see Chapter 4) is most useful. Figure 108 shows the ordinary diameter scale above and the phi scale below. Each Wentworth class limit is an integer, and the phi scale increases with

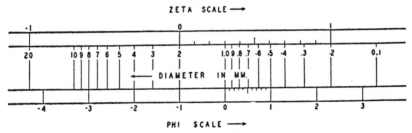

FIG. 108.—Relations between logarithmic grade scales and diameters in millimeters. The "zeta scale" is adapted to Atterberg grades, and the "phi scale" to Wentworth grades.

decreasing grain sizes. Since the phi intervals are equal, ordinary arithmetic graph paper may be used in plotting curves, and the median and quartiles may be read off in phi values directly, to the nearest tenth or hundredth, as desired. Figure 109 illustrates a case; it is the glacial till of Figure 107 superimposed on the phi scale. The curve is in no wise changed; only the independent variable has been changed, and the graph paper is arithmetic instead of logarithmic. The position of the quartiles is conventional also, and since phi increases to the right, Q_3 is greater than Q_1. The three values obtained from this curve, determined in the usual manner of reading the

[1] This discussion is based largely on the paper by Krumbein, *loc. cit.,* 1936.

median and quartiles, and expressed in ϕ terms, are $Md_\phi = 4.00$; $Q_{1\phi} = 1.80$; $Q_{3\phi} = 6.70$. Inasmuch as the independent variable ϕ forms an arithmetic series, it is possible to substitute these values directly in equation (1) for the arithmetic quartile deviation in phi terms, calling the result QD_ϕ to indicate that phi values are used: $QD_\phi = (6.70 - 1.80)/2 = 2.45$.

The geometrical significance of this value in terms of the curve may readily be seen. Since the phi values are expressed in Wentworth grades as units, the difference between the quartiles indicates directly how many Wentworth grades lie between the first and third quartile, and half this value is the quartile deviation. Thus in the glacial till, the first and third quartiles are

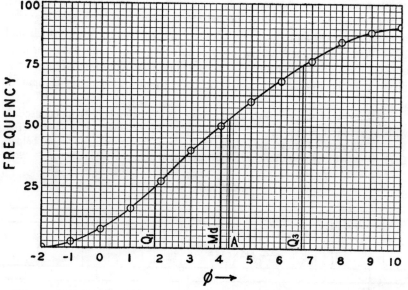

Fig. 109.—Cumulative curve of glacial till (curve C of Fig. 107), plotted with ϕ as independent variable. (After Krumbein, 1936.) See page 237 for ordinate A.

spread over a distance of 4.9 Wentworth grades, and consequently the curve is decidedly drawn out. (This spread of the curve may actually be checked by laying a ruler between the quartiles in Figure 109.)

The value $QD_\phi = 2.45$ may be converted to QD_g by finding its antilog. Because of the reversal of scale direction which follows the use of ϕ, the third quartile in phi terms corresponds to the first quartile in diameter terms in its position on the scale. This means that $QD_\phi = \log_2 QD_g$, rather than the negative log, as one may expect. To find the antilog of QD_ϕ, accordingly, one may use the relation $\log_{10} n = \log_{10} 2 \log_2 n$, where $\log_{10} 2 = 0.301$. Substituting QD_g for n, and QD_ϕ for $\log_2 QD_g$, there results $\log_{10} QD_g = 0.301\ QD_\phi$. Using the value 2.45 for QD_ϕ one obtains $\log_{10} QD_g = 0.738$. The antilog of this is 5.47. This is the value of So shown for the glacial till in Table 20.

Instead of transforming the logarithmic measure to its diameter equivalent by the process outlined, it is convenient to prepare a graph which permits a direct conversion. Figure 110 is such a chart, showing the values of So from 1 to 10 on the vertical logarithmic scale, and $\log_2 So = QD_\phi$ on the horizontal scale. Note that when So equals 2, 4, 8, QD_ϕ equals 1, 2, 3, respectively. The latter values tell how many Wentworth grades are involved in half the spread between the quartiles, for any given value of So. Thus when So = 5.48, QD_ϕ = 2.45, which means that 4.9 Wentworth grades lie between the two quartiles, and the curve is drawn out, as curves of glacial till usually are. Since QD_ϕ is a logarithm, it may be used directly in comparing the relative spread of two or more curves, as noted above, and furthermore this comparison is expressed directly in terms of the number of Wentworth grades involved. The value of the phi notation in this case is that the "sorting" values are expressed in terms of Wentworth grades, which are easily visualized. This result arises from a deliberate choice of ϕ to satisfy these conditions.

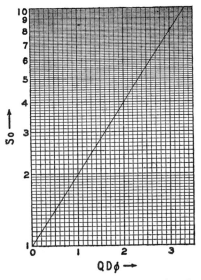

FIG. 110.—Conversion chart for So and $QD\phi$.

Quartile skewness. In a perfectly symmetrical curve, the median exactly coincides with the point half way between the first and third quartiles, but if the curve is skewed, the arithmetic mean of the quartiles departs from the median, and the extent of this departure may be taken as a measure of the skewness. Further, the direction of departure may be included in the measure by assigning positive and negative senses to the two possible directions. The simplest form of this skewness measure is the arithmetic case, Sk_a, which expresses the difference between the arithmetic mean of the quartiles and the median:

$$Sk_a = [(Q_1 + Q_3)/2] - Md = \tfrac{1}{2}(Q_1 + Q_3 - 2Md) \quad . \quad . \quad . \quad (4)$$

This skewness measure may be cast into a geometric form, which is the square root of the ratio of the product of the quartile to the square of the median:

$$Sk_g = \sqrt{Q_1 Q_3 / Md^2} \quad . \quad . \quad . \quad . \quad (5)$$

There is an interesting relation between this measure of skewness and the form introduced by Trask,[1] here referred to simply as Sk. In developing his measure, Trask compared the ratio of the largest quartile and the median to the ratio of the median and the smaller quartile, thus: $Sk = Q_3/Md / Md/Q_1$. This simplifies to $Sk = Q_1 Q_3 / Md^2$, which is obviously the square of Sk_g.

The relation between Trask's measure and Sk_g may be seen readily by considering the third possibility, the log geometric skewness, log Sk_g:

$$\log Sk_g = \tfrac{1}{2}(\log Q_1 + \log Q_3 - 2 \log Md) \quad \ldots \quad (6)$$

This is obviously the log of equation (5) to any base. If one takes logs of Trask's skewness measure, it will be noted that $\tfrac{1}{2} \log Sk = \log Sk_g$.

Table 21 offers a comparison of the skewness values of the four sediments of Table 20, in terms of equations (4), (5), (6). In addition, the last column includes a phi-skewness to be introduced below.

TABLE 21

COMPARISON OF QUARTILE SKEWNESS OF UNDERCLAY, GLACIAL TILL, BEACH SAND, AND BEACH GRAVEL *

Sediment	Sk_a	$Sk = (Sk_g)^2$	$2 \log_{10} sk_g = \log_{10} Sk$	$Skq\phi$
Underclay	+0.001	1.00	0.000	0.000
Glacial till	+0.088	0.74	—0.130	+0.250
Beach sand	0.000	0.96	—0.018	+0.030
Beach gravel ...	—0.053	0.99	—0.004	+0.007

* Data from Figure 107.

The arithmetic skewness, Sk_a, is subject to the same comments that apply to the arithmetic quartile deviation, QD_a, inasmuch as the size factor and the units of measurement enter its values. Thus when QD_a is used to bring out the size factor, presumably the corresponding Sk_a is the best measure to supplement it. In this connection, however, see the comments under kurtosis (page 238) concerning an arithmetic measure of skewness independent of size.

The geometric skewness, or its square, which is identical with Trask's skewness, eliminates the size factor and units of measurement from the resulting values, so that it is a descriptive measure independent of these two factors. When the curve is symmetrical, this skewness is equal to unity, but the values obtained range from numbers less than 1 to numbers larger. As Trask himself points out, the significance of this depends on

[1] P. D. Trask, *op. cit.* (1932).

STATISTICAL METHODS

the fact that numbers less than 1 present a reciprocal relation to numbers greater than 1, so that actually the spread is greater on one side of the curve in the one case, and on the other side in the other case. Thus Sk is itself not an easily visualized measure, because reciprocals are often hard to visualize. For this reason Trask introduced $\log_{10}$Sk, which is positive when Sk is greater than unity, and negative when Sk is less than unity.

Again, however, a measure based on $\log_2$Sk will yield easily visualized values, because such a measure will directly express the skewness in terms of its definition, namely, the extent to which the mean of the quartiles departs from the median. When the median and quartiles are expressed in ϕ values,

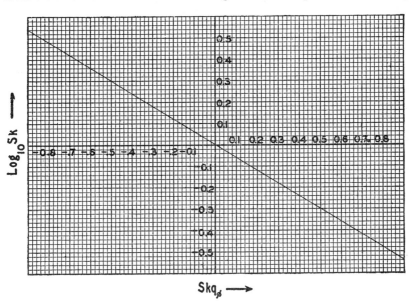

Fig. 111.—Conversion chart for $\log_{10}$Sk and Skq_ϕ.

equation (4) may be used. The ϕ-values for the till in Figure 109 were, it will be recalled, $Md_\phi = 4.00$; $Q_{1\phi} = 1.80$; $Q_{3\phi} = 6.70$. By substitution of these values in equation (4) there results $Skq_\phi = (1.80 + 6.70 - 2[4.00]) = + 0.25$, where Skq_ϕ is the symbol for the phi quartile skewness.[1]

Figure 109 illustrates the geometrical meaning of Skq_ϕ. The ordinate A marks the position of the arithmetic mean of Q_1 and Q_3 expressed in phi terms. Its value is $(Q_1 + Q_3)/2 = (1.80 + 6.70)/2 = 4.25$. Note that $4.25 - 4.00 = + 0.25$, the value of Skq_ϕ. Thus the mean of the quartiles lies 0.25

[1] Elsewhere the symbol Sk_ϕ is used to indicate the phi skewness based on the third moment of the distribution, hence Skq_ϕ is used for the present measure.

of a Wentworth grade to the right of Md_ϕ, and the curve is skewed in the direction of the positive ϕ axis. (This direction of skew is indicated by the $+$ sign; a negative value would indicate that the mean of the quartiles lies toward smaller values of ϕ; i.e., to the left.)

The relation of Skq_ϕ to Trask's Sk may be found directly from the relation between Trask's Sk and equation (6) above: $Skq_\phi = -\log_2 Sk_g = -\frac{1}{2} \log_2 Sk = \frac{1}{2} \log_2 (1/Sk)$. In words, Skq_ϕ is one-half the log to the base 2 of the reciprocal of Trask's Sk. This relation yields to a conversion chart, however, if $\log_{10} Sk$ is used instead of Sk. Figure 111 is such a chart, based on the equation $Skq_\phi = (\log_{10} Sk)/-0.602$, the necessary transformation equation for the conversion of one symbol to the other.

In Table 21 the phi quartile skewness of the four sediments is included in the last column. These values are directly interpretable in terms of Wentworth grades. Thus, the underclay is symmetrical because Skq_ϕ is zero; the beach sand is skewed 0.03 to the right; and the gravel is skewed in the same direction, but to a less extent. It is to be noted that the sign before Skq_ϕ is opposite to that before $\log_{10} Sk$. This is a convention in terms of the change of variable, and is consistent with the phi notation. The curve, whether expressed in phi terms or in diameters, is of course skewed in the same direction; merely the sense assigned to the direction is changed.

Quartile kurtosis. The degree of peakedness of a curve is measured by its kurtosis, which may be defined in various ways. Essentially the kurtosis involves a comparison of the spread of the central position of the curve to the spread of the curve as a whole. To obtain a measure of kurtosis, one may adopt Kelley's equation,[1] which is the ratio of the quartile deviation to that part of the size range which lies between the 10-per cent and 90-per cent lines. The latter values may be referred to as the tenth percentile, P_{10}, and the ninetieth percentile, P_{90}. The arithmetic quartile kurtosis may then be written as

$$Kq_a = \frac{\frac{Q_3 - Q_1}{2}}{P_{90} - P_{10}} = (Q_3 - Q_1)/2(P_{90} - P_{10}) \quad . \quad . \quad . \quad (7)$$

Equation (7), it will be noted, is independent of size or the units of measurement used, inasmuch as it represents a ratio of two spreads. In this manner it differs from the quartile deviation (equation 1), and the quartile skewness (equation 4), both of which are influenced by size factors.[2]

It is not feasible to introduce a simple geometrical measure of kurtosis based on equation (7); however, the corresponding phi quartile kurtosis, Kq_ϕ, may be obtained merely by using phi values in equation (7). The geometrical picture of the measure is the same as in the arithmetic case, except that it applies to the logarithmic curve instead of to the arithmetic grain diameter distribution.

[1] T. L. Kelley, *Statistical Methods* (London, 1924), p. 77.
[2] In conventional statistical practice, especially in connection with moment analysis, both skewness and kurtosis are chosen to be independent of size. An arithmetic quartile skewness having this attribute, and expressed essentially in units of quartile deviation, is discussed by F. C. Mills, *op. cit.* (1924), p. 167.

STATISTICAL METHODS

The kurtosis as defined above, as well as its phi analogue, yields values which decrease with increasing peakedness, in the sense that as the cluster of values in the central part of the curve becomes more pronounced, without a corresponding decrease in the total spread of the curve, the ratio of $(Q_3 - Q_1)/2$ to $(P_{90} - P_{10})$ decreases also. If one prefers a kurtosis value which increases, the reciprocal of equation (7) appears to be suitable.

MOMENT MEASURES

Despite the wide usage of moments in conventional statistics, they have found relatively little application in sedimentary analysis until recently. The earliest applications of moments to sedimentary data were made by Van Orstrand [1] in 1924; and by Wentworth [2] and Hatch and Choate [3] in 1929. Van Orstrand discussed the possibility of representing frequency distributions of sediments by means of Pearson frequency functions. He computed the mode, the arithmetic mean, standard deviation, and skewness of sands, arguing in favor of arithmetic measures based on equal class intervals. Wentworth used logarithmic methods of computing his moments, but referred to his measures as though they were arithmetic instead of logarithmic in nature. Hatch and Choate defined their measures as log geometric moment measures and confined their theory largely to the moments of logarithmically symmetrical curves which could be treated graphically. In 1936 Krumbein [4] developed a series of logarithmic moment measures by means of a logarithmic transformation equation and showed the relation of his measures to the body of statistical theory.

A consideration of moments as applied to sediments should distinguish between arithmetic and logarithmic measures, because the geometrical and physical significance is considerably different in the two cases. As in the discussion of quartile measures, the treatment of arithmetic, geometric, and logarithmic measures will be carried on simultaneously, so that similarities and differences may be brought out as the discussion proceeds.

[1] C. E. Van Orstrand, Note on the representation of the distribution of grains in sands: *Researches in Sedimentation in 1924*. Nat. Research Council, 1925.
[2] C. K. Wentworth, Method of computing mechanical composition types in sediments: *Geol. Soc. America, Bulletin*, vol. 40, pp. 771-790, 1929.
[3] T. Hatch and S. Choate, Statistical description of the size properties of non-uniform particulate substances: *Jour. Franklin Inst.*, vol. 207, pp. 369-387, 1929.
[4] W. C. Krumbein, Application of logarithmic moments to size frequency distributions of sediments: *Jour. Sed. Petrology*, vol. 6, pp. 35-47, 1936.

MEASURES OF THE CENTRAL TENDENCY

The arithmetic mean of the diameter distribution. The arithmetic mean [1] of the diameter distribution is most conveniently calculated from the frequency distribution, although it may be determined graphically.[2] In computing the arithmetic mean, a procedure such as that shown in Table 22 may be used. The actual grades in millimeters are listed in the first column, and the weight percentage frequency is placed in the second column. The third column has the actual midpoints (m) of each grade, regardless of whether the grades are equal or unequal in interval. In the fourth column the frequency has been multiplied by the midpoint, (fm), and the sum of the numbers in this column is 43.97. This sum, divided by the total frequency (100), yields the arithmetic mean diameter in millimeters, $M_a = 0.440$ mm., approximately.

TABLE 22

COMPUTATION OF THE ARITHMETIC MEAN DIAMETER OF A BEACH SAND*

Grade Size (mm.)	Weight Percentage Frequency (f)	m	fm
4—2	0.5	3.0	1.50
2—1	5.6	1.5	8.40
1—1/2	11.7	0.75	8.78
1/2—1/4	53.7	0.375	20.15
1/4—1/8	26.4	0.187	4.94
1/8—1/16	2.1	0.093	0.20
Totals	100.0		43.97

* Sample 22 of F. J. Pettijohn and J. D. Ridge, A textural variation series of beach sands from Cedar Point, Ohio: *Jour. Sed. Petrology,* vol. 2, pp. 76-88, 1932.

The arithmetic mean diameter in millimeters represents the diameter-value of the center of gravity of the frequency distribution. The arithmetic mean is affected by every grain in the distribution, and in some respects it is therefore more typical of the grain distribution than the median. The arithmetic mean may be manipulated algebraically.

The geometric mean.[3] The geometric mean diameter of sediments has not been used extensively in sedimentary work, although in significance it appears to rank as more important than some other means that have

[1] F. C. Mills, *op. cit.* (1924), pp. 113 ff.
[2] See page 255, under Baker's equivalent grade.
[3] F. C. Mills, *op. cit.* (1924), pp. 135 ff.

been used. The geometric mean diameter is defined as the nth root of the product of n items, and its direct computation is a tedious process. By means of a simple logarithmic device, however, a fair approximation of the geometric mean may be obtained from the frequency data.[1] The method used is similar to that used in computing the arithmetic mean, except that the logarithms of the midpoints of each grade are substituted for the midpoint itself. Table 23 illustrates the method. The first column lists the grade sizes, the second column shows the weight percentage frequency (f), the third column shows the midpoint (m) of each grade, and the fourth column has the log [2] of the midpoint to the base 10. The

TABLE 23

COMPUTATION OF THE GEOMETRIC MEAN DIAMETER OF A BEACH SAND*

Grade Size (mm.)	f	m	log m	f log m
4—2	0.5	3.0	+ .477	+ 0.238
2—1	5.6	1.5	+ .176	+ 0.097
1—1/2	11.7	0.75	— .125	— 1.46
1/2—1/4	53.7	.375	— .426	—22.82
1/4—1/8	26.4	.1875	— .727	—19.20
1/8—1/16	2.1	.0937	—1.028	— 2.16
Totals	100.0			—45.40

* The frequency data are the same as in Table 22.

last column has the products fm. The algebraic sum of the products is —45.40. This is divided by the total frequency, 100, to yield —0.454, which is the log of the geometric mean. To convert it to a value which may be found in common log tables, it is added to 10.000 — 10, yielding 9.546 — 10, the antilog of which is 0.352 mm.

The geometric mean diameter is noticeably smaller than the arithmetic mean diameter, as computed for the same sediment in Table 22. This means that the geometric mean lies to the right of the arithmetic mean as usually plotted, in the cluster of grains near the higher part of the frequency curve. It is thus associated with the most abundant grains in

[1] For a graphic method applicable in some cases, see page 254.
[2] For numbers smaller than unity, the logarithm as obtained in ordinary log tables must be converted to its cologarithm. For example, log 0.750=9.875—10. By adding +9.875 and —10.000, one obtains —0.125, the value used in the computations in Table 23.

an asymmetrical distribution. The geometric mean, like the arithmetic mean, is affected by every grain in the distribution, but the geometric mean is not affected to the same degree. The geometric mean may be manipulated algebraically.

The logarithmic mean. The use of a logarithmic mean, expressed and used directly as a logarithm, has received relatively little attention in sedimentary work. Such a mean is defined as the arithmetic mean of the logarithmic frequency distribution, and it is most conveniently computed by transforming the grade scale in millimeters to the logarithmic phi scale or zeta scale mentioned in Chapter 4 and, in connection with quartile measures, in the present chapter. The logarithmic mean may be computed in a manner exactly analogous to that of the arithmetic mean, using, however, the midpoints of the logarithmic grades.

Table 24 shows the limits of the diameter classes in the first column. The corresponding phi values from table 10 (page 84) are listed in the second column. The frequency (f) is shown in the third column, and the midpoints of the phi classes are shown in the fourth column. The products fm are listed in the fifth column. The algebraic sum of the numbers in the last column, +156.15, divided by the total frequency (100), yields the arithmetic mean of the phi distribution, called the "phi mean," $M_\phi = 1.651$.

TABLE 24

COMPUTATION OF THE LOGARITHMIC MEAN OF A BEACH SAND*

Grade Size (mm.)	ϕ	f	m	fm
4—2	−2— −1	0.5	−1.5	− 0.75
2—1	−1— 0	5.6	−0.5	− 2.80
1—1/2	0— 1	11.7	+0.5	+ 5.85
1/2—1/4	1— 2	53.7	+1.5	+80.50
1/4—1/8	2— 3	26.4	+2.5	+66.00
1/8—1/16	3— 4	2.1	+3.5	+ 7.35
Totals		100.0		+156.15

* The frequency data are the same as in Table 22.

The phi mean is the center of gravity of the logarithmic frequency curve, expressed with ϕ as the independent variable. It thus has exactly the same relation to the logarithmic curve as the arithmetic mean of the diameters has to the frequency curve drawn with diameters in millimeters as the independent variable. One should not confuse the two, however. The phi mean, when transformed to its diameter equivalent, becomes the geometric mean of the size distribution. In words the phi mean is the negative log to the base 2 of the geometric mean of the grain diameters.[1]

The phi mean may be used directly in describing sediments, in connection with other logarithmic measures.

The method used in computing M_ϕ may also be used as a more accurate method of finding the geometric mean of the diameters. To convert the phi mean to its diameter equivalent, the antilog of 1.561 must be found. To convert any value in the phi notation to its corresponding diameter equivalent in millimeters, the relations $\phi = -\log_2 \xi$ and $\log_{10} \xi = \log_{10} 2 \; \log_2 \xi$ are used, where $\log_{10} 2 = 0.301$. Substituting $-\phi$ for $\log_2 \xi$, the relation $\log_{10} \xi = -0.301 \phi$ is obtained, and from this equation the antilog of $\log_{10} \xi$ may be found in logarithmic tables. In the example given, $M_\phi = 1.561$. Multiplying this by -0.301 yields -0.469. The colog of -0.469 is obtained by adding this value to $10.000 - 10$, which yields $9.531 - 10$. The antilog of the latter expression, from any common table of logs to the base 10, is 0.340 mm.[2]

It is more convenient to use a graphic method for converting phi values to their diameter equivalents. Figure 112 is such a graph, showing ϕ as ordinate and diameters in millimeters as abscissae. The value $\phi = 1.561$ is chosen along the vertical scale, and where this value intersects the diagonal line an ordinate is dropped to the millimeter scale, yielding the value 0.340 mm.

The mode. An average which is used rather frequently in conventional statistics is the mode.[3] The modal grain diameter may be defined as that diameter which is most frequent in the distribution. The mode, therefore, lies directly at the peak of the curve, and it may be determined graphically either by locating the highest point of the frequency curve, or by finding the point of inflection of the cumulative curve. The modal diameter of the beach sand in Table 22 is 0.300 mm. The mode has not been extensively used in sedimentary work, but it is an average which represents the most abundant, and therefore the most typical, grain in the distribution. The mode, like the median, is an average of position, and is not affected by extreme grain sizes in the distribution.

---

[1] The arithmetic mean of a series of phi values is $M_\phi = \dfrac{\phi_1 + \phi_2 + \ldots + \phi_n}{n}$ and by substituting $-\log_2 \xi$ for ϕ, one obtains $M_\phi = -\dfrac{\log_2 \xi_1 + \log_2 \xi_2 + \ldots + \log_2 \xi_n}{n}$

Such a sum of logarithms is equal to the nth root of the product of the antilogs, which yields $\sqrt[n]{\xi_1 \xi_2 \ldots \xi_n}$. By the definition of the geometric mean, however, this last expression is seen to be the geometric mean itself. Hence the arithmetic mean of the phi distribution is a log of the geometric mean of the diameters.

[2] The value of the geometric mean obtained by the phi method is about 3 per cent smaller than that found by the earlier method of computing the geometric mean. This difference depends upon the precise midpoint used in the computation. The phi method uses the midpoint of the logarithmic classes directly and yields a more rigorous value. However, the approximation furnished by the first method is sufficiently close for most work.

[3] F. C. Mills, *op. cit.* (1924), pp. 124 ff.

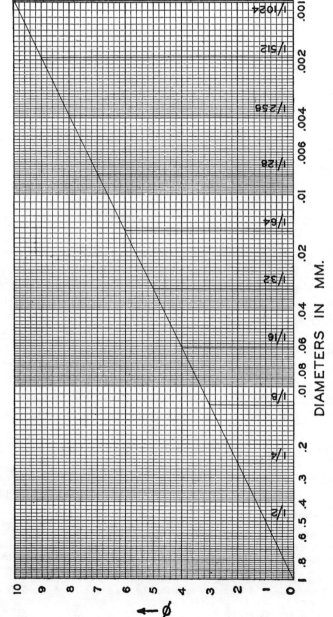

FIG. 112.—Conversion chart for ϕ and diameters in mm., for the range 1/1024-1 mm.

The modal diameter cannot be manipulated algebraically, and it has its most precise meaning, perhaps, in a unimodal distribution, with a single peak. Curves which display several peaks have a corresponding number of modal diameters, but it is not appropriate in such cases to refer to the mode unless one of the peaks predominates strikingly over

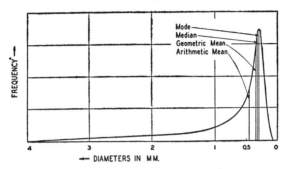

Fig. 113.—Frequency curve of beach sand with *diameter in mm.* as independent variable. The several averages are from Table 25.

the others. Frequency curves of glacial till, either on arithmetic or logarithmic scales, commonly display several modes.

Comparison of average values. The variety of averages discussed in preceding sections indicates the wide choice of measures available for

Table 25

Comparison of Average Values Computed in Tables 22 to 24

Average	Value
Arithmetic Mean, M_a	0.440 mm.
Geometric Mean, M_g	0.352 mm.
Logarithmic Mean, $M\phi$	1.561*
Median Diameter, Md.	0.320 mm.
Modal Diameter, Mo	0.300 mm.

* The diameter equivalent of this value, 0.340 mm., is the geometric mean. See footnote 2 on page 243.

sedimentary work.[1] It is instructive to compare the geometrical significance of the averages in the example used for computation. The several values found for the beach sand are shown in Table 25, which also includes the median and modal diameters for comparison. Figure 113

[1] Other averages, such as the surface reciprocal mean, are discussed under the later heading "Fineness factor."

is the frequency curve of this sediment drawn with diameter in millimeters as independent variable. The diameter values of the several averages are indicated by labeled ordinates. The median, mode, and geometric mean are clustered near the high point of the curve, whereas the arithmetic mean is to the left. As Table 25 indicates, the arithmetic mean is the largest value. In all cases the averages are drawn away from the center of the range (4 mm. — .06 mm.). Figure 114 shows the same curve plotted on a logarithmic scale. (The phi scale is indicated below for convenience.) The curve has become much more symmetrical,

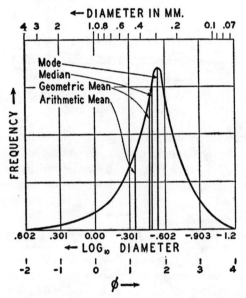

Fig. 114.—Frequency data of fig. 113 plotted with $log_{10} diameter$ as independent variable. Note shift in position of average values. The diameter scale and phi scale are added for comparison. The figure has been enlarged to show the relations clearly.

and the log of the median, the log of the mode, and the log of the geometric mean (the phi mean) are essentially at the center of the curve. The log of the arithmetic mean, however, has been drawn relatively to the left, and no longer occupies a central position. In terms of logarithmic frequency curves (or logarithmic histograms) the arithmetic mean is not so truly a measure of the central tendency as the other averages.

A comparison of values such as the foregoing is important in deciding upon the appropriate average to use in a given case. The physical and geometrical significance of the values changes when logarithmic curves

STATISTICAL METHODS

are drawn, and the practical worker should understand the necessity of making his physical or geological interpretation of the curve conform to its geometrical picture.

Measures of Dispersion

Measures of spread or dispersion about a central value may be set up with respect to the median, the arithmetic mean, or any arbitrary central point in the distribution. As in the case of other measures they may be arithmetic, geometric, or logarithmic in nature.

Mean deviation. The mean deviation, which is used to some extent in conventional statistics, is a measure of the average spread of the data about a mean value.[1] In this case the mean value is chosen either as the arithmetic mean or the median. In the example to be given the arithmetic mean will be used. In words, the mean deviation is $1/N$ of the sum of the deviations from the mean, without regard to whether the deviations are to the right or left of the mean. In computing this measure, Table 26 is set up. In the first column the grades in millimeters are shown. The second column has the percentage frequency, (f), and the third column shows the mid-point (m) of each grade. The fourth column gives the absolute value of the deviation of each mid-point from the arithmetic mean, $M_a = 0.440$ mm. The fifth column shows the frequency multiplied by the deviation, and the sum of the products from this column, 21.91, is written below. By dividing this sum by the total frequency, the mean deviation, $d_a = 0.219$, is found.

TABLE 26

COMPUTATION OF THE MEAN DEVIATION OF A BEACH SAND*

Grade Size (mm.)	f	m	$\|m - M_a\|$[†]	$f\|m - M_a\|$
4—2	0.5	3.0	2.56	1.28
2—1	5.6	1.5	1.06	5.93
1—1/2	11.7	0.75	0.31	3.62
1/2—1/4	53.7	0.375	0.07	3.75
1/4—1/8	26.4	0.187	0.25	6.60
1/8—1/16	2.1	0.093	0.35	0.73
Totals	100.0			21.91

* The frequency data are the same as in table 22.
† The symbol $|m - M_a|$ refers to the absolute value of the difference, regardless of sign. In this example, $M_a = 0.440$ mm., from Table 22.

[1] F. C. Mills, *op. cit.* (1924), pp. 149 ff.

The mean deviation has been used to only a very limited extent with sedimentary data, but a modification of it is used as Baker's grading factor (page 255), which, however, was introduced as an arbitrary measure and not related to its statistical background by any of its users, as far as the authors are aware.

It is not likely that the mean deviation will achieve the wide usage of the standard deviation in sedimentary work, although logarithmic measures analogous to the arithmetic case given above may readily be developed.

Arithmetic standard deviation. The standard deviation of a distribution is a measure of the average spread of the curve about its arithmetic mean, and it is perhaps the most widely used measure of dispersion in conventional statistics.[1] In sedimentary data, the arithmetic standard deviation is computed with respect to the arithmetic mean of the diameter distribution, i.e., the independent variable is diameter in millimeters. The data are usually obtained in terms of unequal class intervals, which are inconvenient for the computation of the standard deviation. The value may be found, however, as shown in Table 27. In the first column are given the grades in millimeters, in the second column is the frequency (f) in each grade. The third column has the mid-point (m) of each grade, and in the fourth column is given the deviation of this mid-point from the arithmetic mean of the grain diameters. This value was found to be 0.440 mm., in Table 23. The fifth column has the deviations squared, and in the sixth column the deviations squared are multiplied by the frequency in each grade. The sum of this column is 12.75. Finally, the square root of 1/100 of the summed value is extracted, yielding the standard deviation: $\sigma_a = \sqrt{\dfrac{12.75}{100}} = 1.13$.

TABLE 27

COMPUTATION OF THE ARITHMETIC STANDARD DEVIATION OF A BEACH SAND*

Grade Size (mm.)	f	m	$m - M_a$	$(m - M_a)^2$	$f(m - M_a)^2$
4—2	0.5	3.0	2.56	6.55	3.27
2—1	5.6	1.5	1.06	1.12	6.27
1—1/2	11.7	0.75	0.31	0.09	1.05
1/2—1/4	53.7	0.375	—0.07	0.005	0.27
1/4—1/8	26.4	0.187	—0.25	0.062	1.64
1/8—1/16	2.1	0.093	—0.35	0.122	0.25
Totals	100.0				12.75

* The frequency data are the same as in Table 22.
[1] F. C. Mills, *op. cit.* (1924), pp. 154 ff.

If the frequency curve is perfectly symmetrical, the standard deviation is a measure of spread such that about 68 per cent of the distribution is contained in the interval $(M_a - \sigma_a)$ to $(M_a + \sigma_a)$.[1] If the frequency curve is not symmetrical, the exact relationship becomes less clear, and the geometrical picture of the measure becomes clouded, especially with curves as asymmetrical as the average sediment plotted with diameters in millimeters as independent variable.

Logarithmic standard deviation. When the logarithmic frequency curve is used for inferences about the sediment, a logarithmic measure is appropriate, inasmuch as it may be used directly in describing the average spread of the logarithmic curve. A logarithmic standard deviation may most readily be obtained by means of the phi or the zeta scale, either of which converts the unequal geometrical diameter grades into equal logarithmic classes. Hence conventional methods of computing the standard deviation may be applied to the data. In Table 28 the method of computation is shown. The first column contains the diameter limits of the Wentworth grades, the second column has the equivalent phi classes, and the third column shows the percentage weight frequency of the data. In column four an arbitrary d scale is chosen with its zero value opposite the greatest frequency. The fifth column has the products fd, which yield the algebraic total $+6.2$. This is divided by 100 to yield $n_1 = +.062$, the first moment about the d origin. Column six shows the square of the d value, and column seven has the products fd^2. The total of this column is $+73.4$; it is divided by the total frequency, 100, to yield $n_2 = +0.734$, the second moment about the d origin. To convert the measure to the second moment about the mean, use is made of the standard equation $\sigma = \sqrt{n_2 - (n_1)^2}$. By substituting the corresponding values one obtains:

$$\sigma_\phi = \sqrt{.734 - (.062)^2} = \sqrt{.730} = 0.855.$$

TABLE 28

COMPUTATION OF THE LOGARITHMIC STANDARD DEVIATION OF A BEACH SAND*

Grade Size (mm.)	ϕ		f	d	fd	d^2	fd^2
4—2	—2—	—1	0.5	—3	— 1.5	9	4.5
2—1	—1—	0	5.6	—2	—11.2	4	22.4
1—1/2	0—	1	11.7	—1	—11.7	1	11.7
1/2—1/4	1—	2	53.7	0	0	0	0
1/4—1/8	2—	3	26.4	+1	+26.4	1	26.4
1/8—1/16	3—	4	2.1	+2	4.2	4	8.4
Totals			100.0		+ 6.2		+73.4

* The frequency data are the same as in Table 22.

[1] B. H. Camp, *op. cit.* (1931), pp. 61 ff.

This computation has been carried out by the "short method of computing the standard deviation" as explained in every statistics textbook.[1] The significance of σ_ϕ in terms of the logarithmic frequency curve is shown in Figure 115. The horizontal scale shows ϕ as the independent variable, and the area under the curve represents the total frequency. At the point $\hat{\phi} = 1.561$ an ordinate has been erected. This is M_ϕ, the arithmetic mean of the phi distribution, and it passes through the center of gravity of the distribution. This is the central value about which the standard deviation is computed. From the example, σ_ϕ was found to be 0.85, and two additional ordinates have been erected at $M_\phi + \sigma_\phi$ and $M_\phi - \sigma_\phi$, or at the phi values 2.41 and 0.71, respectively. Between these two ordinates lies the central part of the distribution, which in a symmetrical curve would include about 68 per cent of the distribution.

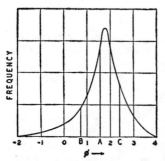

Fig. 115.—Frequency data of Figure 113 plotted directly on phi scale. Ordinate A is the phi mean, Mϕ, and ordinates B and C are at a distance $\sigma\phi$ from the mean.

Inasmuch as σ_ϕ is expressed in phi units, each of which represents one Wentworth grade, the significance of σ_ϕ may readily be seen. The value $\sigma_\phi = 0.85$ means that in the distance $M_\phi - \sigma_\phi$ to $M_\phi + \sigma_\phi$ there are 1.70 Wentworth grades, entirely independently of whether the sediment is coarse or fine. In terms of "sorting," as the term is commonly used, this indicates statistically that the sediment is well sorted.

Geometric standard deviation. Corresponding to σ_ϕ, there is a "diameter-equivalent" which is a geometric measure of spread of the curve with respect to the geometric mean of the grains. This value is called σ_ξ, and is found by converting σ_ϕ to its "diameter-equivalent" by means of the chart of Figure 112.[2]

SKEWNESS AND KURTOSIS[3]

The moment skewness of size frequency distributions is a more difficult concept than the average size or the standard deviation, partly because of the complexity of the concept in terms of grain distribution curves and partly because the physical significance of skewness in sediments is not adequately known. Kurtosis, even more than skewness, represents an essentially unexplored field in sedimentary analysis. For the sake of completeness, however, these two statistical measures will be discussed briefly, but the discussion will be confined to the logarithmic measures, in which the geometrical meaning can be illustrated more effectively than in the arithmetic case.

---

[1] The use of the d scale results in the choice of an arbitrary mid-point about which all the moments are computed. During the computations the correct moments are found in terms of corrections applied to the arbitrary origin by means of the equation for σ. Thus one may compute several moments in one operation, and the method is much shorter than some others which have been used.

[2] Details of this measure are given in W. C. Krumbein, *loc. cit.*, 1936.

[3] B. H. Camp, *op. cit.* (1931), pp. 28 ff.

STATISTICAL METHODS

Table 29 illustrates the method of computing the values needed for skewness and kurtosis. The example is not the same as that formerly used. Instead, another beach sand is used which nearly approximates a normal ϕ curve, about which more will be said later. For completeness the entire set of moments is computed in a single table, to illustrate the full sequence of steps. Table 29 contains all needed data. The diameter classes are shown in the first column, the corresponding phi intervals in the second column, and the frequency in the third. The d scale is chosen with its origin opposite the largest class, as before, and shown in column four. In the fifth column the values fd are shown; the sixth and seventh columns contain d^2 and fd^2 respectively. The eighth and ninth columns show d^3 and fd^3, and columns ten and eleven show d^4 and fd^4. The several moments about the d origin, found by dividing the algebraic totals of the product columns by 100, are

$$n_1 = 0.315$$
$$n_2 = 0.417$$
$$n_3 = 0.327$$
$$n_4 = 0.441$$

The phi mean, M_ϕ, is computed by adding n_1 to the mid-point of the d scale: $M_\phi = 1.500 + 0.315 = 1.815$. Similarly, the phi standard deviation is obtained by the equation

$$\sigma_\phi = \sqrt{n_2 - (n_1)^2} = \sqrt{.417 - (.315)^2} = 0.563$$

TABLE 29
COMPUTATION OF THE FIRST FOUR PHI MOMENTS OF A BEACH SAND*

Grade Size (mm.)	ϕ	f	d	fd	d^2	fd^2	d^3	fd^3	d^4	fd^4
1—1/2 ..	0—1	4.9	—1	— 4.9	1	4.9	—1	— 4.9	1	4.9
1/2—1/4 ..	1—2	58.9	0	0	0	0	0	0	0	0
1/4—1/8 ..	2—3	36.0	+1	+36.0	1	36.0	1	+36.0	1	36.0
1/8—1/16 .	3—4	0.2	+2	+ 0.4	4	0.8	8	1.6	16	3.2
Totals ..		100.0		+31.5		+41.7		+32.7		+44.1

* Sample 7a of W. C. Krumbein, The probable error of sampling sediments for mechanical analysis: *Am. Jour. Sci.*, vol. 27, pp. 204-214, 1934.

For the skewness, the third moment about the d origin must be converted to m_3, the third moment about the mean, by using the standard equation[1]

$$m_3 = n_3 - 3n_2n_1 + 2n_1^3$$

This yields $m_3 = .327 - 3(.417)(.315) + 2(.315)^3 = -0.004$. The third moment, m_3, is used in any of several formulas for skewness. A common and convenient one is based on $\alpha_3 = m_3/\sigma^3$. Skewness is taken as $Sk = \alpha_3/2$. In this example $\sigma_\phi = 0.563$, so that $Sk_\phi = -.004/0.356 = -0.011$.

[1] B. H. Camp, *op. cit.* (1931), p. 26.

Kurtosis requires that the fourth moment about the mean, m_4, be first found, by using the standard equation [1]

$$m_4 = n_4 - 4n_1n_3 + 6n_1^2 n_2 - 3n_1^4$$

This yields $m_4 = .441 - 4(.315)(327) + 6(.315)^2(.417) - 3(.315)^4 = +0.247$. The kurtosis itself is computed in terms of β_2, where $\beta_2 = m_4/\sigma^4$. Kurtosis may then be defined at $\beta_2 - 3$. On this basis, $\beta_2 = 0.247/0.101 = 2.5$; $K_\phi = 2.5 - 3 = -0.5$. In statistical usage β_2 is often used as a test for the normal curve, because in normal curves it is equal to 3. In the present example, β_2 is less than 3, and the curve is designated as "platykurtic." [2]

Zeta moments. A set of moments similar to those just computed may be used with Atterberg's grade scale by applying the zeta notation developed in Chapter 4. Each of the zeta moments has a simple relation to the corresponding phi moments, and may be converted to its equivalent by means of the following equations:

$$M\zeta = 0.301 \ (M_\phi + 1)$$
$$\sigma\zeta = 0.301 \ \sigma_\phi$$
$$\alpha_3\zeta = \alpha_{3\phi}$$
$$\beta_2\zeta = \beta_{2\phi}$$

It may be noted that moments higher than the second are identical in both notations, without the necessity of conversion. Further details of the zeta notation are given by Krumbein.[3]

THE NORMAL PHI CURVE

The normal curve in conventional statistics is defined as the function

$$y = \frac{1}{\sigma_x \sqrt{2\pi}} e^{-(x - M_x)^2 / 2\sigma_x^2}$$

where x is any value of the independent variable, M_x is the arithmetic mean of the x's and σ_x is the standard deviation. The normal curve is completely described by two parameters, M_x and σ_x. That is, the third moment is zero, and the fourth moment has a value of $\beta_2 = 3$. The normal curve is of considerable importance statistically because its properties have been so thoroughly studied, and tables have been prepared for evaluating the frequencies and other characteristics over its entire range.

The tendency for sedimentary curves to become symmetrical on a logarithmic size scale suggests that logarithmic (phi) parameters be substituted for the x-values in the function above, to obtain a normal phi curve, analogous to the conventional curve. In terms of the phi mean, M_ϕ and the phi standard deviation, σ_ϕ, the function is

$$y = \frac{1}{\sigma_\phi \sqrt{2\pi}} e^{-(\phi - M_\phi)^2 / 2\sigma^2_\phi}$$

where ϕ is the value of the independent variable at any point. The importance

---

[1] B. H. Camp, *op. cit.* (1931), p. 26.
[2] F. C. Mills, *op. cit.* (1924), p. 545.
[3] W. C. Krumbein, Korngrösseneinteilungen und statistische Analyse: *Neues Jahrb. f. Min., etc.*, Beil.-Bd. 73, Abt. A, pp. 137-150, 1937.

of this concept is that among the asymmetrical curves of sediments which are commonly encountered when diameters are used as the independent variable, some may be approximately "normalized" by a simple mathematical transformation equation. By means of this normalizing process the curve may be completely described with two parameters, M_ϕ and σ_ϕ. Furthermore, the usual tables of probability developed for the conventional normal curve of statistics may be directly applied to analysis of the phi curve, with no changes of technique whatever.

Significance of Higher Moments

The normal phi curve affords a basis for furnishing the geometrical picture of higher moments. This curve, illustrated in Figure 116, is symmetrical. It is possible to consider this normal curve as the first member of a series of successive derivative curves,[1] such that each succeeding curve represents a higher moment. For example, the third moment is such a function that a

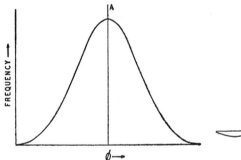

Fig. 116.—Normal phi curve.

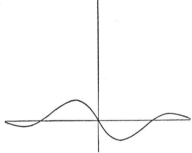

Fig. 117.—Graph of third moment. (Data from Camp, Appendix, Table III.)

graph of its effect on the direction of skewness appears like that in Figure 117. If a curve is skewed, the values of the parameters will govern the exact shape of the curve, but the net effect will be equivalent to adding algebraically the ordinates of Figures 116 and 117, to obtain the curve of Figure 118. Similarly, the fourth moment contains a function of the type shown in Figure 119, and if the ordinates of this curve are added to Figure 116 or Figure 118, the net effect will be an increased or decreased "peakedness," depending upon the exact values of the fourth moment.

In statistical practice it is a common procedure to analyze curves statistically in terms of the moments. The resulting function is called a Gram-Charlier series, and the ϕ analogue of this series may be called the ϕ-Gram-Charlier series. Details of the conventional procedures may be found in advanced statistics texts. An elementary treatment is given by Camp.[2] The de-

---

[1] B. H. Camp, *op. cit.* (1931), pp. 225 ff.
[2] *Op. cit.* (1931), Chap. 3.

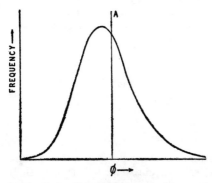

Fig. 118.—Graph of skewed phi curve, prepared by adding the ordinates of Figures 116 and 117 algebraically.

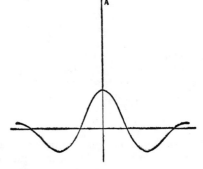

Fig. 119.—Graph of fourth moment. (Data from Camp, Appendix, Table IV.)

termination of the analytical functions of sedimentary curves may in some cases pave the way for a more complete understanding of the environmental factors which control the nature of the frequency distribution.

Graphic Computation of Geometric Mean and Geometric Standard Deviation

For curves which are symmetrical when plotted on a logarithmic size scale, there is a convenient graphic method of determining the log geometric mean and the log geometric standard deviation. This method, developed by Hatch and Choate,[1] involves plotting the cumulative curve on logarithmic probability paper. (In order for the graphic method to apply, this graph must be approximately a straight line.) The diameter-value corresponding to the 50 per cent line is the geometric mean, and the standard deviation is found by computing the ratio between the geometric mean and the diameter-value corresponding to the 15.8 per cent line. This latter determination follows from the fact that in a symmetrical curve the interval between the mean and σ is approximately 34.2 per cent of the distribution. Complete details may be found in the reference cited.

SPECIAL STATISTICAL MEASURES

In addition to standard statistical devices, in which the relation of the measures to statistical theory is known, a number of special devices have been introduced by sedimentary petrologists to describe the characteristics of sediments. These empirical devices include an average size and one or more measures of spread of the data, some of which are related to the skewness of the distribution. In some cases it has been possible to reconcile the measures with standard statistical devices, and

[1] *Loc. cit.*, 1929.

these relations will be brought out in the discussion. In some instances the original authors have not examined the exact geometrical or physical significance of their measures, and apparently some followers of the methods have been content to accept them without scrutiny.

Baker's equivalent grade and grading factor. Perhaps the best known of the empirical measures are those introduced in 1920 by Baker.[1] In Baker's method two measures are used, one representing an average grain size and the other a measure essentially of the spread of the data about the average. The method involved is largely graphic and involves by definition the use of a cumulative curve drawn on an arithmetic scale of diameters, as shown in Figure 120. It will be noted that Baker chose frequency as his independent variable, presumably because his graphic approach is more convenient on that basis. In computing the *equivalent grade,* or average size, the area under the cumulative curve is determined with a planimeter, and the area so found is divided by the length of the frequency line from 0 to 100 per cent. The quotient, which is the length corresponding to the equivalent grade, is then laid off on the vertical axis, and the diameter value at that point is the equivalent grade, which is indicated by a heavy horizontal line in Figure 120, at the value 0.430 mm. Essentially, the equivalent grade is the mean ordinate of the curve, and it may be shown that it is in fact the arithmetic mean diameter of the grain distribution.[2]

Baker's *grading factor* is a measure designed to indicate how nearly the degree of grading approaches perfection. A perfectly graded sediment, according to Baker, is one in which all grains are of the same size, so that there is no deviation from the average size. In that case the computed value of his grading factor would be unity.

Figure 120 also indicates the manner of computing the grading factor. The two shaded areas, one below the curve and above the line (V_a), and the other above the curve and below the line (V_b), are called the *variation areas* of the curve, and the sum ($V_a + V_b$) is called the total variation area. These are also found by means of a planimeter.

Baker defined his grading factor as

$$\text{G.F.} = \frac{\text{Total area under the curve} - \text{total variation area}}{\text{total area under the curve}},$$

[1] H. A. Baker, On the investigation of the mechanical constitution of loose arenaceous sediments by the method of elutriation, etc.: *Geol. Magazine,* vol. 57, pp. 366-370, 1920.
[2] As far as the authors are aware, Baker did not indicate that his grading factor is a graphically determined arithmetic mean. This identity may be proved by a rather tedious mathematical demonstration.

and in the actual computation, the grading factor is found by subtracting the total variation area from the total area, and dividing the difference by the total area.

The relation of the grading factor to statistical theory is less direct than that of the equivalent grade. It may be shown, however, by a somewhat laborious process, that this measure is related to the mean deviation of conventional statistical practice. The relation is

$$\text{G.F.} = 1 - \frac{\text{mean deviation}}{\text{arithmetic mean size}}.$$

To obtain Baker's equivalent grade and grading factor as he defined them, it is necessary to use an arithmetic scale for diameters. Some workers have drawn their cumulative curves on a logarithmic size scale and followed Baker's graphic procedure, calling the result the equivalent grade. What is actually obtained in this case is the geometric mean diameter of the grains—an entirely different value. In similar fashion, if the logarithmic cumulative curve is used for the computation of the "grading factor," the result is not Baker's grading factor, but a geometric measure in terms of the logarithmic frequency distribution.

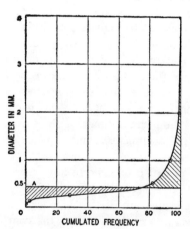

Fig. 120.—Cumulative curve of beach sand, showing Baker's equivalent grade (A), and areas used in computing grading factor.

Baker's two measures have been widely used in the examination of sediments, and his equivalent grade has been chosen as the average value for other empirical statistical methods also. It is probably preferable to indicate that the arithmetic mean size is used in such instances, and indeed the equivalent grade may be computed directly by the methods shown in Table 22.

Niggli's statistical method. Most recent of the empirical statistical methods is that introduced in 1935 by Niggli.[1] Niggli's method combined Baker's equivalent grade with various approximate quartile measures as well as with the maximum and minimum grain sizes. The result is a

[1] P. Niggli, Die Charakterisierung der klastischen Sedimente nach der Kornzusammensetzung: *Schweiz. Min u. Pet. Mitt.,* vol. 15, pp. 31-38, 1935.

STATISTICAL METHODS

method of characterization which Niggli did not attempt to relate to conventional statistical practice. Zingg [1] applied the technique to a number of sediments and discussed the further implications of a wide range of devices for describing sediments. In detail, Niggli's method involves the following steps:

Three fundamental values are chosen, which are called d_{max}, d, and d_{min}. The first of these is the size of the largest particle in the sediment, the second is Baker's equivalent grade, and the last, d_{min}, is the smallest grain in the sediment, usually taken as zero by Niggli, unless the smallest size is appreciable. Niggli points out that in a perfectly symmetrical curve (an arithmetic size scale is used), the value of d would be exactly equal to $\frac{d_{max} + d_{min}}{2}$, which is true in an arithmetic distribution. Inasmuch as few curves are symmetrical, an important measure is thus found in the relation between the value of d and the value $\frac{d_{max} + d_{min}}{2}$; hence a measure is set up from the ratio of these values, and called delta: $\delta = \frac{2d}{d_{max} + d_{min}}$. Next, the total percentage of material lying between d_{min} and d is called p, and a second measure, pi, is defined from this as follows: $\pi = p/50 = 2p/100$. The two measures delta and pi are used as the first characteristic of the sediment, by noting the extent to which they depart from the value 1.

By means of d and p the grain distribution is divided into two parts, one fine and the other coarse. For each of these portions, accordingly, a similar set of measurements is obtained, analogous to d and p. For the fine fraction the average value is called d' and for the coarse d''. Likewise, the corresponding p values are called p' and q'' respectively. Thus Niggli obtains a series of values distributed along the curve, which serve to define it by fixed values. As Niggli points out, if the curve were symmetrical, p' would be the first quartile and q'' the third quartile, whereas d would be the median grain size. Sediments are characterized in part by the relations among d, d', and d''. The greater the interval between d' and d'', the more widespread is the curve; the smaller the difference between d' and d'', the more nearly uniform the sediment is.

Niggli also introduced a sorting index, uninfluenced by size, which he defined as follows: $a = \frac{3(d'' - d')}{d}$, which yields values approximately equal to unity for sediments the bulk of whose grains are well sorted, and values greater than 1 for poorly sorted sediments. The sorting index is essentially the ratio of the difference between the average size of the coarse and fine portions of the curve and the average grain size. The value 3 was chosen as a constant because, according to Niggli, the ideal relation in the Udden (Wentworth) grade scale is $(d'' - d')/d = \frac{1}{3}$.

[1] Th. Zingg, Beitrag zur Schotteranalyse: *Schweiz. Min. u. Pet. Mitt.* vol. 15, pp. 39-140, 1935.

Fineness factor. In 1902 Purdy[1] introduced a *surface factor* or *fineness factor* for describing the texture of clay and ceramic materials. The fineness factor is computed by multiplying the reciprocal of the mid-point of each grade size by the weight percentage of material in the grade, expressed as a decimal part of the total frequency. The sum of the resulting products is the fineness factor. Purdy based his measure on the assumption that the surface areas of the two powders are inversely proportional to their average grain size. Roller[2] examined Purdy's factor in terms of statistical theory and showed that the factor is essentially the reciprocal surface mean diameter of the powder, providing the average size of each grade is defined in terms of a surface mean diameter, d_s, and a percentage weight of material, W, as:

$$\frac{1}{d_s} = \Sigma \left(\frac{W \times 10^{-2}}{d_s} \right),$$

where the right-hand side of the equation indicates the operations used in computing Purdy's factor. Full details of the theory may be found in Roller's paper.

The fineness factor has not been extensively used in sedimentary petrology, but in the light of Roller's work it would appear to offer an excellent approach to the study of properties of finer sediments in terms of their surface area. For the study of pigments, where surface is perhaps the most significant attribute, the measure is of considerable importance.

Sorting indices. A number of writers have introduced various measures of the sorting of sediments, but apparently no one has thoroughly investigated the subject of sorting itself, to determine the most suitable measure of this attribute. The generally accepted definition of sorting is that the more nearly a sediment approaches a single size in its frequency distribution, the better it is sorted. Thus most measures of sorting are statistical in nature and measure essentially the spread of the curve. This is true of Baker's grading factor and of Niggli's index of sorting. Trask's measure of sorting, So, a geometric quartile deviation, is also a measure of spread. In similar fashion the standard deviation, either logarithmic, geometric, or arithmetic, may be used as a measure of statistical sorting. There is a field for investigation on the physical significance of sorting, as well as of the possible influence of sorting on the skewness of the sediments or vice versa.

[1] R. C. Purdy, Qualities of clays suitable for making paving brick: *Ill. State Geol. Survey, Bull. 9*, pp. 133-278, 1908.
[2] P. S. Roller, Separation and size distribution of microscopic particles: *U. S. Dept. Commerce, Bur. Mines Techn. Paper 490*, 1931.

CHOICE OF STATISTICAL DEVICES

The authors are not prepared to commit themselves on any single method of statistical analysis as being the best; the field of statistics as applied to sediments will require mathematical statisticians to investigate all the ramifications of the problem. For the present the individual worker must choose the method which appears best adapted to the end he has in view. In general, three things may guide his choice: the relations of the measures to the body of statistical theory, the relative simplicity of the mere mechanical process of arriving at the numbers, and the simplicity of the geometrical meaning of the measures. It is largely on the latter that one bases his conclusions, and whatever measures are chosen should at least be readily visualized.

The decision between arbitrary methods of description and conventional statistical devices, related to the background of statistical theory, must depend upon the objects of the study. If description and classification are an end in themselves, any measures designed to summarize the data are adequate. If description and classification are only a means to an end, on the other hand, then the measures chosen should serve other purposes as well. Every sedimentary deposit has characteristics which depend upon the conditions of its formation, and these characteristics appear to be most effectively expressed in terms of their statistical parameters. If the relation between sedimentary characteristics and environmental conditions is to be elucidated, it seems reasonable to suppose that the body of theory behind conventional statistical procedures will afford a more direct relationship than the use of measures designed without regard to that body of theory.

The objection is occasionally raised that standard statistical devices yield too few values for the classification of sediments. It was partly this reason that impelled Niggli to develop his measures. However, the argument may be met by proponents of the quartile measures by extending the devices to the deciles or intermediate points. A satisfactory parallel of Niggli's method may be developed by choosing the median and the first and third quartiles as fundamental values, and intermediate deciles where needed. In this manner as many as ten values may be had if desired.

A choice between arithmetic, geometric, or logarithmic measures must depend upon the type of results which are most immediately useful to the investigator. If the influence of size is to be included in the study, measures should be chosen in which the size factor is explicit. When size is to be eliminated, geometric or logarithmic measures may be used. It is important in this connection that the worker understand the dependence or independence of his measures on size, and to that end this information has been given in the body of the chapter. A choice between quartile or moment measures may depend upon the analytical data at hand. It may be stated as a general rule that moment measures are much more sensitive to "open ends" on the sedimentary data than quartile measures are. Hence in working with very fine-grained sediments, where present methods of analysis require grouping a considerable amount of material in the smallest class, the higher moments are perhaps not very reliable, although the first and second moments are usually not greatly affected. Quartile measures, on the other hand, are usually not affected at all by open ends beyond the 25 per cent line in one direction and the 75 per cent line in the other. For partial analyses, therefore, the quartile measures are excellent. However, the disadvantage of quartile measures is often that the behavior of the extreme parts of the curve is not at all reflected, and in studies where departures from the average are to be studied the quartile measures may be of limited use.

The final test of any statistical measure is its mathematical convenience, and this, combined with its relation to the background of statistical theory, enables the worker to gain the maximum value from his study. Beyond standard devices for size distribution studies, there remains much to be done with statistical devices in mineral studies and in connection with shape and surface texture. Likewise problems of sampling and statistical correlation, from the viewpoint of sedimentary data, have not been investigated extensively. The field may therefore be considered wide open for appropriately trained research men.

STATISTICAL CORRELATION

In many scatter diagrams the points are dispersed more or less widely, and a question arises whether there is any definite relation between the two variables.

Statisticians have developed methods of testing data of this nature by means of a *coefficient of correlation*. This is not to be confused with the term correlation as used in a geological sense; statistical correlation is a mathematical procedure which yields a coefficient whose value extends from -1 through zero to $+1$. If the correlation coefficient is equal to $+1$, there is a direct relation between the variables; if its value is -1, there is an opposite (inverse) relation between the variables; and if the coefficient is zero, there is in all likelihood no fixed relation between them. For values other than zero, but neither $+1$ nor -1, the significance of the correlation coefficient depends partly on the nature of the data being examined.

Statistical correlation[1] between two variables is called simple correlation; it may be either linear or non-linear. It is also possible to test the relations among more than two variables by multiple correlation, but the methods become somewhat tedious, and the interested reader is referred to standard texts for methods of computation. An example of linear correlation will be given here to illustrate the method and to indicate some of the advantages and disadvantages of applying this statistical technique to sedimentary data.

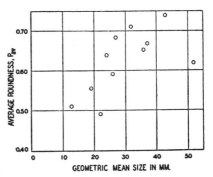

Fig. 121.—Scatter diagram of average roundness and geometric mean size of beach pebbles.

A study of beach pebbles from Little Sister Bay, Wisconsin, showed that in general there was a relation between large average size and average degree of roundness. Eleven samples[2] were studied, and by plotting average size against average roundness the scatter diagram of Figure 121 was obtained. Size was chosen as the independent variable, and roundness as the dependent variable. It will be noted that the points scatter too widely to justify drawing a straight line through them without considerable qualification. It is in such cases that the correlation coefficient may shed light on the problem of possible relations between the variables.

There are several methods of computing the coefficient of correlation, depending upon whether the data are grouped or ungrouped. In the present case they are ungrouped. The following method of computation was chosen so that each step in the process would be explicit, especially the transformation of variables that occurs during the computation and final fitting of straight lines. In Figure 121 the horizontal axis is chosen as X and the vertical axis Y. Hence the "raw data" are given in X and Y units, the variables being

[1] F. C. Mills, *op. cit.* (1924), Chapter 10.
[2] Strictly speaking, eleven samples are not sufficient for a detailed study of correlation. It is preferable to use at least twenty-four samples. Likewise, the direct correlation of average values involves more complex theory, but the example will at least indicate the method of computation.

$X = M_g$, the geometric mean size of the pebbles, and $Y = P_{av}$, the average roundness of the pebbles. In computing the coefficient it is convenient to consider the deviations from the mean values of X and Y. The steps are indicated in Table 30. The first two columns list the X and Y values, and the averages obtained from each of these columns is $X_m = \frac{331}{11} = 30$ mm., $Y_m = \frac{6.86}{11} = 0.62$. The third and fourth columns represent the differences between X and X_m and Y and Y_m; specifically the values are $x = X - X_m$, and $y = Y - Y_m$. These new variables, x and y, are called the deviation values. In columns 5 and 6 the individual values of x and y are squared, and in column 7 the products of x and y are indicated. Note that these last values may be either positive or negative. The figures in the several columns are added and the sum indicated in the last line of the table.

TABLE 30

COMPUTATION OF THE CORRELATION COEFFICIENT OF AVERAGE ROUNDNESS AND GEOMETRIC MEAN SIZE OF BEACH PEBBLES FROM LITTLE SISTER BAY, WISCONSIN

Geometric Mean Size (mm.) X	Average Roundness Y	$X-X_m$ x	$Y-Y_m$ y	x^2	y^2	xy
52	0.62	+22	0	485	0.0000	0.00
43	.74	+13	+0.12	169	.0144	+1.56
36	.65	+ 6	+ .03	36	.0009	+0.18
32	.71	+ 2	+ .09	4	.0081	+0.18
27	.68	− 3	+ .06	9	.0036	−0.18
26	.59	− 4	− .03	16	.0009	+0.12
22	.49	− 8	− .13	64	.0169	+1.04
37	.67	+ 7	+ .05	49	.0025	+0.35
24	.64	− 6	+ .02	36	.0004	−0.12
19	.56	−11	− .06	121	.0036	+0.66
13	.51	−17	+ .11	289	.0121	+1.87
	6.86			1278	0.0634	+5.66

The correlation coefficient, r, is defined as follows [1]

$$r = \frac{p}{\sigma_x \, \sigma_y}$$

where p is the product moment of x and y, and σ_x, σ_y are respectively the standard deviations of the x and y values about X_m and Y_m. Inasmuch as the x and y values are expressed directly as deviations from X_m and Y_m, the

---

[1] F. C. Mills, *op. cit.* (1924), pp. 385 ff.

three needed values are readily found from the values in the table and the following relations

$$p = \frac{\Sigma(xy)}{N} = \frac{+5.66}{11} = 0.52$$

$$\sigma_x = \sqrt{\frac{\Sigma(x)^2}{N}} = \sqrt{116} = 10.8$$

$$\sigma_y = \sqrt{\frac{\Sigma(y)^2}{N}} = \sqrt{0.0057} = 0.07$$

Hence $r = \frac{p}{\sigma_x \sigma_y} = \frac{0.52}{(10.8)(0.07)} = +0.68.$

The correlation coefficient thus has a value between zero and $+1$, indicating that the expected relation is present, but is by no means perfect. This means, essentially, that size alone is not the controlling factor in roundness, a conclusion that the geological evidence itself affords. However, the correlation coefficient at least indicates that the general relation between large average size and high average roundness does hold on the beach in question.

If one wishes to indicate the degree of correlation graphically, he may plot lines of regression on the scatter diagram.[1] The lines of regression represent the straight lines of approximate best fit, both of y on x and x on y. If the correlation coefficient is equal to unity, both of these lines will be identical (i.e., have the same slope and intercepts), whereas if $r = 0$ the indication is that no line fits the data better than any other. For values of r between 0 and $+1$ the angle between the lines is a function of the correlation coefficient.

The correlation coefficient has not been extensively applied to sedimentary data, but it is a common statistical device which may well be used in appropriate situations. Certain precautions should be followed in drawing inferences from the correlation coefficient, however. That is, the correlation coefficient is applicable directly only if the attributes being correlated are continuous variables, expressible as numbers on a continuous scale. The usual case involves n samples, each of which has two variables in common, as in the illustration used. Appropriate cases would include not only size and shape attributes, but various mineral attributes as well. For example, it is proper to use the correlation coefficient with n samples of heavy minerals for the correlation of garnet and hornblende in each.

An illustration of an unconventional use of the correlation coefficient is given by a method of mineral correlation introduced by Dryden.[1] Instead of using n samples of two attributes each, he used 2 samples with

---

[1] F. C. Mills, *op. cit.* (1924), pp. 393 ff.
[2] L. Dryden, A statistical method for the comparison of heavy mineral suites: *Am. Jour. Sci.*, vol. 29, pp. 393-408, 1935.

n attributes each (see page 487). In this case the underlying postulates on which the correlation coefficient are based were not satisfied, because the attributes used do not form a continuous series. However, it is perfectly appropriate to use the *process* of computing the correlation coefficient in this case; but perhaps another term, as "coefficient of mineral association," should have been applied to the result. The point to this discussion is that it is suitable to make the computations as Dryden did, providing no attempt is made to fit r into the extensive background of theory with which the correlation coefficient has hitherto been associated, because it does not satisfy the postulate of this existing theory. However, there seems to be no reason why a second body of theory may not be developed based on postulates suggested by Dryden's use of r.

CHI-SQUARE TEST

The correlation coefficient, as usually employed in statistical work, involves a number of samples. In geological problems it is often desirable to "correlate" two samples, to determine whether they came from the same or different deposits. It is very unusual to find two samples having exactly the same size frequency distribution, or the same percentage of heavy minerals, and the question is how much variation is permissible without rendering invalid the inference that the two samples are from the same parent deposit.

Eisenhart[1] attacked this problem in 1935 by means of the chi-square test, which is applicable to a number of problems in sedimentary work. The theory of χ^2 is beyond the scope of this book, but the essential features of the test may be described by a simple example, as used by Eisenhart. Two samples of sediment have the following numbers of limestone and shale pebbles:

Sample	Limestone	Shale
1	103	794
2	109	781

The question is whether two samples drawn at random from the same parent deposit could show such observed variations due purely to chance. In other words, what is the probability that two random samples would show variations as great or greater than those observed? The chi-square test, applied to these data, shows that one would obtain these or greater variations in 62 per cent of the cases. Thus there is little risk in assuming that the samples are from the same deposit.

[1] C. Eisenhart, A test for the significance of lithological variations: *Jour. Sed. Petrology*, vol. 5, pp. 137-145, 1935.

STATISTICAL METHODS

In applying the chi-square test, one sets up a table showing the samples in vertical columns, and the attributes to be tested in horizontal rows. Each observed frequency is subtracted from an expected frequency (or an "independence frequency" if the former is not known), and the difference is squared and divided by the expected frequency. A series of values are obtained, the sum of which equals χ^2. The observed value of χ^2 is located in a table opposite an appropriate number for the "degrees of freedom," of the table, and the probability desired is found. Details of the test and an introduction to the theory are given in Eisenhart's paper. The complete χ^2 tables are to be found in Fisher's book;[1] a partial table is given by Camp.[2]

The chi-square test may also be used in testing such assumptions as were made in connection with the data of Table 29, which suggested that the sediment approximated a normal ϕ curve. By comparing the observed frequencies and the theoretical frequencies of a normal curve from tables of probability integrals, the "goodness of fit" of the data may be tested by χ^2.

The chi-square test promises to be of considerable importance in the theory of sampling sediments, but it requires detailed study to determine whether the conditions of sampling sediments satisfy, in all cases, the postulates on which the chi-square test is based.

THEORY OF CONTROL

The theory of control, as developed by Shewart,[3] affords a powerful method for testing data to determine whether observed variations are due purely to chance causes or whether they may be attributed to assignable causes of variation. The method was used by Otto[4] in testing the performance of a Jones sample splitter. His tests indicated that certain subjective errors, due to differences in operators, were present. These errors were largely eliminated by developing an improved type of splitter (see page 45), in which lugs required that all operations be standardized.

The theory of control rests fundamentally on the fact that in a normal probability function 99.7 per cent of all cases fall within the range $M_a + \sigma$; in other words, the chances are that no more than three out of a

---

[1] R. A. Fisher, *op. cit.* (1932).
[2] B. H. Camp, *op. cit.* (1931), p. 265.
[3] W. A. Shewart, *Economic Control of Quality of Manufactured Product* (New York, 1931).
[4] G. H. Otto, The use of statistical methods in effecting improvements on a Jones sample splitter: *Jour. Sed. Petrology*, vol. 7, pp. 110-132, 1937.

thousand items will depart from the arithmetic mean by more than three times the standard deviation. Tests are devised to determine whether this relation holds; if not, further tests seek to assign causes to the observed departures from the normal law.

The computations involved in applying the theory are quite tedious, which is an attribute of most detailed methods of analysis. The reader is referred to Shewart's book and Otto's paper for the details of the method.

THE PROBABLE ERROR

Among statistical devices long used in the evaluation of errors is the probable error,[1] defined as that error which will not be exceeded in one half of the observed cases. The probable error bears a constant and simple relation to the standard deviation, σ. This relation is expressed by the equation P.E. $= 0.6745\sigma$.

The probable error was applied to the problem of sampling sediments by Krumbein,[2] who investigated the error in terms of its effect on the median grain diameter.[3] In theory the method applied depends first on the fact that independent errors (sampling errors as opposed to laboratory errors) are related to the total observed error E by the expression [4] $E = \sqrt{(e_1)^2 + (e_2)^2}$, where e_1 and e_2 are the sampling and laboratory errors respectively. A number of samples are collected and separately analyzed to determine the total error E. The samples are then combined into a single composite which is analyzed a number of times, to obtain e_2. This permits the computation of e_1 from the equation above.

The probable error of the mean, PE_m, is defined as $PE/\sqrt{n}$, where PE is the probable error of a single observation, and n is the number of samples. This may be expressed as $PE_m/PE = 1/\sqrt{n}$. This was the equation used by Krumbein. Some writers prefer to use the standard error of the mean, σ_m, defined as $\sigma_m = \sigma/\sqrt{n}$, where σ is the standard deviation. The relation between σ and PE_m is $PE_m = 0.6745\sigma$, from the definition of the probable error. In general, one may express the error of

---

[1] F. C. Mills, *op. cit.* (1924), p. 160.
[2] W. C. Krumbein, The probable error of sampling sediments for mechanical analysis: *Am. Jour. Sci.*, vol. 27, pp. 204-214, 1934.
[3] It is preferable, perhaps, to use the arithmetic mean in such studies, or to approach the problem logarithmically in terms of the phi mean. Fortunately the median deviated about its mean value in a normal manner, so that the method was applicable.
[4] A. Fisher, *The Mathematical theory of probabilities* (New York, 1915), vol. I, p. 106.

the mean as E_m, whereupon the relation is $E_m/E = 1/\sqrt{n}$. This function is discussed in Chapter 2.

Further discussion of the theory of probable errors and details of the method for evaluating the error are given in Krumbein's paper.

SUMMARY OF STATISTICAL METHODS

The preceding sections on correlation, the χ^2 test, the theory of control, and the probable error indicate that there is a growing recognition of the importance of statistical analysis in sedimentary problems. One cannot ignore the contributions which such studies have made and will make to a fuller understanding of the complex study of sediments. As methods of sampling and laboratory analysis are improved, and as more precise methods of evaluating errors are developed, the data furnished by sedimentary studies will become more reliable, and consequently the inferences drawn from the data may be expected to be more sound. Meanwhile, parallel studies of sediment genesis, in terms of the controlling environmental factors, may ultimately lead to an understanding between conditions of deposition and statistical parameters, which will pave the way for more quantitative reconstructions of past environments in historical geology.

CHAPTER 10

ORIENTATION ANALYSIS OF SEDIMENTARY PARTICLES

INTRODUCTION

UNDER certain conditions of deposition sedimentary particles may assume a given orientation with respect to the surface of deposition. The imbrication of stream pebbles is a common example, but it is only one of a large number of similar cases. Numerous writers have described oriented deposits, but comparatively little has been done in a quantitative manner with the large field of study available in the investigation of the primary orientation of sedimentary particles. By *primary orientation* is meant the arrangement in space of the component particles during deposition, regardless of subsequent changes in position.

The fertile fields of research which have been opened in the study of igneous and metamorphic rocks by the techniques of petrofabric analysis suggests that similar results may accrue from a wider application of like methods to sediments.[1] Among sedimentary materials the techniques of analysis may often be more conveniently applied than to igneous or metamorphic rocks. In the latter instances it is necessary to work with oriented thin sections, often composed of small grains; among sediments one may study unconsolidated gravels, for example, in which the particles may readily be examined individually. Smaller particles, such as sand grains, may of course require special techniques, especially among unconsolidated deposits. Artificial induration with bakelite or similar material may preserve original relations among the grains.

Among orientation studies of large particles is that of Wadell,[2] who investigated the orientation of pebbles in an esker and an outwash delta, to determine whether eskers are necessarily the result of deposition in subglacial streams. His results were extremely interesting, inasmuch as they showed that the long axes of the pebbles in the esker gravel were in general parallel to the direction of dip of the bedding planes, whereas in the foreset beds of the delta the orientation of the long axes was more or less diametrically opposite

[1] Suggestions of the possibilities afforded by such studies were given by E. B. Knopf, Petrotectonics: *Am. Jour. Sci.,* vol. 25, pp. 433-470, 1933.
[2] H. Wadell, Volume, shape, and shape-position of rock fragments in open-work gravel: *Geografiska Annaler,* 1936, pp. 74-92.

to the direction of dip of the beds. These relations are shown in Figure 122, adapted from Wadell's paper. In both deposits the dip of the beds was about the same and the sizes of the particles were of the same order of magnitude. Miner [1] had previously noted that fragments in talus were arranged with their long axes parallel to the dip of the slope, and on this basis Wadell offered a tentative conclusion that the accumulation of the pebbles in the esker was essentially subaërial in nature.

Among other studies involving particle orientation may be mentioned Richter's studies of the pebbles in glacial till [2] as a statistical device to determine the direction of ice movement. Richter plotted the orientations of the

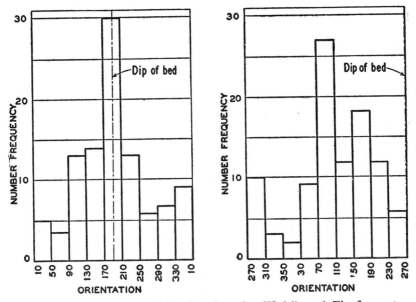

Fig. 122.—Histograms of pebble orientation, after Wadell, 1936. The figure at the left represents esker gravel, that at the right is from an outwash delta.

long axes of the pebbles in terms of compass direction and used the orientation of the modal group as an index of direction of ice movement. The study was supported by actual examinations of the relation of the long axes in cases where the direction of ice movement was known. Richter also pointed out the further implications of his studies in terms of pebble orientation as a function of the nature of glacial movement. According to his view, the arrangement of the pebbles argued for streamline motion of the ice.

[1] N. A. Miner, Talus slopes of the Gaspé Peninsula: *Science*, vol. 79, pp. 229-230, 1934.
[2] K. Richter, Die Bewegungsrichtung des Inlandeises, rekonstruiert aus den Kritzen und Längsachsen der Geschiebe: *Zeits. f. Geschiebeforschung*, vol. 8, pp. 62-66, 1932.

The orientation of roller-shaped pebbles on beaches was studied by Fraser.[1] He found statistically that the most common position was for the pebbles to lie with their long axes parallel to the shore line. On the average, only 16.5 per cent were oriented with their long axes more than 45° from parallelism to the water's edge. Fraser attributed his findings to the tendency for waves to swing pebbles into that position or for roller-shaped pebbles to roll with their long axes perpendicular to the direction of movement.

These examples serve to illustrate the types of quantitative data which may be obtained in orientation analysis. The study of particle positions may properly supplement size, mineral, and shape data in the complete study of sediments.

The study of pebble orientation is no less a statistical operation than mechanical analysis. The orientation of a single particle may be relatively meaningless, but the "average" direction of orientation may be found from a study of the frequency distribution of the individual orientations. Perhaps the most significant "average" to use in such cases is the mode. Whether other statistical devices such as measures of spread or asymmetry of the distribution have significance in these studies is largely a matter for further investigation.

COLLECTION OF ORIENTED SAMPLES

The following details for the collection of oriented sedimentary particles, adapted from Wadell,[2] apply to fairly large pebbles in unconsolidated deposits.

The face of the outcrop is cleaned and a rectangular sampling area enclosing about 100 pebbles is marked on the gravel face. Instruments required for sampling include a Brunton pocket transit, a soft pencil, two fine brushes, and cans of quick-drying black and red lacquers. The direction and dip of the gravel bed are determined, and the compass trend of the face of the outcrop is read.

For collecting an individual pebble, the Brunton is held at eye-level as a prismatic compass, with the mirror in position to reflect the leveling bubble and the compass needle. The compass is held in one hand and the pencil in the other. The back edge of the compass is set parallel to the trend of the gravel face, and the instrument is leveled. A vertical pencil line is now drawn on the pebble parallel to the etched line in the Brunton mirror, and a horizontal line is drawn parallel to the top edge of the mirror. The lines are checked by holding the compass in both hands. When they are found correct, the horizontal line is drawn with black enamel and the vertical line with red enamel. A black dot is also

---

[1] H. J. Fraser, Experimental study of the porosity and permeability of clastic sediments: *Jour. Geology*, vol. 43, pp. 910-1010 (esp. pp. 978 ff.), 1935.
[2] H. Wadell, *loc. cit.*, 1936.

ORIENTATION ANALYSIS

placed in the lower right-hand portion of the pebble to indicate dip-direction and orientation position.

After the lacquer has dried, the pebbles are individually removed. As a general rule the pebbles may be carried in an ordinary container, inasmuch as the lacquer will not be affected by ordinary rubbing of one pebble against the next.

If the sample to be collected is sand or finer material, where the particles cannot be handled individually, the sand must be impregnated with a binder, or a sample removed with a device which does not disturb the relations of the grains to one another. Sampling apparatus similar to that used for porosity determinations may be suitable (see Chapter 20). In all cases the exact orientation of the sample as a whole must be recorded. In indurated sediments orientation lines may be drawn directly on the specimen to be removed from the outcrop.

LABORATORY ANALYSIS OF PARTICLE ORIENTATION

Methods of measuring and tabulating the data in orientation analysis depend upon the nature of the sediments being studied. The technique is fairly simple for unconsolidated material coarse enough to be handled individually; for consolidated materials no simple procedures are known to the authors. The determination of primary particle orientation, as opposed to the determination of crystallographic axes, presents problems of locating the long axis of particles in thin section. The actual long axis of the particle may not lie in the plane of the thin section, and it probably is not sufficient to determine the apparent orientation of the longer axis of the grain section. Methods of analysis analogous to those used in petrofabric analysis [1] may be developed, however, for grains in which there is a more or less fixed relation between the orientation of the long diameter of the grain and a crystallographic axis. For example, it is obvious that in zircon grains the long diameter is parallel to the c-axis; for rounded grains the relations may be less clear. Pettijohn has observed from his own work that statistically the long diameters of quartz grains tend to lie along the c-axis of the original crystals; for other minerals similar characteristics may be found.

Samples of pebbles may most conveniently be measured with an ordinary two-circle contact goniometer. The method is shown in Figure 123, adapted from Wadell's paper. A vertical red line and a horizontal

[1] B. Sander, *Gefügekunde der Gesteine* (Vienna, 1930). H. W. Fairburn, *Structural Petrology* (Queen's University, Kingston, Can., 1937).

black line are drawn on a plate of glass, which is supported before the goniometer. The pebble is mounted on the goniometer with putty, in such a manner that the painted lines on the pebble coincide with the corresponding reference lines on the glass. In this manner the pebble is in the same position as it occupied in the outcrop. The longest axis of the pebble is then determined by inspection, and the dip of this axis is read by rotating the horizontal goniometer circle until the long axis

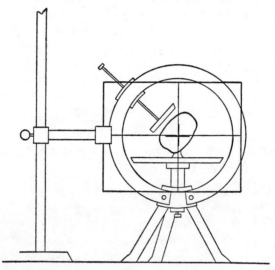

FIG. 123.—Goniometer and glass plate used in measurement of pebble orientation. After Wadell, 1936.

coincides with the plane of the vertical circle, whereupon its dip is found by means of the goniometer ruler.[1] Meanwhile, the amount of horizontal rotation necessary to swing the pebble into position is read. This angle is a measure of the deviation of the pebble from the trend of the gravel face, as represented by the glass plate, and from this information the compass trend of the long axis may readily be computed. The process outlined yields two values for each pebble which may be tabulated or presented in a graph.

PRESENTATION OF ANALYTICAL DATA

The results of an orientation analysis may be expressed as a frequency distribution of the direction of dip of the pebbles (see Figure 122), or the distribution may show the actual angle of dip of the long axes. As a

---

[1] It is assumed in Figure 123 that the long axis of the pebble is properly oriented when the lines on the glass and pebble coincide. Usually the lines will not coincide after the pebble is rotated.

first approximation the mode may be taken as the mid-point of the modal class, although it may be computed, or determined graphically from the inflection point of a cumulative curve of the data.

In addition to histograms of strike or dip direction, orientation studies may include polar coördinate diagrams of the dip and strike of the pebbles. Such charts are circles divided into degrees around the circumference (for strike) and have concentric circles dividing their radii from 0 to 90°, for angle of dip. A dot is placed at the appropriate point for each pebble, which results in a circular scatter diagram. The data are readily visualized, and from the concentration of the dots it is possible to estimate the modal trend. For a more formal determination of the mode the dots on the diagram may be assembled into classes and a histogram prepared.

The choice of the mode as the significant average is suggested by its nature, i. e., it is a measure of the most abundant individuals in the population. If a number of pebbles are being deposited under a set of controlled conditions, it would appear that most pebbles may tend to conform to the set conditions, but complexities of size and shape would cause some deviations from the fixed orientation. The deviations themselves may be significant, however, in terms of their distribution on one side or the other of the mode. Deviations are most conveniently studied in terms of the arithmetic mean of the distribution, by the conventional methods of moment analysis. The standard deviation affords a measure of the average spread, and higher moments afford measures of asymmetry and peakedness. Improper sampling methods may reflect themselves in a skew distribution, or there may be an actual genetic significance to such measures.

Several samples of till pebbles were studied by Krumbein,[1] using essentially the technique of Richter (page 269), in an attempt to apply conventional statistical methods to orientation data. One set of pebbles was collected from a till exposure overlying striated bedrock, so that the known direction of ice movement could be used as a control. The striae varied from N 5° E to N 30° E, a range of 25°. The distribution of strikes of the long axes of the pebbles showed a pronounced mode at N 20° E, well within the range of the striae. The arithmetic mean strike was N 8° E, also within the range. The standard deviation of strikes about the mean strike was 23°. In this sample either the mode or the mean could be used as an index of ice movement, despite the difference of 12° in their trends. Other till samples showed a close agreement between mode and mean, usually within 5°. Moreover, a composite of all samples (400 pebbles) showed an essentially normal (or at least symmetrical) distribution about the mean, suggesting that the deviations are random.

[1] Unpublished data at the University of Chicago.

Studies of serial sets of oriented pebbles along streams or other traverses from source to final disposition may shed important light on orientation changes in the direction of transport. These may reflect an increasing control of strikes as a function of distance (by a decrease in the standard deviation), or complexities may be introduced by changes of size and shape. In any event, there is a fertile field for further research, not only on orientation, but also on the relations between orientation, size, and shape. In all such studies statistical analysis will play an important part; moreover, improved laboratory techniques will enlarge the scope of possible studies.

Among techniques there appears to be a need for a simpler method of collecting the pebbles. The authors have developed a wooden frame measuring about 5 by 7 in. (a small picture frame will do), in which two brass rods are mounted at right angles. A spirit level is set in the frame, and by holding the frame upright, level, and parallel to the face of the exposure, lines may readily be drawn on the pebbles. In this manner 100 pebbles may be collected in an hour or two.

PART II

SHAPE ANALYSIS, MINERALOGICAL ANALYSIS, CHEMICAL ANALYSIS, AND MASS PROPERTIES

CHAPTER 11

SHAPE AND ROUNDNESS

INTRODUCTION

THE shape of sedimentary particles, large and small, is one of the fundamental properties of these particles and is the most recent to be studied quantitatively and statistically. Observers early noted the modification of shape that took place by transportation, and the master experimenter Daubrée himself studied the results of attrition of gravel in a revolving cylinder—an experiment later to be repeated by Wentworth in America and Marshall in New Zealand. Geologists, moreover, noted the characteristic forms imparted to cobbles and pebbles by ice action and those developed by wind abrasion (*Einkanter, Dreikanter*). Others went so far as to generalize that marine and lacustrine pebbles are round and oval or roller-shaped, but not wedge-shaped; that fluviatile pebbles are flat and wedge-shaped.[1] These generalizations and others were based on qualitative data only and that none too certain. H. E. Gregory, as a result of his study of many exceptions to these beliefs, went so far as to say that "of the many factors whose evaluation is essential in establishing distinctions between modes of origin of conglomerate, that of shape of pebbles has perhaps the least significance. No constant difference between the constituents of marine, lacustrine and river gravel is likely to be established."[2] This, too, is a generalization based on little if any quantitative data.

The statement has been made[3] and often repeated and also denied[4] that wind is capable of rounding smaller grains than water. If so, the aqueous or æolian origin of an ancient sandstone could be determined by noting the lower limit of rounding of the grains.

So many geologic factors are involved in the development of shape

[1] R. Hoernes, Gerölle und Geschiebe: *Verhandl. K-K. Geol. Reichsanstalt*, no. 12, pp. 267-274, 1911.
[2] H. E. Gregory, Note on the shape of pebbles: *Am. Jour. Sci.*, vol. 39, pp. 300-304, 1915.
[3] Wm. Mackie, On the laws that govern the rounding of particles of sand: *Trans. Edinburgh Geol. Soc.*, vol. 7, pp. 298-311, 1897. Victor Ziegler, Factors influencing the rounding of sand grains: *Jour. Geology*, vol. 19, pp. 645-654, 1911.
[4] G. E. Anderson, Experiments on the rate of wear of sand grains: *Jour. Geology*, vol. 34, pp. 144-158, 1926.

and roundness that any criterion of origin based on a single principle is likely to be unreliable. Factors that control shape and roundness are: (1) the original shape of the fragment, (2) the structure of the fragment, as cleavage or bedding, (3) the durability of the material, which is in turn a vector property of the rock or mineral fragment, (4) the nature of the geologic agent, (5) the nature of the action to which the fragment is subject and the violence of that action (rigor), and (6) the time or distance through which the action is extended.

It seems clear that if the problem of development of shape is to be studied in any other way than casually, some rigorous definition of shape and roundness must be established. Such qualitative expressions as "angular," "subangular," "subrounded," and "rounded" are vague. No two observers, moreover, can agree on the proper designation of a given sediment. A simple scheme of measuring objectively and a method of expressing numerically the shape and the roundness of a grain is necessary not only for descriptive purposes, but for the prosecution of quantitative studies of the several factors involved in the evolution of the shape of a particle or fragment.

The shape of fragments and grains has a bearing on several other problems. It has been generally assumed—not entirely correctly, as will be shown later—that sand grains and pebbles become progressively more round as they are transported, so that theoretically the direction from which a sediment came could be determined if a progressive increase in roundness, or roundness gradient, were detectable.

A by-product of the study of shape and roundness has been the use made of these characteristics for correlation purposes. Certain horizons have been marked by specific degrees of roundness and sphericity.[1]

Certain other properties of sediments, notably porosity and permeability,[2] are related to the shape of the component grains of the sediment.

REVIEW OF QUANTITATIVE METHODS

Sorby in 1879 classified sand grains into five groups:[3]

1. Normal angular fresh-formed sand, as derived almost directly from granitic or schistose rocks.

---

[1] A. C. Trowbridge and M. E. Mortimore, Correlation of oil sands by sedimentary analysis: *Econ. Geology*, vol. 20, pp. 409-423, 1925. Tor. H. Hagerman, Some lithological methods for determination of stratigraphic horizons: *World Petroleum Congress, Proc. 1933* (London), vol. 1, pp. 257-259.

[2] H. J. Fraser, Experimental study of the porosity and permeability of clastic sediments: *Jour. Geology*, vol. 43, pp. 934-938; 962-964, 1935.

[3] H. C. Sorby, On the structure and origin of non-calcareous stratified rocks: *Quart. Jour. Geol. Soc. London*, vol. 36, *Proc.*, pp. 46-92, 1880.

SHAPE AND ROUNDNESS

2. Well-worn sand in rounded grains, the original angles being completely lost, and the surface looking like fine ground glass.

3. Sand mechanically broken into sharp angular chips, showing a glassy fracture.

4. Sand having the grains chemically corroded, so as to produce a peculiar texture of the surface, differing from that of worn grains or crystals.

5. Sand in which the grains have a perfect crystalline outline, in some cases undoubtedly due to the deposition of quartz over rounded or angular nuclei or ordinary non-crystalline sand.

It may be seen from the above that Sorby's scheme of classification is both descriptive and genetic and involves surface character as well as shape.

Wentworth seems to have been the first to develop a quantitative system of measurement of the shape of individual rock particles independent of origin. Wentworth[1] expressed the shape of pebbles by a *roundness* and a *flatness ratio*. The roundness ratio is r_1/R, where r_1 is the radius of curvature of the sharpest developed edge and R is the mean radius of the pebble.

The mean radius (one half the mean diameter) is sometimes difficult to determine. The mean diameter may be the arithmetic mean of the principal diameters, or $\frac{A+B+C}{3}$, where A, B, and C are the three major diameters of the solid, the length, breadth, and thickness, respectively. The geometric mean may also be used. In that case the mean value is $\sqrt[3]{ABC}$. A major difficulty arises from the fact that no agreement has ever been reached in defining the three diameters of a non-spherical object. Some workers require that the three diameters meet at right angles; others stipulate that the lines of measurement must be at right angles, but do not require a common point of crossing. On pebbles with reëntrant angles, the terms *length*, *breadth*, and *thickness* become ambiguous. Wadell has used the *nominal diameter*, derived from volume measurements of the pebbles, to avoid these confusions.

The flatness ratio is expressed by r_2/R where r_2 is the radius of curvature of the most convex direction of the flattest developed face and R is the mean radius of the pebble. In his study of beach pebbles Wentworth expressed the flatness ratio as the arithmetic mean of the length and breadth divided by twice the thickness, or $\frac{A+B}{2C}$. The radii r_1 and r_2 were first measured by a gage similar to that used by opticians for

---

[1] C. K. Wentworth, A laboratory and field study of cobble abrasion: *Jour. Geology*, vol. 27, pp. 507-521, 1919; The shapes of pebbles: *U. S. Geol. Survey, Bull. 730-c*, pp. 91-114, 1922; The shapes of beach pebbles: *U. S. Geol. Survey, Prof. Paper 131-c*, pp. 75-83, 1922.

measuring the curvature of lenses. Wentworth later developed a flat-type convexity gage. This instrument consists of a low-angle measuring wedge sliding in a split profile block. The profile block is so constructed that the radius of curvature of the corners or edges of pebbles ranging from 1 to 100 mm. may be measured conveniently and rapidly to within 2 or 3 per cent. The readings are made through a reading slot in the split block on a scale which indicates the position of the measuring wedge (Fig. 124). R is computed either as half the arithmetic mean of length, breadth, and thickness of the pebble, or as half the geometric mean of the same dimensions.

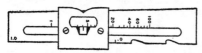

Fig. 124.—Flat type of convexity gage of Wentworth.

Trowbridge and Mortimore[1] used a visual method of determining "roundness" by comparison of the material under question with a more or less arbitrary set of standards.

Lamar[2] devised a mechanical means of determining the relative "roundness" or "angularity" of sand grains in the bulk. The method consists of determining the minimum porosity of sand obtained by compacting. A cylindrical metal tube, 1¼ in. in diameter, working in two guide sleeves, was raised a half-inch from below by a plunger operating on an eccentric and allowed to drop about 100 times a minute. The cylinder struck a piece of felt so as to produce a nearly "dead" fall, thus reducing to a minimum the amount of rebound imparted to the sand in the cylinder. The machine was motor-driven. The percentage of porosity in the sand was determined from the formula $P = \dfrac{100\,(C-V)}{C}$ where C is the volume of the sand and voids measured in the cylinder, V is the actual volume of the sand grains determined by displacement of water in a graduate, and P is the percentage of porosity, with maximum compaction. The relative angularity for the sand was determined by dividing 25.95, the theoretical minimum for spherical sand grains, by the porosity of the compacted sample. The nearer the quotient to 1.0, the less angular the sand. In order that the angularity be thus calculated it is necessary to use sand of one size only, hence the sample to be studied must be screened and the angularity of each screened separate individually determined.

[1] A. C. Trowbridge and M. E. Mortimore, Correlation of oil sands by sedimentary analysis: *Econ. Geology*, vol. 20, pp. 409-423, 1925.
[2] J. E. Lamar, Geology and economic resources of the St. Peter sandstone of Illinois: *Ill. Geol. Survey, Bull. 53*, pp. 148-151, 1927.

SHAPE AND ROUNDNESS

The first attempt to express the "roundness" of individual *sand grains* was that made by Pentland in 1927.[1] He determined the percentage area of the grain projection to that of a circle with diameter equal to the longest diameter of the grain. The area of the grain projection was determined from camera lucida drawings. As Wadell[2] has shown, it is possible to have two different plane figures of equal areas and equal major diameters but of distinctly different shapes (Figure 125).

Cox also studied the projection figures of individual grains.[3] He projected the image of the grains on a screen and from drawings

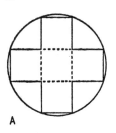

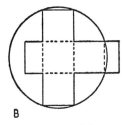

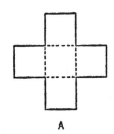

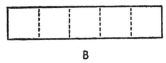

Fig. 125.—Figures with same major diameter and areas but with different shapes. The long diameter of the grain "B" is not the diameter of the circumscribing sphere. The difference between the method of Pentland and that of Wadell is thus demonstrated.

Fig. 126.—Figures with same perimeters and areas but of different shapes.

made calculated the roundness or circularity according to the formula $K = \frac{A \times 4\pi}{P^2}$, where A is the area measured by a planimeter, and P is the perimeter measured by a map measurer. K is the value of the roundness and depends on the shape of the projection figure. For a circle it

---

[1] A. Pentland, A method of measuring the angularity of sands: *Royal Soc. Canada, Proc. and Trans.* (Ser. 3), vol. 21, 1927, Appendix C, Titles and Abstracts, p. xciii.

[2] Hakon Wadell, *Volume, Shape and Roundness of Rock Particles*. A dissertation submitted to the Faculty of the Division of Physical Sciences in candidacy for the degree of Doctor of Philosophy. University of Chicago, June 1932. MS.

[3] E. P. Cox, A method of assigning numerical and percentage values to the degree of roundness: *Jour. Paleon.*, vol. 1, pp. 179-183, 1927.

is 1. Wadell has also shown that it is possible to have two figures of the same perimeter and the same area but of quite different shape (Figure 126).

Tickell[1] more recently used the ratio of the area of the projected grain to the area of the smallest circumscribed circle to express "roundness." This is very nearly the same method as that used by Pentland. For pebbles Tickell recommends the ratio of the volume of the pebble to the volume of the smallest enveloping sphere. The volume of the pebble would be determined by weighing in air and weighing in water. Tickell thus made an improvement over Pentland and Cox, yet he too failed to differentiate between shape and roundness.

Tester,[2] in 1931, proposed a very different method of expressing "roundness." He determined the ratio of the length of the original edge

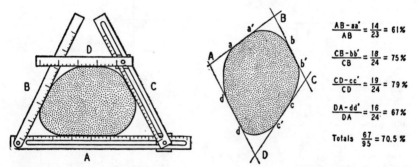

FIG. 127.—Shapometer of Tester and Bay.

FIG. 128.—Example of Tester's method of roundness determination.

or edges of a grain to the portion or portions worn away. Tester and Bay[3] devised a "shapometer" to facilitate the measurements required by his method (Figure 127).

For example, in Figure 128, the outline of an abraded fragment, the remnant edges aa', bb', cc', and dd' are shown and are extended to form polygon, A, B, C, and D. A, B, C, and D are believed to outline the original shape of the block from which the pebble was derived. The length of the lines representing the projected edges, as line AB, and the length of the remnant edges, as aa', are measured. The total length of AB is taken as denominator and the part of the line AB not in contact

---

[1] Frederick G. Tickell, *The Examination of Fragmental Rocks* (Stanford University Press, 1931), pp. 6-7.
[2] A. C. Tester, The measurement of the shapes of rock particles: *Jour. Sed. Petrology*, vol. I, pp. 3-11, 1931.
[3] A. C. Tester and H. X. Bay, The shapometer: a device for measuring the shapes of pebbles: *Science*, vol. 73, pp. 565-566, 1931.

SHAPE AND ROUNDNESS

with the grain (or difference between AB and aa') is the numerator. The ratio (AB-aa')/AB multiplied by 100 gives the percentage of the original edge worn away. The value for each edge is separately determined, and the average for the whole pebble is then computed (see Figure 128).

On the basis of the values obtained, a pebble may be classified in one of the five groups set up by Tester:

Percentage Abraded	Class Name
81-100	Rounded
61- 80	Sub-rounded
41- 60	Curvilinear
21- 40	Sub-angular
0- 20	Angular

Much difficulty is found in application of this method, since it is based on the ratio of an assumed factor, the original shape (largely unknowable), to a known factor, the present shape.

Wadell[1] appears to be the first to differentiate between *shape* (sphericity) and *roundness* and to show that these are two independent variables. Wadell pointed out that roundness was a matter of the sharpness of the corners and edges of a grain, whereas shape has to do with the form of the grain independently of the sharpness of its edges. The several geometrical solids, for examples, cube, tetrahedron, dodecahedron, etc., are clearly of different shapes, yet their respective edges or corners are equally sharp, i.e., the radius of curvature of the edges is 0.

Since the sphere has the smallest surface area in proportion to volume of any solid, it has a higher settling velocity for a given volume than any solid of any other shape.[2] Wadell, therefore, used the sphere as a standard of reference and spoke of the "degree of sphericity" as a measure of the approach of other solids to the sphere in form. Hence in the accumulation of sediment from suspension the degree of sphericity of the component grains is an important factor. So also in the transportation of debris by traction is the spherical form a suitable standard of reference, since a sphere will roll more easily than solids of other shape. An expression which approximately reflects the behavior of a particle in suspension is the ratio of the surface area of a sphere of the

---

[1] Hakon Wadell, Volume, shape and roundness of rock particles: *Jour. Geology*, vol. 40, pp. 443-451, 1932; Sphericity and roundness of rock particles: *Jour. Geology*, vol. 41, pp. 310-331, 1933; Volume, shape and roundness of quartz particles: *Jour. Geology*, vol. 43, pp. 250-280, 1935.

[2] Except only some pear-shaped solids with displaced centers of gravity.

same volume as the particle to the actual surface area of the particle expressed by the formula $\frac{s}{S} = \psi$, where s is the surface area of the sphere of the same volume, S is the actual surface area of the particle, and ψ is the true sphericity. The difficulty of determining the actual surface area and volume of a small grain led Wadell to adopt a working formula similar to that proposed by Tickell. This gives a close approximation of the true sphericity and may be stated: $\frac{d_c}{D_c} = \phi$, where d_c is the diameter of a circle equal in area to the area obtained by planimeter measurement of the projection of the grain when the grain rests on its larger face, D_c is the diameter of the smallest circle circumscribing the projection, and ϕ is the shape value thus obtained.

Wadell's method, like that of Pentland, Cox, and Tickell, which involves the use of projection areas for the study of grain shape, falls in error in the case of very flat grains. Such grains tend to lie upon their flattest developed face. Under these conditions a circular disk and a sphere give the same projection image and therefore the same shape value. Some writers, therefore (Tickell, for example), have specified that the grain be in random position and have taken precautions to insure such orientation. Wadell, on the other hand, specified that the grains be oriented more or less parallel to the largest and intermediate diameters. He found from actual experiment with quartz particles that discrepancies due to this cause were not great. Wadell not only found the discrepancies to be small, but also gave reasons for using the projection area containing the longest and intermediate diameters. Such reasons were related to the behavior of quartz particles to fracturing, chipping, etc.

For large materials, pebbles, etc., Wadell[1] developed a different formula: $d_n/D_s = \Psi$, where d_n is the true nominal diameter of the pebble or the volume of a sphere of the same volume and D_s is the diameter of the circumscribing sphere—usually the longest diameter of the pebble.

As stated above, roundness is a function of sharpness of edges; hence it is possible to have solids with perfect roundness independently of shape. As roundness increases the radius of curvature of the corners increases. An object of cylindrical form with hemispherical ends would be as perfectly rounded as a sphere. Such an object may eventually be worn down to a sphere, but the radius of curvature of its ends must, during the process of wear, remain always equal to the radius of the maximum inscribed circle in the longitudinal section of the solid (Figure 129). Roundness was therefore defined by Wadell as a value computed

---

[1] Hakon Wadell, Shape determinations of large sedimental rock-fragments: *Pan-American Geologist,* vol. 61, pp. 187-220, 1934.

SHAPE AND ROUNDNESS

from a plane figure, either projection or cross-section, in which the radius of the individual corners is divided by the radius of the maximum inscribed circle. The roundness of the individual corners thus obtained is added up and divided by the number of corners. This result is expressed by the formula: $\dfrac{\Sigma \frac{r}{R}}{N} = P$, where r is the radius of curvature of the corner, R is the radius of the maximum inscribed circle, N is the number of corners, and P is the total degree of roundness. The actual manipulative technique has been detailed by Wadell and is here given elsewhere (see page 298).

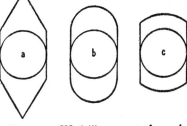

Wadell also used, for reasons given by him, a formula which gave slightly different roundness values than the one given above. The same items are measured as before and the degree of roundness by the expression: $\dfrac{N}{\Sigma (R/r)}$
The maximum value for roundness by this formula is also 1.

FIG. 129.—Wadell's concept of roundness. a, original fragment (with inscribed sphere); b, figure with maximum roundness (radius of curvature of ends equal to radius of inscribed circle); c, figure with low roundness resulting from radius curvature of the ends greater than inscribed circle.

In order to obtain comparable values a standard size was adopted. Large objects such as boulders must be reduced, and small ones like sand grains must be magnified to approximately the same size. Wadell's standard size is 7 cm.

Wadell has given examples of grains differing from one another in roundness and sphericity values and introduces the term *image* as a "binomial" expression of shape made up of the values for roundness and sphericity (see Figure 130).

Wadell has also used the term "degree of circularity" defined as c/C, in which c is the circumference of a circle of the same area as the plane

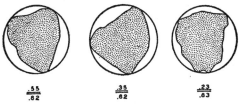

FIG. 130A.—Grains of same sphericity but differing roundness (after Wadell).

figure and C is the actual circumference of the plane figure. This expression is used to describe the shape of a plane figure, presumably the projection image of a sedimentary particle.

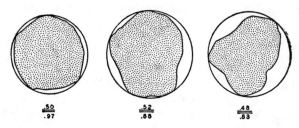

Fig. 130B.—Grains of about same roundness but of differing sphericities (after Wadell).

It is clear, then, from what has been said about the difference between shape and roundness that Wentworth measured roundness, whereas the Pentland, Cox, and Tickell measures express the shape of the grain projections. Wentworth's "flatness ratio," on the other hand, involving the ratio of the mean of the length and breadth divided by thickness, is more of a shape expression. Tester's method is somewhat akin to that of Szadeczky-Kardoss, to be described later, and is a type of roundness measure.

Wadell studied the same material as was investigated by Lamar, namely, the St. Peter sandstone at Ottawa, Illinois, and the sphericity values for each of the grades involved as determined by Wadell agree very closely with the "angularity" values given by Lamar for essentially the same grades. It is evident, therefore, that a person can obtain an average sphericity value for a given grade by Lamar's method. Lamar's method, moreover, measures shape and not "angularity" or roundness in a strict sense.

Other workers have in recent times studied quantitatively the shape of sedimentary particles and fragments. Szadeczky-Kardoss[1] in 1933, after criticizing the method of expressing roundness proposed by Wentworth and also that of Cox, devised a new scheme of measurement and presentation of data. The method applies especially to grains of diameters 2-100 mm. The pebble to be investigated is placed in an apparatus (Figure 131) whereby the profile of the fragment in one plane is mechanically traced without change in size on a sheet of paper. The outline thus obtained is analyzed and the percentage of concave, C, convex, V, and plane, P, parts of the profile is determined. A number of roundness grades were defined as shown in the table below:

[1] E. v. Szadeczky-Kardoss, Die Bestimmung des Abrollungsgrades: *Centralbl. f. Min., Geol., u. Paläon.*, Abt. B, pp. 389-401, 1933.

TABLE 31
SZADECZKY-KARDOSS ROUNDNESS CLASSES

Grade 0	C = 100%	
Grade 1a	C > (V + P)	P > V
Grade 1b		V > P
Grade 2a	(V + P) > C > V	(C + V) > P
Grade 2b		P > (C + V)
Grade 3a	(C + P) > V > C	P > (C + V)
Grade 3b		(C + V) > P
Grade 4a	V > (C + P)	P > C
Grade 4b		C > P
Grade 5	V = 100%	

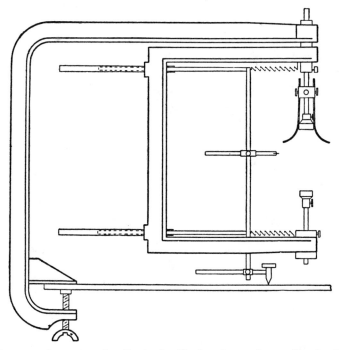

FIG. 131.—Apparatus of v. Szadeczky-Kardoss for tracing profile of pebble.

On a triangle diagram with C, V, and P the three corners (C + V + P = 100%) were plotted the values obtained for each pebble. Differences in roundness due to rock structure and mode of origin of the deposit are thus readily shown (Figure 132).

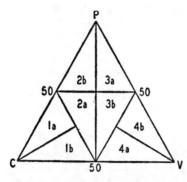

FIG. 132.—Triangle diagram of roundness grades of v. Szadeczky-Kardoss. P = plane, V = convex, and C = concave parts of profile. C + V + P = 100 per cent.

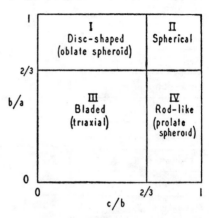

FIG. 133.—Shape classes of Zingg. a, length; b, breadth; and c, thickness.

Zingg[1] in a monographic study used the Szadeczky-Kardoss method of expressing roundness but recognized, following Wadell, the clear difference between roundness and sphericity. Zingg, however, measured the three principal diameters of a pebble, a, b, and c. On the basis of these measurements he set up four classes:

$$\begin{array}{lll} \text{I} & b/a > 2/3 & c/b < 2/3 \\ \text{II} & b/a > 2/3 & c/b > 2/3 \\ \text{III} & b/a < 2/3 & c/b < 2/3 \\ \text{IV} & b/a < 2/3 & c/b > 2/3 \end{array}$$

where $a > b > c$. These are set in a table and given specific names (Figure 133):

Hagerman[2] utilized grain shape to mark different stratigraphic horizons. He carefully split down the sample, mounted a few hundred grains on a microscope slide in such manner that the grains lie on their greatest developed face. The length (l) and breadth (b) of each grain were then measured under the microscope with micrometer ocular or with a microprojector and the b/l ratio computed. The values for each grain

[1] Th. Zingg, Beitrag zur Schotteranalyse: *Schweiz. Min. u. Pet. Mitt.*, Bd. 15, pp. 39-140, 1935.
[2] Tor H. Hagerman, Some lithological methods for determination of stratigraphic horizons: *World Petroleum Congress, Proc. 1933* (London), vol. 1, pp. 257-259.

SHAPE AND ROUNDNESS

(b/l ratio and l) determine a point on the diagram (Figure 134). The b/l ratio is diagrammed as ordinate and l is plotted as abscissa. The grains fall into four groups:

        Small, equiaxial grains   Large, equiaxial grains
        Small, oblong grains   Large, oblong grains

As plotting continues the boundaries of the distribution field begin to appear. A search for limit grains soon fixes the boundary of the field quite closely. The limit observations are marked with heavy dots through which a curve may be drawn. The shape of the distribution field is related to sedimentation conditions at time of deposit (turbulence, current velocity, etc.) and was found to be

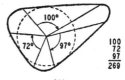

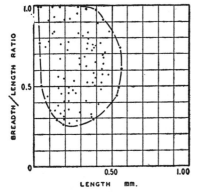

Fig. 134.—Hagerman plot of quartz grains of a sandstone. Limiting or boundary grains shown by heavy dots.

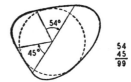

Fig. 135.—Fischer's method of angularity computation. The ratio of the worn (convex) to unworn (plane) portions of the profile, measured in terms of a central angle, determines the angularity.

characteristic of certain stratigraphic horizons.

Hagerman[1] has more recently published detailed results of his method applied to certain Argentine formations.

Fischer, in a recent paper on graywackes, devised a method of studying rounding of grains in thin sections.[2] There are 360° about any point. Fischer chose a central point within the sectional outline of the grain and measured the angles around this point which were governed or subtended by the straight or non-curved parts of the profile. The ratio

---

[1] Tor H. Hagerman, Granulometric studies in northern Argentine: *Geografiska Annaler*, vol. 18, pp. 125-213, 1936.
[2] Georg Fischer, Die Petrographie der Grauwacken: *Jahrb. d. Preuss. Geol. Landesanst.* (Berlin), Bd. 54, pp. 322-323, 1933.

of the sum of these angles to the whole angle of 360° gave the angularity value of the grain. In Figure 135a, for example, the angularity is 70.5 per cent, while in Figure 135b it is 21.9 per cent. Fischer gives no specific instructions for choosing the central point. The authors suggest the center of the inscribed circle.

Recently Wentworth[1] has fallen back on a verbal schedule for description of cobble shapes in which the shape is compared to some well known geometrical form. Terms such as *prismoidal, bipyramidal, pyramidal, wedge-shaped, parallel tabular*, etc., are used. The form of the margin, as viewed from the top was likewise described as hexagonal, pentagonal, trapezoidal, oval, rhombic, etc. The major diameters were also measured, and ratios between these were used in statistical study of the collected materials.

SUMMARY

It is clear from the historical review just given that there is no accepted standard at the moment for measuring the shape of irregular solid particles. The studies reviewed show several lines of approach to this problem:

Method	Attribute Measured	Author
A. Methods based on measurements of individual particles or fragments		
1. Visual comparison with arbitrary set of standards	Roundness and/or shape	Trowbridge and Mortimore
2. Diameter measurements: Ratio between longest and intermediate and/or shortest diameters or mean of length and breadth divided by thickness	Shape	Hagerman Zingg Wentworth
3. Measurements of projection or sectional areas:		
a) ratio of plane to convex and/or convex parts of profile	Roundness	Tester Szadeczky-Kardoss
b) ratio of perimeter to area of grain projection	Shape	Cox

[1] C. K. Wentworth, An analysis of the shapes of glacial cobbles: *Jour. Sed. Petrology*, vol. 6, pp. 85-96, 1936.

SHAPE AND ROUNDNESS

Method	Attribute Measured	Author
c) ratio of area of projection to smallest circumscribing circle	Shape	Pentland Tickell Wadell
d) ratio of radius of curvature of sharpest corner or mean radius of curvature of all corners to half the diameter through the corner or to radius of maximum inscribed circle	Roundness	Wentworth * Wadell
4. Volume measurements: Ratio of diameter of sphere of same volume to diameter of circumscribing sphere; volume ratios of same spheres	Shape	Wadell Tickell
5. Surface area measurements: Ratio of surface area of sphere of same volume to actual surface area	Shape	Wadell
B. Methods based on measurements on the aggregate		
1. Measurement of minimum porosity on uniformity sized materials	Shape	Lamar

* Measurements made on actual pebble rather than plane figure.

PROCEDURES FOR ANALYSIS

Choice of method. The choice of method depends on the use to be made of the results, on the time available, and on the size of the material to be studied. Where the results are to be used for correlation purposes or for comparison of somewhat similar materials, it is probable that some of the less rigorous and more rapid methods of Zingg, Hagerman, and others will suffice. Should these methods fail of the purpose intended, it is possible that the more strict methods of Wadell will succeed. The Wadell methods are very time-consuming and most likely will not be of value for rapid work demanded in oil-field laboratories until they are abridged and made shorter by use of appropriate tables and computing charts or alinement diagrams. Moreover, a method suitable for the study of sands may be unsuited or awkward when applied to pebbles and cobbles, and vice versa. Wentworth's methods, for example, are scarcely applicable to fine materials, though for many purposes they may be satisfactory for coarse sediments. In any event, the worker should have clearly in mind the distinction made by Wadell between shape and roundness. Whatever the method chosen it will give in some instances misleading or erroneous values. The worker should be aware of these limitations and interpret his data accordingly. The results obtained by one method are not usually convertible into values obtained by other methods.

Sphericity. For large fragments Wadell devised a very simple and rapid method. The true sphericity, ψ, is defined by Wadell as s/S, where s is the surface area of a sphere of the same volume as the pebble and S is the actual surface area of the solid. Owing to difficulties of measuring the surface area of an irregular solid, Wadell proposed a "practical method" for actual analysis.[1] In the practical formula $\Psi = d_n/D_s$, where d_n is the true nominal diameter, i.e., diameter of a sphere of the same volume as the pebble, and D_s the diameter of a circumscribing sphere. The value of d_n is computed from a measurement of the pebble volume determined by dropping the pebble in a graduated cylinder and noting the volume of the water displaced.

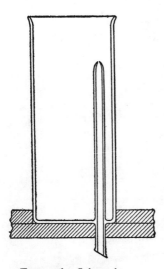

FIG. 136.—Schurecht overflow volumeter. For use either with water or kerosene. Overflow is caught and measured in a burette.

For small pebbles a small graduate (25 or 50 c.c.) may be used and the volume determined to within 0.5 c.c. For large pebbles and cobbles a cylinder with side spout for overflow may be used and the overflow caught and measured with a small graduate. Schurecht has devised an overflow volumeter (Figure 136) in which a burette is used in place of graduates.[2] Other types of volumeters suitable for measurement of the bulk volume of pebbles and cobbles are described in Chapter 19 in connection with the measurement of the volume of rock samples for porosity determination.

Determination of volume should be made to 0.5 c.c. on pebbles of 10 to 20 c.c. in volume, to the nearest 1 c.c. on pebbles of 20 to 50 c.c. in volume,

[1] Hakon Wadell, Shape determinations of large sedimental rock-fragments: *Pan-American Geologist*, vol. 61, pp. 187-220, 1934. A number of typographical errors occur in this paper. The integrational symbol $\int$ is used throughout in place of the symbol f, for *function*. 0.7 in formula 8, p. 205, should be 0.1.

[2] H. G. Schurecht, A direct reading overflow volumeter: *Jour. Am. Ceram. Soc.*, vol. 3, p. 731, 1920.

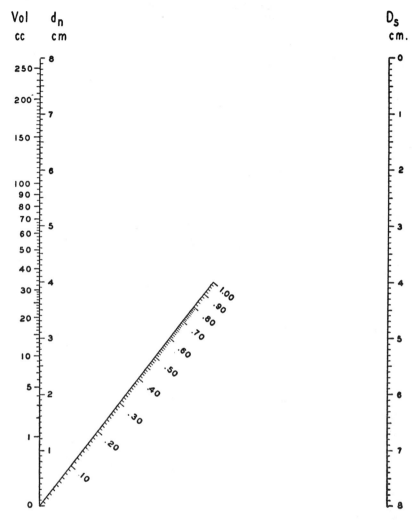

Fig. 137.—Nomograph for computation of sphericity by Wadell's method. For pebbles, the volume value, determined by displacement method, is connected with the D_s (maximum diameter) value measured by gage. Point of crossing, of line thus determined, on slanting scale, gives sphericity. For sand grains, the d_n value computed from area measurements by means of Fig. 139, is connected with D_s value obtained from grain drawing and sphericity read on center scale as before.

and to the nearest 2 c.c. on pebbles of 50 to 200 c.c. in volume. This degree of accuracy will insure a sphericity value correct to the second decimal place.

The longest diameter of the fragment is assumed to be the diameter of the circumscribing sphere, D_s.[1] This may be measured by simple gauge (Figure 37).

The computations on each pebble involve computing d_n from measured volume and evaluating the ratio d_n/D_s. The accompanying monograph simplifies this procedure (Figure 137). On this there are three scales, Vol-d_n, D_s, and center scale. To solve the above problem, locate the volume of the pebble on the left-hand or volume scale and the maximum diameter on the D_s scale. Connect these two points with a straight-edge. Where the straight-edge intersects the center scale read the value of sphericity.

Allowing a minute for measurement of maximum diameter and volume, and a half-minute for calculation with the alinement chart, it is evident that a gravel sample of about 50-100 pebbles can be readily analyzed and that a suite of samples can be studied in the course of a few days.

The use of the alinement diagram does not reduce the accuracy of the method, since errors introduced by the displacement method of volume and errors made by measurement of the long diameter (to the nearest 0.1 cm.) are greater than those involved in reading the scales.[2]

An example of the use of the method is given below:

Data concerning early pre-Cambrian conglomerate at locality M.B. 1, Manitou Lake, District of Kenora, Ontario.

Location: Lahay Bay, Manitou Lake, District of Kenora, Ontario, Canada. About 50 chains due east of narrows between Lahay Bay and Manitou Straits on narrow neck of land separating Lahay Bay from small bay south of the same.

Note on Exposure: Vertical conglomerate beds, strike N 58° E, containing numerous pebbles and cobbles and a few boulders up to 20 inches in diameter. A count in one square yard disclosed 37 black cherts and iron-formation, 36 granites, 60 greenstones and metadiorites, 12 felsites and porphyries, 1 vein quartz. Bedding pronounced with pea-like grits alter-

---

[1] As noted by Wadell, this is not always the case. Exceptions are, however, believed to be too rare to materially affect the analysis.

[2] Wadell calculated the values obtained by this method with those obtained by the formula s/S for certain definite geometric forms and found the values of the practical method here given to be about 0.10 lower, on the average, than the true sphericity. He corrected for this difference by adding 0.10 to all values less than 0.80. This correction, however, as noted by Wadell, produced some difficulties of overlapping of values in the 0.70-0.90 range.

nating with pebble and cobble conglomerate. Pebbles readily weather out of matrix.

TABLE 32

Shape Analysis: $\Psi = d_n/D_s$

Pebble No.	Rock Type	D_s (cm.)	Vol. (c.c.)	Ψ	Remarks
1	Vein quartz	5.31	34	0.762	Slightly broken
2	Felsite	4.62	19	.718	Slightly broken
3	Greenstone	6.31	29	.604	
4	Granite	6.42	60	.759	A little matrix attached
5	Greenstone	5.00	23	.708	A little matrix attached
6	Greenstone	4.00	19	.823	
7	Felsite	3.80	15	.803	
8	Felsite	3.20	5	.664	
9	Granite	8.50	174	.815	

Wadell's method of determination of volume, sphericity, and roundness of sedimentary particles. Wadell has described in great detail the various steps involved in analysis of a sediment so as to determine volume, sphericity, and roundness of quartz particles according to his definitions of these properties.[1] Since on the whole the concepts of shape and roundness presented by Wadell seem to the authors to have the soundest theoretical basis, a brief outline of the procedure used by Wadell is here given.

The sample is split down and screened into several fractions, based on the Wentworth-Udden grade scale, which are then weighed.[2] Each screened fraction is further split down to an amount small enough to be spread over a microscopic slide. The grains on each slide are counted and also weighed. Knowing the number of grains and their weight makes possible the computing of the total number of grains in each sieve separate, since the latter is weighed also. Thus the frequency of grains by number (rather than by weight) is known for the sediment. A few drops of clove oil ($n = 1.560$) serve as mounting medium for the grains.

The slide is next placed under the microscope and, by means of a camera lucida, the outlines of each grain are drawn. It is necessary for

[1] Hakon Wadell, Volume, shape and roundness of quartz particles: *Jour. Geology*, vol. 43, pp. 250-280, 1935.
[2] Wadell, in fact, removed all minerals except quartz by bromoform. Such procedure is probably advisable in those sediments in which minerals other than quartz are abundant.

determination of roundness that the average diameter of the reproduced grains be about the same size. Wadell chose 7 cm. as the "standard size." It is evident, therefore, that different objectives will be needed for the different grades in order to achieve this standard. In order to avoid distortion, especially with high-power objectives, the grains to be drawn should be placed in the center of the field. About 50 grains in each grade size should be drawn.

The area of each grain reproduction is then determined by a polar planimeter (Figure 138). From this value the *nominal sectional diameter*, i.e., the diameter of a circle with the same area as the projection, is computed. This may be done graphically by use of a chart (Figure

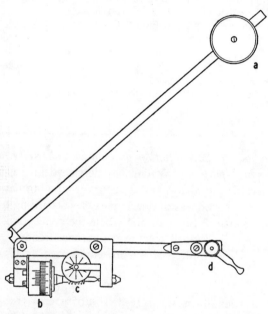

FIG. 138.—Polar planimeter. *a*, weight and pin; *b*, vernier and revolving drum recording 0 to 1 sq. in.; *c*, revolving recording disk, 0 to 10 sq. in.; and *d*, tracing stylus and thumb-hold.

139). The diameter of the smallest circumscribing circle is also measured, usually the long diameter of the grain, and from this and the nominal sectional diameter the sphericity is obtained, as per formula $\phi = \dfrac{d_c}{D_c}$ where d_c is the diameter of a circle equal in area to the area obtained in the standard size when the grain rests on one of its larger faces and D_c is the diameter of the smallest circle circumscribing the grain reproduction.[1]

[1] Wadell defines the nominal sectional diameter as that of the *non-magnified* grain. No distinction between the non-magnified grain and the image as drawn is here made, since the sphericity as calculated is a ratio in which size effects are canceled out.

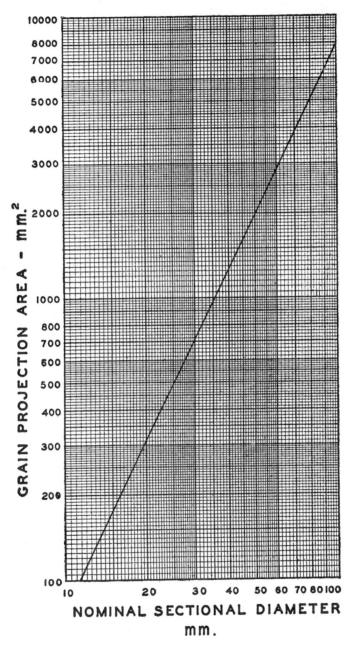

Fig. 139.—Chart for computing nominal sectional diameter (of the magnified grain) from measurement of the projection area.

While this formula or definition of sphericity differs from that earlier given by Wadell for large fragments (see page 284), he has shown that since the nominal sectional diameter as above defined approximates quite closely the true nominal diameter, the result obtained from grain projections is quite comparable to that obtained on large fragments.[1] He has shown also that the values for the sphericity thus obtained approach fairly closely the actual sphericity in which surface areas are taken into account.[2]

Fig. 140.—Celluloid circle scale of Wadell for radii measurements of grain drawings.

Roundness is also obtained from the grain projections according to Wadell's formula for roundness of plane figures (see page 285). The radius of curvature of each corner is obtained by placing a transparent celluloid scale (Figure 140) over the image of the grains as drawn and magnified to the standard size of about 70 mm. This scale, on which are drawn some 35 concentric circles differing from one another in radius by 2 mm., is adjusted so that the radius of curvature of each corner may be obtained. The radius of the maximum inscribed circle is likewise obtained. The roundness is then computed according to Wadell's formula. Example:

$R = 19$ mm. = the radius of the maximum inscribed circle
r = the radius of curvature of a corner. The values for r are shown in the table below.
$N = 10$ = the number of corners, see Figure 141B.
Table of values of r:

$r_1 - 3$; $r_2 - 2$; $r_3 - 10$; $r_4 - 3$; $r_5 - 2$; $r_6 - 7$; $r_7 - 13$; $r_8 - 4$; $r_9 - 4$; $r_{10} - 6$.

$\Sigma r = 54$. $\dfrac{\Sigma r}{N} = 54/10 = 5.4$ $5.4/19 = .28+$

To measure the roundness of pebbles and larger objects a different scheme of obtaining the projection image is required. Hough,[3] in a study of material 3 mm. to 100 mm. in diameter, photographed or photostated the pebbles, using appropriate lens combinations to achieve a standard size. The pebbles were first separated into groups of approximately uniform size and placed in their most stable positions on a black background. A white square bearing identification number and a white cellu-

---

[1] Hakon Wadell, loc. cit., pp. 259-263, 1935.
[2] Idem, pp. 264-266.
[3] J. L. Hough, personal communication, 1937.

SHAPE AND ROUNDNESS

loid pocket ruler were also placed in the field. If the images were less than standard size, further enlargement by projection printing is necessary. The authors have used a simpler scheme for obtaining projections of pebble images. The pebbles were placed on a glass tracing table with light source beneath. With a "pointolite"[1] or similar illumination the shadow of the pebble is sharply defined on a paper placed on a glass plate supported an appropriate distance above the pebbles themselves.

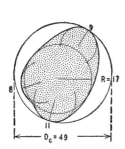

Fig. 141A.—Wadell method of determination of shape and roundness of sand grains. D_c, diameter of circumscribing circle; R, radius of inscribed circle.

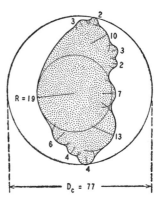

Fig. 141B.—Wadell method of determination of shape and roundness of sand grains. D_c, diameter of circumscribing circle; R, radius of inscribed circle.

Short method for determination of size and shape of sand grains. Wadell's procedure as given above is very time-consuming. The following is suggested as a modification for the worker who wants to retain the principles developed by Wadell, but who, for lack of time, cannot give every sample exhaustive treatment.

The sample is carefully split down with a Jones splitter to about 1 or 2 g., which in turn is split down by some means—preferably a microsplit—until some 300 to 500 grains remain. These are then mounted on a glass slide, in piperine ($n = 1.68$) if possible, to bring the grains into clear view. The slide is then placed under the microscope and suitably magnified so that the outlines of the grains may be drawn some 100 times their actual size with camera lucida, or they are placed on a microprojector and the images as projected on a drawing board are traced.

[1] "Pointolite" is the trade name of a bulb which emits rays from a very small, white-hot bead of tungsten. It is essentially a point-source of illumination and gives, therefore, very sharp shadows. Obtainable from James G. Biddle Co., Philadelphia, Pa.

If one wishes to know the average size of the grains, it is only necessary, as pointed out by Krumbein,[1] to measure the maximum horizontal intercepts of randomly oriented grains. If, however, one is concerned about the sorting of the grains as measured by the standard deviation, it is better to measure the grain areas, as projected, with a planimeter. This is, of course, absolutely necessary if one wishes to determine their shape or sphericity. From the measured areas the nominal sectional diameter, or the diameter of a circle of the same area, is computed. The diameter of the circumscribing circle, usually the maximum diameter of the grain, is also measured, and the sphericity for each grain is computed by taking the ratio of these two measurements as stated above. It is not possible to measure roundness on the images as drawn unless they are all of about the same order of size and preferably of the

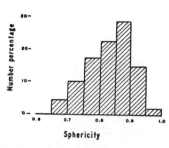

FIG. 142.—Histogram of sphericity analysis of sample of St. Peter sandstone based on microscopic measurements of 100 grains.

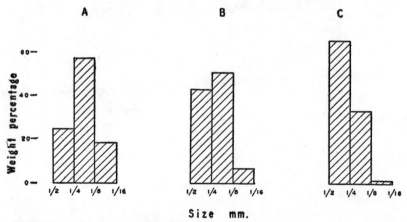

FIG. 143.—A, histogram of sieve analysis of St. Peter sandstone; B, histogram of same based on microscopic measurements of intermediate diameter of grains; C, histogram of same based on microscopic measurements and computations of nominal diameter; 100 grains used for B and C.

"standard size," namely, about 7 cm. over all. If these conditions are fulfilled the roundness may be measured as outlined in the preceding section.

[1] W. C. Krumbein, Thin-section mechanical analysis of indurated sediments: *Jour. Geology*, vol. 43, pp. 489-496, 1935.

TABLE 33
DATA DERIVED FROM MICROMETER ANALYSIS AND COMPUTATION

1	2	3	4	5	6	7	8	9	10
Grain No.	Proj. Area (in.²)	Proj. Area (mm.²)	Proj. d_c ($d_c = \sqrt{4A/\pi}$)	Proj. D_c (mm.)	ϕ (d_c/D_c)	Actual d_{cn} ($d_c/100$)	d_{cn}^3 (mm.)	Volume (c.c.³) ($\pi/6 d_{cn}^3$)	Weight (g.) (V × 2.65)
1.....	.97	635	28.4	31.0	.92	.284	.022906	.011994	.031784
2.....	.27	177	15.0	17.0	.88	.150	.003351	.001754	.004648
3.....	.87	569	26.9	32.0	.84	.269	.019476	.010197	.027022
etc.									

Abbreviations: Proj. = projection
d_c = diameter of circle of same area as magnified grain projection
D_c = diameter of smallest circle circumscribing grain projection
d_{cn} = non-magnified nominal sectional diameter of grain
V = volume
ϕ = sphericity

NOTE.—In practice it is usually unnecessary to compute the values of columns 9 and 10. The value given in column 8 is proportional to weight (K d_{cn}^3 = weight). The class or size grade to which the grain belongs is known from column 7. Add the column 8 values of all grains in a given class, and divide by the total of column 8, to find the percentage for that class.

SEDIMENTARY PETROGRAPHY

The computations involved are considerable, and as an illustrative example the determinations made on a sample of St. Peter sand are here given. The example is more elaborate than is required by routine analysis, since the writers wish to point out several ways in which the results may be expressed and to show differences in the results when the composition is expressed on a weight basis from those when it is expressed on a number basis and the differences between an analysis made by screening and one made by micrometric methods on a very small sample. (See Figures 142 and 143.)

Sieves classify grains according to their intermediate diameter. Hence in order to obtain strictly comparable results by microscopic analysis, it is necessary to measure the shortest diameter of the projection area. Since grains tend to lie on their greatest developed face, parallel to their longest and intermediate diameters, it is clear that the shortest diameter of the projection is really the intermediate diameter of the grain. The true shortest diameter will be perpendicular to the projection. That this scheme of measurement gives values more nearly identical with those obtained by screening is seen by inspection of Figure 143. Here the data obtained by screening and that obtained by microscopic measurement and recomputation to a weight basis may be compared. As seen in 143B, the data from measurement of the intermediate diameter is closer to that obtained by screening.

CHAPTER 12

SURFACE TEXTURES OF SEDIMENTARY FRAGMENTS AND PARTICLES

INTRODUCTION

The intimate details of the grain surface, independent of size, shape, or mineral composition, are termed the *surface texture* of the grain. A grain, for example, may be polished, frosted, or etched. A pebble may be marked with striations or by percussion marks. Such features are here defined and described.

These detailed characters have genetic significance and may be criteria of value. The frosting on sand grains has, for example, been said to denote æolian action,[1] while striations are most usually attributed to glaciation.[2] While these generalizations are open to question and while the origin of many surface textures is not at all clear, it is evident that as our knowledge increases these external characteristics of sedimentary particles and fragments deserve careful attention.

Just as a sand grain or pebble may inherit its shape from an earlier deposit of different origin, so too may a particle or fragment inherit the surface markings that it bears. However, a geological agent will, if time be sufficient, impose its own unique character on the particle or fragment. Some considerable time must elapse before size and shape modifications are evident, but the surface characteristics are more readily destroyed or modified. It is perhaps the sensitiveness of these features to change that makes them all the more important.[3] They record most faithfully the effect of the last cycle of transportation.

---

[1] W. H. Sherzer, Criteria for the recognition of the various types of sand grains: *Geol. Soc. Am., Bull.*, vol. 21, p. 640, 1910.

[2] C. K. Wentworth, An analysis of the shapes of glacial cobbles: *Jour. Sed. Petrology*, vol. 6, p. 85, 1936. Wentworth describes glacial striations in some detail and points out also the similarities between those of glacial origin and those produced by river ice. See: The shape of glacial and ice jam cobbles: *ibid.*, vol. 6, p. 97, 1936.

[3] Wentworth, for example, found by experimental study that a travel of but 0.35 mi. was necessary to remove striæ on hard limestone and greenstone pebbles. See: The shapes of pebbles, *U. S. Geol. Survey, Bull.* 730-c, p. 114, 1922.

But as has been pointed out it is not impossible to find sand grains whose surfaces may be described as smooth, rough, glassy, frosted, pitted, and stained, all in the same sand. If the character of the surface of the grain bears any relation to the origin of the sand, it is clear that the mixture of the above surface types in the same sand indicates that the grains were derived originally from several different types of deposits and that they have not been worked over sufficiently to have these inherited surface textures destroyed. However, in a great many sands there is one type of surface texture that predominates, and it is assumed that in such deposits an understanding of the surface texture may be extremely helpful in determining the conditions of deposition of the sediments.

The study of these features has lagged much behind that of the study of other fundamental properties—size, shape, and mineral composition. Even now no quantitative method of measuring these features is known. Our discussion then is confined to a brief definition of each feature and a purely descriptive classification of these surface characters.[1]

Surface characters are most conveniently discussed with reference to the size of the particle or fragment on which they appear. There are many features known on cobbles and pebbles which cannot or do not appear on sand grains. This is due to the fact that the pebbles and other large fragments are generally rock fragments—often of more than one mineral—whereas the sand grains are largely single minerals and are microscopic in size. Moreover, the pebbles and cobbles are studied megascopically, whereas the sand grains are examined with the microscope.

SURFACE TEXTURE OF LARGE FRAGMENTS ($>$ 2 MM. DIAMETER)

The surface features of large fragments fall into three categories, namely, degree of smoothness, degree of polish or gloss, and surface markings.

Polish or gloss has to do with the degree of luster of the surface. This property is primarily related to the regularity of reflection. Much scattering or diffusion of light produces a *dull* surface. The presence of high-lights indicates a good polish. A polished or glossy surface may or may not be smooth, it may be striated, grooved, or pitted.[2]

---

[1] For these definitions and classification, the writers are indebted to Miss Lou Williams, who prepared a summary of the literature on this subject, while at the University of Chicago, for the Committee on Sedimentation.

[2] Miss Williams finds, for example, that a pebble with a polished and apparently smooth surface reveals microscopic striations under high magnification. In fact, the presence of such microstriations may distinguish between polish induced by abrasion, which exhibits such striations, and gloss or chemical polish, which lacks such microstriations.

SURFACE TEXTURES

TABLE 34

SURFACE TEXTURES OF FRAGMENTS OVER 2 MM. IN DIAMETER

May be smooth, or scratched, furrowed and grooved, pitted or dented	A. DULL versus B. POLISHED (Gloss)
May be dull or polished	C. SMOOTH versus D. SURFACE MARKINGS 1. FURROWED AND GROOVED 2. SCRATCHED 3. RIDGED 4. PITTED OR DENTED

Smoothness is the evenness of the surface. A *smooth* surface is one on which no striations, pits, ridges, or other features are observable. A smooth surface may be either polished or dull. The antithesis of *smooth* is *rough*. Roughness may be due to pits or to nondescript small irregularities of the surface.

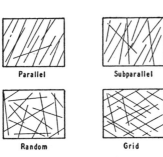

FIG. 144.—Patterns of arrangement of striations on pebble or cobble face.

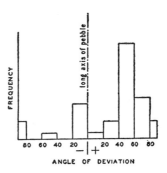

FIG. 145.—Histogram of deviations of striations from long axis of glacial pebble.

Much attention has been given to the nature of surface markings and blemishes when these are present. Many pebbles are *scratched* or striated. Such striations may occur in *parallel* or *subparallel* sets, *grid* patterns, or in *random or scattered* fashion (Figure 144). The relation of the striations to the long axis of the pebble may be important and the deviations in direction from such long axis may be measured and studied statistically.

Krumbein,[1] for example, studied the striæ on a glacial cobble and measured their deviations from the long axis of that cobble. The angle between the long axis and each striation was determined. The data obtained were then divided into nine classes based on ten-degree intervals and a frequency curve was obtained (Figure 145). Conceivably such a statistical study might reveal differences between cobbles striated by one agent or another.

Such detailed studies may not always be warranted, but may, in the case of certain glacial deposits, be important. A distinction between ice-jam river cobbles and true glacial cobbles may be made on the basis of arrangement of striations.[2] Some pebbles are *furrowed* or *grooved*. The opposite of grooved is *ridged*.

Pebble surfaces may also be *pitted* or dented. Pits vary much in form and size. Crescentic impact scars or *percussion marks* are notable on some pebbles, particularly quartzite pebbles. Conceivably *chatter marks* may appear on glacial cobbles. Indented pebbles, pebbles with shallow oval-shaped depressions, due to solution at point of contact between pebbles under pressure, are also known.

SURFACE TEXTURES OF SMALL FRAGMENTS (<2 MM. DIAMETER)

The surface texture of a sand grain may be described as dull or polished and as smooth or rough. As suggested in the table on page 307, various combinations of these qualities are possible.

The dullness or polish, as in the case of pebbles, is a quality of luster. A *dull* surface is one lacking in brilliance or luster, whereas a *polished* surface is one of high gloss. A grain surface may have a low polish, often pearly, or it may have a high polish, vitreous or brilliant.

A grain surface may be smooth or rough independent of the luster of the surface. A *smooth* surface lacks relief when seen under the microscope,[3] whereas a *rough* surface has inequalities, projections or pits. Where the irregularities are linear, the term *striated* may be used. When they are of geometric form and of chemical origin, the term

---

[1] W. C. Krumbein, unpublished data.
[2] C. K. Wentworth, *loc. cit.*, 1936.
[3] A pebble may have a smooth surface to the naked eye but under the microscope appear minutely rough. As stated, the definitions for surface characters of pebbles are given for the unaided eye, while those for sand grains are given as observed under the microscope.

etched is used. When the irregularities are very minute a *frosted* surface results, but where the irregularities are larger and scattered the surface is termed *pitted*. Grains of some minerals are subject to secondary enlargement. The secondary growth is deposited from solutions in optical and crystallographic continuity with the original material. Such grains exhibit microscopically small *facets*.

TABLE 35
SURFACE TEXTURES OF FRAGMENTS UNDER 2 MM. IN DIAMETER

May be smooth or rough	A. DULL	versus	B. POLISHED
May be dull or polished	C. SMOOTH	versus	D. ROUGH 1. STRIATED (usually glacial action) 2. FACETED (secondary growth) 3. FROSTED ("ground-glass" surface) 4. ETCHED (solvent action) 5. PITTED

NOTE.—The term *mat surface* is not well defined but is perhaps most often used in place of *frosted surface,* though perhaps the surface described by the former term is one of finer texture.

It is not within the province of this book to discuss the genetic significance of these characteristics. The reader is referred to the report by Miss Williams for a résumé of current opinion on this subject.[1] It is worth while to recall that some workers have given considerable weight to surface texture. It is noteworthy also that some of these textures can form in very different ways. Polish or gloss, for example, may be produced by gentle attrition or wear or it may be induced by solution or, as in the case of the pebbles, by deposition of a vitreous film, exemplified by desert varnish. Likewise the frosted surface may be formed by the rigorous action of wind, chemical etching, or incipient secondary enlargement.[2]

As is evident from the foregoing summary, our knowledge of the sur-

[1] *Report of the Committee on Sedimentation, 1936-37,* pp. 114-128, Nat. Research Council, 1937.
[2] R. Roth, Evidence indicating the limits of Triassic in Kansas, Oklahoma, and Texas. *Jour. Geology,* vol. 40, pp. 718-719, 1932.

face characters of grain is very incomplete and inexact. Even our definitions and our classification of the surface textures are not wholly satisfactory. We have yet to be content with verbal descriptions, though, as pointed out under the discussion of striæ, some attempt has been made to quantify our observations and to express the results in statistical terms. Perhaps ultimately the polish on grains will also be described more exactly—perhaps even measured by a photometer.

The interpretation of these features is even more unsatisfactory, but when we have fully organized and adequately defined what we know about the surface characters, we may hope that experimental and observational data will accumulate that will make for a better understanding as well.

CHAPTER 13

PREPARATION OF SAMPLE FOR MINERAL ANALYSIS

INTRODUCTION

THE section of this book which follows is devoted to the techniques of study which deal with the mineral composition of the sediments. The various steps involved are dependent on the type of study chosen. When the mineral grains are to be studied (as in the case of sub-surface correlation), it is necessary, after collection of a sample, to prepare it for analysis. Such preparation involves a disaggregation or breaking-down of the sample followed by a separation of the sample into two or more fractions which are more or less homogeneous mineralogically and mounting of the separates in a suitable way for microscopic study. The separations are usually achieved by panning, or by use of heavy liquids, the electromagnet, or some special method or device. Since the average sand is largely quartz, the separation methods are usually designed to segregate the minor accessory minerals from the quartz (and occasionally abundant feldspar). The minor accessories, or so-called "heavy minerals," even though present in very small amounts (tenth of 1 per cent or less), have proved of most worth both in correlation and in provenance studies.

The next step involves identification of the minerals. This is best accomplished optically by means of the polarizing microscope, though microchemical and other methods are of occasional value. Following identification, the mineral frequencies are determined by actual count or by estimation and are then recorded in tabular or graphic form. Statistical analysis of the data obtained may then be made if desired.

The non-clastic sediments are more generally studied in thin sections in which both the mineral components and the textures may be identified. Consequently we have given instruction in the preparation of such thin sections from both consolidated and feebly coherent materials as well as methods of identification suitable to the study of thin sections. Planimetric analysis of the section makes possible quantitative results, if such are desired.

Both the thin section and the mineral concentrates of a sediment should be studied if full information is required. The thin section gives the only method of careful study, of textures and structure, whereas the mineral concentrates are most ideal for the study of the mineral composition.

Both methods have their limitations. In the case of exceedingly coarse materials microscopic study is awkward and unnecessary, and in the case of very fine materials it is quite barren of results owing to the limits of visibility. Other methods, such as chemical analysis, x-ray and spectroscopic investigations, must then be resorted to for information concerning the mineral composition of the rock. In the case of sedimentary materials, other than clastic rocks, such as the phosphates, dolomites, gypsum, and coal, the thin section is the most usual method of study, though it has been found advantageous to study the detrital components of these rocks which have been isolated by some method suitable to the material in question.[1]

In all cases an interpretation of the results obtained from the mineral grains or the thin section is a matter of large scope and beyond the purpose of this volume. For such information the reader must refer to the voluminous literature on these aspects of sedimentary petrology or to such larger works as the *Treatise on Sedimentation,* published under the auspices of the Committee on Sedimentation. Since the industrial value or use of the various sedimentary materials is closely tied up with their mineral composition, the study of their composition becomes economically important. But a discussion of this too is beyond the scope of this book, and the reader is therefore referred to special papers and works on the subject.

PREPARATION OF SAMPLE

Disaggregation

Assuming the most usual case, namely a mineral grain study of a detrital rock (supplemented by thin section), the authors will consider the preparation of the material for examination. Since this involves, as pointed out above, some methods of separating the whole sample into fractions more or less simple in composition, and this in turn implies a state of complete disaggregation, it will be necessary to give an ac-

---

[1] J. E. Lamar, Sedimentary analysis of the limestones of the Chester series: *Econ. Geology,* vol. 21, pp. 578-585, 1926.

count of the treatment preliminary to making the mineral separations and then discuss the separation methods themselves.

Mineral analyses and preparation of the sample for analysis can be intelligently pursued only if the general make-up of the clastic rocks be kept in mind. The following groups of materials may be present:

(After Holmes)

I. *Allogenic constituents*
   1. *Pebbles* or other fragments of preëxisting rocks
   2. *Composite grains* consisting of more than one mineral
   3. *Simple grains* or particles of unaltered minerals such as quartz, muscovite, garnet, etc.
II. *Organic remains*
III. *Authigenic constituents*
   1. *Alteration products* or synthetic recrystallization products formed *in situ* from any of the allogenic constituents: e.g., clay from feldspar; limonite, leucoxene, glauconite, etc.
   2. *Infiltration products,* or materials introduced from external sources, generally present as a cement
   3. *Recrystallization products,* formed by the recrystallization of materials already present in the sediment

NOTE.—New materials produced during hydrothermal alteration or contact metamorphism may also be considered authigenic constituents.

A preliminary examination of the material is advisable in order to select intelligently the method of treatment most likely to be successful. Deverin,[1] following Cayeux,[2] gives six possible cases involved in disaggregation problems. These are:

(1) Material very coherent and unattacked by weak acid, for example quartzite.
(2) Very coherent material partially soluble in dilute acid, for example calcareous sandstone.
(3) Material slightly coherent, unattacked by weak acid, for example clay-shale.
(4) Material slightly coherent, but attacked by dilute acid, for example chalk.
(5) Unconsolidated materials, unattacked by weak acid, for example siliceous sand.
(6) Unconsolidated materials attacked by weak acid, for example carbonate sand.

[1] L. Deverin, L'Étude lithologique des roches sédimentaires: *Schweiz. Min. u. Pet. Mitt.*, Bd. H, pp. 29-50, 1924.
[2] Lucien Cayeux, Introduction a l'étude pétrographique des roches sédimentaires: *Texte* (Imprimerie Nationale, Paris, 1931, re-impression), pp. 4-8.

For (1) the thin section is still the principal method of study.[1] Thin sections are of much value for (2), (3), and (4), but are rarely used for (5) and (6). In (3) and (4) disintegration is accomplished by slaking or softening in water and by rubbing with a stiff brush. When the rock is attacked by weak acid (1 HCl:4 H_2O), it may be broken down by acid treatment. It is advisable, however, to restrict the quantity and strength of the acid and the time of treatment as far as possible, since the acid does in part destroy certain minerals either by solution or by decomposition. Such loss may be anticipated by the preliminary examination suggested above, and due consideration must be given it. In some cases an alkaline digest, either KOH or NaOH solution, will facilitate disintegration of the rock. Goldman[2] found this to be of use in sandstones with opaline silica cement. The alkaline digest is of course the only one that can be used if calcareous microfossils are to be looked for. Certain siliceous forms may, however, be destroyed by its use. Several other methods have been suggested for the disaggregation of rocks that do not yield to the simpler methods. Freezing and thawing have been recommended by Hanna and Church[3] for the purpose of disintegrating shale containing microfossils. Tolmachoff[4] soaked shaly rocks in a hot solution of "hypo" (sodium hyposulphite). Since hypo is several times less soluble in cold water than in hot, crystallization begins at once as soon as the solution has been sufficiently cooled, accompanied by disintegration of the shale. Should the liquid become supersaturated and fail to crystallize, inoculation with a little of the solid will start crystallization at once and produce disruptive pressures. Tolmachoff also used sodium sulphate and ordinary washing soda. The hypo was found to be the most satisfactory. Saturation by heating the sample with sodium acetate[5] has also been tried, as has saturation with sodium carbonate followed by quenching with HCl with attendant evolution of CO_2, which also has the effect

---

[1] W. Wetzel, in Sedimentpetrographische Studien, *Neues Jahrb. f. Min.*, B.B. 47, pp. 39-92, 1922, examined splinters and flakes of chert instead of thin sections. Even such materials as the quartzites may be *crushed* and the fines screened out and the remainder treated in the same manner as that given for the loose sands. This treatment, however, is likely to introduce changes in the mineral frequencies.

[2] M. I. Goldman, Petrographic evidence on the origin of the Catahoula sandstone of Texas: *Am. Jour. Sci.* (4), vol. 39, pp. 261-287, 1915.

[3] G. D. Hanna and C. C. Church, Freezing and thawing to disintegrate shales: *Jour. Paleon.*, vol. 2, p. 131, 1928.

[4] I. Tolmachoff, Crystallization of certain salts used for the disintegration of shales: *Science*, vol. 76, pp. 147-148, 1932.

[5] M. Guinard, The disintegration of diatomaceous deposits: *Jour. Queckett Micros. Club*, ser. 2, vol. 3, p. 188, 1888. G. D. Hanna and H. L. Driver, The study of subsurface formations in California oil-field development: *Summary of Operations, Calif. Oil Fields*, vol. 10, No. 3, pp. 5-26, 1924.

PREPARATION OF SAMPLE

of breaking up the rock in some cases.[1] Most methods of treatment are greatly promoted if the rock is first crushed. This is usually accomplished with an iron mortar and pestle. A crushing action, rather than a grinding or abrasive action, should be used, since the latter gives an objectionable amount of fine dust and tends to destroy the original form of the grains. Occasional sifting through a sieve with openings of 0.5 mm. (about 30 mesh) and repeated crushing of the oversize is recommended, otherwise an undue amount of dust is developed as a consequence of abrasion. Tickell suggests that for crushing small amounts of rock fragments may be broken with a hammer and assayer's anvil (3 x 3 x 1 in.).[2] A 1½-in. pipe will prevent the fragments from scattering. Tickell[3] also recommends the use of a screen made by soldering a piece of wire mesh to the bottom of a sheet-iron cylinder about 2½ in. in diameter and 2 in. long. If it has been thoroughly cleaned by brushing and jarring before use, the crushed sample can be gently sifted through without contamination by grains that would otherwise collect in the meshes. A diamond mortar (Figure 146) may be used where small quantities or mineral grains are to be crushed.

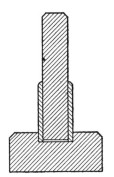

FIG. 146.—Mortar for crushing mineral or rock fragments (about 3 in. high).

More recently disaggregation of clastic rocks by means of a pressure chamber has been tried and found successful.[4] Although various solvents and solutions were forced into rock specimens under high pressure, none was found to be as effective as a supersaturated solution of sodium sulphate. Each sample to be treated is placed in a beaker, covered with the solution, and placed on a rack in the pressure chamber. The latter, see Figure 9, was constructed of an ordinary 10-in. steel casing some 12 in. long. A plate of ½-in. steel was welded to one end to form a bottom, while at the other end a flange of the same material, 1½ in. wide, was welded to the outside of the casing. A cover of ½-in. steel, 14 in. in diameter, was bolted to the flange with ½-in. bolts. A heavy composition boiler gasket between cover plate and flange prevented escape of any vapors. Holes in the cover permitted attachment of a ¼-in. stop-

[1] Albert Mann, Suggestions for collecting and preparing diatoms: *Proc. U. S. Nat. Mus.*, vol. 60, pp. 1-8, 1922.
[2] Frederick G. Tickell, *The Examination of Fragmental Rocks* (Stanford University Press, 1931), p. 35.
[3] Frederick G. Tickell, *op. cit.*, p. 36.
[4] G. L. Taylor and N. C. Georgesen, Disaggregation of clastic rocks by use of a pressure chamber: *Jour. Sed. Petrology*, vol. 3, pp. 40-43, 1933.

cock and pressure gauge. A pressure of 350 lbs. was developed.

When the sample has been reduced to grains consisting mainly of single minerals, the fine dust can be removed by screening through a sieve with openings of about 0.061 mm. (about 250 mesh). The finer material may also be removed by washing in water and decanting. In this case settling in a 15-in. column of water for 5 min. will eliminate everything below about $\frac{1}{32}$ mm. in diameter. Repeated washings, of course, are necessary to reduce the quantity of fine material to a negligible quantity. It is desirable to screen the dried residue in order to divide it into three or four fractions, each quite uniform in size, which can then be mounted for study under the microscope.

Clarification of Grains

It is convenient for identification purposes to clear the grains which are coated with iron oxide, etc., or with weathering products and to dissolve those grains which have been weathered beyond recognition, particularly if the latter are very numerous. This is accomplished to some extent during treatment of the sample during disaggregation. The hydrochloric acid treatment, for example, used in the solution of the carbonate cements, removes iron-oxide stains in addition to removing coatings of the cement.[1] In the case of HCl, however, certain detrital minerals are likely to be partly or wholly dissolved, and a microscopic check is necessary to determine whether or not this has been the case. Reed, however, says,[2] "In spite of published assertions to the effect that this treatment destroys apatite, hypersthene, and other minerals of a similar degree of stability, in several experiments these minerals were not visibly affected by boiling from as much as an hour in 50 per cent acid."

Hydrogen sulphide has also been used to remove the iron oxide coatings.[3] A water suspension of the minerals is treated with hydrogen sulphide. This treatment changes the finely divided iron oxide to iron sulphides which dissolve quickly in 0.05N hydrochloric acid. Other minerals, such as silicates and apatite, are not appreciably affected, although the carbonates would be.

Mackie[4] made it his procedure to examine each heavy mineral frac-

[1] The action of HCl in removal of iron-oxide coatings is greatly accelerated if a little stannous chloride is added to the acid.
[2] R. D. Reed, Some methods of heavy mineral investigation: *Econ. Geology*, vol. 19, pp. 320-337, 1924.
[3] M. Drosdoff and E. Truog, A method for removing iron oxide coatings from minerals: *Am. Mineralogist*, vol. 20, pp. 669-673, 1935.
[4] Wm. Mackie, Acid potassium sulphate as a petrochemical test and solvent: *Trans. Edinburgh Geol. Soc.*, vol. 11, pp. 119-127, 1915-1924.

PREPARATION OF SAMPLE 315

tion—separated from the light minerals—after the following treatment:

(1) the fraction as originally separated, without treatment
(2) hydrochloric acid treatment (heat with dilute HCl)
(3) fusion with $KHSO_4$ and solution in water

Mackie found that the $KHSO_4$ (acid potassium sulphate) fusion in a platinum crucible was very effective in removing iron oxides, quite superior to HCl, and he found, moreover, that it only slowly dissolved apatite. It completely dissolves anatase, chromite, magnetite, ilmenite, pyrite and marcasite.

The solvent action of other acids or combinations of acids, such as nitric acid, sulphuric acid, hydrofluoric acid, hydrofluoric and sulphuric acids, and hydrofluoric and nitric acids, has been investigated,[1] and the results can be had by consulting the literature on this subject. The reader is also referred to Table 39, page 355.

In some heavy mineral concentrates certain minerals appear in such quantity as to mask the less frequent mineral species that may be present. It is then desirable to remove the overabundant mineral, if possible, by some simple method. Minerals that play such a rôle include pyrite, gypsum, anhydrite, and barite. Pyrite can be removed by heating with 15 per cent nitric acid. Gypsum is usually not present in the heavy residues due to its low specific gravity, but when present due to abundant inclusions of iron oxide, it may, according to Milner,[2] be eliminated by digesting with a strong ammoniacal solution of ammonium sulphate. Barite can be removed by concentrated sulphuric acid. Strong hot HCl will dissolve the anhydrite. In all cases it must be remembered that some of the other minerals in the residue may be adversely affected by the solvents and may be partly or wholly dissolved (see section on Chemical Separation Methods).

Bituminous sands may be clarified by treatment with a mixture of petroleum ether and carbon disulphide followed by a wash in alcohol or with benzol, chloroform, or ether. Milner[3] gives a method for the removal of hydrocarbon materials which can be used for the quantitative determination of the amount present in the sample if desired. The extraction apparatus is a Soxhlet extractor (Figure 147) or simply a reflux condenser attached to an Erlenmeyer flask, in which the crushed

---

[1] C. L. Doelter, *Handbuch der Mineralchemie* (1914). A. A. Noyes, *Qualitative Analysis of the Rare Elements.*
[2] H. B. Milner, *Sedimentary Petrography,* 2nd ed. (Thos. Murby and Co., London, 1929), p. 65.
[3] H. B. Milner, *op. cit.,* pp. 70-73.

sample is suspended in a "thimble" (porous perforated porcelain cup). The whole apparatus is then placed on a water-bath, and after some solvent, such as benzol, is placed in the flask to a depth of about an inch, heat is applied. When clarification is complete the sample is treated as usual. The evaporation of the solvent will leave a hydrocarbon residue, the amount of which can thus be determined.[1]

Special Preparation Problems

The techniques just described apply in the main to the arenaceous sediments. The mineralogy of these rocks has been most important in problems of correlation, and hence these rocks have been more widely studied. In some cases it has been found useful to use some of these techniques for study of the mineralogy of other types of sediments. This is usually done when minor accessory minerals are to be studied, and, as usual, this study must be preceded by some concentration of these lesser constituents. Here again there is no standard procedure and the problem presented by each sediment must be met in a different way.

Fine-grained argillaceous rocks. No special problems of preparation are involved with these materials that have not been treated at some length in the section dealing with mechanical analysis (page 51). Action with acid is, however, permissible in some cases where the mineralogy is to be studied without mechanical analysis, though the destructive effect of the acid on certain minerals should be

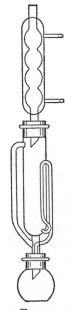

Fig. 147.—Soxhlet extractor.

kept in mind. Treatment with alkaline reagents may give better results with the argillaceous materials. Owing to difficulty of readily identifying the minerals in the clays, it is common practice to remove these finest materials and discard them, leaving a small residue of fine sand for study with the microscope. This is achieved by means of a special elutriation apparatus, such as that devised by Eichenberg[2] or more simply by elutriation or decantation, the latter following a 5-min. settling in a 15-in. column of water which will remove everything less than $\frac{1}{32}$ mm. in diameter, or a like period of settling in a 5-in. column which will per-

---

[1] See also E. M. Spieker, Bituminous sandstone near Vernal, Utah: *U. S. Geol. Survey, Bull. 822-c,* pp. 77-98, 1930.

[2] W. Eichenberg, Ein Schlämmapparat für Tone: *Centralbl. f. Min. Geol. u. Pet.,* Abt. A, pp. 221-224, 1932.

mit washing out of everything under 1/64 mm.¹ The principal difficulties of the study of the fine-grained sediments relate to the separation into like mineral groups and to identification. These problems are treated elsewhere (see section on study of clay minerals and section on heavy mineral separation in clays and silts).

Crystalline rocks. Owing to the development of many new minerals by metamorphism, the suite of minor accessories originally present loses somewhat of its identity. Nevertheless, it is occasionally important to investigate these minor constituents, as they may be the sources of these minerals in sediments of later geologic age or may be important in determination of the igneous or sedimentary origin of the rock. In these cases the rock is crushed to a size about that of the constituent minerals, and, following sieving out of all the oversize and fine dust, then subjected to the usual concentration methods. For a more detailed discussion of the treatment of crystalline rocks the reader is referred to the literature on the subject.²

Calcareous rocks. Since these are nearly all calcite or dolomite, the remaining constituents can be concentrated by crushing the rock and leaching with acid. Except in cases of very abundant residue, it has been the practice to examine the minor minerals as a whole rather than separated fractions. In the more impure calcareous rocks where the residue is large, mineral separation methods may be applied, the procedure to be used will depend on whether the residue is mainly arenaceous or argillaceous. If the residue is composed of authigenic minerals, such as pyrite, or chert, the separations made are for convenience of study and do not have the same significance, as for example, the mechanical analysis of the arenaceous residues. For details on the method of study of insoluble residues in calcareous rocks the reader is referred to page 494.

Coal. The minor mineral constituents of coal have rarely been investigated. To concentrate these minerals is a special problem. Two methods have been proposed. The first involves burning of the coal and washing the ash to eliminate the finer dust. Such a method is drastic in its action. The second method, the solution of the organic matter with caustic potash, pyridine, phenol, chloroform, selenium oxychloride,³ or

[1] Marcellus H. Stow, Washing sediments to obtain desirable size of grain for microscopic study: *Am. Mineralogist,* vol. 16, p. 226, 1931.
[2] *Report of the Committee on Accessory Minerals for 1936–37* (National Research Council, Division of Geology and Geography, Washington, D. C.).
[3] E. Stach, *Kohlenpetrographisches Praktikum* (Gebrüder Borntraeger, Berlin, 1928). Stach describes the chromic acid method, "Schulze's method" (potassium chlorate and nitric acid), and the diaphonol method.

sodium hypochlorite, is to be preferred if it can be made to work. Crushing before treatment with the solvent is necessary.

TABLE 36

SUMMARY OF DISAGGREGATION PROCEDURE FOR MINERALOGICAL STUDIES

Condition of Sample	Non-calcareous Cement	Calcareous Cement
Unconsolidated	#5 No problem involved *	#6
Slightly coherent †	#3 (a) Crush with fingers, wood block, or rubber-covered pestle (b) Soak in water and use stiff brush	#4
	(c) Digest in alkaline solution (KOH, NaOH, or NH$_4$OH)	(c) Digest in weak acid (HCl)
Very coherent	#1 (a) Thin-section (b) Crush in iron mortar (c) Sodium thiosulphate treatment	#2 (a) Thin section (b) Digest in HCl

* It is sometimes desirable to remove carbonate, which is troublesome in bromoform separations due to nearness of specific gravity (2.7 — 2.9) to that of bromoform (2.85). Acid clarifies iron oxide-stained grains also.

† May be thin-sectioned after special treatment to indurate sufficiently for cutting and grinding. Indurate with Canada balsam or bakelite.

CHAPTER 14

SEPARATION METHODS

PRELIMINARY ENRICHMENT OF SAND IN HEAVY MINERALS

Owing to the high cost of heavy liquids, some workers have used other methods of concentrating the heavy fraction of the sediment. Most often these other methods are but rough concentrations and are preliminary to actual separation by other means. Such preliminary enrichment reduces the amount of heavy liquid required for a separation. Such methods are sizing, panning, and vibration.

It is known from observation that the heavy minerals of a sand are largely concentrated in its finer grades. This is due to the fact that for a certain dominant size of quartz and feldspar there is a smaller size of magnetite and like heavy minerals, which are deposited together because they have what Schöne called the same hydraulic value or the same settling rates. Consequently by screening out the coarse, light fractions the heavy minerals will be materially concentrated.

Heavy minerals can be separated from the light ones by repeated panning as in panning for gold. The method is used mostly for separating small amounts of heavy accessory minerals from a large amount of rock or sediment and works best if there is a large difference in gravity. A conical pan is preferred by some to the common miner's pan.[1] There are rather serious objections to preliminary panning, as pointed out by Smithson,[2] if exact frequencies of the minerals are to be determined. The proportions are materially affected by panning.

Panning is carried out in a shallow circular vessel with wide flaring rim. A common size is some 18 in. in diameter at the top and 8 in. in diameter at the base. The depth is 4 in. The pan is made of thin black sheet steel. It should be entirely free from grease.

[1] O. A. Derby, On the separation and study of the heavy accessories of rocks: *Proc. Rochester Acad. Sci.*, vol. 1, pp. 198-206, 1891.
[2] Frank Smithson, The reliability of frequency-estimations of heavy mineral suites: *Geol. Magazine*, vol. 67, pp. 134-136, 1930. See also C. J. C. Ewing, A comparison of the methods of heavy mineral separation: *Geol. Magazine*, vol. 68, pp. 136-140, 1931.

The pan is filled with the sediment to be studied and set in water a few inches deeper than the pan. The material is thoroughly wetted, stirred up and disintegrated. At the same time any large pebbles are washed from the pan. From time to time the pan is sharply swirled in a horizontal plane until nothing is left but clean sand. The pan is then tipped gently forward, is held level with the surface of the water, and is given a circulatory motion. This alternates with stirring so that the lighter materials are gradually spilled over the edge. Concentration is carried as far as desired; the first stages are completed rapidly, but the later stages must be done carefully to avoid loss of heavy minerals. A further concentration is best effected by means of heavy liquids. Otherwise the pan is so rocked (jerk and flow) at the end to spread the "colors."

Salmojraghi[1] found that when a handful of dry sand is agitated on a sheet of paper there is a marked enrichment in heavy minerals in that part which works to the bottom and which may be seen by inclining the sheet. He believed that with three or four such operations all the components of a sand may be determined with the possible exception of the extremely rare ones. Salmojraghi, moreover, found that a relationship could be established between the proportions of the minerals in the sand enriched by dry agitation and their true proportions in the natural sand. The authors believe, however, that while this method is an aid to study, it can hardly take the place of the more complete and easy separation with heavy liquids.

SEPARATION OF MINERALS ON BASIS OF SPECIFIC GRAVITY

Heavy Liquids

A heavy liquid substance to be satisfactory should be (1) inexpensive, (2) easily prepared or purchased, (3) transparent, (4) liquid at ordinary temperatures, that is, having a low melting point, (5) non-corrosive, (6) chemically inert, (7) without odor, (8) fluid, not viscous, and (9) easily concentrated or diluted.

Many liquids have been used or investigated, but no one has all these desirable qualities. Several, however, are decidedly superior and are to be recommended. Some liquids that have undesirable features, notably potassium mercuric iodide (Thoulet solution), have been widely used. Hence details concerning the properties, preparation, and uses of most

---

[1] F. Salmojraghi, Sullo studio mineralogico delle sabbie e sopra un modo di rappresentarne i risultati: *Atti soc. ital. sci. nat.*, vol. 43, pp. 54-89, 1904.

liquids, including those heretofore generally used, have been omitted and only five of the most useful and satisfactory fluids are described in detail.

As Davies[1] has suggested, the double nomenclature of the heavy liquids, that is, the use of both the chemical name and the name of the first user or discoverer of the fluid, is confusing. The usage is further often inconsistent. Consequently the liquids or solutions are here given according to their chemical name. The chemical name has the double advantage of giving the composition of the liquid as well as recalling to mind some of the properties of the substance.

Bromoform (tribrom-methane). Schroeder van der Kolk used bromoform in 1895 as a heavy liquid for the separation of the minerals of a sand.[2] It has since been used rather widely by investigators and is at present perhaps the most commonly used heavy liquid.

Bromoform (tribrom-methane), $CHBr_3$, is a halogen substitution product of methane. It is a highly mobile liquid at ordinary temperatures and has a specific gravity of 2.89 at 10° C. Sullivan also gives the melting point at 9° C. and the boiling point at 151.2° C.[3]

Bromoform may be more readily purchased than prepared in the laboratory. Commercial bromoform, however, is usually low in gravity, often below that of quartz (2.66) due largely to dissolved alcohol. For work, therefore, where it is desired to separate the quartz from the heavier minerals it is necessary to remove the alcohol by the procedure given below.[4] Bromoform changes in specific gravity with temperature, about .0023 per degree Centigrade. Consequently pure bromoform will have a specific gravity of 2.87 at ordinary laboratory temperatures (20° C.). This is quite sufficient to affect a separation of quartz and feldspar from the heavier minerals. Since this separation is usually all that is desired, bromoform is likely to be one of the most useful heavy liquids.

Bromoform is miscible in all proportions with carbon tetrachloride (CCl_4), benzene (benzol) (C_6H_6), alcohol (C_2H_5OH), and acetone (CH_3COCH_3). From carbon tetrachloride and benzene mixtures it may be received by fractional distillation, using an Engler-type flask and

[1] G. M. Davies, Nomenclature of the heavy liquids: *Geol. Magazine,* vol. 57, p. 287, 1920.
[2] J. L. C. Schroeder van der Kolk, Beitrag zur Kartirung der quartaren Sande: *Neues Jahrb. f. Min., etc.,* vol. 50, Bd. I, pp. 272-276, especially p. 274, 1895.
[3] John D. Sullivan, Heavy liquids for mineralogical analyses: *U. S. Bureau Mines, Technical Paper 381,* p. 10, 1927.
[4] The authors, for example, took commercial bromoform, specific gravity of 2.638, and were able with three washings to raise the specific gravity to 2.838.

Liebig condenser. Benzene boils at 80.5° C. and carbon tetrachloride boils at 76.7° C. Since bromoform boils at 151.2° C., the tetrachloride or benzene would come over first. Bromoform diluted with alcohol (as commercial bromoform often is) or acetone could be similarly recovered, but a much simpler and more satisfactory method is that suggested by Ross.[1] To the bromoform-alcohol or bromoform-acetone mixture is added a large volume of water (fifteen times as large) in a two-liter bottle. After vigorous shaking the heavy bromoform phase separates out with but very little alcohol or acetone and the water phase contains most of the alcohol or acetone. Most of the water is decanted and the process repeated two or three times. After the third decantation,[2] the bromoform-water mixture is poured into a separatory funnel and the bromoform almost free of water is drawn off and run into a funnel fitted with several thicknesses of filter-paper which will absorb any dispersed water. The bromoform filtered should be clear and have a specific gravity of about 2.85 at room temperature. A little bromoform is lost due to dispersion in the decanted water but the loss is not large (3 to 6 per cent). Cohee, using this method of recovery, has described an apparatus set-up.[3]

Another method of recovery and of clarification in the case of separation of free bromine is that of fractional crystallization by freezing-out of the bromoform. If a dish of bromoform is placed out of doors on a cold winter day, the bromoform will soon crystallize out as clear, transparent platy crystals. (Pure bromoform freezes at 9° C. but a somewhat lower temperature is required for solutions containing alcohol.) If the residual liquid be strained off from the crystals and the latter melted, it will be found that the bromoform will be clear and of high specific gravity. A second crystallization may be needed to completely clear the substance.

Sometimes the bromoform becomes discolored due to petroleum present in the sands undergoing separation. Hanna[4] found fuller's earth useful in the

---

[1] Clarence S. Ross, Methods of preparation of sedimentary materials for study: *Econ. Geology*, vol. 21, pp. 454-468, 1926. The method of recovery described by Ross was later redescribed by Bracewell. See S. Bracewell, Recovery of bromoform: *Geol. Magazine*, vol. 70, p. 192, 1933, and F. Smithson, The recovery of bromoform: *Geol. Magazine*, vol. 71, p. 240, 1934.

[2] Ross recommends shaking with water in excess and allowing the material to stand 24 hr. If the volume of water is *greatly* in excess and if the shaking and decantation are repeated two or three times, it will be unnecessary to allow the mixture to stand. By two or three such decantations the writers concentrated bromoform, S.G. 2.590, to 2.835 with no great loss.

[3] George V. Cohee, Inexpensive equipment for reclaiming heavy liquids: *Jour. Sed. Petrology*, vol. 7, pp. 34-35, 1937.

[4] Marcus A. Hanna, Clarification of oil-discolored bromoform: *Jour. Paleon.*, vol. 1, p. 145, 1927.

SEPARATION METHODS

clarification of the bromoform in such cases. The fuller's earth, however, will not clarify bromoform discolored by free bromine. Shaking with NaOH or alcoholic KOH will remove free bromine.

Carbon tetrachloride may be used as diluent as indicated above. It must be recovered by fractional distillation, however, and therefore is less desirable than acetone or alcohol. It has one advantage, namely that, since carbon tetrachloride has a specific gravity of 1.58, a larger volume of solution of given specific gravity will be obtained if it is used instead of acetone, alcohol, or benzene. In case a permanent series of liquids differing by small intervals in specific gravity is desired, for the purpose of specific gravity determination, perhaps carbon tetrachloride is the most economical diluent. Sullivan[1] has investigated the properties of this combination and the following table is given by him. Sullivan used bromoform of an original gravity of 2.61 (commercial bromoform) and hence concluded that the solution had but a restricted use in the separation of heavy minerals from gangue but that it was satisfactory for separating coal from bone.

Specific Gravity of Mixtures of CCl_4 and $CHBr_3$ at 25° C.

Per Cent $CHBr_3$ (by Vol.)	S. G.	Per Cent $CHBr_3$ (by Vol.)	S. G.
100	2.61	25	1.84
75	2.35	0	1.58
50	2.09		

Bromoform is "commercial"

The mineral grains collected on the filter-paper after gravity separation may be collected in a bottle labeled "Bromoform washings" from which the bromoform may be recovered when the quantity becomes sufficiently large. The method of recovery is that of washing with large volumes of water as described by Ross.

Normally bromoform is used without dilution for the separation of minerals with specific gravity greater than 2.85 from those with a gravity less than that figure. As indicated by Hanna,[2] however, there

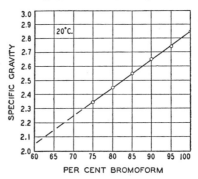

Fig. 148.—Curve showing specific gravity of various bromoform-acetone mixtures.

are separations in which a liquid of less density is needed. Bromoform with acetone will give such a liquid. (See Figure 148.)

---

[1] John D. Sullivan, *op. cit.*, p. 11.
[2] Marcus A. Hanna, Separation of fossils and other light materials by means of heavy liquids: *Econ. Geology*, vol. 22, pp. 14-17, 1927.

Hanna outlines the following separation:[1]

(1) Bromoform-alcohol liquid. S. G. 2.685. Quartz, etc., floats, calcite and aragonite fossils sink as do also the "heavy minerals."
(2) Bromoform-alcohol liquid. S. G. 2.60. Glauconite, orthoclase, glauconite-filled shells, and air-filled shells float. Quartz sinks.
(3) Bromoform-alcohol liquid. S. G. 2.20. Air-filled shells float. Glauconite, glauconite-filled shells, and orthoclase sink.

Hanna estimates that 90 per cent or more of the fossils are thus separated from the quartz. Possibly a quantitative separation of quartz from orthoclase in arkosic sands can be made in this way.

TABLE OF SPECIFIC GRAVITY MIXTURES OF $CHBr_3$ AND CH_3COCH_3
(BROMOFORM AND ACETONE) AT 20° C.

Per Cent $CHBr_3$ (by Vol.)	S. G.	Per Cent $CHBr_3$ (by Vol.)	S. G.
100	2.846	85	2.551
95	2.743	80	2.450
90	2.652	75	2.346

Bromoform is not entirely free from objectionable features. These are:

(1) It is decomposed by strong light. Prevent by keeping in brown bottles in the dark.
(2) It also deteriorates with heat. Store in a cool place.
(3) It evaporates readily, hence vessels containing bromoform should be kept stoppered or covered.
(4) Cost is sufficiently high, though not excessive,[2] to make it profitable to conserve the acetone washings and to recover the bromoform therefrom.
(5) Bromoform is subject to strong convection currents. Milner[3] states that this difficulty may be overcome by carrying out the separations in a fume cupboard or constant temperature chamber or a room free from warm air currents.
(6) Bromoform is toxic. While it is not corrosive nor are its vapors strongly toxic if inhaled, the fumes are mildly so if one works with the liquid in the open laboratory for a long time without interruptions. If the separations are carried out in a well-ventilated hood this objection will be overcome.

Acetone as a diluent has some objectionable features, notably its rapid evaporation, necessitating constant checking of the specific gravity of the bromoform-acetone mixture, and also its inflammability, which makes it impossible to use near an open flame. The rapid evaporation is an ad-

[1] In this table "alcohol" has been substituted for "benzol" of Hanna.
[2] Bromoform was obtainable in 1936 at about $2.25 per pound (about 157 c.c.).
[3] Henry D. Milner, *op. cit.*, p. 40.

vantage, however, in that minerals washed free of bromoform with acetone dry almost instantly and are ready for mounting and examination. Alcohol has much the same disadvantages.

Acetylene tetrabromide (tetrabrom-ethane). The most promising alternate to bromoform is apparently acetylene tetrabromide. It is rather widely used. Muthmann[1] seems to have been the first one to use it for mineral separations.

Acetylene tetrabromide (tetrabrom-ethane), $C_2H_2Br_4$, is a colorless mobile liquid with a specific gravity of 2.96 at 20° C. It is miscible in all proportions with carbon tetrachloride or benzene, giving a range of gravities from that of the pure liquid, 2.96, to the gravity of the respective diluents, 1.6 or 0.9. Either diluent may be used for washing the mineral grains free of the tetrabromide, though for rapid washing benzene is probably the better. The acetylene tetrabromide may be recovered by fractional distillation. Since the compound is also miscible with alcohol and insoluble in water, alcohol may be used as diluent and the tetrabromide recovered by washing with water as is the case with bromoform. The alcohol used, however, must be absolute alcohol otherwise the acetylene tetrabromide becomes clouded upon dilution.

Sullivan,[2] who used commercial acetylene tetrabromide with a specific gravity of 2.89, gives the following data on mixtures of the tetrabromide and carbon tetrachloride:

Specific Gravity of Mixtures of CCl_4 and $C_2H_2Br_4$ at 25° C.

Per Cent $C_2H_2Br_4$ (by Vol.)	S. G.	Per Cent $C_2H_2Br_4$ (by Vol.)	S. G.
100	2.89	25	1.91
75	2.58	0	1.58
50	2.24		

Acetylene tetrabromide may be purchased (at about $1.50 per pound in 10-lb. lots) or prepared in the laboratory according to the method outlined by O'Meara and Clemmer.[3]

A thousand grams of liquid bromine is placed in three or four gas-washing bottles, and a small amount of water is added to prevent excessive volatilization of the bromine. Acetylene gas, produced by the action of water on calcium carbide, is bubbled slowly through the series of bottles until the reaction is complete. This is indicated by a change of

[1] W. Muthmann, Ueber eine zur Trennung von Mineralgemischen geeignete schwere Flüssigkeit: *Zeits. Kryst. Min.,* Bd. 30, pp. 73-74, 1899.
[2] John D. Sullivan, *op. cit.,* pp. 11-12.
[3] R. G. O'Meara and J. Bruce Clemmer, Methods of preparing and cleaning some common heavy liquids used in ore testing: *U. S. Bureau Mines, Rept. of Investigations No. 2897,* pp. 1-3, 1928.

color, since the acetylene tetrabromide is light amber. Excessive temperature rise should be prevented. The resulting heavy solution is placed in a separatory funnel and agitated with a dilute caustic solution to remove the uncombined bromine. When the water and the acetylene tetrabromide have separated into two layers, the latter is drawn off and dehydrated with calcium chloride. O'Meara and Clemmer report a yield of 90 to 96 per cent.

Recovery of the acetylene tetrabromide from the benzene or carbon tetrachloride may be effected by fractional distillation. Simple evaporation will, however, produce the same result. If the washings are permitted to stand open at room temperature, the diluent evaporates slowly. When the residual liquid is up to the desired gravity, it is filtered and is then ready for further use. Gentle heating facilitates the concentration, but rapid or excessive heating results in partial decomposition. If decomposition occurs, as indicated by a dark color, the acetylene tetrabromide can be restored by shaking with a small amount of bromine in a separatory funnel, then adding a small amount of sodium hydroxide to remove the excess bromine. The solution should become light straw-colored. Dehydration with calcium chloride follows. The liquid is then allowed to stand in an open vessel until it loses the peculiar odor acquired during the cleaning process, when it is ready for further use. If the liquid was very dark to begin with, a repetition of the cleaning process may be necessary. Sulphurous acid may be substituted for the caustic, according to O'Meara and Clemmer, and the danger of formation of ethylene dibromide during the caustic treatment is thus eliminated.

Methylene iodide. Brauns is credited with the first use of methylene iodide for mineral separation.[1] O'Meara and Clemmer[2] give the most recent published statement of the preparation and reclamation of this fluid. Owing to its high cost, the authors do not recommend it for heavy mineral work. It is, however, one of the most useful immersion liquids for refractive index work, hence something of its properties is given here.

Pure methylene iodide, CH_2I_2, has a light straw color and a specific gravity of 3.32 at 18° C. It has an index of 1.74. It is miscible in all proportions with benzene or carbon tetrachloride, from which it may be recovered by fractional distillation at reduced pressure or simply by allowing the more volatile lighter liquid to evaporate. The products of separation, therefore, may be washed with these substances and the iodide recovered by slow evaporation. Heating to hasten evaporation should be avoided, owing to decomposition of the iodide. Ross[3] states that the

---

[1] R. Brauns, Ueber die Verwendbarkeit des Methylenjodids bei petrographischen und optischen Untersuchungen: *Neues Jahrb. f. Min.*, etc., Bd. 2, pp. 72-78, 1886.
[2] R. G. O'Meara and J. Bruce Clemmer, *op. cit.*, pp. 3-4.
[3] Clarence S. Ross, Separation of sedimentary materials for study: *Econ. Geology*, vol. 23, p. 334, 1928.

SEPARATION METHODS

methylene iodide may be diluted with acetone or alcohol and recovered by washing with water in the same manner as with bromoform. For refractive index work, methylene iodide is diluted with alpha monobromonaphthalene to give a series of liquids ranging in index from 1.66 to 1.74. Methylene iodide is decomposed by strong light and should be kept in tin-foil covered or brown bottles. Mercury or copper foil may be used to restore normal color and remove the free iodine liberated, or clarification may be achieved by shaking with dilute KOH or NaOH solution in a separatory funnel followed by dehydration of the iodide with calcium chloride.

Methylene iodide cost $15 per pound in 8-lb. lots according to O'Meara and Clemmer (1928). It may, however, be prepared in the laboratory by the method given by these authors.

Thallous formate. Clerici[1] used thallium formate ($TlCO_2H$) for the separation of minerals in 1907. Since then this substance has come to be more widely used, and solutions of this salt together with the double thallium formate-malonate ("Clerici's solution") are probably the most satisfactory substances known for use where liquids of high specific gravities are desired.

Thallous formate ($TlCO_2H$) is an organic salt very soluble in water. Sullivan[2] gives the melting point as 94° C. and the specific gravity as 4.95 at 105° C. The aqueous solutions have the following gravities:

Per Cent H_2O (*by Weight*)	*Temperature*	*S. G.*	*Melting Point*
0	105° C.	4.95	94° C.
5	60° C.	4.19	54° C.
10	42° C.	3.72	31° C.
15	25° C.	3.39	22° C.
20	25° C.	3.09	—
25	25° C.	2.86	—

The solubility of the salt in water increases rapidly with increasing temperature. The formate does not decompose at the boiling point of water but does slowly decompose at higher temperatures. The diluted solutions may be concentrated by evaporation.

The salt may be purchased, but is readily prepared from thallium carbonate by adding an equivalent weight of formic acid. Sullivan[3] gives the procedure for preparing the formate from thallium sulphate and quotes the procedure given by Clerici for making the material from

---

[1] Enrico Clerici, Preparazione di liquidi per la separazione dei minerali: *Atti. Rend. R. Accad. Lincei. Roma,* ser. 5, vol. 16, 1 semestre, pp. 187-195, 1907.
[2] John D. Sullivan, *op. cit.,* pp. 20-21.
[3] John D. Sullivan, *op. cit.,* p. 20.

metallic thallium. The preparation of the salt is also described by Vhay and Williamson.[1]

Thallium formate-malonate (Clerici's solution). Clerici,[2] who worked with thallium formate solutions, also discovered that solutions of the double salt, thallium formate-malonate, were suitable for mineral separations. Clerici's solution has also been investigated by Vassar,[3] from whose work the following data are taken.

Clerici solution is a mixture of thallium malonate, $CH_2(COOTl)_2$, and thallium formate, $HCOOTl$, which at ordinary room temperatures has a density of 4.25. Clerici is quoted as stating that increasing heat will increase the density of the solution because of increased solubility of the salts at higher temperatures. A concentrated solution at 35° C. has a specific gravity of 4.4 and at 50° C. a specific gravity of 4.65; at 90-100° C., pyrite floats. It is possible to dilute with water in any quantity and to reconcentrate. The solution is more mobile than Thoulet's solution (potassium mercuric iodide), is odorless, and has a slight amber color. At ordinary temperatures the solution appears to be stable and inert, but sulphides should not be left too long in hot solutions.

The solution is prepared by neutralizing two equal parts of thallium carbonate with equivalent parts of formic acid and malonic acid, each separately, and then mixing, filtering, and evaporating until almandite floats. A weight of 111 g. of malonic acid dissolved in a little water will neutralize 500 g. of thallium carbonate, and 115 g. of 55.5 per cent formic acid will neutralize the same amount of thallium carbonate. One kilogram of thallium carbonate will make approximately 300 c.c. of Clerici's solution. If the solution is made from the dry salts, thallium formate and thallium malonate, 7 g. of each will dissolve completely in 1 c.c. of water, but 10 g. of each will leave a part undissolved. Vassar gives the method of preparing Clerici's solution from metallic thallium, but it is omitted here because thallium carbonate can be purchased and the procedure for its manufacture from the metal is quite laborious. Most recently Rankama[4] has given detailed instructions for the purifying of Clerici's solution, as well as acetylene tetrabromide, by treatment with bone charcoal.

[1] J. A. Vhay and A. T. Williamson, The preparation of thallous formate: *Am. Mineralogist,* vol. 17, pp. 560-563, 1933.

[2] Enrico Clerici, *op. cit.,* pp. 187-195; Ulteriori ricerche sui liquidi pesanti per la separazione dei minerali: *Atti. Rend. R. Accad. Lincei. Roma,* ser. 5, vol. 31, pp. 116-118, 1922.

[3] Helen E. Vassar, Clerici solution for mineral separation by gravity: *Am. Mineralogist,* vol. 10, pp. 123-125, 1925.

[4] Kalervo Rankama, Purifying methods for the Clerici solution and for acetylene tetrabromide: *Bull. Comm. Geol. de Finlande, No. 115,* pp. 65-67, 1936.

SEPARATION METHODS

TABLE 37

PROPERTIES OF SOME COMMON HEAVY LIQUIDS

Heavy Liquid	Comp.	S.G. 20° C.	S.G. Change per °C.	F.P. °C.	B.P. °C.	Approx. Cost 1,000 g.	Diluent	S.G. Diluent	Recovery	Remarks
Bromoform	$CHBr_3$	2.87	.0023	9°	151.2°	$5.00	Acetone or alcohol	0.79	Washing with water	Rather volatile
Acetylene tetrabromide	$C_2H_2Br_4$	2.96	.0023	0.1°	151°	$3.40*	"	"	"	
Methylene iodide	CH_2I_2	3.32	.0021	5°	180° decomp.	$34.00†	Benzol or alcohol	0.88	Evaporation	Sensitive to light
Thallium formate sol.	$TlCO_2H$	3.39‡		94°		$55.00	Water	1.00	Evaporation	
Thallium formate-malonate sol.	$TlCO_2H$ and $CH_2(COOTl)_2$	4.25				$55.00	Water	1.00	Evaporation	

* If purchased in 10-lb. lots, according to O'Meara and Clemmer (1928).
† If purchased in 8-lb. lots, according to O'Meara and Clemmer (1928).
‡ At 25° C.

Choice of liquid. An aqueous solution of thallium formate is probably the most satisfactory liquid for a gravity range of 2.89 (bromoform) to 3.39 at ordinary temperatures (25° C.). Where gravities lower than 2.89 are required, either bromoform or acetylene tetrabromide may be used; where gravities higher than 3.39 are desired, an aqueous solution of the formate-malonate is more suitable. The formate-malonate solution, of course, could be used for all ranges except that it is more expensive than the other liquids mentioned. The composition, cost, and properties of these most useful liquids are summarized in Table 37.

Other heavy liquids. As pointed out at the beginning of the section on heavy liquids, there are some other liquids which may be or have been used for mineral separations and specific gravity work. Of the many which have been investigated or proposed but a few ever achieved wide popularity. As noted, even these are now largely replaced by the organic liquids described above. Since a few laboratories still have a supply of the earlier-used liquids, it seems advisable to indicate the nature of these liquids and to point out where instructions for their use and recovery may be obtained.

Thoulet solution (or Sondstadt's solution), potassium mercuric iodide, was one of the most widely used heavy liquids.[1] It is an aqueous solution and therefore recovered by evaporation and diluted with water. It is very corrosive, rather viscous, and reacts with certain minerals [2] and becomes dark on the liberation of iodine. For these reasons its use should be discouraged. It has a specific gravity of about 3.0-3.2.

Klein solution,[3] an aqueous solution of cadmium borotungstate, has a density of 3.3-3.4. This liquid is poisonous, though not corrosive; it is also decomposed by certain minerals, notably the carbonates, and on exposure to light becomes very dark.

A benzol solution of the double iodide of tin and arsenic with a density of 3.6 was used by Retgers.[4] It is very toxic, readily decomposed in the presence of water, and has a dark red color. Retgers [5] also used a double salt of thallium mercuro-nitrate which may be dissolved in water. This is a quite fluid substance, is transparent, and does not react with the metallic sulphides and is therefore generally more satisfactory.

---

[1] E. Sondstadt, Note on a new method of taking specific gravities, adopted for special cases: *Chemical News,* vol. 29, pp. 127-128, 1874. J. Thoulet, Séparation des éléments non ferruginuex des roches, fondée sur leur différence de poids specifique: *C. R. Acad. Sci. Paris,* vol. 86, pp. 454-456, 1878. V. Goldschmidt, Ueber Verwendbarkeit einer Kaliumquecksilberjodidlösung bei mineralogischen und petrographischen Untersuchungen: *Neues Jahrb. f. Min., etc.,* B.B. I, pp. 179-238, 1881.

[2] T. L. Walker, Alteration of silicates by Sondstadt's solution: *Am. Mineralogist,* vol. 7, pp. 100-102, 1922.

[3] D. Klein, Sur une solution de densité 3.28, propre à l'analyse immédiate des roches: *C. R. Acad. Sci. Paris,* vol. 93, pp. 318-321, 1881.

[4] J. W. Retgers, Die Bestimmung des specifischen Gewichts von in Wasser löslichen Salsen, III. Die Darstellung neuer schwerer Flüssigkeiten: *Zeits. phys. Chem.,* Bd. 11, pp. 328-344, 1893.

[5] J. W. Retgers, Versuche zur Darstellung neuer schwerer Flüssigkeiten zur Mineraltrennung: *Neues Jahrb. f. Min., etc.,* Bd. II, pp. 183-195, 1890.

SEPARATION METHODS

Rohrbach[1] used an aqueous solution of barium mercuric nitrate. This solution is, however, difficult to prepare, very easily decomposed, and very poisonous.

For a brief review of the large literature on heavy liquids, the reader is referred to Sullivan's paper.[2] Sullivan described the properties of carbontetrabromide-carbon tetrachloride, stannic bromide-carbon tetrachloride, stannic chloride, stannic iodide, antimony tribromide, antimony trichloride, thallous silver nitrate, mercurous nitrate, thallous mercurous nitrate, mercuric chloride-mercuric iodide-antimony trichloride, all in addition to those of bromoform, acetylene tetrabromide, and thallous formate.

Reasons for imperfect separations with heavy liquids. Mineral separations are often incomplete or imperfect owing to convection currents in the separating fluid, to entrapment of grains of one density within the bulk of the fraction of opposite density, to inclusions with the mineral grain or to attachment to other mineral grains of either lower or higher density or to alteration products which cause density to differ from theoretical value, to extreme fineness of size which causes the material to "ball-up" or settle with extreme slowness, or to smallness of difference in density between grains and liquid. Careful control of temperature will prevent convection, while repeated separations or frequent stirring will usually overcome the normal incomplete separation. The settling rate of very small grains and grains of density near that of the fluid may be accelerated by use of a centrifuge. Separation of the fine materials is promoted by the use of a vacuum pump.[3]

Standardization of heavy liquids. A simple and approximate method of checking the specific gravity of a liquid is to put a small crystal of a mineral of known specific gravity in the liquid and determine whether the grain floats or sinks. In this way one can readily test a liquid for the separation of a given mineral, as quartz for example. If it is desired to make a liquid of a given specific gravity, this can be done by diluting the heavy liquid with the proper fluid until a mineral of the gravity desired remains suspended in the solution.

If 20 c.c. or any other convenient volume of the liquid be pipetted into a previously weighed container and then weighed, the specific gravity of the liquid can be calculated by dividing the weight found (in grams) by

---

[1] Carl Rohrbach, Ueber eine neue Flüssigkeit von hohem specifischen Gewicht, hohem Brechungsexponenten und grosser Dispersion: *Wildem. Ann.*, N. F., 20, pp. 169-174, 1883.

[2] John D. Sullivan, Heavy liquids for mineralogical analyses: *U. S. Bur. Mines, Technical Paper 381*, pp. 5-9, 1927.

[3] R. C. Emmons, On gravity separation: *Am. Mineralogist*, vol. 15, p. 536, 1930.

the volume (in cubic centimeters). Goldschmidt,[1] using this principle, gives the details for careful determination of the specific gravity of a liquid which will just suspend a mineral grain. His method involves the accurate weighing of a liquid in a 25-c.c. measuring flask. The average of three weighings is assumed to be the correct value and is used in calculations. Greater accuracy, however, is achieved if some special container, such as the Sprengel tube,[2] is used (Figure 149). In this case the liquid is sucked directly into the tube and weighed. Then, since the weight of the tube is known and the weight of the tube filled with distilled water can also be determined, it follows that

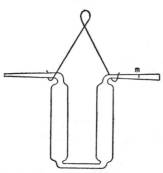

FIG. 149.—Sprengel tube for weighing a precise volume of liquid. (After Sollas.)

$$G = \frac{w' - w''}{w - w''}$$

where w is the weight of tube and water, w' is the weight of tube and liquid whose specific gravity is desired, and w'' is the weight of tube alone.

The index of refraction may be used to check the specific gravity of a fluid composed of mixtures of two independent liquids. To do so, however, requires an accurate method of measuring the index of refraction of liquids, preferably a refractometer, and this is likely to be a more circuitous way of obtaining specific gravity than some other method. It is necessary also to determine the index of known mixtures of the liquids (with specific gravities of the mixtures also known) so that a curve showing the relation between gravity and index can be constructed. With such a curve, however, it is possible to determine the specific gravity of a combination of the liquids with but a single drop of the mixture. Merwin[3] has published the data for a solution of barium-mercuric iodide (Rohrbach's solution). Vassar[4] has prepared a table showing in like manner the relation between the index of refraction and the specific gravity of a solution of thallium formate-malonate (Clerici's solution):

[1] V. Goldschmidt, Ueber Verwendbarkeit einer Kaliumquecksilberjodidlösung bein mineralogischen und petrographischen Untersuchung: *Neues Jahrb. f. Min., etc.*, B.B. I, pp. 196-199, 1881.
[2] W. J. Sollas, On a modification of Sprengel's apparatus for determining the specific gravity of solids: *Proc. Roy. Dublin Soc.*, n.s., vol. 5, pp. 623-625, 1886-1887.
[3] H. E. Merwin, A method of determining the density of minerals by means of Rohrbach's solution having a standard refractive index: *Am. Jour. Sci.* (4), vol. 32, pp. 425-432, 1911.
[4] Helen E. Vassar, *op. cit.*, p. 125.

SEPARATION METHODS

Indices of Refraction	Specific Gravity
1.6761	4.076
1.6296	3.695
1.6154	3.580
1.5990	3.434
1.5815	3.280
1.5693	3.184
1.5620	3.114
1.5515	3.024
1.5363	2.884
1.5156	2.692

Measurements at 19.5° C.

The Westphal balance is perhaps the most accurate and most satisfactory device of all, though its use involves a little more time than that of the hydrometer described below.[1] This apparatus is essentially

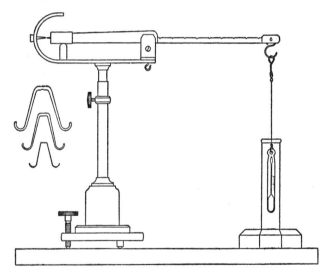

FIG. 150.—Westphal balance. Riders to be suspended on beam are shown at left. Largest rider marks one unit when placed at end of beam and marks tenths when placed in appropriate notch. Next largest rider indicates hundredths, and the smallest rider marks thousandths. Liquid to be measured is placed in cylinder at right.

a beam balance (Figure 150). From one end of the beam is suspended, by a platinum thread, a weighed sinker, usually a short glass thermometer tube (thus enabling one to make a determination of the temperature of the liquid in question at the same time). The balance is made so that

[1] E. Cohen, Ueber ein einfache Methode das specifische Gewichteiner Kaliumquecksilberjodidlösung zu bestimmen: *Neues Jahrb. f. Min.*, etc., Bd. II, pp. 87-89, 1883.

when the sinker is immersed in water the beam is horizontal as indicated by a pointer on the left end. If the liquid is heavier than water the sinker rises, and to restore balance riders must be hung upon the right arm of the balance, which is graduated into ten equal parts. Since there are three sizes of riders, each division has a different value for each rider. In terms of specific gravity, each division represents 0.1 for the largest or unit rider, 0.01 for the rider of intermediate size, and 0.001 for the smallest rider. The specific gravity of a liquid is, therefore, given by summing up the readings given by the position of the three kinds of riders plus one (since the beam alone is in balance when the sinker is immersed in water).

A simple and accurate method is that involving a hydrometer which is especially calibrated for heavy fluids [1] (Figure 151). This gives the specific gravity by direct reading, accurately to two decimal places (and approximate to three). The principal objection to such a hydrometer is that it needs a rather large quantity (200 c.c. or more) of liquid to float it.

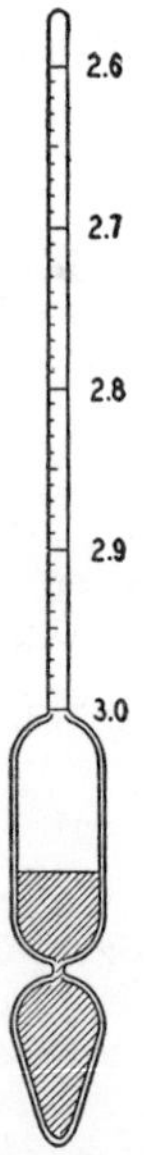

Fig. 151.—Heavy-liquid hydrometer (Blake).

Tester [2] has devised a hydrometer for measuring the specific gravity of a heavy liquid. The hydrometer requires but 5 c.c. liquid and has a range of values from 2.0 to 5.0. It is made of glass tubing and consists of a ball float (a) above which is a glass stem (b) graduated from 10 to 25 g., subdivided into tenths, and below which is a liquid chamber (c) (Figure 152). A small amount of mercury, used as a balancer, is sealed in the ground glass stopper (d) which fits the liquid chamber. To use, the liquid chamber is filled to a 5 c.c. mark and then stoppered, inverted, and immersed in a column of distilled water. The scale is read, and the value noted is divided by five to give the specific gravity. The instrument may be used as a means of measuring the specific gravity of a solid.

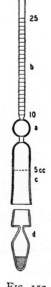

Fig. 152.—Tester's heavy-liquid hydrometer. a, ball float; b, graduated stem; c, liquid chamber, and d, ground glass stopper.

---

[1] Such a hydrometer, 20 cm. long, with bulb diameter of 2 cm., and range from 2.6 to 3.0 with scale graduated every 0.01, can be obtained from Messrs. T. O. Black, 57, Hatton Garden, London, E. C. 1.

[2] A. C. Tester, A convenient hydrometer for determining the specific gravity of heavy liquids: *Science*, n.s., vol. 73, pp. 130-131, 1931.

Separation Apparatus for use with Heavy Liquids

Gravity separations by means of heavy liquids have been carried out in various kinds of apparatus varying from a simple evaporating dish on the one hand to the elaborate Penfield apparatus for use with low-melting solids on the other. The apparatus should be inexpensive, not too fragile, and so constructed that little loss occurs with volatile liquids. It should permit the grains to be agitated or stirred, should be so constructed that the mineral crops are readily removed from the apparatus, and should not be so narrow that clogging occurs during separation. Moreover, the separating vessel should be as broad as possible so that the floating minerals will have but a slight thickness, should have no abrupt reëntrants on which the grains may lodge, and should have valves with a diameter as large as the outlet tube.

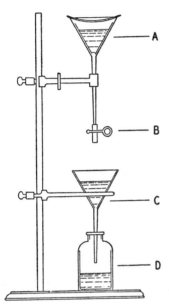

FIG. 153.—Simple apparatus for heavy-liquid separation. A, funnel with heavy liquid; B, rubber tube and pinch-cock; C, funnel fitted with filter-paper; D, bottle to collect used heavy liquid.

FIG. 154.—Special funnel for heavy-liquid separation.

The several types of separating devices are briefly described. Simplest of all is a simple filtering funnel, the stem of which is fitted with a short length of rubber tubing and pinch-cock (Figure 153). A watch-glass may be used to cover the funnel to reduce evaporation losses. Several workers have used funnels of this type but have modified them by increasing the steepness of the funnel walls to prevent lodgement of grains and have introduced stop-cocks into the stem [1] (Figure 154). One of the earliest devices was such a funnel, the

[1] Clarence S. Ross, Methods of preparation of sedimentary materials for study: *Econ. Geology*, vol. 21, pp. 454-468, 1926. Such a funnel with steep walls and stopcock is obtainable from Emil Greiner of Fulton and Cliff Streets, New York, N. Y. L. Van Werveke, Ueber Regeneration der Kaliumquecksilberjodidlösung und über einen einfachen Apparat zur Trennung mittelst dieser Lösung: *Neues Jahrb. f. Min.*, etc., Bd. 2, pp. 86-87, 1883.

upper part of which was cylindrical, with a short stem, with valve, which fitted by a ground glass joint into a lower bottle [1] (Figure 155).

Other workers have used bulb- or, better, pear-shaped separatory funnels. Earliest of these is the Harada tube [2] which was closed at the upper end by a tight-fitting stopper and at the lower end by a valve, the opening in which was of the same diameter as the tube in which it is located (Figure 156). After shaking and complete separation, the lower end is placed in an accessory vessel and a stopcock opened. Brögger [3] built a similar tube but added a second large stop-cock above the first (Figure 157). Separations are rarely complete after one

Fig. 155.—Apparatus of Church.

Fig. 156.—Harada tube.

Fig. 157.—Brögger apparatus.

settling period, but with the Brögger apparatus it is possible to make a clean separation by first closing the middle valve—after the preliminary settling—and then inverting the apparatus, allowing a second separation to take place in both portions, then slowly returning the apparatus to normal position and reopening the middle valve so that the separated portions of the light and heavy minerals unite.

Another device which makes such repeated separations possible is that

---

[1] A. H. Church, A test of specific gravity: *Mineral. Mag.*, vol. 1, pp. 237-238, pl. 8, fig. 7, 1876-1877.

[2] K. Oebbeke, Beiträge zur Petrographie der Philippinen und der Palau-Inseln: *Neues Jahrb. f. Min., etc.*, B.B. 1, pp. 451-501 (esp. p. 457), 1881.

[3] W. C. Brögger, Om en ny Konstruktion af et isolations-apparat for petrografiske undersögelser: *Geol. Fören. i. Stockholm Förh.*, vol. 7, pp. 417-427, 1884.

devised by Laspeyres[1] in which two pear-shaped vessels are connected by a stop-cock and at each end are closed with large ground glass stoppers (Figure 158). A modification of this device is described by Hauenschild[2] in which the two stop-cocks are modified or converted into two small vessels (Figure 159). They are large enough so that when they are detached the heavy liquid and mineral grains can be contained therein.

Wülfung[3] also devised an apparatus in which repeated separations are possible. It differs from the others in that it is a linked-shaped affair composed of two branches—each one a curved tube—which are connected to each other at both

Fig. 158.—Laspeyres separating vessel.

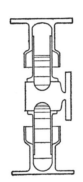

Fig. 159.—Hauenschild apparatus (after Cayeux).

Fig. 160.—Wülfung separation tube.

ends by two valves (Figure 160), and by means of which the two parts can be shut off. To make repeated separations the valves are closed and heavy liquid and minerals are introduced, through a glass-stoppered opening, into one half only. After partial separation the lower valve is opened to allow the heavy grains and some liquid to pass into the other half. The two halves are then again separated by closing the valves, and, after some liquid is added to both sides, a second separation takes place on each side. By opening the lower valves and tilting the apparatus, the heavy fractions unite, while opening the upper valve permits the light portions to join.

[1] H. Laspeyres, Vorrichtung zur Scheidung von Mineralien mittelst schwerer Lösungen: *Zeits. f. Kryst. Min.*, vol. 27, pp. 44-45, 1897.
[2] A. Hauenschild, *Zeitschr. f. Baumaterialenkunde*, März, 1898. Description and figure in Cayeux, *L'Etude pétrographique des roches sédimentaires. Texte* (1931), p. 59.
[3] E. A. Wülfung, Beitrag zur Kenntnis des Kryokonit: *Neues Jahrb. f. Min., etc.*, B.B. 7, pp. 164-165, 1891.

Smeeth[1] developed an apparatus, later modified by Diller,[2] in which an upper pear-shaped separatory vessel could be entirely detached from a lower vessel and base of candlestick form (Figure 161). The junction is a ground glass joint. The upper vessel could be closed at its lower end by a glass stopper on the end of a glass rod which extends upward and out through the top opening. A ground glass cap in turn closed the top opening. Luedecke also made a separating device on much the same order [3] (Figure 162).

Thoulet,[4] in 1879, used a long, narrow, burette-like separating tube, the outlet stem of which, how-

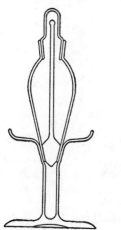
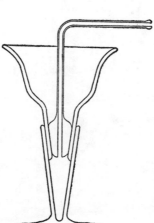
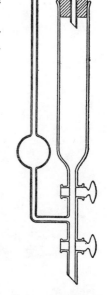

Fig. 161.—Diller modification of Smeeth apparatus.

Fig. 162.—Luedecke apparatus.

Fig. 163.—Thoulet tube (slightly modified).

ever, has two valves. A parallel side tube of smaller diameter was joined to the main apparatus between the two valves (see Figure 163). Air forced through the side tube and up into the main separating vessel agitated the heavy liquid. The upper valve is then closed to allow quiet separation. Afterwards the valve is opened to allow the heavy minerals

---

[1] W. F. Smeeth, An apparatus for separating the mineral constituents of rocks: *Proc. Roy. Dublin Soc.*, vol. 6, pp. 58-60, 1888.
[2] J. S. Diller, The Smeeth separating apparatus: *Science*, n.s., vol. 3, pp. 857-858, 1896.
[3] O. Dreibrodt, Trennungsapparat nach Prof. Dr. O. Luedecke: *Centralbl. f. Min.*, etc., pp. 425-426, 1911.
[4] J. Thoulet, Séparation mécanique des divers éléments minéralogique des roches: *Bull. Soc. Min. France*, vol. 2, pp. 17-24, 1879.

to drop into the space between the valves, after which it is closed and the lower valve is opened to release the heavy crop. Oebbeke[1] simplified the Thoulet apparatus but owing to its fragile nature and difficulty of cleaning it has been but little used.

Fraser[2] suggested the simplest of all special devices. His apparatus is a tube, intermediate between a V- and a U-tube, which tapers from one end to the other (Figure 164). The tube is filled two-thirds full of heavy liquid and so held that the wider limb is vertical and so that the liquid rises to the open end of the smaller limb. The sand is introduced at the wide end. The light minerals thus accumulate at the top of the liquid in this wider portion while the heavy minerals settle to the bend in the

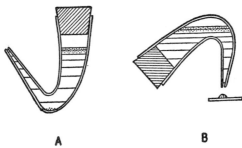

FIG. 164.—Fraser tube. A, initial position; B, inverted position. Slight pressure on cork forces out drop of heavy liquid and heavy mineral onto slide.

tube. The tube should now be inverted in such a way that the heavier minerals fall into the narrow tube (B, Figure 164), where they collect in the jet and may be expelled on a glass slide for observation by pushing in the cork. The apparatus is easily made, without fragile or complicated parts, uses little liquid, and is inexpensive. It is probable, however, that it is best suited to rapid qualitative work rather than complete quantitative study.

Woodford[3] suggested the use of the Spaeth sedimentation glass for heavy liquid separation. This is a conical-shaped glass with wide base. Near the base of the cone is a large stop-cock, a part of which is cut away to make a relatively large cup-shaped hole of dimensions continuous with the upper part of the vessel. If the stop-cock is turned, its

[1] K. Oebbeke, Beitrage zur Petrographie der Philippinen und der Palau-Inseln: Neues Jahrb. f. Min., etc., B.B. I, pp. 450-501 (esp. p. 456), 1881.
[2] F. J. Fraser, A simple apparatus for heavy mineral separation: Econ. Geology, vol. 23, pp. 99-100, 1928.
[3] A. O. Woodford, Methods for heavy mineral investigations: Econ. Geology, vol. 20, pp. 103-104, 1925.

content of heavy minerals are isolated from the glass above (Figure 165).

Penfield[1] designed an apparatus to be used with low-melting solids. In its improved form (Figure 166) it consists of a tubular separating vessel fitted, by means of a ground glass joint at its lower end, to a small hollow glass cap. A hollow stopper connected to a glass tube which extends the entire length of the upper tube shuts off this upper tube from the lower hollow cap. The whole apparatus is put in a large test-tube in a beaker of hot water. An air stream through the tubular stopper agitates the melt and contained mineral grains. After separation is complete the stopper is inserted and isolates the heavier minerals in the lower hollow cap, from which they are then removed. The double nitrate of silver and thallium is used with this device. Since, however, this salt has a specific gravity of 4.5 at its melting point of 75° C., while Clerici's solution has a gravity of 4.25 at room temperature (20° C.), it seems to the authors unnecessary to use the former and the complicated Penfield separator except in rare cases.

FIG. 165.—Spaeth sedimentation glass.

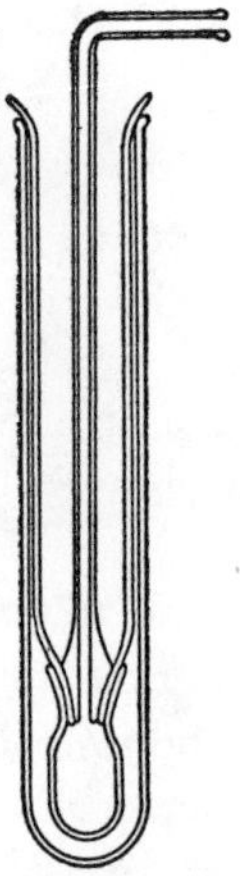

FIG. 166.—Penfield separation apparatus for heavy melts.

HEAVY-LIQUID FRACTIONATION WITH CENTRIFUGE

It may be necessary, in the case of fine sands and silts, to use the centrifuge to accelerate the separation. An inexpensive but permanent centrifuge tube which makes a rapid and efficient separation of grains smaller than 0.5 mm. is described by Taylor.[2] It consists of a tube 120 mm. long with a cylindrical upper part having a diameter of 17 mm. and a similar lower part with a diameter of 9 mm. A cork-tipped plunger, slightly larger than the constricted lower part of the tube, is inserted in the upper wider part to the point of narrowing, where it effectively separates the two parts of the tube. A cork is provided for the opening of the centrifuge tube, and this fits tightly on the stem of the plunger so that the latter can be fixed at any desired position (Figure 167). The

---

[1] S. L. Penfield and D. A. Kreider, On the separation of minerals of high specific gravity by means of the fused double nitrate of silver and thallium: *Am. Jour. Sci.*, vol. 48, pp. 143-144, 1894. S. L. Penfield, On some devices for the separation of minerals of high specific gravity: *Am. Jour. Sci.*, vol. 50, pp. 446-448, 1895.

[2] G. L. Taylor, A centrifuge tube for heavy mineral separations: *Jour. Sed. Petrology*, vol. 3, pp. 45-46, 1933.

heavy minerals collect in the lower constricted portion during centrifuging. They are stoppered in the plunger while the light minerals are washed out of the upper part of the tube.

Brown [1] has also described the use of the centrifuge in separation of heavy minerals. Brown used an ordinary centrifuge tube and removed

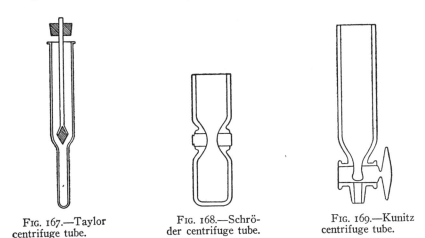

FIG. 167.—Taylor centrifuge tube.

FIG. 168.—Schröder centrifuge tube.

FIG. 169.—Kunitz centrifuge tube.

the heavy minerals by means of an ordinary pipette fitted with a rubber bulb.

Müller [2] and Schroeder [3] both apparently used the centrifuge in mineral separations, Müller used a modified sling tube, anticipating in principle that later described by Taylor, while Schroeder used a tube with valve in the middle, which upon closing separated the light and the heavy portions (Figure 168). Kunitz [4] also used a sling tube modified with a large stop-cock at the lower end for separation purposes. The stop-cock, however, had only a cup-like depression in which the heavier constituents collected. Turning the valve 180° permitted the escape of the heavy crop at the lower end of the tube (Figure 169).

[1] Irvin C. Brown, A method for the separation of heavy minerals of fine soil: *Jour. Paleon.*, vol. 3, pp. 412-414, 1929.
[2] Heinrich Müller, Neues Zentrifugenglas zum quantitativen Trennen von Kornigen und pulverigen Gemengen verschiedenen spez. Gewichts mit Hilfe von schweren Lösungen: *Mitt. Min.-Geol. Staatsinstitut. Hamburg*, H. 11, pp. 1-6, 1929; Uber ein angeandertes Zentrifugenglas zum Trennen nach dem spez. Gewicht: *Centralbl. f. Min., etc.*, Abt. A, pp. 90-91, 1932.
[3] Fritz Schroeder, Scheidetrichterzum Einsetzen in die Zentrifuge beim Trennen von Mineralgemischen mit schweren Flussigkeiten: *Centralbl. f. Min., etc.*, Abt. A, pp. 38-46, 1930.
[4] W. Kunitz, Eine Schnellmethode der gravimetrischen Phasenanalyse mittels der Zentrifuge: *Centralbl. f. Min., etc.*, Abt. A, pp. 225-232, 1931.

The centrifuge effects a very complete separation between minerals differing but slightly in specific gravity. Dolomite and calcite [1] may be quantitatively separated by its use as also may anhydrite and rock salt.[2]

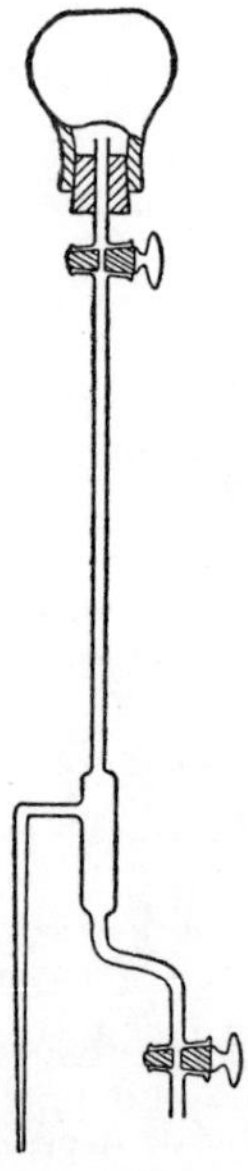

None of the various devices for heavy liquid separation of minerals in the centrifuge is wholly satisfactory. The stop-cocks will not hold heavy liquids when subject to centrifugal force. It is necessary, therefore, to employ only those with sealed ends. With these, however, it is difficult to separate the light and heavy mineral crops. To overcome these difficulties Berg [3] devised a pipette consisting of (1) a small glass tube (D) 16 cm. long and a 2-mm. bore but tapered to a 1¼-mm. opening at the lower end, (2) a trap (C) which collects the heavy minerals carried through the small tube by the rising heavy liquid, (3) a stop-cock (B) which permits drainage of the trapped minerals into a filter-paper, and (4) a rubber bulb which can be closed off from the rest of the pipette by a stop-cock (E) (see Figure 170).

Berg used about 1 g. of sediment in an ordinary 15-ml. centrifuge tube with about 15 ml. of heavy liquid. After thorough shaking to disperse the sediments, the material is centrifuged until the liquid is clear. To remove the heavies, the rubber bulb is depressed, the bulb stop-cock is closed, as is also the lower stop-cock, and the small glass tube (D) is lowered to the bottom of the centrifuge tube. The bulb stop-cock is then opened and the heavy liquid and the heavy minerals rise into the trap. The bulb stop-cock is then closed and the pipette is withdrawn until it is just out of the liquid, when the bulb stop-cock is opened just enough to clear the small pipette tube of liquid. The material in the trap is then released onto a filter-paper by opening the lower stop-cock. Or, if desired, the material may be released into a second centrifuge tube to effect a second settling to insure complete separation of the heavy mineral fraction.

Fig. 170.—Berg pipette for collecting heavy residue from bottom of common centrifuge tube.

[1] C. B. Claypool and W. V. Howard, Method of examining calcareous well cuttings: *Bull. A. A. P. G.*, vol. 12, pp. 1147-1152, 1928.
[2] F. von Wolff, Die Trennung fester Phasen durch die Zentrifuge: *Centralbl. f. Min., etc.*, Abt. A, pp. 449-452, 1927. Carl W. Correns, Ueber zwei neue einfache Verfahren für das Zentrifugieren mit schweren Lösungen: *Centralbl. f. Min., etc.*, Abt. A, pp. 204-207, 1933.
[3] Ernest Berg, A method for the mineralogical fractionation of sediments by means of heavy liquids and the centrifuge: *Jour. Sed. Petrology*, vol. 7, pp. 51-54, 1937.

Procedure for Separating the Minerals of a Clastic Sediment by Means of Bromoform

As may be seen by referring to the preceding section, there are many types of apparatus designed for separating the heavy minerals from the light, using some heavy liquid. Nevertheless, the set-up shown in Figure 153 is about as satisfactory as any, except for very fine material, since it is readily constructed from materials on hand in any laboratory.

Bromoform of known specific gravity is first poured into the upper funnel, after which a weighed amount of dry sediment (prepared as previously described) is placed in it and thoroughly stirred; and the heavier minerals can then be drawn off by opening the pinch-cock. If the proportion of heavy minerals is small, as is the case with ordinary unconcentrated sediments, a large amount of materials, say about 50 g., can be added at once. When the amount of heavier minerals is large, as in concentrated sediments (and in many crushed igneous rocks), the mixture should be added in several portions, the pinch-cock being opened each time, to prevent the accumulation of too much material in the neck of the funnel. Cover the funnel with a watch-glass to reduce evaporation losses.

It is convenient to have the "used" heavy-liquid bottle at hand, provided with a filtering funnel with filter-paper, and to allow the heavy minerals to drop on the filter-paper. After the heavy liquid has drained off, the filter-paper can be detached and opened, and the heavy grains washed off by placing the paper face downwards in a watch-glass or porcelain dish containing alcohol (or acetone). The washings thus obtained should be put in the bottle marked "Bromoform Washings." (Ultimately, the amount of heavy liquid in the washings will become considerable and it will be recovered.) *After* the removal of the "heavy" minerals, the bromoform in which the light minerals are floating can be run off through the pinch-cock and through a clean filter-paper into the "used" heavy-liquid bottle. After replacing the "used" heavy-liquid bottle with that marked "Bromoform Washings," alcohol may be used to wash down the light minerals remaining in the upper funnel.

Weigh the "lights" and "heavies" and put each in a vial and label. A permanent mount in Canada balsam should be made of the light minerals and a mount of the heavy minerals should also be made either in Canada balsam or piperine (see instructions on same).

Procedure for recovery of bromoform. Add the washings containing bromoform to a large volume (1 gal.) of cold water in a large stoppered bottle. Shake vigorously. Decant most of the water. Repeat twice more. After the last decantation pour the remaining water and separated bromoform into a separatory funnel. Draw the bromoform off from below and run into a funnel fitted with several thicknesses of filter-paper. Collect the bromoform filtrate. If it is not clear run it through a second funnel and filter-paper. (The paper absorbs any dispersed water and any wax that may be formed.) Test the bromoform for specific gravity. Put it in a brown bottle and label "Used Bromoform."

SEPARATION ON THE BASIS OF MAGNETIC PERMEABILITY

Minerals can be classed as *paramagnetic* when the lines of magnetic force pass through them more easily than through air (air 1), and *diamagnetic* when they pass less readily. Common paramagnetic substances are: iron, nickel, cobalt, manganese, chromium, cerium, potassium, platinum, and aluminum. Common dimagnetic substances are: bismuth, antimony, zinc, silver, copper, water, sulphur, phosphorus, boron, the halogens, silica, etc. The grouping of the elements in the compound will influence its permeability. Both paramagnetic and dimagnetic elements may be present, but the compound will not necessarily be either one or the other. When iron is present, the compound is usually magnetic. Ferrous iron is usually more effective than ferric iron. Magnetism of a particular grain may be due to the presence of magnetic inclusions.

Because of the varying magnetic permeability of minerals it is possible to make a separation of one kind from another. Either a permanent magnet or an electromagnet may be used. An ordinary permanent magnet of considerable strength is rather large; it requires remagnetizing at intervals; and it is difficult to detach grains attached to it. On the other hand, it is relatively cheap and can, if fitted with special pole pieces, be used for separating weakly magnetic material. The separation of the weakly magnetic materials is, however, better achieved by using an electromagnet, which, if suitably constructed, is much stronger than a permanent magnet of the same size. The electromagnet, moreover, exerts no attractive force when the current is shut off, and hence recovery of the magnetic crop is rendered easy.

Permanent magnets. Crook[1] recommends an instrument of U-shape, the limbs of which are 6 in. long, made of a steel bar about 1 in. wide and ½ in. thick. (See Figure 171.) Two adjustable pole pieces of soft iron fit against the smooth free end of the limbs to which they are secured by binding screws. These pole pieces are slotted so that the gap between them can be varied. The free ends of the pole pieces taper gradually and are gently curved or bent downwards.[2]

---

[1] T. Crook, A simple form of permanent magnet suitable for the separation of weakly magnetic minerals: *Geol. Magazine,* vol. 5, pp. 560-561, 1908.

[2] The L-shaped pole pieces used by some investigators are unsatisfactory and result in loss of magnetic intensity. The bend should be gradual, not right-angled.

SEPARATION METHODS

Smithson[3] developed a simple method for observing the magnetic properties of single mineral grains. A piece of cardboard, about 1¾ by 1½ in., is bent over at one end to give a square 1½ by 1½ in. Two darning needles are heated to redness and cooled slowly. These are then forced through the fold in the card about ½ in. apart and fixed to the card with seccotine so that their points are close together. The two needles are then placed against the poles of a horseshoe magnet and a strong magnetic field between the points results.

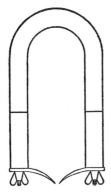

Fig. 171.—Simple form of permanent magnet with adjustable pole pieces.

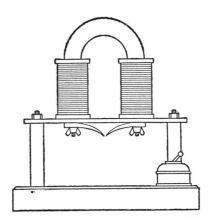

Fig. 172.—Electromagnet for mineral separation.

The grains are immersed in some liquid on a watch-glass, and their behavior observed when the needle points are brought near.

Electromagnetic separation. Delesse[2] was apparently the first to investigate thoroughly the magnetic properties of minerals and to achieve a separation of minerals by the electromagnet. His work was largely unnoticed, and no further work was done on minerals until 1872, when Foque[3] used the electromagnet to isolate the various rock constituents. Other workers soon followed. Mann[4] placed an electromagnet in a horizontal position. Between the poles of the magnet was a vertical glass

---

[1] Frank Smithson, A simple method of observing the magnetic properties of mineral grains: *Mineral. Mag.*, vol. 22, pp. 419-421, 1930.

[2] A. Delesse, Sur le pouvoir magnétique des minéraux et des roches: *Annales des Mines*, vol. 14, pp. 429-486, 1848.

[3] F. Foque, Nouveaux procédés d'analyse médiate des roches et leur application aux laves de la dernière éruption de Santorin: *Comptes Rend. Acad. Sci. Paris*, vol. 75, pp. 1089-1091, 1872.

[4] Paul Mann, Untersuchungen über die chemische Zusammensetzung einiger Augite aus Phonolithen und verwandten Gesteinen: *Neues Jahrb. f. Min., etc.*, Bd. 2, pp. 181-185, 1884.

tube filled with water. As the mineral grains settled slowly downward the more magnetic substances were retained at the poles.

Doelter[1] set up a scale of magnetic permeability based on twelve common minerals. Crook,[2] on the other hand, established four classes of minerals, namely, very magnetic, such as magnetite; moderately magnetic, such as ilmenite; weakly magnetic, for example, monazite; and non-magnetic, such as zircon or quartz. This scheme of classification is here followed and the four classes are numbered or designated I, II, III, and IV respectively in the section dealing with the properties of minerals.

To affect a separation into the several groups either the distance between the pole pieces must be varied or the strength of the current controlled. Several types of pole pieces have been used.

Crook[3] more than anyone else has discussed the use of the electromagnet in petrography.

An electromagnet suitable for petrological work should be made with a soft iron core, 1 in. in diameter, bent into a U-shape with vertical limbs about 6 in. long and poles some 4 in. apart. The soft iron pole pieces should be slotted and attached with wing nuts so that the tips of these pieces can be adjusted over a wide range (Figure 172). The pole pieces should also be bent down at the correct angle to obtain the greatest magnetic intensity. The magnet and adjustable pole pieces should be raised sufficiently above the base to give ample working space. Such a magnet may be operated from a storage battery or from a direct current lighting circuit if sufficient resistance, such as a bank of lamps, is inserted or from an alternating current circuit with suitable rectifier.

Procedure for separation of minerals with common electromagnet: Since, in general, a heavy liquid separation precedes the use of the electromagnet, the size of the sample to be treated is small and therefore suitable for magnetic separation. To make a separation, place the mixed grains on a piece of stiff paper or cardboard and first of all remove by means of the ordinary horseshoe magnet any highly magnetic minerals (I), such as magnetite, etc., that may be present. Then with the tips of the poles of the electromagnet about a centimeter apart, bring the grains into the vicinity of the pole tips and switch on the current. The first portion of the more magnetic grains will entrap some non-magnetic

---

[1] C. Doelter, Ueber die Einwirkung des Elektromagneten auf verschiedene Mineralien und seine Anwendung behufs mechanischer Trennung derselben: *Sitz. d. K. Akad. d. Wiss., Wien,* Bd. 85 Ab. I., pp. 47-71, 1882.

[2] T. Crook, Systematic examination of loose detrital sediments, in F. H. Hatch and R. H. Rastall, *The Petrology of the Sedimentary Rocks* (London, 1913), p. 369.

[3] T. Crook, The use of the electromagnet in petrography: *Science Progress,* vol. 2, pp. 30-50, 1907. See also C. J. Ksanda, An electromagnetic separator for laboratory use: *Jour. Optical Soc. Am.,* vol. 13, 1926.

grains. It is advisable, therefore, by switching off the current to let the grains fall back onto another sheet of paper and then to repeat the operation several times. In this way a group of moderately magnetic minerals (II) can be separated. When the separation of this group is complete, reduce the interval between the tips of the pole pieces to about a half-centimeter and then continue the separation as before in order to isolate another group of minerals which are only weakly magnetic (III). The residue remaining after this group has been extracted is non-magnetic (IV) or nearly so.

Sometimes impurities make any otherwise non-magnetic mineral slightly magnetic. If the impurity is merely a ferruginous cementing material coating the grain, it can be destroyed by acid treatment, but if, as is sometimes the case, the impurity exists as a microscopic inclusion, the separations will be imperfect.

Electromagnetic separators. Several types of electromagnetic separators utilizing the principle of the electromagnet have been devised. One of these is the Davis magnetic separator.[1] This apparatus consists of a powerful electromagnet, a glass tube in which the separation takes place which lies between the poles of the magnet, and an electric motor which oscillates the separation tube. A current of water flows through the tube while the separation is taking place. The machine is designed to operate on a 110-volt D.C. system. Ten grams of material of 100 mesh or smaller are used for analysis. The glass tube is nearly filled with water and the sample is washed into the tube, where it settles down. The magnetic fraction collects at the poles of the magnet. The upper end of the tube is stoppered and the whole machine tipped to an angle of about 45°. After the water supply is adjusted to a flow of about 1 gal. in 10 min., the motor is started. The speed is so adjusted by rheostat control that the tube makes about 60 to 100 strokes per minute. At the end of 10 min., the water in the tube should be clear. The non-magnetic fraction will be washed free from the magnetic concentrate. The latter remains in the tube. After the water flow is stopped, a portion of the water is drawn off and the outlet is closed. Removal from the poles will then release the magnetic crop, which may be washed out of the tube.

A different type of separator is that designed by A. F. Hallimond.[2] The Hallimond separator consists of a horizontally rotating conveyor disk. This disk is carried on a central upright spindle, which is rotated clockwise by a small motor and worm gear. Near the outer edge of the disk is a broad groove, on the floor of which the material is carried. The

[1] Manufactured and sold by Dings Magnetic Separator Company, Milwaukee, Wisconsin.
[2] A. F. Hallimond, An electromagnetic separator for mineral powders: *Mineral Mag.*, vol. 22, pp. 377-381, 1930.

groove is cut away at intervals, leaving six shelves with alternate gaps through which the magnetic material can periodically be discharged. As the conveyor disk rotates, the shelves pass beneath a hopper, the feed-valve of which is magnetically operated to open and close at the right moment and accordingly deposit a small amount of material on each shelf in turn as it passes beneath the hopper. The shelf then passes in succession below (a) a small single-pole magnet which removes the magnetite, (b) a large magnet of moderate strength, and (c) a large magnet of maximum strength. Each magnet lifts a fraction of the magnetic material from the shelf of the conveyor disk and holds it until a gap has come below the poles. At this moment a current interruption releases the magnetic materials, which then drop through the respective gaps into glass dishes below. The shelf with the non-magnetic residue passes beneath a fixed brush which sweeps the bottom of the groove and carries the tailings into a fourth dish. The timing of the current interruption is achieved by a commutator with six vertical copper bars corresponding with the gaps in the conveyor. The separator operates best with evenly sized materials. It can be used successfully on samples as small as a half a gram.

SEPARATION ON THE BASIS OF DIELECTRIC PROPERTIES

This method of separation is based on the principle that two oppositely charged poles attract each other with a force that varies inversely as the square of the distance between them. The force with which the poles are attracted is modified by the surrounding medium, and the ratio of the attraction in a vacuum as compared with the ratio in a given medium is called the *dielectric constant*. A mineral grain in a medium of a given dielectric constant will be attracted to the electric field between the poles if it has a constant greater than that of the medium.

Dielectric separation methods have been studied by Hatfield,[1] Holman,[2] Tickell,[3] and Berg.[4] Berg, the most recent worker in this field, has ap-

[1] H. S. Hatfield, Dielectric separation, a new method for the treatment of ores: *Min. and Met. Inst., Bull. 233*, 8 pp., 1924.

[2] Bernard Holman and St. J. R. C. Shepherd, Dielectric mineral separation: Notes on laboratory work: *Min. and Met. Inst., Bull. 233*, 7 pp., 1924. B. W. Holman, The dielectric constant as a factor in ore concentration: *Mining Mag.*, vol. 28, pp. 267-271, 1923.

[3] Frederick G. Tickell, *The Examination of Fragmental Rocks* (Stanford University, California, 1931), pp. 43-45.

[4] Gilman A. Berg, Notes on the dielectric separation of mineral grains: *Jour. Sed. Petrology*, vol. 6, pp. 23-27, 1936.

TABLE 38

CLASSIFICATION OF MINERALS FOR HEAVY LIQUID AND MAGNETIC SEPARATION

Heavy Liquid	II. Moderately Magnetic	III. Weakly Magnetic	IV. Practically Non-magnetic	
2.89 — Bromoform / Acetylene tetrabromide / Methylene iodide / Thallous formate / Thallous formate-malonate	2.3 Glauconite		2.0-2.4 Zeolites 2.3 Gypsum 2.5 Leucite Kaolin 2.5-2.6 Alkali feldspars 2.61-2.76 Soda-lime feldspars 2.6-2.75 Scapolite group 2.62 Chalcedony 2.63 Nepheline 2.64 Cordierite 2.65 Quartz 2.72 Calcite 2.85 Dolomite 2.85 Phlogopite	"Light" minerals
2.96			2.90 Muscovite 3.0 Tremolite	
3.3	3.1 Iron-rich biotite 3.2 ⎤ 3.3 ⎥ Iron-rich 3.4 ⎬ Amphiboles 3.5 ⎥ and Pyroxenes 3.6 ⎦	3.1 Biotite Amphiboles Pyroxenes 3.2 Tourmaline 3.4 Epidote 3.5 Olivine 3.6-3.7 Staurolite	3.1 Enstatite 3.15 Apatite 3.18 Andalusite 3.2 Fluorite 3.23 Sillimanite 3.5 Sphene 3.52 Topaz 3.6-3.7 Spinel 3.62 Kyanite 3.9 Anatase 3.94 Brookite	
3.4				
4.3	3.8 Siderite 4.1 Almandine	3.8 *Pleonaste* 4.0 Garnet 4.1 *Cr. spinel* 4.2 Ferriferous rutile	4.0 Perovskite Corundum 4.2 Rutile	"Heavy" minerals
	4.3 Melanite 4.4 *Chromite* 4.5 Xenotime 4.8 *Ilmenite* 5.1 Hematite	5.2 Monazite 5.3 *Columbite*	4.5 Barite 4.7 Zircon 5.0 *Pyrite*	

NOTES. *Magnetite*, 5.17, *titanoferrite*, 4.65, and *pyrrhotite*, 4.65, are the only highly magnetic minerals (I).
Good conductors are italicized and may be separated by electrostatic methods.
Table modified after Holmes.

parently developed a satisfactory technique for this method. Unlike the earlier workers, Berg used 880 volts, 60-cycle alternating current secured by use of an ordinary radio transformer. A 1-ampere fuse is placed in the primary circuit, and a resistance of 2,000 ohms in the secondary circuit. Wires from the secondary circuit lead to needles mounted on an insulated pencil-like handle. The space between the needle-points serves as the electric field needed for separation. After experimenting with several liquids, Berg used mixtures of furfural and benzene (as distinct from benzine) which have dielectric constants of 42.0 and 2.28 respectively. For example, 1 part of furfural and 5 parts of benzene give a resulting medium with dielectric constant of 8.9, computed from the formula $V_1K_1 + V_2K_2 = (V_1 + V_2)K$, where V_1 and V_2 are the respective volumes taken and K_1 and K_2 are the dielectric constants of the components, while K is the constant of the mixture.

The separation is carried out in a small, flat glass container, such as a culture dish. Owing to the inflammable nature of benzene and to arcing produced by good conductors, such as magnetite and ilmenite, Berg recommends the removal of such minerals magnetically. After this treatment the minerals are placed in the container and furfural and benzene are drawn from burettes in the proportions desired. Needle-points are immersed in the liquid and placed about 1 mm. apart. The grains with a dielectric constant higher than the liquid move into the area between the needle-points, whereas those with lower constant are repelled. The grains attracted to the points are then transferred to a smaller dish placed in the larger container, the needle-points at all times held in the liquid, by moving the poles over the small dish and *then* withdrawing them from the liquid. When the needles are raised above the liquid the grains fall off into the small dish, which is then removed with forceps. The mineral grains are washed with alcohol and collected on filter-paper.

Berg measured the dielectric constants of a considerable number of minerals. His results are as follows:

Dielectric Constant — *Minerals*

- 14–15 .... Glauconite
- 8–9 .... Opal
- 6–7 .... Biotite (Fe)
- 5–6 .... Actinolite, augite, biotite (Mg), chlorite, epidote, kaolinite, monazite, sphalerite
- 4–5 .... Gypsum, hornblende, kyanite, sillimanite, titanite
- 3.5–4 .... Albite, anorthite, apatite, barite, cerussite, corundum, diopside, dolomite, fluorite, garnet, halite, nephelite, olivine, orthoclase, siderite, spodumene, staurolite, tourmaline, wollastonite
- 3–3.5 .... Andalusite, calcite, quartz

SEPARATION METHODS

Rosenholtz and Smith[1] have also recently studied the dielectric properties of minerals. Current from a 110-volt, 60-cycle A.C. system was stepped up to 220 volts by a small transformer. A 2,000-ohm resistance was connected in series with the transformer. Biological dissecting needle-holders of the type which permits replacement of the needles were utilized. The electrical connections were made to the knurled tightening rings. The needle-points were bent facing each other and placed 1 mm. apart. C. P. carbon tetrachloride, C. P. methyl alcohol, and triply distilled water, of which the dielectric constants are 2.24, 33.7, and 81 respectively, were used.

The mineral powder was pulverized to pass 250 mesh. Three to four cubic centimeters of carbon tetrachloride were run from a 10-c.c. burette into a glass caster cup. A very small portion of the powder was placed in the cup and the current turned on. Methyl alcohol was then added, a few drops at a time, from another burette. When definite repulsion, instead of attraction, of the particles was noted the volume of the two liquids used was recorded. From these values the constant of the mixture and therefore of the mineral was computed as was done by Berg.

The determinations by Rosenholtz and Smith for common sedimentary minerals are summarized and abridged in the table below:

Dielectric Constant	Minerals
5–6	Apatite, corundum, rutile, titanite, tourmaline
6–7	Actinolite, albite, anhydrite, anorthite, augite, calcite, celestite, epidote, garnet (andradite and almandite), gypsum, hypersthene, limonite, microcline, olivine, opal, orthoclase, quartz, siderite, spinel, staurolite, topaz, zircon
7–8	Aragonite, barite, diallage, diopside, fluorite, garnet (grossularite), halite, hornblende, kyanite, monazite, tremolite
8–9	Andalusite, chalcedony, clinozoisite, chlorite, dolomite, enstatite, spodumene
9–10	Biotite, sillimanite
10–12	Glauconite, kaolinite, muscovite
Over 33.7, under 81	Ilmenite, magnetite, marcasite, pyrite
Over 81	Graphite, hematite

Inspection of the data given by Berg and those given in the table above, as well as similar data given in the *International Critical Tables*, will show considerable disagreement. A part of this probably arises from the different procedures (voltages, etc.) used and a part from the different liquids employed.

[1] Joseph L. Rosenholtz and Dudley T. Smith, The dielectric constant of mineral powders: *Am. Mineralogist*, vol. 21, pp. 115-120, 1936.

SEPARATION ON BASIS OF ELECTRICAL CONDUCTIVITY

Minerals of high conductivity may be separated from those of low conductivity by electrostatic methods. If the grains of a sand previously screened are lightly sprinkled over a copper or metallic plate and a rod of ebonite excited by rubbing with fur or flannel is brought near, certain of the grains will be attracted to the rod, whereas other grains, the poor conductors, will remain on the plate. This negatively charged rod induces positive charges on the upper surfaces of the grains and negative charges on the under side of the same. The negative charges of the conducting grains are removed by the metal plate, and since they are then but positively charged they are attracted to the rod, whereas the non-conducting grains lose their negative charges but slowly and remain unattracted.

This method of separation was studied by several workers.[1] It remained for Crook to devise special apparatus for this purpose and to apply the method to sedimentologic studies.

Crook used two copper plates. (See Figure 173.) These are separated a small distance by glass insulators. The lower plate is grounded and the

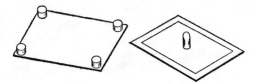

FIG. 173.—Crook electrostatic plates.

upper plate has its lower side shellacked. An induced charge is put on the upper plate by means of an electrostatic machine or simply by use of an electrophorus and cat skin. The conductive minerals will become negatively charged and will be attracted to the upper plate. The shellac keeps the charge from being neutralized. Both grains and the atmosphere must be dry in order to make the method a success. The grains adhering to the upper plate may be removed by brushing.

This method separates the metallic from the non-metallic minerals.

[1] D. Negreano, Procédé de séparation électrique de la partie metallique d'un minerai de sa gangue: *C. R. Acad. Sci. Paris,* vol. 135, p. 1103, 1902. Also vol. 136, pp. 964-965, 1903. L. I. Blake, Electrostatic concentration: *Eng. and Min. Jour.,* vol. 79, pp. 1036-1037, 1905. T. Crook, The electrostatic separation of minerals: *Mineral. Mag.,* vol. 15, pp. 260-264, 1909; Further remarks on the electrostatic separation of minerals: *Mineral. Mag.,* vol. 16, pp. 109-111, 1911. P. Riboni, La separazione elettrostatica dei minerali: *Rend. R. Ist. Lomb.,* vol. 49, pp. 649-660, 1916.

SEPARATION METHODS 353

Thus pyrite, hematite, magnetite, ilmenite, and other metallics are separated from quartz and other non-conducting minerals.

A continuously operating separator was designed by Blake[1] for commercial ore treatments. Very probably a similarly designed separator, on small scale, could be used in laboratory work.

SEPARATION ON THE BASIS OF VISUAL PROPERTIES

It is sometimes necessary for microchemical or optical study to pick out a few grains of a particular mineral. For teaching or reference purposes a permanent mount of a few grains of a particular mineral may be required. For such needs the grains are sorted out by hand usually with the aid of a binocular microscope in much the same manner in which foraminifera and other microfossils are sorted out for study. A small portion of the sample or heavy mineral concentrate is sprinkled thinly and evenly over the bottom of a shallow cardboard tray or on a smooth glossy sheet of paper or on a polished brass plate. The individual grains may be picked up by means of a very fine brush (about 00 size red sable) moistened with glycerine or cedar oil. Some workers recommend a needle on the point of which is a bit of viscous balsam. The needle with adhering grain is dipped into a drop of xylol which will dissolve the balsam and release the grain desired. The mineral grain soon will be left dry by the evaporation of the xylol. Mineral grains thus segregated may be temporarily deposited in the hollows of a glazed porcelain tile (such as is used for color-testing in chemical analysis).

Other means have been devised for handling single grains. See Chapter 16, page 389.

SEPARATION ON THE BASIS OF SHAPE

Most minerals are more or less equidimensional, with the length and breadth not over three to four times the width. Fibrous and flat grains tend to stick to a surface. By placing the grains on an inclined piece of blotting paper, the micas and fibrous minerals may be made to remain on the paper, while the rest will roll off. The same result can be achieved by pouring the grains through a funnel with a damp surface. The method is useful, as it separates almost all the micas.

[1] L. I. Blake, *op. cit.*, pp. 1036-1037.

SEPARATION ON BASIS OF SURFACE TENSION

The commercial process of oil flotation may be applied to rock minerals, but good laboratory methods for rock separation are not developed.[1]

SEPARATION ON BASIS OF CHEMICAL PROPERTIES

Chemical methods for the separation of minerals are based on the differing rates at which different minerals are attacked by reagents. A typical example would be the separation of quartz from calcite. Treatment with dilute hydrochloric acid would dissolve the calcite without affecting the quartz. Different reagents attack different minerals. While it would not be within the scope of this volume to describe all the separations possible, some of the more common separations will be given.

Soluble minerals may be dissolved away from insoluble ones. Remove the calcite from an altered rock by means of acetic acid. Dilute HCl is commonly used for effecting a separation of the carbonate cement and carbonate grains of a rock from the non-carbonate fractions.

To remove olivine and other soluble minerals from common feldspars and insoluble minerals, digest the powder in 1:1 HCl in large volume. Olivine is rarely present in sediments, and hence this method is of little importance.

To separate kaolin from quartz and orthoclase in a potters' clay, decompose the kaolin by 1:1 sulphuric acid; boil till it fumes. Dilute, decant the solution, and finish the solution by warming a few minutes with 5 per cent Na_2CO_3.

Several minerals can be separated from others because they dissolve in HF more slowly than most minerals. Even combinations of HF and H_2SO_4 attack them very slowly. The most common minerals of the group are magnetite, hematite, zircon, rutile, corundum, tourmaline, and staurolite. The rarer insoluble minerals include anatase, brookite, perovskite, andalusite, sillimanite, kyanite, cassiterite, topaz, axinite, and spinel.

Cold hydrofluosilicic acid, H_2SiF_6, will in time decompose the silicate minerals but will not attack quartz. Use has been made of this fact to determine the percentage of quartz or "free silica" in industrial dusts.[2]

Acid potassium sulphate ($KHSO_4$) may be used to advantage to eliminate some of the already partly decomposed mineral grains in a sediment. A great many detrital minerals are unaffected and will remain intact.

A brief summary of the reagents that may be used and the minerals affected is given in the table below. For complete chemical analysis see section on "Chemical Analysis."

[1] Anon., Small-scale flotation test: *Mining Mag.*, vol. 28, pp. 122-123, 1923.
[2] Adolph Knopf, The quantitative determination of quartz ("free silica") in dusts: *U. S. Treasury Dept., Public Health Reports*, vol. 48, no. 8, 1933.

TABLE 39

REACTIONS OF COMMON DETRITALS TO VARIOUS SOLVENTS

Mineral	Fused KHSO$_4$	Boiling Conc. HCl	Boiling Conc. HCl and SnCl$_2$	Conc. H$_2$SO$_4$ and KHSO$_4$	Boiling 60% HClO$_4$	Boiling Conc. H$_2$SO$_4$
Actinolite	..	..	..	..	3	2?
Apatite	2	1	1	..	..	..
Augite	..	6	6	..	4	..
Biotite	..	5 or 6	..	..	1	1
Chromite	2	..	..	..	..	..
Enstatite	..	..	6	..	..	6
Epidote	4 or 6	6	6	4	4	6
Garnet, var. almandite	..	6	6	..	2?	..
Garnet, var. grossularite	6	..	..	..	..	..
Glaucophane	..	..	..	..	6	..
Hornblende	3	5	4	..	3	4
Ilmenite	1	..	..	..	..	..
Magnetite	1	4	1	..	2	..
Pyrite	1	6	5	..	1	..
Staurolite	..	6	6	..	6	6
Titanite	5	6	..	..	..	..
Tourmaline, pink	6	6	6	..	6	6
Tourmaline, black	4	6	6	..	4	..
Tremolite	6	..	..	..	..	6
Zoisite	6	6	6	..	..	6
Muscovite	6	6	6	..	6	6

NOTE.—Degrees of attack:
 1. Completely decomposed
 2. Corners rounded, partial solution
 3. No rounding, but marked loss of pleochroism and birefringence
 4. Traces of rounding and/or some loss of pleochroism
 5. Some bleaching
 6. No action

Table prepared from unpublished experiments of George H. Otto.

CAUSES INTRODUCING ERRORS IN SEPARATION

1. Variation in specific gravity in same mineral as a result of gas bubbles or weathering. Grains may be made up of more than one mineral. Such things easily influence magnetic and specific gravity separations.

2. Separates are not always sharp, especially with poor methods. Minerals lying along the border line may go into either group.

3. In specific gravity methods and in magnetic separates the mass action effect causes certain minerals to be trapped in other separates than that in which they belong. This effect can usually be overcome by making a separation at least twice.

4. Very fine material tends to "ball up" in lumps of various size which are composite and thus cannot be separated. Separation can be made down to individual minerals, but such a process is usable only up to a certain point.

SYSTEMATIC SCHEMES OF SEPARATION

Combinations of methods are possible. After a mechanical analysis of a sediment, or a heavy-liquid separation, the fractions may be divided by magnetic and electrostatic methods with excellent results. Systematic schemes of separation and examination have been outlined by Tomlinson,[1] Holmes,[2] and Hatch and Rastall.[3]

[1] C. W. Tomlinson, Method of making mineralogical analysis of sand: *Trans. Am. Inst. Min. Eng.*, vol. 52, pp. 852-861, 1915.
[2] Arthur Holmes, *Petrographic Methods and Calculations* (London, 1923), p. 103.
[3] F. H. Hatch and R. H. Rastall, *The Petrology of the Sedimentary Rocks* (London, 1913), Appendix by T. Crook, "The Systematic Examination of Loose Detrital Sediments," pp. 377-379.

CHAPTER 15

MOUNTING FOR MICROSCOPIC STUDY

INTRODUCTION

If the mineral crop acquired by some separation method is small it may be mounted entire; but if it is too large and a portion only can be used, care must be used in the selection of this portion. Also in the case of the light minerals care must be exercised in the selection of the small portion for mounting, most especially where quantitative data with respect to size, shape, or mineral composition are to be recorded.

SPLITTING

The quartering-down or splitting-down of the sample so that a representative fraction will be had for mounting on a microscope slide may be accomplished in several ways. One is to place four rectangular sheets of smooth paper, 2 x 4 in., together in such a way that each overlaps one half of one other and so that altogether they form a square. The heavy minerals should be carefully poured into the center of this square and flattened out into a circle. The pieces of paper may then be pulled apart, alternate quarters rejected, and alternate quarters combined and the process repeated as many times as necessary to give the desired size of sample.

Time and effort can be reduced by use of some of the miniature sample-splitting devices known. One such device consists of a conical hopper whose very small opening at the tip is centered over the intersection of two knife edges set at right angles to each other. The sample is thus split into four approximately equal quarters. The alternate quarters are rejected and the remainder again passed through this device until a sample of the proper size is obtained.

Another type of splitter is a miniature Jones ore sample splitter designed and built by George H. Otto.[1] (See Figures 174 and 175.) In the Otto "Microsplit" the inclined chutes (adjacent chutes slope in opposite

---

[1] George H. Otto, Comparative tests of several methods of sampling heavy mineral concentrates: *Jour. Sed. Petrology*, vol. 3, pp. 30-39, 1933.

directions and lead alternately to two pans) are built up from a series of sixteen brass plates, each $\frac{1}{16}$ in. thick, with polished inclined edges; the plates are separated by very thin bronze partitions. The parts are pressed together in a strong vise and screws inserted to hold the parts in place. Otto tested this splitter as well as the two methods outlined

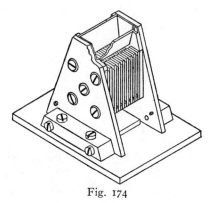

Fig. 175

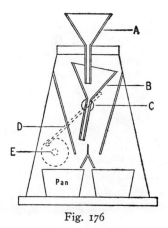

Fig. 176

FIG. 174.—Otto Microsplit.

FIG. 175.—Pan for microsplitter.

FIG. 176.—Appel splitter. A, hopper; B, oscillating funnel which feeds sample over knife edge; C, funnel shaft; D, E, rocker arm and drive shaft.

above, and found that the Microsplit gave the best results for time expended.

Apparently by one unaware of Otto's Microsplit, there appeared recently in the *Engineering and Mining Journal* a description and working drawings of two miniature sample splitters.[1] The larger one is recommended for cutting down samples for assaying and chemical analysis, while the other is designed for quantitative microscopic work. Both are essentially of the same design as that invented by Otto.

Wentworth, Wilgus, and Koch[2] developed a micro-splitter based on a different principle. In their device a circle of small glass vials were mounted upright on a small turntable. A small section of glass tube

---

[1] Anon., Splitting small samples accurately with the Microsplitter: *Eng. and Min. Jour.*, vol. 138, pp. 185-186, 1937.
[2] C. K. Wentworth, W. L. Wilgus and H. L. Koch, A rotary type of sample splitter: *Jour. Sed. Petrology*, vol. 4, pp. 127-138, 1934.

MOUNTING FOR MICROSCOPIC STUDY 359

(which held the sample to be split) was mounted on trundle rollers and so tilted as to allow for a small, steady flow of sand from the lower end of the tube into the glass vials on the rotating turntable. Each vial in turn passed through the sand stream. The turntable was driven by a small crank. A rubber band belt from the crank to the tilted glass tube served also to revolve this part. Tests showed a superiority of a large model of this type of splitter when applied to large samples for mechanical analysis. No marked difference, however, was observed in the results achieved by the rotating micro-splitter and the Otto Microsplit.

Appel[1] developed a still different type of device for sample splitting. It consists of an oscillating funnel, driven by a crank attached to a small induction motor. The funnel is pivoted on a horizontal axis passing through the hopper part. The funnel stem swings alternately on one side and the other of a knife edge and so leads alternately to each of two pans. A variable resistance permits control of the oscillation period. The sand to be split is fed into the hopper of the oscillating funnel at a point below its axis of swing by a second funnel placed above it. (See Figure 176.)

MOUNTING

The residue thus split down may be mounted in some temporary fashion or permanently. For temporary mounts some immersion liquid is used. For study of the light minerals clove oil ($n = 1.53+$) may be used. If a mounting fluid of higher index is desired, bromoform ($n = 1.589$), monobromonaphthalene ($n = 1.658$), or methylene iodide ($n = 1.74+$), or any other liquid desired may be used (see table, page 381). The mineral grains after examination can be washed with a little xylol and dried and reserved for further study.

It is sometimes desirable to change the position of grains mounted in a temporary medium, as for example, in the study of interference figures (page 401). This may be done by the use of very viscous media such as Canada balsam ($n = 1.54$), Peru balsam ($n = 1.59$), bakelite varnish ($n = 1.63$), or AFS mounting medium ($n = 1,685$).[2] Mounting the mineral grains with some coarser glass fragments will prevent the cover glass from holding the grains close to the slide. When it is desired to turn a fragment under the microscope, the motion of the balsam is usually sufficient when the cover glass is touched with a pencil. The

---

[1] J. E. Appel, Unpublished research at University of Chicago laboratory of sedimentology.

[2] Esper S. Larsen and Harry Berman, The microscopic determination of the nonopaque minerals: *U. S. Geol. Survey, Bull. 848*, p. 24, 1934.

glass also helps prevent fibrous and platy minerals from falling down on the slide.

Permanent mounts may be made in media of low index such as Canada balsam ($n=1.54$) or kollolith, a German synthetic resin ($n=1.535$), or in media of high index such as piperine ($n=1.68$).

Workers in recent years have turned to the synthetic resins for permanent mounting media of high refractive index. Alexander,[1] for example, calls attention to several high-index resins ranging from 1.60 to 1.65 and points out the advantages of these materials over Canada balsam. Cameron[2] describes the properties of hyrax, another synthetic product with index of 1.7135. Hyrax serves as a mountant for thin-sections of rocks composed largely of quartz and feldspar, since the relief of these minerals is thus markedly increased and the details of structure and textures are accentuated for study and for photographic purposes. Like piperine, it is a better mounting media for the heavy minerals, since it divides these into two more or less equal groups. Even for light minerals, a better contrast between similar species is sometimes obtained, as between quartz and orthoclase, for example. Keller[3] has also described a synthetic resin for mounting purposes. He points out that such a medium facilitates a distinction between mullite and corundum.

If the minerals have been previously separated into lights and heavies by means of bromoform, it will be found that most of the light minerals have low index and most of the heavy minerals have a high index of refraction. Therefore a medium such as Canada balsam, which divides the range of indices of the light minerals into two more or less equal parts, and a medium such as piperine, which divides the range of indices of the heavy minerals likewise into two parts, seem most useful. It is of course possible to use one medium for all minerals regardless of index.

Several procedures have been given for using balsam.[4]

In order to insure an even distribution, the grains may be put in a drop of distilled water on a clean slide and the water allowed to evaporate; or the slide may be touched with the wet cork from a bottle of

---

[1] A. E. Alexander, Recent developments in high index resins: *Am. Mineralogist*, vol. 19, p. 385, 1934.

[2] Eugene N. Cameron, Notes on the synthetic resin hyrax: *Am. Mineralogist*, vol. 19, pp. 375-383, 1934.

[3] W. D. Keller, A mounting medium of 1.66 index of refraction: *Am. Mineralogist*, vol. 19, p. 384, 1934.

[4] Fanny Carter Edson, Criteria for the recognition of heavy minerals occurring in the Mid-Continent field: *Okla. Geol. Survey Bull. 31*, p. 7, 1925.

R. D. Reed, Some methods for heavy mineral investigations: *Econ. Geology*, vol. 19, pp. 320-337, 1924.

MOUNTING FOR MICROSCOPIC STUDY

solution of gum tragacanth and the grains sprinkled evenly thereon. In either case after evaporation is complete the grains will be firmly attached to the slide. They are then covered with balsam and heated slowly on a hot plate (stove lid on tripod heated by small Bunsen flame) until the gum has melted. Care should be taken to heat slowly to prevent formation of bubbles. Should the latter form, they will rise to the surface, where they may be removed with a toothpick or similar tool. After the balsam has cooled long enough (so that a droplet on a pencil-point is brittle enough to crush) a clean cover glass is placed on the slide, with care not to trap air bubbles. The slide is then removed from the hot plate, and after it has cooled a little the cover glass is gently pressed down. Excess balsam appearing at the edges is removed after it is hard by means of a knife and then by xylene.

Piperine ($n = 1.68$) has been used by only a few workers in sedimentary petrography. It is an alkaloid obtainable in crystalline form and is more convenient for mounting purposes than balsam gum. The mineral grains to be mounted are mixed with piperine and placed on a glass slide which is placed on a hot plate. After the piperine has melted (m.p. = 129.5° C.) a cover glass is added and the mount is complete. The piperine will, however, crystallize after a few days and it is necessary to heat for a moment in order to clarify the slide for reëxamination. Martens[1] says that this tendency may be overcome by heating the slides in an oven at a temperature of 180° C. for one hour. This causes some darkening around the edges of the slide. Excess piperine around the edge of the cover glass may be removed with alcohol. Such removal should be done before the final heat treatment, since the presence of the solvent will start crystallization. As pointed out by Martens, piperine has an advantage over Canada balsam in that it is easier to prepare a mount, since piperine has an index which divides the heavy minerals, whereas the index of most heavy minerals is above that of Canada balsam; and since piperine has high dispersion, which produces a color fringe around the margins of mineral grains of about the same index, which in turn enables the observer to compare indices of grain and piperine without use of the Becke test.[2]

Materials larger than ½ mm. in diameter are mounted with difficulty, since the grains hold the cover glass up. It is possible to use a vulcanite ring, cemented to the slide by balsam, in which the large grains may be placed and on which a circular cover glass is fastened. Since the largest grains are likely to be the light minerals, it is with these that most difficulty is encountered. The space within the ring and between the cover glass and slide may be left empty or filled with balsam. Another method is to use a microscope slide in which a shallow bowl-shaped hollow has been ground on one surface.

---

[1] James H. C. Martens, Piperine as an immersion medium in sedimentary petrography: *Am. Mineralogist*, vol. 17, pp. 198-199, 1932.

[2] Piperine suitable for petrographic work must be crystallized and free of impurities. It is obtainable from Eastman Kodak Company, Rochester, New York.

For large mineral grains and microfossils several mounting slips have been devised.[1] One form is made from a celluloid strip, of thickness greater than grain or fossil, punched at regular intervals with an ordinary paper punch, which is *cemented* to a glass slide. Each hole thus becomes a cell into which a grain or fossil may be placed. A thin cover glass is then cemented on to the entire strip. A suite of minerals or

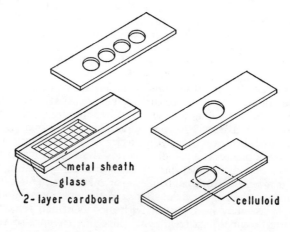

FIG. 177.—Cardboard slides for mounting loose grains or microfossils.

fossils may thus be mounted together. These mounts are usually "air mounts" rather than Canada balsam mounts.

Difficulty was encountered in cementing celluloid to glass. The cement used (celluloid dissolved in acetone) eventually dried and cracked. Possibly this difficulty could be corrected by cementing the punched celluloid strip to a celluloid slide.

For examination under the binocular microscope with dark background of a few loose grains or fossils a similar mounting slip may be made by punching in like manner a cardboard strip glued to a piece of thin black cardboard of the same size. Such slides and similar ones with cover glass are available on the market.[2] The grains or fossils are mounted loose or are glued to the back of the cell in which they are placed. (See Figure 177.)

[1] E. T. Thomas, An aid to the study of foraminifera: *Bull. Am. Assoc. Petrol. Geol.*, vol. 9, pp. 667-669, 1925.
[2] Obtainable from R. P. Cargille, 26 Courtlandt Street, New York, N. Y., or W. H. Curtin and Company, Houston, Texas.

PREPARATION OF THIN SECTIONS

Very detailed instructions for the preparation of thin sections have been published so many times that it seems unnecessary to repeat the details here except in so far as the special problems presented by the sedimentary rocks—especially unconsolidated and friable materials—are involved. Moreover, many individuals no longer prepare their own sections. The making of a good thin section requires considerable skill and time. Labor is usually saved by having such sections as are needed made by experienced technicians.[1]

If, however, thin-sectioning must be done in one's own laboratory, the following condensed procedure is given. Work is started with chips about 1 in. square, as thin as is consistent with the strength, and preferably across rather than parallel to the bedding. These are ground smooth by one of various methods; e.g., on a rotating lead or soft iron lap, with emery or carborundum and water; or by the use of carborundum plates. Field outfits are simply carborundum bricks. The smoothing process moves in succession to finer and finer abrasives. Polishing is done on wet glass alone. The smooth surface thus prepared is cemented to a glass slide with Canada balsam. This is done by pressing the dry, smooth surface into properly cooked Canada balsam on the slide. (The cooking is done on a hot plate and continued until a small globule of balsam removed on a pin-point is no longer sticky but hardens at once.)

On the cemented rock chip there is now ground a second smooth surface parallel to the first. The grinding is carried out as before, but as the chip gets thinner the grinding is done by hand on a glass plate, using only a fine abrasive. Grinding is continued until the slice is wholly transparent and of standard thickness (0.03 mm.). The thickness may be checked optically by noting the order of interference color exhibited by the quartz (see page 393). When proper thickness is attained, a cover glass is cemented over the section. Some balsam, previously cooked as described above, is poured over the rock slice. A cover glass of suitable size is then placed on top of the balsam. Suitable heating will soften the balsam so that the cover glass may be pressed down, at the same time expelling any air bubbles present.

Various precautions and details of preparation are given by Johannsen[2] and others,[3] though they are best learned by experience. Section-

[1] Reed and Mergner, 5519 Nevada Avenue, N.W., Washington, D. C.; W. Harold Tomlinson, 114 Yale Avenue, Swarthmore, Pennsylvania; Gregory, Bottley & Co., 30 Church St., Chelsea, London, S.W. 3, England. F. Krantz, Dr. F. Krantz, Rheinisches Mineralien-Kontor, Bonn am Rhine, Germany.

[2] Albert Johannsen, *Manual of Petrographic Methods*, 2nd ed. (New York, 1918), pp. 572-604.

[3] R. E. Head, The technique of preparing thin sections of rock: *Utah Eng. Exper. Sta., Tech. Paper No. 8*, 1929. Mary G. Keyes, Making thin sections of rocks: *Am. Jour. Sci.* (5), vol. 10, pp. 538-550, 1925. A. Allen Weymouth, Simple methods for making thin sections: *Econ. Geology*, vol. 23, pp. 323-330, 1928. Henry B. Milner and Gerald M. Part, *Methods in Practical Petrology* (Cambridge, England, 1916), pp. 1-14.

cutting machines save much time otherwise required for grinding. Detailed description and methods of use are given in the references referred to.

With friable materials and wholly unconsolidated sediments, some preliminary treatment to indurate the specimen is necessary. This may be done by boiling the chip for five to ten minutes in balsam to which a little shellac has been added.[1] Johannsen has cited and briefly described more than a dozen similar procedures used by earlier workers in preparing thin sections of friable materials. These older methods have certain objectionable features which are overcome by the use of bakelite varnish. The use of this material was first described by Ross[2] and later by Leggette.[3] Bakelite is a synthetic phenol resin product which hardens on heat treatment and has an index of 1.60-1.64. A thin chip or slice is covered with diluted bakelite[4] (B.V. 1305 bakelite varnish diluted with methyl alcohol and a few drops of acetone) and placed in a desiccator which is then evacuated. After a few hours (or longer if necessary) of soaking, the rock chips are placed in an oven and kept at a temperature of 70° C. for 2 to 6 hr. Later the temperature is raised to 100° C. or, if there are no hydrous minerals present, to 200° C., for several days at the lower temperature or for but a few hours at the higher figure. If the specimen is then completely and thoroughly impregnated, it may be ground down with water and abrasive in the usual manner. If colloids are present, Leggette recommends benzene in place of water.

Several writers have found synthetic resins and other products to be of use in the hardening of materials for sectioning purposes. The reader is referred to these papers for details.[5]

[1] Henry B. Milner and Gerald M. Part, *op. cit.*, p. 7. Albert Johannsen, *op. cit.*, pp. 599-602.
[2] Clarence S. Ross, A method of preparing thin sections of friable rocks: *Am. Jour. Sci.* (5), vol. 7, pp. 483-485, 1924.
[3] Max Leggette, The preparation of thin sections of friable rocks: *Jour. Geology*, vol. 36, pp. 549-557, 1928.
[4] Obtainable from the Bakelite Corporation, Chicago, Ill.
[5] Clarence S. Ross, Methods of preparation of sedimentary materials for study: *Econ. Geology*, vol. 21, pp. 454-468, 1926. (Describes use of kollolith.) R. J. Schaffer and P. Hirst, The preparation of thin sections of friable and weathered materials by impregnation with synthetic resin: *Proc. Geol. Assoc.*, vol. 41, pp. 32-43, 1930. W. Ahrens and H. Weyland, Die Herstellung von Dünnschliffen aus locherem Material für petrographische Untersuchungen: *Centralbl. f. Min., Geol., u. Paläon.*, Abt. A., pp. 370-375, 1928. (Ketone resin method.) Eugene N. Cameron, *op. cit.*, 1934. (Synthetic resins.)

NITROCELLULOSE FILM METHOD OF STUDY

A film method for studying textures may sometimes be substituted for the thin-section. This method for studying textures is achieved by coating a smooth or etched surface of a specimen with a nitrocellulose solution and stripping off, or peeling, the subsequently hardened film. The paleobotanists apparently first used these peels to study the structures of fossil plants. The advantages of closely spaced serial sections for the reconstruction of the structure of the fossil is evident. Sedimentary iron ores of the Clinton type and certain manganese ores and phosphate rocks may be examined more easily by this method than by thin-sections. The films have the advantage over thin sections in that they are more easily and rapidly prepared; they are of such thickness that they lie wholly within the focal depth of the microscope lenses and thus yield clear photomicrographs without hazy outlines; closely spaced serial sections are possible, permitting the construction of a three-dimensional concept of rock structures, fossils, etc. But inasmuch as only the textural relations are recorded in the film, it is still necessary to use other methods for mineral identification.

Appel[1] has described the technique of preparing these films for the study of rocks and ores. The specimen to be studied is ground to a plane surface and smoothed with 600F carborundum on a glass plate. Sometimes, when there is insufficient relief on the surface of the specimen, etching is used to bring out the structure. Weak (5 per cent) hydrochloric acid may be used on calcareous rocks, while strong hydrofluoric acid is used for siliceous materials. The calcareous rocks are treated for a few seconds while the hydrofluoric acid treatment takes 1 to 30 min. The surface is washed with water and then coated with nitrocellulose. This solution is prepared by dissolving 20 g. of 20-sec. gun-cotton in 200 c.c. of tech. butyl acetate to which 1 c.c. of a plasticizer such as dimethyl phthalate, castor oil, or tri-cresyl phosphate has been added. The specimen must be perfectly level before the solution is poured on. After several days of drying at room temperature the film is loosened around the edges and stripped with a pair of tweezers. It may then be mounted on a glass slide by means of an uncooked solution of balsam which hardens on evaporation. In some cases a dry mount is more satisfactory.

[1] J. E. Appel, A film method for studying textures: *Econ. Geology,* vol. 28, pp. 383-388, 1933.

CHAPTER 16

OPTICAL METHODS OF IDENTIFICATION OF MINERALS

INTRODUCTION

THE authors are aware that most readers of this volume will probably be somewhat familiar with the polarizing microscope and optical theory involved, but in order to make the book more or less self-contained, brief consideration is given to the manipulative methods. For information beyond that here given, the reader is referred to the standard texts.[1] Moreover, most books on optical crystallography and many courses in petrography are written or taught with a view to the study of minerals in thin sections. Such differences in method as arise from study of mineral grains, rather than thin slices, are therefore worthy of discussion. The writers have found from experience that individuals quite able to identify common rock-making minerals at sight in thin section had to relearn the same minerals in grains.

THE POLARIZING MICROSCOPE

The microscope is a most valuable tool for research on the sedimentary materials. Its use is threefold: (1) to measure and study size, shape, and surface texture of the component grains, (2) to study textures and structures as seen in thin sections, and (3) to identify the minerals by optical means. It is to the third use that this section

[1] Albert Johannsen, *Manual of Petrographic Methods,* 2nd ed. (New York, 1918). Joseph P. Iddings, *Rock Minerals,* 2nd ed. (New York, 1911). A. N. Winchell, *Elements of Optical Mineralogy* (New York): *Part I. Principles and Methods,* 5th ed. (1937); *Part. II. Descriptions of Minerals,* 3rd ed. (1933); *Part III. Determinative Tables,* 2nd ed. (1929). N. H. Hartshorne and A. Stuart, *Crystals and the Polarising Microscope* (London, 1934). W. H. Fry, Petrographic methods for soil laboratories: *U. S. Dept. Agric., Tech. Bull. 344.* 1933. E. S. Dana and W. E. Ford, *A Text-Book of Mineralogy* (with an extended treatise on crystallography and physical mineralogy), 3rd ed. (New York, 1922). F. Rinne and M. Berek, *Anleitung zu optischen Untersuchungen mit dem Polarisationsmikroskop* (Leipzig, 1934). F. E. Wright, The methods of petrographic-microscopic research, their relative accuracy and range of application: *Carnegie Inst. Wash. Pub. 158,* Washington, D. C., 1911. E. M. Chamot and C. W. Mason, *Handbook of Chemical Microscopy,* Vol. I (New York, 1931).

OPTICAL METHODS OF IDENTIFICATION

of this book is mainly devoted. Minerals may be distinguished in a number of ways, but for the non-opaque mineral grains the optical method is the most acceptable.

The polarizing or petrographic microscope is a compound microscope and differs from the ordinary compound instrument in that it is equipped

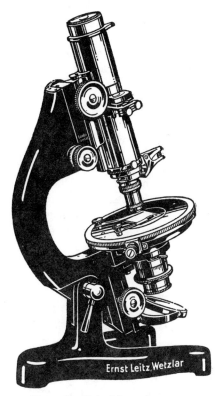

FIG. 178.—Polarizing microscope.

with two prisms which produce plane polarized light. Excepting for this feature it is in principle the same also as the ordinary microscope (Figure 178).

Mechanical components. The microscope consists of certain mechanical arrangements for mounting and adjusting the essential optical system.[1] The mechanical components can be grouped under three principal

---

[1] For full description of the microscope, both mechanical parts and optical system, see one or several of the larger works on microscopy: E. M. Chamot and C. W. Mason, *Handbook of Chemical Microscopy*, Vol. I, *Principles and Use of Microscopes and Accessories, etc.*, 1st ed. (New York, 1931). Albert Johannsen, *Manual of Petrographic Methods*, 2nd ed. (New York, 1918).

classes: (1) the body of the stand with rack-and-pinion coarse motion and micrometer slow motion, together with the tube and objective clutch or nosepiece; (2) the rotating object stage; (3) the lower section composed of the foot and hinge fitting together with the motion devices of the condenser and the iris diaphragm.

The body of the stand carries the tube in which are mounted the optical parts of the microscope proper—i.e., the objectives and eyepieces. The objectives are usually changed either by means of a revolving nosepiece or by means of an objective changing clutch into which the centering collar of the lens fits.

The tube consists of two parts. The upper one or drawtube, which carries the eyepiece, is in some instruments capable of sliding within the lower one to which the objective is attached so that the over-all length of the tube may be varied and the distance between the ocular and objective lenses altered.[1] The displacement of the drawtube is indicated on a scale at the upper end of the tube.

The tube as a whole can be moved up and down by rack and pinion. This is the coarse focusing adjustment and is operated by turning the large milled heads (Figure 178). The fine focusing adjustment is effected by a smaller milled head a little lower down. This screw has vernier attachments to allow reading of complete turns and fractions thereof which may be converted to vertical movements expressed in decimals of millimeters.[2] The fine adjustment is thus a measuring device of importance. With higher magnifications it alone should be used for focusing.

In the tube should be slots and other arrangements for the reception of an auxiliary Bertrand lens, a nicol prism (analyzer), and other optical accessories.

The arm which supports the tube and focusing parts is connected by a joint, which permits tilting of the upper parts of the instrument, to the pillar and base, the latter usually of modified horse-shoe form.

The object stage should be of a horizontal rotating type with periphery graduated in degrees with appropriate verniers for measuring the angle of rotation and with screw clamps for arresting movement. To the revolving type stage may be attached a mechanical stage, which by means of two thumb-screws permits movement of the object slide in two directions at right angles to one another (see Figure 179). Appropriate

---

[1] A variable adjustment is required only when the cover glass deviates greatly from the standard thickness of 0.17 mm. Many microscopes do not have an adjustable drawtube.

[2] The fine adjustment if properly made is without slack.

OPTICAL METHODS OF IDENTIFICATION 369

scales and verniers indicate the amount of each component of movement.

Other special stages used in petrographic work include the integrating stage, with the aid of which planimetric analysis may be carried out conveniently and with precision (Figure 244, page 467). The apparatus has six independent measuring spindles by means of which the proportional amounts of six different constituents may be summed up in one operation. For descrip-

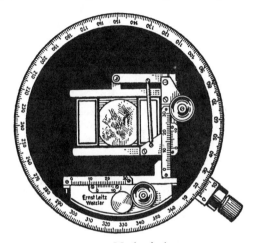

FIG. 179.—Mechanical stage.

tion of same and instructions for use see F. E. Thackwell, Quantitative microscopic methods with an integrating stage applied to geological and metallurgical problems: *Econ. Geol.*, vol. 28, pp. 178-182, 1933.

The universal stage has special uses in research on the optical constants of minerals and on petrofabrics or particle orientation. This stage is so made that the slide or thin section to be examined is placed between two glass hemispheres separated from it by a film of oil. The stage is so constructed that the slide may be tilted into any plane and rotated in that plane. The orientation is read from graduated circles and arcs on the apparatus. These vary from two to five in number.[1]

The substage mechanism is moved vertically by a rack-and-pinion mechanism. In the substage mechanism is mounted the lower nicol prism (polarizer), the condenser lens, and an iris diaphragm. A mirror arm and fork, allowing rotation and turning of the mirror to any desired position is also required.

Optical components. The optical components consist of the mirror,

[1] For fuller description and use see R. C. Emmons, A modified universal stage: *Am. Mineralogist*, vol. 14, pp. 441-461, 1929, and W. Nikitin, *Die Federow-methode* (Berlin, 1937).

the lower nicol prism or polarizer, the condenser lens, the objective lens, the upper nicol or analyzer, and the eyepiece or ocular lens. Certain optical accessories are also used.

The lowest part of the substage equipment is the mirror. This is plane on one side and concave on the other and is so pivoted as to be free to rotate so that either side may be used. The plane mirror is used with lower magnifications, while the concave mirror is used for higher powers which require greater intensity of illumination. When the supplementary condenser lens is used, however, the plane mirror is preferred. Omitting for the moment the analyzer prism, the condenser lens forms the balance of the substage illuminating equipment. This lens, as its name implies, is a lens for condensing and thereby increasing the intensity of the light which passes through it. A cone-shaped bundle of rays converge at the focal point of the lens. The condenser lens (and lower nicol) may be racked up and down to bring the lens into focus, so that the object is most brightly illuminated. The iris diaphragm, located just below the condenser lens, is an almost indispensable item for microscopic work. A two-piece supplementary condenser lens, which can be put in or out of position by a quarter turn of a small attached stud, is used for strongly convergent light as in the examination of interference figures.

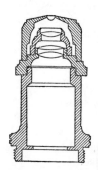

Fig. 180.—Section of objective lens.

The objective is the lens or lens system which is attached to the lower end of the microscope tube (see Figure 180). Several interchangeable objective lenses should be available so that various magnifications can be achieved with the same ocular lens. For ordinary work three objectives are all that are required.

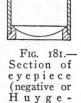

Fig. 181.—Section of eyepiece (negative or Huygenian type).

Objectives are usually corrected to some extent for spherical and chromatic aberration. Achromatic objectives are corrected for aberration of one color and for primary spherical aberration. Semi-apochromatic objectives are spherically corrected for two, while apochromatic lenses are corrected for three colors. Measurements with polarized light should be made only with achromatic objectives. Apochromatic lenses cause some noticeable and undesirable polarization effects. Apochromatic lenses, on the other hand, are best suited for high magnifications and for photomicrographic work with ordinary light.

OPTICAL METHODS OF IDENTIFICATION 371

Some objectives are immersed in a liquid, most usually cedarwood oil ($n = 1.52$) or water ($n = 1.333$). The focal length and the aperture of such immersion objectives are thereby changed.

The ocular which fits the upper end of the tube has for its purpose the magnification of the image formed by the objective. It consists most usually of but two planoconvex lenses. The upper lens of the ocular pair is the eye lens, and the lower one is the field lens (see Figure 181).

It is advantageous to have an adjustable eye lens so that cross-hairs may be brought into sharp focus. Some oculars are equipped with a micrometer scale or with a cross-line micrometer, which presents a field ruled in squares. These eyepieces may be calibrated with a stage micrometer—a glass slide on which is a 1-mm. scale divided into tenths and hundredths.

A valuable though not indispensable addition to the microscope as described is a Bertrand lens. This lens is inserted when needed, just above the analyzer. It, together with the ocular lens systems, forms a "secondary microscope" the purpose of which is to magnify the interference figure formed in the tube just above the objective lens during conoscopic investigations. The Bertrand lens should be focusable and should have an iris diaphragm and be adjustable for centering purposes also.

If no Bertrand lens is present, the figure may be viewed unmagnified by simply lifting out the ocular and looking down the tube. In such cases a small metal cap with a single pin-hole in the center greatly aids in isolation of the figure from small grains. The cap fits on the upper end of the microscope tube.

The nicol prisms for the production of polarized light are most essential. The upper nicol or analyzer and the lower nicol or polarizer are alike in construction. Each is made from a block of calcite, variety Iceland spar, which is so cut and recemented that one of the two polarized rays produced by the double refraction of this substance strikes the cementing balsam layer at an angle greater than the critical angle and is totally reflected, whereas the other ray is transmitted. The lower nicol is usually built into the substage illuminating equipment and is rarely removed. The upper nicol, however, is inserted and removed at will in a slot in the barrel of the instrument just above the objective lens.

The polarizing prisms are called "nicols" even when they are not strictly Nicol prisms. Several types of prisms have been devised since Nicol made his prism in 1828. The Glan-Thompson and the Ahrens prisms are now commonly used.

There is now available a new product, "Polaroid," mounted in plates which may be used instead of a polarizing prism.[1] Such relatively inexpensive accessory plates may thus be added to a common biological microscope and make it serviceable for some petrographic work. One of the plates is mounted below the condenser and serves as polarizer while the other is placed in "crossed" position above the ocular lens. This upper plate is the analyzer and functions much as a "cap-nicol" but since it permits a wide field of view it is without the drawback of a cap nicol. It is necessary, however, for optical work that the microscope to be converted have a revolving stage.

Various accessories are necessary for optical work. The quartz wedge—a mounted wedge of quartz cut from a single crystal—and the gypsum (selenite) plate are indispensable. A mica plate is also desirable. These are inserted into the slot just above the objective lens.

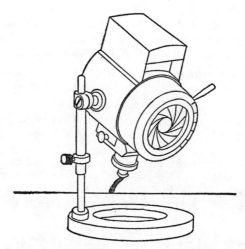

Fig. 182.—Microscope lamp with diaphragm.

A camera lucida or drawing device and a stage micrometer are two valuable accessories. A suitable lamp with "daylight" glass for illumination is also required (Figure 182).

MEASUREMENT OF SMALL PARTICLES

The microscope is a most useful tool for the measurement of small particles. A complete analysis of particle size is even possible by use of the instrument (see Chapters 5 and 6). By means of the microscope and camera lucida the grains observed may be drawn and roundness and

[1] Polaroid microaccessories are obtainable from Bausch & Lomb Optical Co., Rochester, N. Y., in suitable mountings for microscopic use.

OPTICAL METHODS OF IDENTIFICATION 373

sphericity measurements made. It is important to know how the microscope may be used as a measuring tool.

Horizontal linear measurements are sometimes made with a micrometer stage (Figure 179). This is a mechanical stage in which two thumbscrews are capable of movement of the stage in two directions at right angles to one another. Two scales and attached verniers record the amount of movement in each direction. The mechanical stage is calibrated by means of a stage micrometer (glass slide on which are ruled the divisions of a millimeter). The stage is properly aligned and the distances actually passed through as shown by the cross-hairs and the stage micrometer are noted. This calibration should extend throughout the entire length of both screw movements. The length or diameter of any small object may then, after calibration, be measured by placing one end of the object on the appropriate cross-hair and reading the stage vernier and then reading the vernier again after a turn of the thumb screw has brought the other end of the grain to the same cross-hair.

More usually the grain lengths may be measured by means of the eyepiece micrometer (see Figure 72). The eyepiece scale may also be calibrated by using the stage micrometer. All that is necessary is to compare the eyepiece scale with the stage micrometer scale by looking through the microscope. Such calibration is of course good only for a particular combination of lenses. It is only necessary to know how many microns or what fraction of a millimeter one division of the ocular scale represents. The size of a grain is then determined by multiplying the number of divisions intercepted on the ocular scale by the value of each division.

In place of a micrometer eyepiece with scale, an ocular fitted with a Whipple disk or net-grating, which shows the field divided into squares, may be used. These are calibrated in the same manner as the eyepiece with the linear scale.

To measure thickness of a grain, crystal, or other object, the micrometer screw or "slow-motion adjustment" is used. By focusing on a scratch or speck of dust on the upper surface of a glass slide and reading the setting of the micrometer screw, then raising the tube by means of the fine adjustment until the upper surface of the object resting on the slide is in sharp focus and again reading the setting, then noting the difference between the two readings, one can obtain a measure of the thickness of the object. It is necessary, of course, to know what vertical distance is represented by each division on the micrometer screw. This information is usually supplied by the manufacturer, or it can be computed by successive readings on an object of known thickness—measured by micrometer calipers or other means. Since the fine adjustment is often not uniform throughout its range, it may be necessary to utilize only the middle portion of the range which has been previously checked for accuracy. The fine adjustment may moreover have some "lost motion." To overcome this difficulty, the focus should always be obtained by upward motion. The method outlined applies only to measurements in air. If the grain is in some immersion liquid, the apparent thickness will be less than the true thickness. The difference between the true and apparent thickness is a function of the index of the liquid. In fact, this apparent decrease in thickness may itself be applied to the measurement of the index of the liquid itself when the latter is placed in a cell of known thickness (see page 385).

The measurement of areas under the microscope is accomplished by use of the net-ruled eyepiece. The unit squares enclosed within the image boundaries are a measure of the area. Areas may also be measured by use of a camera lucida. The grain image is drawn on ruled graph paper and the area enclosed by the image is computed as for the net-ruled eyepiece, or the area enclosed within the grain outlines is measured with a planimeter, or the images are simply cut out and weighed on a chemical balance. In all cases calibration is first necessary so that the true areas may finally be computed.

To a considerable extent a microprojector of inexpensive sort may be utilized in linear and areal measurements. The image of the grains is projected on paper on the desk or table top or on a drawing board and the linear dimensions are directly measured with a centimeter rule or the areas are measured with a planimeter. The magnification is first determined by projecting the image of a stage micrometer scale. A stand which will permit easy adjustment for different degrees of enlargement is of much aid with the microprojector.

FUNDAMENTAL OPTICAL CONSTANTS

The fundamental optical constants of minerals are the indices of refraction, the crystallographic orientation of the directions of vibration that correspond to the indices of refraction, and the amount of absorption of light vibrating in these directions. All the other properties commonly tabulated—birefringence, optic sign, optic axial angle, dispersion, and extinction angle—are but derived constants which can be computed

OPTICAL METHODS OF IDENTIFICATION 375

from the fundamental constants above. In practice, however, it is often advisable or more convenient to measure them directly.

The student of sedimentary minerals should be skilled in the use of the microscope in the study and measurement of the optical constants, both fundamental and derived, so that he may rapidly and confidently record the characteristics of the minerals or use the optical data on record for the purpose of identifying an unknown mineral.

In the section which follows, the authors briefly review the methods of determination of the indices of refraction, the determination of the vibration axes and their relations to the crystallographic axes and planes of reference, and the determination of color and pleochroism. In the section on convergent light and its use, methods of measuring and recording some of the derived constants are given. A short preliminary section on crystal form, cleavage, and inclusions is also given.

OBSERVATIONS IN ORDINARY LIGHT

Crystal Form and Grain Shape

The *form* or shape of the detrital minerals is a characteristic useful for identification purposes. For this reason the appearance of the common detrital minerals under the microscope is figured (Chapter 17).

The shape of a grain depends on its crystal habit, cleavage, and manner of fracture as well as its resistance to abrasion along the several crystal axes. The mineral as formed in the parent rock may or may not have a well-developed or automorphic crystal habit. Many minerals, such as quartz, are without good initial form, whereas others, such as apatite and zircon, have exceptionally well-developed crystal forms. Such crystal forms are of course modified by abrasion and breakages during transportation, though with the very resistant minerals, such as zircon, such modifications are slight, especially where the time involved in transportation is not great. Many sediments, therefore, show detrital grains with automorphic habit. Even long-enduring transportation, which results in wearing off of the sharp interfacial angles and even of the crystal facets, does not wholly obliterate the original elongation of the grain. The end-product is often of elliptical or elongate rod-like form with rounded ends. The same sediment, of course, might contain two varieties of the same mineral—an automorphic zircon and a well-rounded zircon, for example—which may be significant in tracing the sediment to its source or for correlation purposes.

Detrital minerals, even though well worn, may undergo secondary enlargement. Such secondary enlargement, due to deposition of the same mineral in optical and crystallographic continuity with the original grain, will result in reconstruction of the automorphic form of the mineral. Crystal faces and sharp angles appear. Quartz is especially subject to such enlargement, as are also calcite, feldspar, and even tourmaline.[1] (See Figures 230, 232, pages 443, 445.)

Cleavage, Parting and Fracture

Cleavage is a property of crystalline materials expressed as easy breakage along certain directions. These directions are parallel to simple crystal faces. Cleavage is properly described, therefore, in crystallographic terms or less strictly in terms of the number of directions and the degree of perfection. Such a description summarizes kind and quality. A cleavage may be cubic, for example, if it is parallel to the cube faces. A mineral with such cleavage would show three directions of cleavage mutually at right angles to each other. On the other hand, a mineral with basal cleavage (like mica), that is, cleavage parallel to the basal pinacoid, would show but one direction of cleavage. The table below summarizes the most common cleavages, their terminology, and their characteristics.

The shape assumed by a mineral controlled by cleavage depends on whether a cleavage exists alone or is found in combination with other cleavages. A perfect basal cleavage alone yields plates or sheets, as does any pinacoid cleavage. Two pinacoidal cleavages, both perfect, would yield elongate lath-shaped fragments, whereas three pinacoidal cleavages in combination, all equally good, yield parallelopipeds. In the orthorhombic system such parallelopipeds would be right-angled. In general a prismatic cleavage alone gives long, lath-shaped pieces, but such cleavage plus a good basal cleavage might give squarish or stubby block-like fragments.

Since a grain has the tendency to lie on its largest developed face, it follows that cleavage fragments tend to lie with the best-developed cleavage faces parallel to the mounting slide. Some difficulty therefore arises in determining the number of cleavages present. It is not apparent to a beginner, for example, that a mineral like mica with only one very

---

[1] So many examples of secondary enlargement of quartz are known that no reference need be given. Secondary enlargement of feldspar is not rare. See, for example, Samuel S. Goldich, Authigenic feldspar in sandstones of southeastern Minnesota: *Jour. Sed. Petrology*, vol. 4, pp. 89-95, 1934. Even the secondary enlargement of tourmaline has been reported in several places. For example see Stella West Alty, Some properties of authigenic tourmaline from Lower Devonian sediments: *Am. Mineralogist*, vol. 18, pp. 351-355, 1933.

OPTICAL METHODS OF IDENTIFICATION

perfect cleavage has any cleavage at all. Under the microscope a mica grain shows usually an irregular or ragged outline (Figure 227, page 438). A mineral with two good cleavages, such as hornblende, appears to have but one cleavage, since the grain is marked with straight parallel fractures running the length of the fragment (Figure 205, page 415). Generally, then, the cleavages of a mineral grain as examined microscopically are equal to the number of apparent cleavages plus one.

TABLE 40

COMMON CLEAVAGES

Cleavage	System	Symbol	No. of Directions	Remarks
Cubic	Isometric	100	3	At right angles
Octahedral	Isometric	111	4	Not at right angles
Dodecahedral	Isometric	110	6	Not at right angles
Rhombohedral	Hexagonal	10$\bar{1}$1	3	Yields rhombic blocks
Pinacoidal	Tetragonal	001	1	"Basal" cleavage
	Hexagonal	0001	1	
	Orthorhombic, Monoclinic, Triclinic	001, 010, 100	1 each	Cleavage may be parallel to any one or two or all three pinacoids so that 1, 2, or 3 directions of cleavage appear. All mutually at right angles only in orthorhombic system
Prismatic	Tetragonal	110 or 100 and 010	2	At right angles
	Hexagonal	10$\bar{1}$0	3	Not at right angles
	Orthorhombic, Monoclinic, Triclinic	110	2	Not at right angles (011) and (101). (domes or "horizontal prisms" same in appearance as (110) cleavage)

SEDIMENTARY PETROGRAPHY

The existence of but a single cleavage is best detected by examination of the grain under crossed nicols. The grain, if anisotropic, will exhibit a uniform interference color (because of uniform thickness), or if the color be not uniform, it will change abruptly along a fracture (due to step-like changes in thickness) rather than grade imperceptibly from point to point on the grain surface.

As in the case of degree of automorphism, so the degree of perfection of cleavage varies greatly. Some minerals, such as kyanite, always show cleavage faces and owe their shape to cleavage, while others, such as staurolite or tourmaline, exhibit no traces of cleavage. General descriptive terms such as *perfect, imperfect, distinct,* etc., are employed to record such differences.

Parting is a phenomenon similar to cleavage which marks certain specimens of certain minerals. It is microscopically indistinguishable from cleavage.

Minerals without cleavage or at best with but an indistinct cleavage sometimes show a characteristic *fracture*. Staurolite, for example, is often marked by a good conchoidal fracture, as is also garnet (Figures 221 and 237, pages 432, 451). The extremities of some grains, notably those of hypersthene, diopside, and enstatite, as well as those of some amphiboles, are marked by a ragged saw-toothed appearance (dentate) due to fracture and possibly to solution.

INCLUSIONS AND OTHER VARIETAL CHARACTERS

Many crystals contain *inclusions*. These may be solid, liquid, or gaseous. They may or may not be systematically arranged. Quartz, for example, often shows liquid and gas inclusions arranged in definite planes. In other cases quartz is filled with rutile needles in random orientation.

Mackie, in 1896, in a remarkably fine quantitative study even by modern standards, paid especial attention to the inclusions of quartz. He divided the quartz in the sands studied into four groups, namely (1) with automorphic mineral inclusions, (2) with acicular (rutile?) inclusions, (3) with liquid and gas inclusions, and (4) without any inclusions. Mackie showed that the quartz of the possible source rocks of the sands in question differed materially in the nature of the inclusions found in the quartz of these rocks and he was, therefore, not only able to trace the sands to their sources but to express quantitatively the proportions contributed by each source type.[1] In 1919 Gilligan[2] followed Mackie's scheme of study and was able to reach similar con-

---

[1] Wm. Mackie, The sands and sandstones of Eastern Moray: *Trans. Edinburgh Geol. Soc.,* vol. 7, pp. 148-172.
[2] Albert Gilligan, The petrography of the Millstone Grit of Yorkshire: *Quart. Jour. Geol. Soc. London,* vol. 75, pp. 260-262, 1920.

clusions with regard to the Millstone Grit. More recently Tyler[1] made an intensive study of the inclusions in the quartz of the St. Peter sandstone.

Many minerals are essentially free of inclusions, but others, such as zircon, are often characterized by abundance of inclusions.

One of the very convenient applications of the Becke line is in the study of microscopically small inclusions. When the inclusion has a higher index, as in the case of apatite inclusions in quartz, and the tube of the microscope is raised, the Becke line moves toward the inside of the inclusion, the whole center of which becomes bright. On the other hand, when the inclusion has a lower index than the grain, the inclusion becomes dark upon raising the tube. Liquid and gas inclusions both have an index much less than that of the containing mineral, so that their relief is *high*—so much higher for gases that gas inclusions appear as entirely black dots. When an inclusion of high relief has an index lower than that of quartz or feldspar, it is likely to be liquid or gaseous. When an inclusion consists of both liquid and gas, the gas tends to take the form of a spherical bubble, and the combination can be recognized without any tests of its indices. Liquid and gas inclusions, moreover, are generally irregular in form, though they are often arranged in lines and planes across the mineral.

Some minerals, particularly zircon, exhibit a *zoned* structure (Figure 242, page 455). The zoning is often marked by zones of inclusions, but not always.

Brammall,[2] for example, has described and figured the zoning of zircon. He found the origin of this zoning difficult to determine, but thought it was due to "laminations" and in some cases due to regular arrangement of dust-like suspensions, probably ilmenite.

A varietal feature occasionally noted is *striation* on the crystal faces. This is a notable feature of rutile, for example. Parallel or longitudinal cleavage cracks should not be mistaken for striations.

Determination of the Refractive Index by Immersion Methods

Method. If a colorless isotropic substance such as glass be immersed in a liquid of the same refractive index, the substance will become entirely invisible.[3] Inasmuch as the two have the same index, no reflection or refraction can take place at the boundary between the two sub-

---

[1] Stanley A. Tyler, The St. Peter sandstone in Wisconsin: *Jour. Sed. Petrology*, vol. 6, pp. 72-77, 1936.
[2] A. Brammall, Dartmoor Detritals; A study in provenance: *Proc. Geol. Assoc.*, vol. 39, pp. 28-48, 1928.
[3] It is assumed that the solid is colorless or of the same color as the liquid and that it is free of inclusions of any kind. If these conditions are not fulfilled, the object will remain visible.

stances. Use is made of this phenomenon in the determination of the refraction of a substance by means of the microscope. The material to be studied is powdered, some few grains are placed in a drop of liquid of known index on a glass slide, a cover glass is added, and the preparation is observed under the microscope. Successively different liquids are used until the border effects disappear or are reduced to a minimum. The advantage of the method is that it requires very little of the material to be studied, that it can be done with the microscope without expensive accessory equipment, and that it is both accurate and rapid.

Preparation of immersion liquids. The ideal immersion liquid is colorless, chemically stable, and of low volatility. The liquid must not dissolve nor react with the material to be studied. The liquids are generally made up in sets ranging from about 1.45 to 1.75 with intervals of about 0.01 in refractive index making in all some thirty liquids. For special purposes media of index lower or higher than those afforded by such a set are required. If possible, liquids miscible in one another should be used so that a mixture of a drop each of adjacent liquids may be used for greater accuracy.

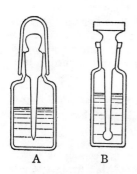

FIG. 183.—Bottles for immersion liquids. Bottle A is better type since it has ground glass cover as well as stopper.

The liquids are kept in small bottles of about 15- to 20-c.c. capacity which are fitted with ground glass stoppers and dropping rods and ground-glass caps (Figure 183).[1] Since each cap is ground to fit a particular bottle, each cap as well as bottle should be labeled, and care should be taken not to interchange caps. The bottles are best placed in a wooden block in which suitable holes have been drilled and then placed in a case which is both light- and dust-proof.[2]

Certain liquids, notably methylene iodide, darken on exposure to light. Some refractive index sets available are without covers, and the bottles of liquids sensitive to light are covered with tin-foil. The writers much prefer a storage box with cover, since this protects against dust as well as light.

The index of an immersion liquid varies slightly with the temperature. Liquids should therefore be standardized and used at the same tempera-

[1] Such bottles are obtainable for about 45¢ each from A. H. Thomas and Co., West Washington Square, Philadelphia, Pa.
[2] Such a case is described and figured by Larsen. See Microscopic determination of the non-opaque minerals: *U. S. Geol. Survey, Bull. 679,* p. 16, 1921.

OPTICAL METHODS OF IDENTIFICATION

ture. Unless the temperature variation is extreme its effect is small and may be neglected unless very accurate results are required. If the liquids are mixtures rather than pure substances, the index may change with age. This is especially true if the two substances differ greatly in boiling point. For this reason they should always be kept tightly stoppered and the index checked periodically, about every six months.

A wide choice of liquids is available. The particular liquid chosen should meet as nearly as possible the requirements of the ideal liquid as stated above. A list of possible liquids and their properties is given in Table 41.

TABLE 41

REFRACTIVE INDICES OF VARIOUS IMMERSION LIQUIDS

Liquid	$t°C.$	N_D	*Liquid*	$t°C.$	N_D
Water	29	1.333	Canada balsam	...	1.54+
Ether	24.8	1.350	Clove oil	...	1.533–
Acetone	20	1.359			1.544
Ethyl alcohol	20.5	1.361	Anise oil	...	1.547
Hexane (*n*)	20	1.375	Nitrobenzene	20	1.552
Heptane (*n*)	17.6	1.388	Dimethylaniline	20	1.559
Amyl alcohol (*n*)	20	1.410	Bromobenzene	20	1.560
Ethylene chloride	20	1.444	Toluidine (*o*)	20	1.570
Chloroform	19	1.446	Aniline	20	1.584
Kerosene	...	1.450	Bromoform	19	1.595
Carbon tetrachloride	...	1.466	Chloroaniline (*m*)	20	1.593
Glycerine	...	1.473	Cassia oil	...	1.586–
Turpentine	...	1.475			1.603
"Nujol"	...	1.475	Cinnamon oil	...	1.605–
Olive oil	...	1.476			1.619
Xylene (*p*)	23.4	1.494	Iodobenzene	17.8	1.621
Toluene	20	1.495	Carbon disulphide	20	1.628
Benzene	20	1.502	Tetrabromoethane	20	1.638
Xylene (*m*)	20	1.498	α bromonaphthalene	20	1.658
Xylene (*o*)	20	1.505	α chloronaphthalene	20	1.633
Sandalwood oil	...	1.507	Phosphorus tribromide	26.6	1.697
Cedarwood oil	...	1.510–1.516	Piperine (m.p. 129°)	...	1.68
			Klein's solution	...	1.70
Chlorobenzene	20	1.523	Thoulet solution	...	1.717
Wintergreen oil	...	1.536	Methylene iodide	15	1.738
Ethylene bromide	20	1.538	Barium mercuric iodide solution	...	1.793
			Sulphur in methylene iodide (sat.)	...	1.778

REFRACTIVE INDICES OF LOW-MELTING SOLIDS FOR INDEX DETERMINATION

Piperine	1.68
Piperine and iodides	1.68–2.10
Sulphur and selenium	1.998 Na–2.716 Li
Selenium and arsenic selenide	2.72–3.17 Li

Iodides prepared by adding 35 g. iodoform, 10 g. sulphur, 31 g. SnI_4, 16 g. AsI_3, and 8 g. SbI_3 to 100 g. methylene iodide. Warm to hasten solution, allow to stand, and filter off undissolved solids.

Of course one can purchase a prepared set of index liquids. They are rather expensive.[1]

F. E. Wright proposed certain liquids which could be mixed in various proportions to obtain a wide range of indices.[2] This is the set made up and used by the authors. The set consists of the following:

Mixtures of	Index
Kerosene and turpentine	1.450–1.475
Turpentine and clove oil	1.480–1.535
Clove oil and α monobromonaphthalene	1.540–1.655
α monobromonaphthalene and methylene iodide	1.660–1.740
Sulphur in methylene iodide	1.740–1.790
Methylene iodide, iodoform, antimony iodide, arsenic sulphide, antimony sulphide and sulphur	1.790–1.960

To determine the approximate index of a mixture of two liquids, the following formula may be used:

$$V_1 N_1 + V_2 N_2 = (V_1 + V_2) N$$

where V_1 and V_2 are the respective volumes of the two liquids and N_1 and N_2 are the indices of the same, while N is the index of the mixture. By use of this formula the proportion of two liquids to produce a mixture of a specific index may also be obtained. However, the index must be verified by some other means, preferably a refractometer.

In order to avoid the change in index due to differential evaporation of such mixtures, R. C. Emmons[3] proposed a set of thirty liquids, most of which are separate compounds. These liquids differ by slight but unequal intervals.

[1] Such standardized liquids for determining refractive index are obtainable from R. P. Cargille, 26 Cortlandt Street, New York, N. Y., and from Ward's Natural Science Establishment, Rochester, N. Y.

[2] F. E. Wright, The methods of petrographic-microscopic research: *Carnegie Inst. Publ. No. 158*, p. 96, 1911.

[3] R. C. Emmons, A set of thirty immersion media: *Am. Mineralogist*, vol. 14, pp. 482-483, 1929.

OPTICAL METHODS OF IDENTIFICATION

TABLE 42

Liquid	B.P. °C.	mm.	$N_D 24°$ C.	Temp. Coeff.	Dark Container
Trimethylene chloride ..	119.5	735	1.446	.00045	
Cineol	53-54	8 (Eastman)	1.456	.00041	
Hexahydrophenal	161	740	1.466	.00044	
Decahydronaphthalene ..	88.5-90.5	38	1.477	.00040	
Isomylphthalate	(Tech.)	Eastman	1.496	.00038	
Tetrachloroethane	143-144.4	740	1.492	.00051	
Pentachloroethane	158-160	740	1.501	.00048	
Trimethylene bromide ..	165-167	Eastman	1.513	.00048	
Chlorobenzene	130.5-130.8	740	1.523	.00053	
Ethylene bromide and chlorobenzene	......	...	1.533	.00054	
o-Nitrotoluene	220-220.4	740	1.544	.00053	
Xylidine	114-118	27	1.557	.00050	X
o-Toluidine	87-88.5	17	1.570	.00047	X
Aniline	66.8-67.2	4.5	1.584	.00045	X
Bromoform	147.5	736	1.595	.00056	
Iodobenzene and bromobenzene	......	...	1.603	.00054	X
Iodobenzene and bromobenzene	......	...	1.613	.00054	X
Quinolin	117-121	16	1.622	.00049	X
α-Chloronaphthalene	128-132	20	1.633	.00044	
α-Bromonaphthalene and α-chloronaphthalene ..	......	...	1.640	.00044	
α-Bromonaphthalene and α-chloronaphthalene ..	......	...	1.650	.00044	
α-Bromonaphthalene and α-iodonaphthalene	......	...	1.660	.00045	X
Ditto	......	...	1.670	.00044	X
Ditto	......	...	1.680	.00044	X
Ditto	......	...	1.690	.00044	X
Methylene iodide and iodobenzene	......	...	1.700	.00060	X
Ditto	......	...	1.710	.00063	X
Ditto	......	...	1.720	.00066	X
Ditto	......	...	1.730	.00068	X
Methylene iodide	80	15	1.738	.00070	X

NOTE.—All of the liquids can be obtained from the Eastman Kodak Co., Rochester, N. Y., or in sets ready for use from Dr. C. W. Muehlberger, Service Memorial Institute, Madison, Wisconsin, or Ward's Natural Science Establishment, Rochester, N. Y.

For minerals less than 1.45 in index, Harrington and Buerger[1] recommend the lower petroleum distillates. They prepared a series from 1.35 to 1.46.

A considerable number of minerals have indices greater than 1.738 (methylene iodide). The embedding media for such minerals are solid at ordinary temperatures. Low-melting solids for the higher index values include piperine in which are dissolved the tri-iodides of arsenic and antimony[2] and melts composed of sulphur and selenium[3] and mixtures of selenium and arsenic selenide and the halogen compounds of thallium.[4] The preparation and properties of such low-melting solid mixtures are summarized by Larsen. To use these materials a little of the embedding medium is melted on a glass slide, a little of the powder to be examined is dusted into the melt, and a cover glass is added.

Standardization of index liquids. Whether the liquids are prepared by oneself or purchased, it is necessary to check their indices from time to time. This may be done in several ways.

For liquids of medium index, a refractometer is most satisfactory.[5] The refractometer should have a range from about 1.450 to 1.840, rather higher than the average instrument has. For some work it is desirable to have refractometer prisms the temperature of which may be controlled by circulating water.

A one-circle goniometer may be used in place of a refractometer for measurement of index of refraction. A goniometer has the advantage over the refractometer that its range of usefulness is not limited by the index of any glass prisms. To use the goniometer, a hollow glass prism is needed, or, if a prepared prism is not available, one can be made from a pair of square cover glasses. The latter should be exactly plane.[6] The cover glasses may be used together along one edge or, more simply, mounted in stiff wax on the goniometer plate in the same way in which a crystal is mounted for measurement of interfacial angles. An angle of about 30° is best suited for the method of vertical incidence or for refraction liquids of a high index with

---

[1] V. F. Harrington and M. J. Buerger, Immersion liquids of low refraction: *Am. Mineralogist,* vol. 16, pp. 45-54, 1931.

[2] H. E. Merwin, Media of high refraction for refractive index determinations with the microscope: *Jour. Wash. Acad. Sci.,* vol. 3, pp. 35-40, 1913.

[3] H. E. Merwin and E. S. Larsen, Mixtures of amorphous sulfur and selenium as immersion media for the determination of high refractive indices with the microscope: *Am. Jour. Sci.* (4), vol. 34, pp. 42-47, 1912.

[4] Tom Barth, Some immersion melts of high refraction: *Am. Mineralogist,* vol. 14, pp. 358-361, 1929.

[5] Such an instrument is the Abbe-Spencer refractometer with water-jacketed prism for temperature control and with rack-and-pinion movement. Graduations permit direct reading of the index to the third decimal.

[6] They may be tested by reflection for imperfections. A distant window cord or some similar straight object is observed by reflection. If it is distorted, the cover glass is faulty. About one cover glass in three or four is satisfactory.

the method of minimum deviation. A prism angle of about 60° is best suited for ordinary liquids using the method of minimum deviation (Figure 184).

Adjust the prism, or pair of cover glasses, in the same manner as with a crystal, so that both plates are vertical and so that the intersection of the two is nearly centered. Measure the angle between them. Insert a drop of liquid so that capillarity will hold it at the intersection of the plates. Measure the deviation caused by the prism of liquid, either when the deviation is at a minimum, or when there is vertical incidence. From the two measured angles the index may be calculated by the formulas:

For vertical incidence, $n = \dfrac{\sin(\alpha + \delta)}{\sin \alpha}$

For minimum deviation, $n = \dfrac{\sin \frac{1}{2}(\alpha + \delta)}{\sin \frac{1}{2}\alpha}$

where α is the angle between the prism faces and δ is the deviation.[1]

A number of methods have been devised whereby the refractive index of a liquid may be measured under the microscope. Several of these methods depend on the displacement of image. When an object is viewed through a medium the surface of which is perpendicular to the line of vision, the image observed will appear to lie in a plane above that of the object. The amount of displacement will depend on the thickness and refractive index of the interposed layer of material. To apply this principle to liquids an object glass with shallow cell of known thickness is placed on the microscope stage.[2] The microscope is carefully focused on the bottom of the empty cell. The cell is then filled with the liquid to be measured and the microscope again focused on the bottom of the cell. The number of divisions on the

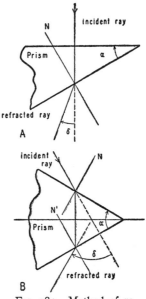

FIG. 184.—Method of refractive index (of liquid) determination by use of hollow prism. A, vertical incidence; B, minimum deviation.

micrometer screw that are required to bring the instrument into sharp focus once more are carefully noted. The index of the liquid is true thickness/apparent thickness. The apparent thickness is the true thickness minus the displacement observed (and measured by means of the micrometer adjustment). This method of index measurement is only approximate.[3] A similar method in which a cell of unknown thickness is employed may also be used. If the displacement of the image by equal thicknesses of liquids of different

---

[1] Complete instructions are given in Dana's *Textbook of Mineralogy*, 3rd ed., pp. 216-219 (1922). See also same reference, pp. 155-157, for instruction in use of the goniometer.

[2] Cells of the type required are obtainable from Zeiss, or they may be constructed by cementing a perforated brass disk to the object slide.

[3] For further details, see Chamot and Mason, *Handbook of Chemical Microscopy*, Vol. I (New York, 1933), pp. 380-383.

refractive indices is known, a curve may be drawn which shows displacement plotted against index of refraction. The index of an unknown liquid is then readily determined. This method is the more reliable of the two.[1]

Other possible methods for measurement of the index of refraction of a liquid under the microscope have been devised. These, however, require special cells not likely to be readily available, and therefore further details are not given here.[2]

A simple though more approximate method which utilizes the Becke effect is to compare the index of a solid of known index with that of the liquid of which the index is desired. To avoid complications isotropic solids are used. Chamot and Mason[3] give a table of thirty-five isotropic crystals ranging from 1.326 to 2.25 in index that may be used. Some of these are relatively rare salts and difficult to obtain, while others, like the alums, may, owing to isomorphic mixing, not have the theoretical index given. Rutherford[4] has described the use of powdered glass of known refractive index. The glass is obtained from lens and optical manufacturers, is screened to 100-150, 150-200, and 200-250 mesh size, and is kept in a series of bottles properly labeled as to index. In the determination of the index of a mineral in a liquid mixture with a volatile component, a bit of powdered glass may be added to check the index of the liquid after or during index determination of the mineral. The isotropic character of the glass will serve to distinguish the mineral and the glass. If the mineral is isotropic, the screened size of the glass may be used to differentiate the two.

Becke method. The method as detailed here is applicable to substances which are isotropic, that is, which do not affect polarized light or which remain dark upon rotation of the stage under crossed nicols. These substances include glass and minerals of the isometric system. Optic axis sections of anisotropic materials may also be studied as here outlined.

The mineral grains are broken in a small agate mortar by gently tapping or crushing (not grinding). A very little of the powder[5] is dusted into a drop of index liquid, and a cover glass[6] is added. The preparation

[1] See Chamot and Mason, *op. cit.*, pp. 383-384.
[2] F. E. Wright, The measurement of the refractive index of a drop of liquid: *Jour. Wash. Acad. Sci.*, vol. 4, pp. 269-279, 1914. E. Clerici, Sulla determinazione dell'indice di rifrazione al microscopio: *Atti. Rend. R. Accad. Lincei. Roma*, vol. 16, pp. 336-343, 1907.
[3] Chamot and Mason. *op. cit.*, p. 387.
[4] Ralph L. Rutherford, A convenient method for checking the index of a liquid: *Am. Mineralogist*, vol. 9, pp. 207-8, 1924.
[5] Some workers prefer to screen material through a 100-mesh screen and to mount the powder retained on a 150-mesh screen. This eliminates troublesome coarse fragments and fine dust too small for easy study.
[6] Cover glasses for this work may be made by splitting the usual square ⅞-in. glasses into four small squares. These may be discarded after each test. Small cover glasses may also be purchased.

OPTICAL METHODS OF IDENTIFICATION 387

is placed on the microscope stage. A medium- or high-powered objective is used. Choose a nearly vertical contact between the mineral and the immersion oil; such contacts are narrow lines that do not swing sideways during focusing. Focus carefully. Reduce the light, by means of the substage diaphragm, so that it is not dazzlingly bright. With the fine-adjustment micrometer screw alternately raise and lower the objective from the position of sharp focus. Look for a bright line (*Becke line*) parallel to the contact of the mineral and oil moving back and forth across the contact as the lenses are moved. The rule is: When the microscope tube is raised, the Becke line moves toward the mineral when the latter has an index higher than the immersion liquid, and conversely. Similar tests with other refraction liquids are made until the index of the mineral is found to be between two adjacent liquids in the set. The index is then expressed as the average of the indices of the liquids in the two adjacent bottles plus or minus one half the difference between them. For example:

$$n > 1.608$$
$$n < 1.620$$
$$n = 1.614 \pm .006$$

Closer determinations of the index may be made by mixing the two liquids. Place one drop of each liquid close together on a slide. Mix with edge of cover glass. Index of mixture is average of the two liquids.

Certain observations will shorten the labor involved and save time. The prominence of the borders of the grain, indicated by the broad dark boundaries and bold outlines, is a measure of the difference in index between immersion liquid and mineral grain. The grains which, therefore, appear to "stand up" prominently are said to have *high relief*. As the indices of the grain and oil approach one another, the grain becomes less conspicuous, until it disappears altogether. When the grain is faintly visible, it is said to have *low relief*. Inspection of the first mount made will tell the observer whether the immersion liquid chosen is close to or far from the index of the mineral. Such observations will govern the selection of other test liquids.

Often when the grain has a low relief there appears about the boundary a fringe of color. The boundary will appear bluish on one side and reddish or orange on the other. This is a dispersion phenomenon and is due to the fact that when white light is used for illumination it is broken up at the grain boundary.

This is known as the *Christiansen effect*. It is noticeable when the material immersed in the liquid has a refractive index the same as the liquid for certain colors or wave lengths of light. The light for which the refractive indices are identical will be transmitted, and the other colors will be deviated by refraction. As a consequence the particle appears faintly colored by transmitted light. This is most especially observed in permanent mounts in piperine. Minerals slightly higher in index appear bluish, while those slightly lower appear yellowish or orange. This accounts for the anomalous bluish tint of some colorless garnets in piperine mounts.

The true value of the index is then determined either by noting carefully which way the yellow portion of the color fringe moves (the values usually recorded in the tables are for yellow light—"D line") or by using monochromatic light of known wave length.

Very often the grain boundaries are not perpendicular. The grain instead is rather lenticular in shape. In this event, the grain behaves as a small biconvex lens: when the index of the mineral is higher than that of the liquid, the light is brought to a focus just above the grain, as with a biconvex lens in air; but when the mineral has the lower index, the

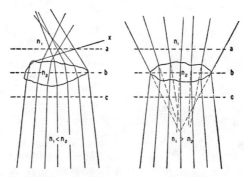

Fig. 185.—Behavior of very small grains. n_1, index of liquid; n_2, index of grain. a, above focus; b, sharp focus; c, below focus.

light rays are refracted and bent away from the grain (see Figure 185). This phenomenon serves to determine the index of very small grains. Those with index higher than the liquid become centrally illuminated and, if extremely small, appear as pin-points of light, whereas those lower in index become dark and appear, if very small, as black dots.

Method of inclined illumination. In this method a low-power objective lens is used. After the grains are in focus, half of the field is darkened or shaded by holding a card or the hand over the mirror. The grains that lie near the edge of the shadow—which should be out of focus—usually show a bright line along one border, and a dark line on the opposite border. If the substance and the immersion liquid are of about the same index, the bright and dark borders are replaced by orange and blue borders. To determine whether the material has a higher or lower index than the oil when the dark border of the grain is on the dark side of the field, check the lens system used, by noting which side of the grain is light when the test is made on some substance in water, i.e., when the index of the material is known to be greater than that of the liquid.

Single and double variation methods. As pointed out above, the index of refraction of solids is not the same for all wave lengths of light. The numerical difference between the values for violet and for red light constitutes a measure of this dispersion. Normally (and in the tables of this volume) the index is given only for yellow (D line) light. The index of the immersion liquids is also measurably different for the different wave lengths of light.

Temperature effects a change in the index of both liquids and solids, though the change per degree for liquids is appreciable while the change for solids is negligible.

Advantage is taken of the phenomena of dispersion and change in index with change of temperature to provide a more exact method of measuring the refractive indices of crystals. Measurements are carried out on the stage of an ordinary petrographic microscope. The material to be studied is mounted in the usual way and the Becke effect observed. The wave length of the light or the temperature of the immersion liquid is then altered until the Becke line disappears, indicating an exact rather than an approximate match of the index of the immersion fluid and that of the solid. The single-variation method of Merwin[1] involves the change of the wave length only by means of a monochromator, while that of Emmons[2] involves a change of temperature of the liquid by use of a special cell to be placed on the microscope stage. Emmons also developed a combination of these two methods, the double-variation method. Since special accessory equipment—refractometer, etc.—is required and since it is rare in practice that such refined methods are required in sedimentary petrography, the reader is referred to the papers by Emmons for further details. For greater accuracy in the determination of the index and other optical constants the method is unexcelled. The universal stage may be used to advantage in conjunction with the Emmons double-variation apparatus.

Methods of handling single grains. If the material is scanty and but a grain or two are available for study, it may be necessary to transfer the grain from one liquid to another. Several methods for handling single grains have been proposed.[3] These consist largely of the use of special types of tweezers for picking up the grain to be studied. Such methods of handling single grains are also of service when their specific gravity is to be measured.

Ross[4] suggests that when temporary mounts, such as refractive index

---

[1] H. E. Merwin, *Jour. Am. Chem. Soc.*, vol. 44, p. 1970, 1922.

[2] R. C. Emmons, The double dispersion method of mineral determination: *Am. Mineralogist*, vol. 13, pp. 504-515, 1928; The double variation method of refractive index determination: *Am. Mineralogist*, vol. 14, pp. 414-426, 1929.

[3] F. C. Calkins, Transfer of grains from one liquid to another: *Am. Mineralogist*, vol. 19, pp. 143-149, 1934. Francis C. Partridge, Methods of handling and determination of detrital grains and crushed rock fragments: *Am. Mineralogist*, vol. 19, pp. 482-487, 1934. Arthur D. Howard, A simple device for the manipulation of individual detrital grains of minute size: *Jour. Sed. Petrology*, vol. 2, pp. 160-161, 1932.

[4] Clarence S. Ross, Some methods for heavy mineral investigations: *Econ. Geol.*, vol. 19, pp. 320-337, 1924.

liquids, are used it is possible to pick out a particular grain for special study by means of a bit of viscous balsam on a needle-point. The needle is dipped down into the immersion fluid and withdrawn before the balsam has a chance to dissolve. By dipping the mineral grain adhering to the balsam into a drop of xylol on another slide the grain can be immediately set free and will soon be left dry by the evaporation of the xylol.

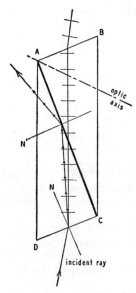

FIG. 186.—Passage of light through nicol prism. ABCD, nicol prism; N, normal to base of prism; N', normal to cementing balsam layer; AC, cementing balsam layer. Note but a single polarized ray is transmitted by the prism. The complementary ray is totally reflected by the balsam layer and turned aside.

OBSERVATION IN PLANE POLARIZED LIGHT

(TRANSMITTED LIGHT AND CROSSED NICOLS)

ISOTROPISM AND ANISOTROPISM

As noted elsewhere, the petrographic microscope differs from the ordinary compound microscope in that it has a pair of "nicol" prisms for the production of polarized light.

Polarized light is produced by the calcite of these prisms. Calcite, like other crystalline materials, except those of the isometric system, has the property of *double refraction*. A single ray of ordinary light entering such a crystal is split in two rays each of which is *plane polarized* with directions of vibration at right angles to each other. In the nicol prism one such complementary ray is turned aside by total reflection at the layer of balsam cementing the two calcite blocks of the prism together (Figure 186).

Since the two nicol prisms, the polarizer and the analyzer, are mounted in the microscope with the planes of vibration of the two at right angles to each other, they are described as "crossed." If both prisms be in the path of the light, that ray which passes the first prism is completely cut out when it reaches the second. Since glass and other amorphous materials and minerals of the isometric system do not themselves either polarize light or rotate a beam of polarized light, it follows that when these *isotropic* materials are placed on the stage of the microscope between crossed nicols the field will remain dark. On the other hand, crystals of the less symmetrical systems

OPTICAL METHODS OF IDENTIFICATION

are doubly refracting and resolve the ray from the polarizer into two complementary plane-polarized rays which are further resolved by the analyzer into four components one pair of which is turned aside by total reflection while the other is transmitted to the eye (Figure 187). The transmitted two rays are polarized in the same plane and because of

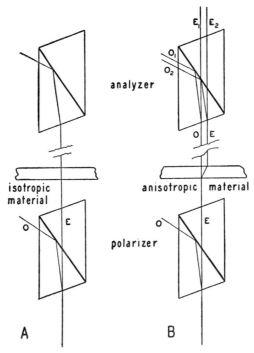

FIG. 187.—Passage of light through the prisms (crossed position). A, with isotropic material on stage; B, with anisotropic material on stage. In A the single transmitted ray from polarizer is cut out by analyzer, while in B the crystal resolves the single ray into two rays which are in turn resolved by the analyzer into four rays, two of which are transmitted to the eye. These two vibrate in the same plane and produce interference effects.

phase differences interfere with one another and produce interference colors. The grain of an *anisotropic* mineral, therefore, appears illuminated against a dark background and as the stage is rotated these grains change from light to dark, from color to black, four times in a rotation. This removal of all light during rotation of anisotropic minerals between crossed nicols is called *extinction*. Extinction occurs when the directions of vibration of the two rays produced by the mineral on the stage of the microscope are brought parallel to the direction of vibration of the polarizer and analyzer prisms of the microscope. (See Figures 188, 189 and 190.)

Anisotropic or doubly refracting substances as noted produce two rays which have two different indices and therefore two different velocities. One may be called the "fast" ray and the other the "slow" ray. It is due to this velocity difference that the rays emerge from the mineral grain out of phase, so that when components of these rays are resolved by the prisms into the same plane, interference results. The amount of re-

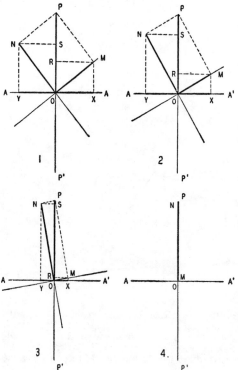

FIG. 188.—Explanation of extinction. Vector OP, representing direction of vibration of light from polarizer, is resolved by mineral on stage into vectors OM and ON which represent direction and intensity of rays transmitted by the mineral. These two rays are resolved by the analyzer into vectors OX and OR and OY and OS respectively. OR and OS are the rays turned aside in the analyzer (see Fig. 187B) while OX and OY are the rays which vibrate in the same plane and produce interference effects.

FIG. 189.—Continuation of Fig. 188. In 3, owing to rotation of the stage, the vectors OX and OY are very small, while in 4, they have vanished altogether. No light, therefore, reaches the eye. The mineral is said to show extinction.

tardation is a vector property of the crystal. Its numerical measure is the difference in the indices of the two rays transmitted.[1] This value is termed *birefringence* (abbreviated B in the tables).

The interference color produced is related to the birefringence. Interference colors are said to belong to a low, medium, or high order and, other things being equal, express a low, medium, or high value of the birefringence.

[1] Actually the "wave index" rather than the "ray index" is involved. These two are not always the same.

OPTICAL METHODS OF IDENTIFICATION 393

Interference colors are the usual spectrum colors. In order of increasing retardation involved these are violet, indigo, blue, green, yellow, orange and red. These may be repeated many times in regular cycles as may be seen by examining the quartz wedge between crossed nicols. The cycles or *orders* are designated *first order, second order,* etc.

The colors of the higher orders, however, merge somewhat and yield pale pinks and greens and in the highest orders produce a "high-order white." The first order is also slightly anomalous in that the first interference colors are gray and white instead of blue and green as in the other orders. The white here seen is "white of the first order."

Not only is the birefringence a property of the specific mineral involved, but it also is a vector property of the crystal itself. In certain directions, for example, the retardation or birefringence is zero; such directions are called *optic axes*. In all other directions the retardation is greater. It is clear, therefore, that different crystallographic orientations result in different colors.

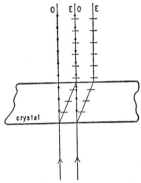

FIG. 190.—Diagram to show how two rays may emerge at the same point and travel the same path. Resolution into the same plane of vibration and retardation of one ray with respect to the other lead to interference.

A basal section or plate of a uniaxial crystal, i.e., a section or plate cut perpendicular to the single optic axis of these minerals, an axis which is coincident with the *c* crystallographic axis, appears dark under crossed nicols and remains dark throughout the rotation of the stage. Such isotropic-appearing minerals of course are readily proved to be anisotropic, since they yield an interference figure if examined in convergent polarized light.

The birefringence is also related to the thickness of the grain examined. A greater thickness gives also a greater retardation of the slow ray behind the fast and different colors as a result.

FIG. 191.—Interference bands on quartz grain.

Mineral grains, either crushed or detrital, also show the relation between thickness and interference color. Many grains are more or less lens-shaped. When such grains are examined between crossed nicols, they exhibit bands of interference color which are roughly concentric (Figure 191). The lowest colors are along the thin edges and the highest colors are found in the thicker middle part of the grain. One can determine how high the color is in the middle part by counting, from the margin towards the center, the red bands —each red band marks one order of color. Grains which show but one interference color throughout do so because they are everywhere of the same thickness. Such an observation is in fact indicative (in the case of crushed fragments) of a cleavage parallel to the slide on which the grain rests.

Colored minerals also yield interference colors, but these may be masked or blended with the natural color of the mineral. If the mineral has very high birefringence it will exhibit the same appearance under crossed nicols as in ordinary light. Rutile, for example, shows the same reddish-brown color between crossed nicols as in ordinary light.

Anomalous polarization and incomplete extinction. The student should guard against anomalous polarization, occasionally exhibited by certain minerals otherwise isotropic. Garnet, for example, may exhibit a weak birefringence. Such anomalous polarization is usually very feeble and irregular. The phenomenon is produced by strain within the crystal. The beginner is also apt to confuse the polarization effects of numerous large inclusions with that of the containing crystal itself. An effect related to anomalous polarization and also produced by internal strain in the crystal is wavy extinction. Dynamic pressure often induces wavy extinction. The strain shadows are the portions at extinction position. As the stage is rotated the shadows move in a "wavy" or irregular fashion over the crystal as the extinction position is reached. The quartz of quartzite, greywackes, and similar sediments is prone to show such extinction irregularities.

Incomplete extinction is also shown by certain minerals of notably high dispersion, especially titanite. As the extinction position is approached the mineral becomes bluish, but fails to extinguish completely. Such behavior is of diagnostic value.

Aggregates of very small polarizing particles are sometimes confusing. Such aggregates (leucoxene, chert, etc.) most often fail to extinguish at any point during the rotation of the stage. In some cases, however, the aggregate will behave approximately as a single crystal and will show partial extinction at certain positions. In the latter case an interference figure may even be obtainable, though as a rule the aggregate fails to yield a figure.

SIGN OF ELONGATION

The *sign of elongation* is a very useful and easily made test on minerals of anisotropic character. Many minerals have a tendency to an elongate habit when they crystallize. Thus apatite and zircon are commonly elongated in the direction of the vertical axis. This habit (acicular, prismatic, etc.) persists even in worn grains, and where pronounced the sign of elongation of the mineral should be determined.

To do so the nicols are crossed, the grain is turned to a position midway between points of extinction, and the quartz wedge is inserted thin end first. If the interference colors of the grain rise, that is, move toward the higher orders, the slow rays of the grain and the wedge coin-

OPTICAL METHODS OF IDENTIFICATION 395

cide; if not, they are opposed (at right angles to one another). If the slow ray is parallel (or approximately so) to the length of the grain, the sign is *positive;* otherwise *negative.*

Some writers prefer to say "length slow" or "length fast" which means in effect that the slow ray is parallel to the length of the grain (positive elongation) or otherwise. Nearly all uniaxial crystals have their elongation parallel to the vertical axis. In these cases the optic sign and the sign of elongation are the same. This is not true of the biaxial minerals. In some few biaxial minerals the sign of elongation becomes ambiguous and may be either positive or negative.

Optic Orientation

The geometrical relation between the crystallographic axes and the principal optical directions or vibration axes is the *optic orientation.* The relations between the two are fairly simple in uniaxial crystals but they are rather complex in the biaxial crystals.

Uniaxial crystals. Uniaxial crystals, those of the tetragonal and hexagonal crystallization, are crystals which have but one optic axis or direction of isotropy. Except in this direction light is doubly refracted and the two transmitted rays, the *ordinary ray* (O) and the *extraordinary ray* (E), are each polarized in planes at right angles to one another. The index of the O-ray is ω while that of the E-ray is ε.[1] When $\varepsilon > \omega$ (or when the E-ray is the slow ray) the crystal is said to be *positive;* when $\varepsilon < \omega$ (E-ray is the fast ray) the crystal is *negative.*

The velocity (or index) of the ordinary ray is constant irrespective of its direction of travel while that of the extraordinary ray is variable and is a vector property of the crystal. The vibration direction of any O-ray always lies in the plane of the *a* crystal axes while the vibration direction of the E-ray always lies in a plane containing the *c* crystallographic axis. The *c* axis is also the optic axis or direction along which light may travel through the crystal without any double refraction.

Any randomly oriented grain of a uniaxial crystal, then, will yield two transmitted rays. One of these will have the index ω, while the other will have an index greater or less, depending on whether the crystal is positive or negative, than ω. The value of the other index (that of the E ray) will equal the limiting value, ε, only when light travels through the crystal at right angles to the optic axis (*c* crystallographic axis). Such crystals will exhibit the maximum birefringence and therefore the highest orders of interference colors. For that reason grains showing highest colors should be chosen in refractive index determination (see page 398).

Uniaxial crystals normally exhibit *parallel* or *symmetrically inclined* extinction. A zircon crystal, for example (Figure 192), shows extinction

[1] Index of the *E-wave* rather than the *ray* is more technically correct.

parallel to the prism faces but has extinction symmetrically inclined to the pyramid faces.

The extinction angle is observed by centering the microscope (page 402), aligning the crystal, crystal face or cleavage direction parallel to the vertical cross-hair, noting the stage reading, and after rotating the crystal on the stage (between crossed nicols) to the extinction position, reading the stage again. The difference in the two readings is the extinction angle or technically the angle between the vibration direction of one of the transmitted rays and the crystal direction chosen.

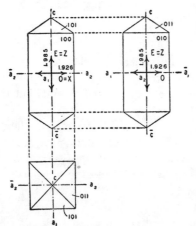

Fig. 192.—Orientation diagram of zircon.

Biaxial crystals. Passing through any point in a biaxial crystal there are three planes at right angles to one another, each of which is characterized by the fact that one of the two rays which can travel in any direction in the plane has a constant refractive index. These planes are defined by three mutually perpendicular axes which are, respectively, the vibration directions of rays having the minimum, maximum, and a particular intermediate refractive index.[1] The axes are designated X, Y, and Z and the corresponding refractive indices of the rays vibrating along these axes are α, β and γ. The plane of X and Z is the *principal optic plane*. Y is the *optic normal*. The two optic axes lie in the optic plane. X and Z bisect the angles between these axes. The bisector of the acute angle is the *acute bisectrix* and may be either X or Z. The bisector of the obtuse angle is the *obtuse bisectrix*. If X is the acute bisectrix, the crystal is *negative*; if not, it is *positive*. The angle between the acute bisectrix and an optic axis is designated V. 2V is twice the angle V and is the angle between the two optic axes or the *optic axial angle*.

Uniaxial crystals are really a special case of a biaxial crystal in which the optic axial angle, 2V, is zero and in which $\beta = \alpha$ or γ. Likewise isoaxial (isometric) crystals represent a special case in which $\alpha = \beta = \gamma$. In biaxial crystals $\gamma > \beta > \alpha$.

Several possible relations between the vibration axes X, Y, and Z and the crystallographic axes *a*, *b*, and *c* are possible. In the orthorhombic

[1] In these particular planes ray index and wave index are the same.

OPTICAL METHODS OF IDENTIFICATION 397

system X, Y, and Z coincide with *a, b,* and *c.* Any one of the vibration axes may coincide with any one of the crystal axes. In the monoclinic system only the *b* axis coincides with a vibration axis. Most often $Y = b$, though *b* may coincide with either X or Z. Normally, however, the optic plane is (010). In the triclinic system no systematic relation exists between the vibration and crystal axes.

Biaxial crystals may exhibit parallel or inclined extinction. The latter may be symmetrical or *oblique.* As a consequence of the relations described above, the crystals of the orthorhombic system show parallel or symmetrically inclined extinction; those of the monoclinic system usually show parallel or oblique extinction while those of the triclinic system have oblique extinction. (See Figures 193 and 194.)

Extinction angles and orientation diagrams. The experimental determination of the relations between the vibration axes and the crystallographic axes involves much labor and requires considerable skill. It is best accomplished with the universal stage. An understanding of the relationships existing, however, is necessary to intelligently interpret the observed relations between cleavage and extinction position, to determine what indices of refraction are likely to be exhibited by certain cleavage fragments or certain crystal orientations, etc. One will find in this book and in others in which the optical data are tabulated abbreviated statements of the relationship between the optical and crystallographic directions.

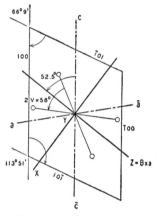

Fig. 193.—Gypsum orientation diagram. Cleavage fragment. An example of the optic orientation of a monoclinic substance.

For example, anhydrite has indices $\alpha = 1.570$, $\beta = 1.575$, $\gamma = 1.614$. It is biaxial, and it is positive (a fact which may be determined from the indices since when β approaches α a crystal is positive, while when β approaches γ, the crystal is negative). The optic plane is parallel to (010), $X = c$, and $Z = a$. $2V = 42°$. Anhydrite is orthorhombic with cleavage perfect parallel to (001) and to (010) and distinct parallel to (100). From this information we would expect the mineral to yield cleavage blocks of rectangular form all of which would exhibit parallel extinction. The more common cleavage flakes would be parallel to (001) and (010). The former would show a well-centered acute bisectrix figure and yield indices of 1.575 and 1.614, while the latter (parallel to 010) would yield only a flash figure but would give indices of 1.570 and 1.614. See Figure 194.

DETERMINATION OF THE INDEX OF REFRACTION OF AN ANISOTROPIC SUBSTANCE

A slightly modified procedure from that given on page 386 is required for the determination of the index of refraction of an anisotropic substance. A sprinkling of grains mounted in an immersion liquid is examined under crossed nicols. Attention is first given to the grains with the highest interference colors. A grain showing high colors is turned to

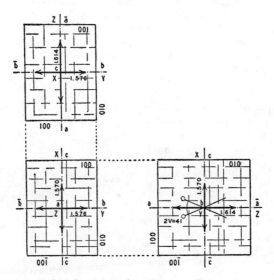

FIG. 194.—Anhydrite orientation diagram. Cleavage fragments.

extinction. The upper nicol is removed and the index of the mineral is then compared with that of the liquid by means of the Becke test. The grain is then rotated 90° and the other index is in like manner compared to that of the immersion liquid. This comparison is repeated on at least a half-dozen grains in each preparation, for with grains in random position, many will be so placed as not to give the extreme or limiting values. Grains with cleavage or markedly tabular or acicular crystal form do not have random orientation, and the final results have therefore to be taken with reservations.

Similar tests are made with the different index liquids until a liquid is formed with an index just a little higher than any obtainable index from the mineral; and another with an index just a trifle lower than any index shown by the mineral grain. The values of the extreme indices (α and γ or ε and ω) are then estimated as for a single index. The difference between the maximum and the minimum index is the birefringence ($\gamma - \alpha$).

OPTICAL METHODS OF IDENTIFICATION 399

The underlying theory of the method of refractive index determination outlined above is not complicated. Each anisotropic mineral is doubly refracting and yields two transmitted rays (except when the light travels along an optic axis). These two rays (and associated waves) differ in velocity and therefore in index and are polarized in planes at right angles to one another. The maximum difference or maximum birefringence is obtained from grains which lie with their principal optic plane—the plane containing the bisectrices and the optic axes—parallel to the stage of the microscope. Such sections exhibit the highest order of interference color. The nicol prisms of the microscope offer the means for studying each transmitted ray separately. The mineral is first turned to the extinction position so that it will yield but a single ray (the polarized ray from the lower nicol passes without further resolution by the mineral grain); with the upper nicol out of the path of light, the index may be studied by the Becke method. A rotation of 90° makes similar study of the complementary ray possible.

The index of β is always intermediate between α and γ. Select a grain with low interference colors. This grain will have both rays with similar indices and equal to β.

Such grains will be so oriented that the observer will be looking down the optic axis. This orientation may be verified by means of an interference figure. Such an orientation if exact will give but a single ray which is the β index. In the case of uniaxial minerals the value obtained is that of the ordinary ray, which will be equal to either the fast or the slow ray. (Such an index is designated ω.)

In practice the determination of all three indices may be readily carried out at the same time so that tests on a half-dozen immersion liquids with as many mounts will be sufficient. The results should be systematically recorded. See example below.

Record of Index of Refraction Tests

1. Test with liquid 1.565
   $\gamma > 1.565$
   $\alpha > 1.565$
   $\beta > 1.565$
2. Test with liquid 1.603
   $\gamma > 1.603$
   $\alpha < 1.603$
   $\beta < 1.603$
3. Test with liquid 1.580
   $\gamma > 1.580$
   $\alpha < 1.580$
   $\beta > 1.580$
4. Test with liquid 1.572
   $\gamma > 1.572$
   $\alpha > 1.572$
   $\beta > 1.572$
   Therefore
   $\alpha = 1.576 \pm .004$
   The other indices are determined and recorded in a like manner.

In all cases, whether the material is prepared by crushing of the specimen or is natural sand, it is advisable to screen to fairly uniform size. This facilitates selection of grains of proper orientation, since the high colors will then be due solely to orientation and not to size.

Color and Pleochroism

Color is a residual effect. That is, a mineral illuminated with white light may selectively absorb certain wave lengths and transmit or reflect others. The subtraction of certain wave lengths from white light leaves a residue which to the eye appears colored. The color perceived is the complement of the color absorbed. The color of a mineral by transmitted and reflected light is usually, though not always, the same.

It is important to note that the intensity of color of a mineral viewed by transmitted light is in part a function of the thickness. Hornblende appears black in thick grains but may be blue-green under the microscope. Many minerals which appear deeply colored in the hand specimen may be essentially colorless under the microscope. For this reason the color given under the description of the mineral is the color seen microscopically.

Certain minerals, when viewed in plane polarized white light, transmit different colors in different directions. Such selective absorption, which is a vector property of the crystal, is called *pleochroism,* and a statement of the nature and intensity of the pleochroism is called the pleochroic formula.

Isometric crystals and amorphous substances exhibit no pleochroism. Crystals of the tetragonal and hexagonal systems are dichroic, while crystals of the less symmetrical systems are trichroic.

The pleochroic effect is observed by examination of the mineral grains or crystals on the microscopic stage. As the stage is rotated (lower nicol only in the path of the light) the mineral may show different colors in different positions. The pleochroism may be one of change from colorless to colored (example, andalusite), or one of change in intensity (example, tourmaline, from light to dark brown), or one of change in quality of color (example, hypersthene, from pink to green). Hence a complete statement of pleochroism involves both intensity and quality. Thus pleochroism may be weak or strong. Pleochroism is described in terms of the colors exhibited and the associated optical directions as well. Just as an anisotropic biaxial substance has three indices of refraction with corresponding vibration directions X, Y, and Z, so also may each vibration direction be associated with a different intensity or quality of color. It is necessary, therefore, to record the color exhibited in each of these principal optical directions.

OPTICAL METHODS OF IDENTIFICATION 401

To record the pleochroism of a mineral, it is necessary to study a number of grains scattered in random fashion in the microscope field. A grain is chosen and turned to extinction, and the color in plane polarized light is recorded. The grain is turned 90° and a second recording is made. Which of the two rays thus recorded is fast and which is slow is also determined by means of the quartz wedge (see page 394). Other grains are examined in like fashion. The several observations are combined, and from the data the true pleochroic formula may be deduced. For example, examination of a slide on which are scattered crushed fragments of glaucophane will yield the necessary information to record the pleochroism of that mineral. A given grain may show X' = colorless, Z' = violet blue. Another grain may show X' = violet blue and Z' = dark blue, while still another may show X' = colorless, Z' = dark blue. From this data it is clear that X = colorless, Y = violet blue, and Z = dark blue.

The opaque mineral should be examined in reflected light. This is done by shading or tilting the mirror to exclude transmitted light. A strong light is then brought near the grains on the stage. The color seen by reflected light is then recorded. "Dark-field" illuminators of a special sort for examination by reflected light are available for the microscope.[1]

OBSERVATIONS IN CONVERGENT LIGHT

Interference Figures

Convergent light is needed for the study of optic axes as elsewhere defined. An optic axis is a direction through an anisotropic crystal along which light may travel and suffer no double refraction. When viewed in this direction, the crystal appears no different from an isotropic crystal. Minerals that have but one such direction of isotropy are *uniaxial*, whereas minerals with two such directions are *biaxial*. Uniaxial minerals include all those which crystallize in the tetragonal and hexagonal systems and include such species as zircon and calcite, while the minerals crystallizing in the orthorhombic, monoclinic, and triclinic systems are biaxial. Staurolite, gypsum, and kyanite are examples.

When birefringent minerals are placed in a cone of strongly convergent light and properly examined, they may be made to yield an "interference figure." From such figures it is possible to distinguish between a uniaxial and a biaxial substance.

[1] See Chamot and Mason, *Handbook of Chemical Microscopy*, Vol. 1 (New York, 1931), pp. 86-94.

To obtain an interference figure with the microscope the accessory condenser lens is added to the optical system.[1] An objective lens for high magnification is attached, and the microscope is centered and focused on a good-sized grain which shows low interference colors. After crossing the nicols the observer either removes the eyepiece and looks down the tube, or inserts the Bertrand lens. In either case, if a properly chosen grain is examined, an interference figure will be seen. To make certain of its nature, the stage is rotated and the effect on the figure noted.

FIG. 195.—Centered uniaxial interference figure.

To center the microscope: Some microscopes are centered by means of two centering screws located on the objective collar, while on others two thumb screws for centering are attached to the stage. The procedure for centering is the same in either case. A speck of dust or other similar small object on a glass slide and in the field of view is placed at the point of intersection of the cross-hairs. The stage is rotated until the dust particle, which describes a circle, is at its greatest distance from the center of the field. Then by means of the centering screws it is brought about half way back to the center of the field. The particle is then again placed at the center (or point of intersection of the cross-hairs) by moving the slide and the whole procedure as given above is repeated. Two or three such adjustments should suffice.

Uniaxial figure. Uniaxial crystals give a right-angled dark cross more or less completely in the field which may or may not be accompanied by concentrically placed color rings that center about the intersection of

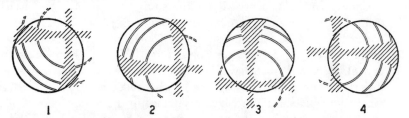

FIG. 196.—Off-center uniaxial interference figure showing successive positions upon rotation of stage.

the two arms of the cross (Figure 195). As the stage is rotated the figure remains wholly unaffected if it is perfectly centered, or, in an off-center figure, the cross as a whole moves, without rotation, in such a way that the center of the cross describes a small circle (Figure 196). The arms

[1] This lens is usually placed in line by a quarter-turn of the stud below the stage and above the condenser lens proper.

OPTICAL METHODS OF IDENTIFICATION 403

or bars of the cross do not pivot or swing but remain unchanged in direction.

Biaxial figure. Biaxial minerals give a figure in which two dark hyperbolas or *isogyres* are conspicuous. Upon rotation of the stage each of these hyperbolas pivots about a center (the points of emergence of the

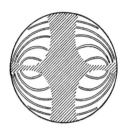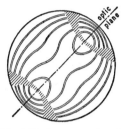

FIG. 197.—Acute bisectrix (biaxial) figure, centered.

optic axes). At certain positions they are straight and are joined to form a cross not unlike that formed by uniaxial minerals. The colored rings, if present, are not circular, but are symmetrically curved about the two centers. (See Figure 197.) Some grains are so centered as to yield an *optic axis figure* instead of the *bisectrix figure* described. In this case one of the two points of emergence of the optic axes is nearly in the center of the field and but a single isogyre is visible (Figure 198). As the stage is rotated this dark bar pivots much as does a compass needle (hence the term "compass needle figure") and usually becomes alternately straight and curved. In such a figure the colored rings, if present, enclose the optic axis but are not quite circular in plan.

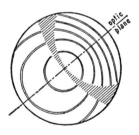

FIG. 198.—Optic axis figure.

In the case of the well-centered figures described, the student will have little difficulty in distinguishing between uniaxial and biaxial materials. Markedly off-center figures are more difficult to diagnose, and it is sometimes impossible, even for an experienced worker, to make certain of the optical nature of the mineral. The beginner will do well to study figures obtained from a thin sheet of biotite (pseudo-uniaxial) and muscovite (biaxial), which will be well centered, before attempting analysis of off-centered and less complete figures.

If the grains are abundant and possess no cleavage which would result in unfavorable orientation, examination of a half-dozen grains of promising appearance, i.e., with low interference colors, should give a usable figure.

Good figures are not always obtainable. This is due to unfavorable orientation due to prominent cleavage, to interference by twinning, to interference of alteration products or inclusions, to smallness of the grain, and to interference of other nearby grains in the field. Effect from other grains can be reduced or eliminated by use of the iris diaphragm of the Bertrand lens or by use of a metal peep-sight (page 371).

Determination of the Optic Sign

Uniaxial minerals. Obtain a uniaxial interference figure, preferably well centered. Insert the gypsum plate in the slot above the objective lens. The cross will turn red with alternately yellow and blue quadrants; the yellow and blue color is located at the apex of each quadrant. (See Figure 199.) If the yellow quadrants are on opposite sides of the arrow on the test plate (indicating the slow ray of that plate), the mineral is

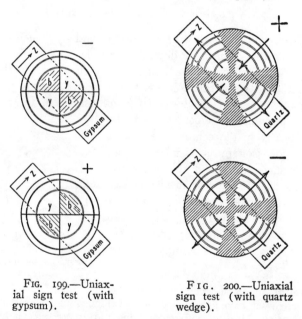

Fig. 199.—Uniaxial sign test (with gypsum).

Fig. 200.—Uniaxial sign test (with quartz wedge).

optically positive (+), otherwise it is negative (−). If the cross is a little outside of the field, a determination may often be made by first rotating the stage so that the observer, with a little experience, can tell which quadrant is observed at a given moment. When the proper relation of the quadrant is understood and the color exhibited known, the optic sign may be inferred with certainty.

Sometimes the sign may be more conveniently determined with the quartz wedge—especially so when owing to high birefringence the colored rings are very numerous. The wedge is inserted, thin end first. The

OPTICAL METHODS OF IDENTIFICATION   405

colored rings appear to migrate or shift as the wedge is pushed in. The rings appear to expand or move in two opposite quadrants and to contract or move in the alternate pair of quadrants (Figure 200). If they move out of the quadrants on opposite sides of the arrow (slow ray) on the wedge, the mineral is positive (+); otherwise it is negative (−).

Biaxial minerals. The optic sign of a biaxial mineral is determined in a somewhat similar way. If the acute bisectrix figure has many colored rings, the quartz wedge may be used. Rotate the stage until a line connecting the image of the axes (melatopes) is in the 45° position. Push

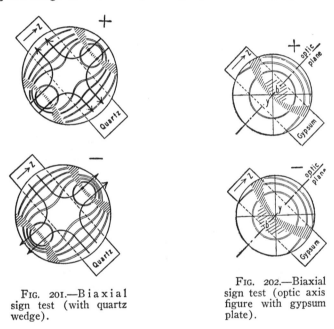

Fig. 201.—Biaxial sign test (with quartz wedge).

Fig. 202.—Biaxial sign test (optic axis figure with gypsum plate).

in the quartz wedge, thin end first, with its slow ray (arrow) *parallel* to the trace of the axial plane (or line connecting the melatopes). If the colored rings contract from the area between the images of the axes, moving from the acute bisectrix to the axes, the mineral is negative (−); otherwise it is positive (+) (Figure 201).

If a figure contains few or no colored rings, but has the points of emergence of the optic axes in the field, the stage should be rotated until the line connecting the melatopes is "horizontal" or "east-west" and a biaxial type of cross is formed. Insert a gypsum plate. The cross turns red and on opposite sides of the isogyre near the pivoting point there will appear yellow and blue spots. If the yellow areas are located—as in the case of a uniaxial cross—in opposite quadrants on either side of the arrow or slow ray of the gypsum plate, the mineral is positive; otherwise it is negative.

For optic axis figures, one in which a single bar or isogyre stays in the field during rotation, use the gypsum plate. The stage is first rotated so that the bar is curved and the optic plane—which bisects the arc formed by the isogyre—is in the 45° position. Insert the gypsum plate with arrow parallel to the optic plane. The hyperbola turns red and has near the point of emergence of the optic axis a blue spot on one side and a yellow spot on the other (Figure 202). If the yellow spot is on the *convex* side of the bar, the mineral is positive (+); otherwise it is negative (—).

Minerals with high optic angles have a very nearly straight isogyre in all positions. On such minerals the optic sign is obtained with difficulty or not at all.

The optic sign may sometimes be told by inspection of the values for the three indices of refraction of a biaxial mineral. If $\beta - \alpha$ is decidedly greater than $\gamma - \beta$, the mineral is negative; if decidedly less, the mineral is positive.

ESTIMATION OF THE OPTIC ANGLE (2V)

From study of an optic axis figure it is possible to estimate the optic axial angle (angle between the two optic axes) of a mineral. The degree of curvature of the hyperbola varies as the stage is rotated. At its position of *maximum* curvature, however, it varies in shape from a right angle, through arcs of successively less sharp curvature, to a straight line. The limiting values of the optic angle (or 2V) which correspond to the limiting curvatures are 0° and 90° respectively. Figure 203 is useful in estimating the angle. It is usually sufficient to know whether it is small, moderate, or large.

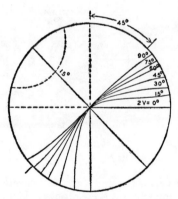

FIG. 203.—Estimation of 2V from centered optic axis figure (after Wright). For 2V = 90° the axial bar is practically a straight line.

Another means of estimation is afforded by the acute bisectrix figure. As the stage is rotated the biaxial cross breaks up into two hyperbolas. These pivot or swing and may or may not leave the field of view. If they entirely leave the field of view as the stage is turned, the optic angle exceeds 55° for most microscopes.

In an obtuse bisectrix figure—not unlike the acute bisectrix figure of a mineral with a large optic angle—the isogyres remain in the field only through a small angle of rotation of the stage. On such figures the rules for the determination of the optic sign as given above do not hold unless the words *positive* and *negative* be interchanged. It is important therefore to know whether

OPTICAL METHODS OF IDENTIFICATION

the figure observed is that of the acute or the obtuse bisectrix. The amount of rotation of the stage necessary to bring the isogyres from the crossed position to the margin of the field is roughly greater than 30° and less than 15° for the two types respectively.

Dispersion and Dispersion Formula

The optical properties of a crystal vary with the wave length of the light used. As we have seen, the index of refraction of a mineral can be completely stated only when the wave length of light is specified. Since each index of a biaxial mineral is affected, so also the birefringence is modified and is different for each different color of light, so that the optic axial angle is not the same for all wave lengths. As a result the position of the isogyres may be different for red light than for blue light. These differences are very small for most substances, but for some, for example titanite, they are distinctly visible under the microscope and are often diagnostic of the mineral. Very high dispersion of the optic axes is detected by study of the isogyres in an interference figure. If the hyperbolas are marked by a red fringe on one side and a bluish fringe on the other, the dispersion is said to be "very high" or "extreme." When the red fringe is on the convex side and the blue is on the concave side, the dispersion is recorded as $\rho > v$, or red greater than violet. The opposite case is $\rho < v$.

One other noticeable effect of very high dispersion is incomplete extinction. The extinction position of a crystal varies with the different colors or wave lengths of light. When the crystal reaches the extinction position for one color, the complement is seen if white light was used as source of illumination. As a result at no time is darkness achieved. This is called incomplete extinction and is characteristic of certain minerals, as for example, titanite.[1]

SPECIAL METHODS FOR THE STUDY OF CLAYS

Most clay materials, including kaolin and the other clay minerals, occur in nature as very impure mixtures that contain quartz, feldspar, micas, and in many specimens a large variety of other minerals. Only within the last ten years have methods been perfected which permit detailed study of minute crystal aggregates which provide reliable data on the mineral composition of these sedimentary materials. This field, although of great importance, has been studied mainly by a few workers in this

[1] Not to be confused with lack of extinction shown by compound polarizing aggregates.

country and abroad and has developed into a very specialized branch of sedimentary petrography. A brief résumé of the methods of investigation is here given.

Two steps are involved in the study of the clays and related materials. These are (a) preparation of materials and (b) identification.

A. Preparation

Marshall,[1] in an excellent review of the whole subject of the mineralogy of clays, gives directions for preparing the sample. He first removed the calcium ions, especially calcite and gypsum, by treatment with dilute acid followed by washing in distilled water. The material thus prepared is dispersed by shaking with ammoniacal solution for some hours followed by sedimentation and decantation so that the material is separated into three fractions, namely 20-5 mu, 5-2 mu, and 2-1 mu. For fractions less than 1 mu (.001 mm.), Marshall used the centrifuge and was able to obtain fractions as small as 100 mumu (.0001 mm.).

Ross and Kerr,[2] who were primarily interested in the kaolin minerals in the clay, used several methods of purification. The large, clean crystals were separated by hand-picking, the kaolin crystals were separated by washing free of the still more finely divided associated material, or the clay minerals were separated from quartz and feldspar by use of heavy liquids and the centrifuge. In only exceptional materials could the kaolin crystals be removed by hand-picking or washing. In any event the purity of the final product was checked optically.

As is elsewhere mentioned, the sand fraction may be isolated by washing with water and treated by ordinary methods.

B. Identification

For materials of 1 to 20 microns (.001 to .020 mm.) in size optical methods may be used for identification purposes. The approximate refractive index, determined by the immersion method, the approximate double refraction determined by measuring the thickness of the grain and observing the order of interference color, and the observation of the interference figure suffice. These observations can readily be made on material 20 mu or larger in size. For material less than this the main reliance is placed on the refractive indices. An oil-immersion lens extends the range of usefulness of the microscope.

For material less than 1 micron in size several methods are available. These are (1) optical, (2) X-ray, (3) chemical analyses, (4) dehydration tests, and (5) base-exchange capacity.

[1] C. E. Marshall, Mineralogical methods for the study of clays: *Zeitsch. f. Krist*, vol. 90, pp. 8-34, 1935.
[2] Clarence S. Ross and Paul F. Kerr, The kaolin minerals: *U. S. Geol. Survey, Prof. Paper 165-E*, pp. 151-180, 1931.

OPTICAL METHODS OF IDENTIFICATION 409

Optical methods. Marshall found that by use of proper immersion liquids and a *dark-field* illumination the refractive index could be determined (Figure 204). Small particles of slightly lower refractive index than the liquid emit a purplish or bluish light, while those slightly higher give a yellowish or orange light.

Hendricks and Fry[1] and later Grim[2] discovered a method whereby

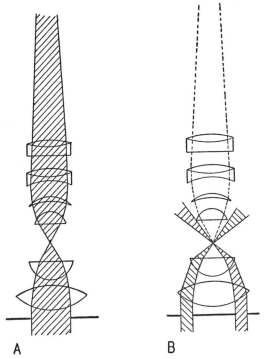

Fig. 204.—Methods of object illumination. A, normal transmitted bright-field illumination; B, dark-field illumination.

clay aggregates are formed in which all the individual particles have about the same optical orientation, so that the optical constants of the clay mineral can be measured easily and accurately. Grim prepared his sample by shaking with distilled water for some hours and then making a thin suspension which was allowed to stand until the non-clay constituents had settled to the bottom. A microscope glass slide is then sus-

[1] S. B. Hendricks and W. H. Fry, The results of X-ray and microscopical examination of soil colloids: *Soil Science,* vol. 29, pp. 457-479, 1930.
[2] Ralph E. Grim, The petrographic study of clay minerals—A laboratory note: *Jour. Sed. Petrology,* vol. 4, pp. 45-47, 1934.

pended by means of a wire sling in a horizontal position in the upper part of the suspension. The accumulated film of clay is dried, and flakes of the same are mounted in refractive index liquids as are larger grains of any minerals. These aggregates behave optically as a single crystal and may even yield interference figures from which the optical character and the value of 2V can be ascertained. So also the indices of refraction and birefringence may be obtained. Grim warns against the use of a deflocculating agent to secure a satisfactory suspension because clays have base-exchange capacity and adsorptive power which produce changes in optical characters. Fine-grained quartz causes some difficulty, as does calcite. The latter, however, may be removed by passing a stream of carbon dioxide gas through the suspension. The carbonate is thereby converted to the soluble bicarbonate. It seems to the authors that the Pasteur-Chamberland filter might be used to advantage to secure good dispersion where interfering ions are present.

Marshall accomplished the same result by dielectric means. A dilute clay suspension is placed in a parallel-sided cell with platinum foil electrodes. An alternating current is passed through the cell, and, owing to the fact that the dielectric constant varies with crystal direction, the clay crystals assume a common orientation when the current is on so that measure of the double refraction is possible. For details the reader is referred to Marshall's original paper on this technique.[1]

X-ray methods. Ross and Kerr found that by working with pure materials standard diffraction patterns showing distinct differences between the various kaolin minerals could be obtained. The film-holder used was modified by the construction of an inclined slit system set at 45° to the path of the incident rays, and the sample was placed before the final slit. This modification produced diffraction lines with a sharper resolution.

Chemical analyses. These have the greatest value only when they are made on purified materials, the purity of which has been checked optically. Since they are carried out in the standard way, no comment is required.

Dehydration tests. Dehydration tests may be carried out in an electric furnace, the temperature of which may be determined by a calibrated thermocouple and may be varied from about 100° C. to 800° C. The sample should be maintained at constant temperature until repeated weighings show that there is no further loss of weight. The results should be plotted as a "dehydration curve" with loss of water in percentage as

---

[1] C. E. Marshall, in *Trans. Faraday Soc.,* vol. 26, p. 173, 1930.

OPTICAL METHODS OF IDENTIFICATION 411

the vertical ordinate and temperature as the abscissa. Ross and Kerr showed that a distinction between such closely allied clay minerals as kaolin, nacrite, and dickite could be made on the basis of such curves.

Base-exchange capacity. The bases in clay vary in proportion and character. Changes in the character of the bases reflect geologic differences under which the clays were deposited. In a large variety of clay materials magnesium is the dominant base, but calcium, potassium, and sodium are also important. Under certain conditions bases of clays are exchangeable. That is, one base may be driven out and another take its place without any breakdown of the clay molecule as a whole. The measure of this capacity is accomplished by titration of an electrodialysed clay fraction with a standard alkali. Details are given in the paper by Marshall.[1]

Other methods. Other methods of separation of the several components of clays have been tried. Drosdoff[2] tried several physico-chemical methods, including centrifuging with heavy liquids, electrodialysis, differential flocculation, and crystallization in a bomb. He also tried certain chemical solvents, notably a sodium carbonate solution, sodium acid oxalate, and ammonium acetate.

Urbain[3] obtained the finest colloid fraction by centrifuging the clay suspension. By suitably placed electrodes and proper technique he was able to precipitate the negatively charged colloid and to keep the positively charged crystalline fraction in suspension.

The study and identification of the clay minerals is still in an unsatisfactory state. It is the opinion of Ross and Kerr that "descriptions of clay minerals written before 1900 or even later are with a few notable exceptions of little value judged by present-day methods." Among the recent papers summarizing our present knowledge of the clay minerals, including optical and other properties of these materials, are those above mentioned by Ross and Kerr and Marshall. The reader interested further in this subject should consult these and other recent papers.[4]

[1] C. E. Marshall, *loc. cit.,* p. 21, 1935.
[2] Matthew Drosdoff, The separation and identification of the mineral constituents of colloidal clays: *Soil Science,* vol. 39, pp. 463-478, 1935.
[3] Pierre Urbain, Sur la séparation des divers constituents des argiles: *Comptes Rendus,* vol. 198, p. 964, 1934.
[4] Clarence S. Ross and Paul F. Kerr, The clay minerals and their identity: *Jour. Sed. Petrology,* vol. 1, pp. 55-65, 1931. R. E. Sommers, Microscopic examination of clays: *Jour. Wash. Acad. Sci.,* vol. 9, pp. 113-126, 1919. W. H. Fry, Petrographic methods for soil laboratories: *U. S. Dept. Agric., Tech. Bull. 344,* 1933. C. E. Marshall, Clays as minerals and colloids: *Trans. Ceram. Soc.,* vol. 30, pp. 81-96. R. H. Bray, R. E. Grim, and P. F. Kerr, Application of clay mineral technique to Illinois clay and shale: *Bull. Geol. Soc. Am.,* vol. 46, pp. 1909-1926, 1935.

CHAPTER 17

DESCRIPTION OF MINERALS OF SEDIMENTARY ROCKS

(With Emphasis on the Common Rock-making Minerals and Detritals)

INTRODUCTION

SINCE it is a somewhat different problem to identify and describe the minerals of a sedimentary rock than it is to do the same for an igneous or metamorphic rock, tables of identification and description of minerals are here included. There are several reasons for the differences. Clastic sediments, for example, may theoretically contain any mineral present in any of the rock existing within the drainage area from which the sediment was derived. It is not uncommon for a sand to contain twenty to thirty mineral species. This is in contrast to an igneous rock, which has a mineral suite more or less restricted by the equilibrium conditions under which it formed. Great emphasis moreover is placed on the composition of the feldspar present in an igneous rock, and great pains are taken to determine the variety or subspecies inasmuch as the feldspar is a key mineral in classification of the rock. In a sand, on the other hand, the precise determination of the plagioclase is usually not of special significance. In fact, a half-dozen different kinds might be present. The methods useful in the case of an igneous rock, based on the assumption that but a single kind is present, are obviously valueless in the study of a sedimentary rock. A second difference arises in consequence of the method used in microscopic study. The study of the mineral grains as such rather than the thin section plays a more important rôle in the study of the sedimentary materials, especially in the medium-grained clastics. A student familiar with the common minerals in thin section may be quite at a loss to recognize the same minerals as detrital grains, most especially if they are mounted in piperine or some other medium of high index. Lastly, in the study of detrital sediments, great weight is given to the minor accessory minerals. While it is true, particularly within the last decade, that students of the igneous rocks have turned also to the minor accessories, in general only the common rock-making minerals are

DESCRIPTION OF MINERALS 413

familiar to the student. When such an individual turns to the sediments, he must learn the microscopic appearance and optical properties of many generally unfamiliar minerals such as allanite, monazite, xenotime, spinel, anatase, etc.

While it is theoretically true that any mineral present in the possible parent rocks is to be anticipated in the sediments, it is only the very abundant parent rock minerals or the very stable minerals that survive and appear in the sediments. Consequently we have seen fit to list and describe only some fifty-odd minerals—those which are commonly encountered. Each of these is described in terms of its physical, optical, and detrital characteristics in the order named. The physical and optical properties are taken from standard sources, especially from Larsen and Berman.[1] Since the detrital form of the mineral is of diagnostic value, the microscopic appearance of the mineral is both described and figured. The relief shown is that presented by a balsam mount. Inasmuch as many minerals are not sharply defined species but rather members of an isomorphous system, it has been necessary to generalize the variable optical properties to cover the group or possible range of variability. Precise determination of composition does not carry the same significance in the clastic sediment that it does in the case of the igneous or metamorphic rock. A half-dozen kinds of amphibole, several kinds of pyroxene, etc., may be encountered in a sediment. To determine the precise nature of each kind is often effort wasted. On the other hand, minor varietal characteristics, such as color, form, inclusions, zoning, or any other distinguishing marks will be noted by the careful observer. These minor characteristics may be of value in both correlation and studies of provenance. The worker should be alert to note such differences and make a record of them. The presence, for example, of both a rounded and an unrounded zircon, or an amber-colored and a colorless garnet, or a quartz with rutile inclusions and a quartz with only liquid and gas inclusions may be diagnostic of a particular horizon or be the clue to important drainage changes or diastrophic movements within the area of provenance.

Inasmuch as sediments are also studied by means of thin sections, a short description of such characters as prove of value in section studies have also been given for those few minerals likely to be seen in thinsections. The study and identification of clay minerals is rather a special

---

[1] E. S. Larsen and Harry Berman. The microscopic determination of the nonopaque minerals: *U. S. Geol. Survey, Bull. 848, 1934.*

field. To introduce the reader to the subject, a short table of the physical and optical characters of the better-known clay minerals is given. This is taken from the well-known published writings of Ross, Kerr, Grim, and Marshall (see page 407 for discussion of the technique of clay mineral studies).

The minerals described herein are arranged in alphabetical order. The description of each mineral faces the figured grains of the same. Some minerals, particularly varieties and closely related species, are for convenience grouped together because they have so many properties in common. To facilitate finding such minerals cross-references are inserted.

In the section which follows, G denotes specific gravity, Dc is the dielectric constant as given by Rosenholtz and Smith (see page 351), and Mag is the magnetic permeability class (see page 346). The last paragraph in each description is an attempt to summarize the significant or diagnostic properties and detrital characteristics of each mineral.

MINERAL DESCRIPTIONS

ACTINOLITE (*see* Amphiboles)

AMPHIBOLE GROUP
  Tremolite-actinolite, hornblende
$Ca_2(Mg, Fe)_5Si_8O_{22}(OH)_2$ = Tremolite-actinolite $G = 2.9$-3.5
$Ca_4(Na_2)(Mg, Fe)_{10}(Al, Fe)_2SiO_{14}O_{44}(OH, F)_4$ = Hornblende $Dc = 6$-8
Monoclinic $Mag = II$-III
Perfect (110) cleavage at about 55° and 125°; less
  distinct || to (100) and (010)

Colorless to light and dark green and brown
Pleochroism weak to marked. $X < Y < Z$
    Examples: X = colorless, Y = yellow-green, Z = blue-green
              X = yellow, Y = green, Z = olive-green
              X = light brown, Y = brown, Z = green-brown
Tremolite-actinolite:
$\alpha = 1.599$-1.628, $\beta = 1.613$-1.644, $\gamma = 1.625$-1.655. $B = .026$-$.027$
Hornblende:
$\alpha = 1.658$-1.698, $\beta = 1.670$-1.719, $\gamma = 1.679$-1.722. $B = .019$-$.026$
Biaxial, negative. (Pargasite is positive. Not common.) Optic plane is (010).
    $Y = b$, $Z \wedge c$ 14-24° (smaller in basaltic hornblende)
$2V$ = moderate to large (60-90°)
Elongation, positive.

DESCRIPTION OF MINERALS

Grains elongate, prismatic, marked by longitudinal cleavage and by strong diagonal cross fractures. Also as irregular fractured fragments. Gradations between all members of the amphibole group make it impossible to assign each grain of a heavy mineral residue to its proper place. The actinolite and tremolite members may be combined and counted as one. They are marked by lower index of refraction, lack of color (tremolite) or very pale color (actinolite). "Common hornblende" may be defined as those grains in shades of brownish-green or greenish-brown. "Blue-green hornblende" on the other hand is the soda-rich type with Z greenish-blue or bluish-green, Y green or bluish-green and X yellowish-green. Most of the blue-green hornblendes are related to the glaucoamphiboles and are therefore more characteristic of metamorphic rocks. "Basaltic hornblende" shows the typical deep pleochroism in red-brown, small extinction angle and high birefringence. The hornblendes are often nearly opaque and appear translucent only on the thin edges.

Glaucophane:
$\alpha = 1.621$, $\beta = 1.638$, $\gamma = 1.639$. $B = .018$
Colored, lavender blue with distinct pleochroism. $X =$ colorless, $Y =$ violet blue, $Z =$ dark blue
$Z \wedge c$ $4°$ to $6°$

Marked by its blue color and pleochroism and by positive elongation. Derived from schists and gneisses.

The amphibole group as a whole is characterized by cleavage-controlled prismatic form, inclined extinction (smaller angle than monoclinic pyroxenes), marked pleochroism (contrasted with general lack of pleochroism of monoclinic pyroxenes). Basaltic hornblende, igneous; common hornblende, igneous and metamorphic; tremolite, actinolite and glaucophane, metamorphic.

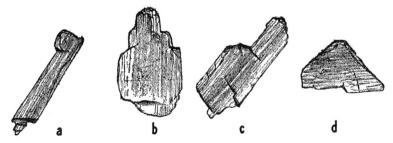

FIG. 205.—Amphibole: *a-c*, actinolite; *d*, hornblende. Lake Erie Shore sand, Cedar Point, Ohio.

ANATASE (Octahedrite)
TiO_2

Tetragonal: Commonly pseudo-octahedral; also tabular and prismatic $G = 3.82\text{-}3.95$

Distinct basal (001) and pyramidal cleavage (111) $Mag = IV$

Pale yellow or indigo blue

Pleochroism weak: X = yellow to light brown or blue, Z = orange to brown or deep blue, $X < Z$.

$\omega = 2.554$, $\epsilon = 2.493$. $B = .061$

Uniaxial, negative (frequently anomalously biaxial)

Tabular grains markedly rectangular in outline with corners sometimes truncated (by 111 faces) are common. These grains, which appear isotropic, yield well-centered uniaxial cross in convergent light. (001 faces striated parallel to borders (100, 010) and to possible beveling of corners (111). Pyramidal (pseudo-octahedral) forms also known. Occasional parallel groups of crystals noted.

May be either authigenic or detrital. The latter grains usually marked by wear. Characterized by yellow or blue color, rectangular grains marked by "geometrical patterning" (striations) and negative uniaxial character.

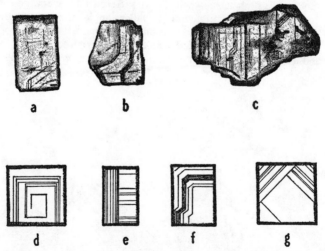

Fig. 206.—Anatase: *a-b*, Lower Estuarine series, Yorkshire; *c*, Pliocene, Cornwall; *d-g*, geometrical patterning on basal plates (after Brammall).

ANAUXITE (*see* Clay minerals)

DESCRIPTION OF MINERALS 417

ANDALUSITE (and Chiastolite)
Al_2SiO_5 $G = 3.15$
Orthorhombic: Nearly square prisms $Dc = 8\text{-}9$
Good prismatic (110) cleavage at about right angles; $Mag = IV$
 traces || to (100) and (010)

Colorless to pink
Pleochroism marked. X = rose pink, Y and Z = nearly colorless
$a = 1.634$, $\beta = 1.639$, $\gamma = 1.643$. $B = .009$
Biaxial, negative. Optic axial plane parallel to (010); $X = c$; $Z = a$
 Parallel extinction
Elongation negative
$2V = 85°$. Dispersion weak

Elongate worn or broken prism to irregular grains showing only a conchoidal fracture. Inclusions common, sometimes carbonaceous. They may be concentrated in the crystal in geometrically symmetrical areas; var. chiastolite.

FIG. 207.—Andalusite var. chiastolite: a, Oligocene Bovey Beds, Heathfield, E n g l a n d (after Boswell); b, gravel at Riddaford, England (after Boswell).

a   b

Distinguished by form, negative sign and elongation, and pleochroism. Alters readily to colorless mica which renders crystal turbid. Derived from contact metamorphic zones in shaly material.

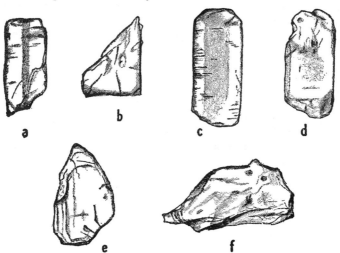

FIG. 208.—Andalusite: a-f, Pliocene, Cornwall.

Anhydrite
$CaSO_4$
Orthorhombic
Perfect (001, 010) cleavage; distinct (100) cleavage;
  yielding rectanguar cleavage flakes

$G = 2.93$
$Dc = 6-7$
$Mag = IV$

Colorless
$a = 1.570, \beta = 1.575, \gamma = 1.614.$ $B = .044$
Biaxial, positive. Optic plane parallel to (010); $X = c, Z = a$.
  Extinction parallel
$2V = 42°$

Grains markedly rectangular to irregular. Also euhedral, dentate, rounded to very ragged or etched. Rectangular grains show "flash figure" or less commonly well-centered acute bisectrix figure. Dark inclusions and zonal structure sometimes shown.

An authigenic mineral associated with limestones, rock salt, and gypsum. Soluble in hot HCl. Distinguished by the rectangular cleavages in three directions, by strong birefringence and distinguished from barite and celestite by low specific gravity.

Apatite
$Ca_5(F, Cl)(PO_4)_3$
Hexagonal: Long to short prismatic, terminated by base
  or pyramid
Imperfect basal (0001) cleavage

$G = 3.16-3.22$
$Dc = 5-6$
$Mag = IV$

Colorless
$\omega = 1.630-1.644,$ $\epsilon = 1.633-1.649.$ $B = .003-.005$
Uniaxial, negative (anomalously biaxial, 2V small)
Elongation, negative

Grains oval or nearly circular in plan to slightly worn elongate prismatic form. Often contains inclusions arranged in rows or planes.

Marked by detrital form and low birefringence. May be wholly removed from a sediment by acid digest since it is soluble in HCl (though see comment, page 314). Derived from acid igneous rocks and pegmatites.

DESCRIPTION OF MINERALS 419

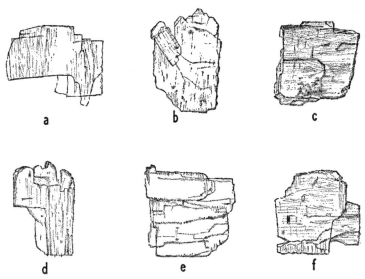

Fig. 209.—Anhydrite: *a-f*, Miocene, Persia.

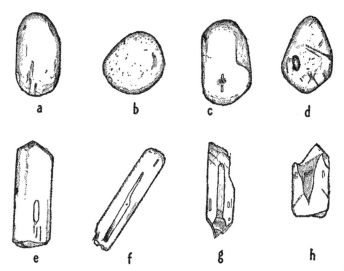

Fig. 210.—Apatite: *a-d*, Trias, Yorkshire; *e-h*; Nutfield fuller's earth, England (After Newton).

ARAGONITE
CaCO$_3$ G = 2.94
Orthorhombic; acicular $\parallel$ to c Dc = 7-8
Cleavage distinct $\parallel$ to (010), traces $\parallel$ to (011) and (110)   Mag = IV

$\alpha = 1.530, \beta = 1.680, \gamma = 1.685$ B = .155
Biaxial, negative. Op. pl. = (100)   X = c
Extinction parallel; negative elongation
2V = 18°

Fragments prismatic, irregular. Rare as detrital mineral. Largely as shell fragments.

Distinguished from calcite by higher specific gravity, acicular form, absence of rhombohedral cleavage, parallel extinction, and biaxial character. Soluble in dilute acid. Aragonite found in gypsum deposits; also in certain fossil shells and corals.

AUGITE (*see* Pyroxenes)

BARITE and Celestite
BaSO$_4$ G = 4.5
Orthorhombic; tabular (001)   Dc = 7-8
Perfect (001) (110) and less perfect (010)   Mag = IV

Colorless
$\alpha = 1.636, \beta = 1.637, \gamma = 1.648$. B = .012
Biaxial, positive. Optic plane is (010); X = c, Z = a
2V = 37½°

Grains very fantastically irregular to ragged. Sharply angular cleavage fragments and diamond-shaped grains also known. Rounded grains rare. With numerous inclusions. Diamond-shaped grains show obtuse bisectrix figure; such grains show first-order colors and symmetrical extinction. Prismatic grains exhibit parallel extinction and bright colors.

A mineral of authigenic origin. Grain shape apparently due either to shape of interstice in which it was deposited or by cleavage. Marked by high index, high specific gravity, and low optic angle.

Celestite (SrSO$_4$) very similar to barite but less common. Also authigenic. Has lower index ($\alpha = 1.622$, $\beta = 1.624$, $\gamma = 1.631$. B = .009), larger optic angle (51°), and lower specific gravity (3.96).

DESCRIPTION OF MINERALS

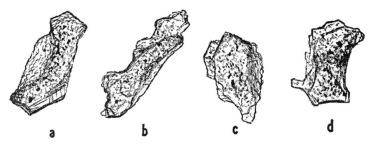

FIG. 211.—Barite: *a-d*, Trias, Cheshire.

BEIDELLITE (*see* Clay Minerals)

BIOTITE
$K_2O.4(Mg, Fe)O.2(Al, Fe)_2O_3.6SiO_2.H_2O$
Monoclinic: Pseudo-hexagonal plates
Perfect (001) basal cleavage

$G = 2.7-3.1$
$Dc = 9-10$
$Mag = III$

Brown. Also green.
Marked pleochroism in thin sections. $X < Y$ and Z
$\alpha = 1.573-1.584$, $\beta = 1.620-1.648$, $\gamma = 1.620-1.648$. $B = .050-.064$
Biaxial, negative. X near *c*. $Y = b$.
2V is very small.

In flakes varying from hexagonal to rounded irregular. Grains always yield perfectly centered pseudo-uniaxial cross. Inclusions (around zircon, xenotime, or allanite) with dark halos are characteristic. Some biotites have rutile needles as inclusions.

Biotite is a detrital grain marked by deep brown color, lack of pleochroism, perfectly centered negative pseudo-uniaxial interference figure, and inclusions with halos. Strong pleochroism, "bird's-eye maple" appearance under crossed nicols and inclusions with pleochroic halos mark biotite in thin-sections. Alters to chlorite and by diagenetic action is changed to glauconite. Derived from igneous and metamorphic rocks.

BROOKITE [1]
TiO_2
Orthorhombic; tabular parallel to (100)
Cleavage: (110) indistinct; also (001)

$G = 3.9$
$Mag = IV$

Straw-yellow, amber, and shades of orange and brown. Weakly pleochroic.
$\alpha = 2.583$, $\beta = 2.586$, $\gamma = 2.741$. $B = 0.158$
Biaxial, positive. $X_{red} = b$. $X_{blue} = c$. $X = a$.
$2V_{Na} = 30°$, $0°$ for yellow-green. Dispersion very strong.

Squarish grains, with truncated corners, flattened parallel to (100) very common. Markedly fluted and striated parallel to vertical axis. A weaker second set for striations at right angles to the first is seen on some grains. Brookite is often much corroded. Incomplete extinction in white light. Interference figure uniaxial in yellow-green light; biaxial in yellow light. In white light figure is anomalous, being composed of two highly dispersed biaxial figures at right angles to each other.

Brookite is characterized by its color, tabular habit, marked striations parallel to edge, incomplete extinction, and anomalous interference figure. Differs from anatase in its lack of diagonal striations and in its very high dispersion effects. Derived from acid igneous rocks and from crystalline metamorphic rocks.

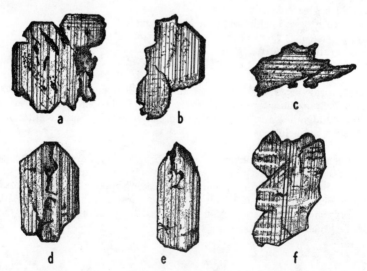

FIG. 212.—Brookite: *a-c*, Recent sand, Devonshire; *d-f*, Dartmoor, England (after Brammall).

[1] For a good detailed description of brookite see: Alfred Brammall, Dartmoor detritals: A study in provenance. *Proc. Geologists Assoc.*, vol. 39, pp. 31-34, 1928.

CALCITE and Dolomite
$CaCO_3$
Hexagonal-rhombohedral: Scalenohedra, prisms with rhombohedral termination and less commonly rhombohedra.
Perfect rhombohedral ($10\bar{1}1$) cleavage

$G = 2.71$
$Dc = 6-7$

$Mag = IV$

Colorless
$\omega = 1.658$, $\epsilon = 1.486$. $B = .172$
Uniaxial, negative
Symmetrically inclined extinction (to $10\bar{1}1$ cleavage)

Soluble with effervescence in cold HCl. Marked by a sharp change in relief ("twinkling") upon rotation of the stage. Detrital grains irregular or marked by cleavage boundaries. In thin section calcite exhibits cleavage which appears as very fine straight lines, discontinuous, unequally spaced, and in obliquely intersecting sets. Calcite shows also polysynthetic twinning marked by a series of parallel bands. Under crossed nicols, calcite exhibits iridescent interference colors of very high orders.

Detrital grains rare except in recent sediment owing to solubility, cleavage, and softness of the mineral. Characterized by "twinkling," high-order interference colors, and marked cleavage. It occurs in sediments as (1) interlocking mosaic of anhedral crystals, (2) small euhedral crystals, and (3) fibrous groups or spherulites. Fibers may have parallel extinction (elongated parallel to c) or oblique extinction.

Distinguished from dolomite only with difficulty. Principal distinctions are: (1) Calcite rarely crystallizes as rhombohedra; dolomite nearly always has this habit (exceptions are known, however). (2) Selective stains may be used (see page 495). (3) Calcite has lamellar twinning. (4) Calcite effervesces in cold dilute HCl. (5) Calcite has lower specific gravity (calcite 2.71, dolomite 2.87).

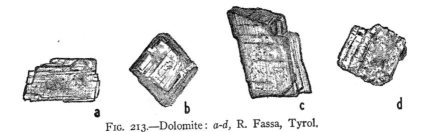

FIG. 213.—Dolomite: *a-d*, R. Fassa, Tyrol.

Dolomite

$Ca(Mg, Fe)(CO_3)_2$
Hexagonal-rhombohedral; rhombohedra
Cleavage: rhombohedral ($10\bar{1}1$)

$G = 2.83\text{-}3.00$
$Dc = 8\text{-}9$
$Mag = IV$

Colorless
$\omega = 1.682$, $\epsilon = 1.503$ $B = 0.179$
Uniaxial, negative

Dolomite occurs as euhedral rhombs, as subangular to rounded grains, and as microcrystalline aggregates.

Distinguished from calcite with difficulty (*see* Calcite).

Cassiterite

SnO_2
Tetragonal; prisms terminated by low pyramids; twinned on (101)
Cleavage: (100) (111) imperfect

$G = 4.35\text{-}4.53$
$Mag = IV$

Colorless, yellow, amber, to reddish brown (near blood-red). Pleochroic.
$\epsilon = 2.093$, $\omega = 1.997$ $B = 0.096$
Uniaxial, positive

Typical grains irregular, showing conchoidal fracture or, at best, indistinct cleavage. Slightly elongate worn euhedra also known. Zoning prevalent parallel to crystal form or to vertical axis, also diagonal and irregular. Irregular blotch or parti-coloring common. Striations on prism faces both parallel and oblique to vertical axis. Horizontal striations less common.

Minute grains of cassiterite are easily confused with rutile and even with anatase, zircon, and sphene. Cassiterite grains placed on sheet zinc and moistened with dilute hydrochloric acid become coated in a few seconds with a dull gray film of metallic tin. Cassiterite derived from veins, pegmatites, and granite.

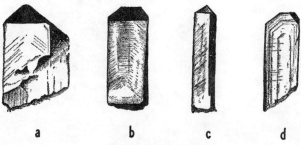

Fig. 214.—Cassiterite: *a-d*, Dartmoor, England (after Brammall).

DESCRIPTION OF MINERALS

CELESTITE (*see* Barite)
CEYLONITE (*see* Spinel)
CHALCEDONY (*see* Quartz)
CHERT (*see* Quartz)
CHIASTOLITE (*see* Andalusite)

CHLORITE—SERPENTINE GROUP
$4R''O \cdot R'''_2O_3 \cdot 3SiO_2 \cdot 4H_2O$, where $R'' = $ Mg, Fe''; $G = 2.5$-3.0
$R''' = $ Al, (Fe''')
Monoclinic (?): Micaceous, pseudo-hexagonal plates $Dc = 8$-9
Perfect basal (001) cleavage   Mag $=$ III

Dirty yellow-green to green
Pleochroic in thin section. X and Y > Z or reverse
β — varies from 1.56-1.65. B $=$ low (.003-.009), exceptionally higher
Biaxial, negative and positive. Bx_a nearly perpendicular to (001)
$2V = $ small (0°-30°)
Dispersion strong in some cases

Grains flat, rounded, irregular cleavage flakes. May show "ultra-blue" abnormal interference color (penninite) or may exhibit compound polarization (aggregate). Color blotchy, in part colors owing to bleaching. May show zones following hexagonal plan.

Marked by pale green color, weak birefringence, micaceous habit, well-centered psuedo-uniaxial interference figure, and strong dispersion of some species. Chlorite is not a single species but is a complex group of closely related minerals the members of which are distinguished only with difficulty. "Serpentine": term used for antigorite member of chlorite group. Marked by different optical orientation ($Bx_a \perp$ to 100), large 2V, fibrous to irregular. Irregular grains, compound polarizing. Alteration product of amphiboles, pyroxenes, or micas. Produced by low-grade anamorphism of clays, shales and present in cement of graywackes.

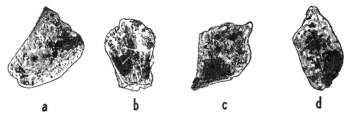

FIG. 215.—Chlorite: *a-d,* variety "serpentine," shore sand, Cornwall.

TABLE 43

IMPORTANT CLAY MINERALS

CLAY MINERALS

Mineral	Composition	Crystal System	Refractive Indices			B	Optic Sign	General Appearance and Remarks	Spec. Gr.
			α	β	γ				
KAOLINITE-ANAUXITE (Leverrierite)	$2H_2O.Al_2O_3.\pm SiO_2$	Probably mono.	1.561	1.565	1.566	.005	−	Translucent to opaque; greatly elongated $\parallel$ to c (vermicular) Absorbs dyes very strongly; pleochroic	2.59
DICKITE	$2H_2O.Al_2O_3.2SiO_2$	Mono.	1.560	1.562	1.566	.006	+	Clear, transparent thin crystal plates. Not strongly stained by dyes; non-pleochroic	2.62
NACRITE	$2H_2O.Al_2O_3.2SiO_2$	Mono.	1.557	1.562	1.563	.006	−	Transparent trilled crystal plates; wedge-shaped cleavage plates. Not readily stained	2.50

KAOLIN GROUP

Data from Ross and Kerr (*U. S. Geol. Survey, Prof. Paper 165-E*) and by Marshall (*Zeitsch. f. Krist.*, vol. 90, p. 8-34, 1935).
See Figure 216.

Group	Mineral	Formula	Crystal system	α	β	γ	Birefringence	Sign	Habit	G
	Halloysite	$Al_2O_3 \cdot 2SiO_2 \cdot nH_2O$	......	1.553 (variable)			0	..	......	2.0–2.2
	Allophane	$Al_2O_3 \cdot SiO_2 \cdot 5H_2O$	......	1.47–1.50			0	..	Irregular plates	1.85–1.89
Montmorillonite–Beidellite Group	Montmorillonite	$Al_2O_3 \cdot 5SiO_2 \cdot (Ca,Mg)O \cdot nH_2O$	Mono.	1.492	.....	1.513	.021	−	......	2.25 ± .03
	Beidellite	$Al_2O_3 \cdot 3SiO_2 \cdot nH_2O$	......	1.517	.....	1.549	.032	−	Plates, short blades	2.30 ±
	Nontronite	$Fe_2O_3 \cdot 3SiO_2 \cdot nH_2O$	......	1.580	.....	1.615	.035	+ or −	Blades and fibers. Pleochroic; yellow ∥ to fiber; brown or green ⊥ to fibers	2.495
Potash-Bearing Clays	......	......	......	1.528–1.543	.....	1.550–1.565	.022	..	......	......
	Pyrophyllite	$Al_2O_3 \cdot 4SiO_2 \cdot H_2O$	......	1.552	1.558	1.600	.048	..	Platy habit. Some long blades with Z ∥ to length	2.8–2.9

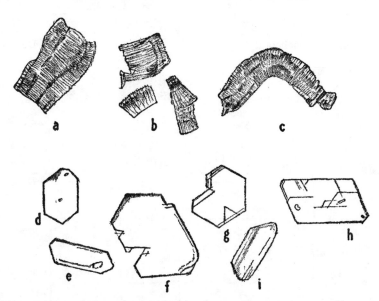

Fig. 216.—Clay minerals: *a-b*, kaolin (residual), Joel, Idaho; *c*, kaolin, Franklin, N. C. (after Ross and Kerr); *d-i*, dickite. See pp. 426 and 427 for description of mineral properties.

COLLOPHANE
$3Ca_3(PO_4)_2 nCa(CO_3,F_2)(H_2O)_x$ $G = 2.6$-2.9
Amorphous: Colloform, i.e., bone structure, incrustations,
    oölitic, concretionary, nodular, and banded

White, yellow, brown, gray, black
$n = 1.57$-1.63
Isotropic (may be anisotropic low birefringence)

Fragments are irregular, colorless to brown, translucent with high relief. Between crossed nicols the mineral is either dark (isotropic) or has very low first-order interference colors (double refraction due to strain). Fossil bone shows Haversion canals and lacunæ which appear as dark lines and dark spots.

Collophane crystallizes to dahlite. Is rare as a detrital mineral. It occurs (1) as a constituent of bones, teeth, and other organic structures; (2) in yellow to brown grains, ovoid to irregular, often enclosing micro-fossils, or it may have concentric or oölitic structure or may be composed of radial fibers which have low birefringence; (3) as a cement; (4) as a constituent of phosphatic nodules; and (5) as a constituent of coprolites.

DESCRIPTION OF MINERALS

CORUNDUM
Al_2O_3
Hexagonal, scalenohedral
Cleavage: Rhombohedral ($10\bar{1}1$) parting; also perfect basal (0001) parting

G = 4.0
Dc = 5-6
Mag = IV

Colorless, pink, blue. Sometimes pleochroic; ϵ = blue, ω = bluish green
$\omega = 1.769$, $\epsilon = 1.760$. B = .009
Uniaxial, negative
Elongation negative

Grains commonly irregular, fracture-controlled, to worn; crystal form obliterated. Less common are basal flakes which give good uniaxial figure. Color irregular, blotchy. Colorless grains most common.

a   b

FIG. 217.—Corundum: *a-b*, Eocene, Haldons, E n g l a n d. (After Boswell.)

Characterized by high refractive index, low birefringence, blotchy coloring (if any). Derived from igneous (syenites) or metamorphic rocks (contact metamorphosed limestones).

DIALLAGE (*see* Pyroxenes)
DICKITE (*see* Clay Minerals)
DIOPSIDE (*see* Pyroxenes)
DOLOMITE (*see* Calcite)

Dumortierite
$8Al_2O_3 \cdot B_2O_3 \cdot 6SiO_2 \cdot H_2O$ $\qquad G = 3.3$
Orthorhombic; acicular $\parallel$ to c
Cleavage: (100) distinct

Blue, greenish, reddish violet. Pleochroic, X = deep blue or violet, Y = yellow to red or nearly colorless, Z = colorless or very pale blue.
$\alpha = 1.678$, $\beta = 1.686$, $\gamma = 1.689$. B = .011
Biaxial, negative. X $\wedge$ c = 0°$\pm$. Z = a. Parallel extinction
2V = 30-40°. Negative elongation.

Found as unevenly terminated prismatic grains; often with conspicuous striations parallel to prism edge.

Resembles indicolite (blue tourmaline), but has a higher index, a distinctive pleochroism. Derived from pegmatite, gneisses, and schists.

    a   b   c   d

FIG. 218.—Dumortierite: *a-d*, crushed grains from Dartmoor, England. (After Brammall.)

Enstatite (*see* Hypersthene)

Epidote
$Ca_2(Al, Fe)_3Si_3O_{12}(OH)$ $\qquad G = 3.36$
Monoclinic; elongated parallel to b; striated $\parallel$ to b. $\qquad Dc = 6\text{-}7$
Cleavage perfect basal (001); imperfect (100) $\qquad Mag = III$

Pale greenish yellow to lemon-yellow
Pleochroism distinct. X = colorless, Y = pale greenish yellow, Z = colorless
$\alpha = 1.722\text{-}1.729$, $\beta = 1.742\text{-}1.763$, $\gamma = 1.750\text{-}1.780$. B = .028-.051
Biaxial, negative. Optic plane (010). Y = b, Z $\wedge$ c 2-5°. One optic axis nearly perpendicular to basal cleavage
2V = 69-80°

Equidimensional sharply angular to subrounded grains of characteristic greenish yellow. Grains very often yield centered optic axis or one bar ("compass-needle") interference figure. Bar nearly straight. Such grains are non-pleochroic. Most grains exhibit high-order interference color and therefore appear much the same under crossed nicols as in ordinary light.

DESCRIPTION OF MINERALS

Marked by weak but distinct pleochroism, greenish yellow color, high index of refraction, and optic axis interference figure. Derived from metamorphosed igneous rocks.

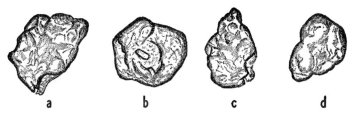

FIG. 219.—Epidote: *a-d*, shore sand, Lake Michigan, Indiana.

FLUORITE
CaF_2
Isometric: Cubic habit
Perfect octahedral (111) cleavage.

$G = 3.18$
$Dc = 7-8$
$Mag = IV$

Colorless
$n = 1.434$
Isotropic

Colorless with irregular to cleavage-controlled triangular form are typical. Planes of liquid and gas inclusions sometimes noted.

Recognized by high relief, very low index, isotropic appearance, and octahedral cleavage. Derived from acid igneous rocks, pegmatites, and veins. Also found in sedimentary and igneous rocks.

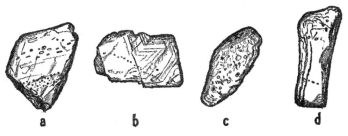

FIG. 220.—Fluorite: *a-d*, shore sand, Cornwall.

Garnet

$R''_3 R'''_2 (SiO_4)_3$, where $R'' = $ Mg, Fe'', Ca, Mn $\qquad$ G = 3.5-4.3
$\qquad R''' = $ Al, Fe''', Cr
Isometric: Dodecahedra, trapezohedra $\qquad$ Dc = 6-8
Conchoidal fracture $\qquad$ Mag = III

Colorless, pale pink, red, orange, apricot-yellow, amber; rarely green
$n = 1.70$-1.90
Isotropic. Anomalously anisotropic

Garnet is variable in color, form, and inclusions. It is commonly in well-rounded grains, but in many cases it appears in angular, sharp-cornered, irregular grains bounded by conchoidal fracture surfaces. Euhedra rare. On the better-rounded grains the surface is marked by percussion scars. Others are etched into a "mosaic" of facets, or pits. Anisotropic inclusions often of large size, some circular or rounded in form, others of acicular (rutile) habit, still others lath-shaped. Apatite, zircon, biotite, magnetite, feldspar and quartz known as inclusions.

Garnet is characterized by high relief, isotropism, conchoidal fracture, and color. The several species of garnet cannot be distinguished except by careful determination of index and specific gravity. The varietal characters, especially form and color, should be noted, as difference may be of significance in provenance studies. (See also spinel, page 449, with which garnet is confused.) Derived from igneous and metamorphic rocks, especially the crystalline gneisses and schists.

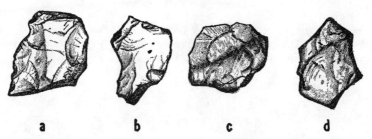

Fig. 221.—Garnet: *a-d*, Lake Erie shore sand, Cedar Point, Ohio.

DESCRIPTION OF MINERALS

GLAUCONITE [1]

$(OH)_{6-8} \cdot K_3 (Mg, Fe'', Ca)_{1-3} \cdot (Fe''', Al, Si)_{8-6}$ $(Si_{13-14}Al_{2-3})O_{38-40}$
Monoclinic: Platy, parallel to (001) to formless
Perfect (001) cleavage to indistinct fracture

$G = 2.2-2.8$
$Dc = 10-12$
$Mag = II$

Olive to grass-green
Pleochroism variable. X = dark bluish green, pale yellowish green,
 Y and Z = lemon-yellow, grass-green
$\alpha = 1.597-1.609$, $\beta = 1.618-1.630$, $\gamma = 1.619-1.630$. $B = .013-.027$
Biaxial, negative. X near c
$2V$ = small ($0-20°$)

Grains rounded, often polylobate, papilliform, ovoid, spherical, subcylindrical. Surface often polished, sometimes cracked. Polarization varies from grains that behave optically as units to grains showing only aggregate polarization so that determination of optical constants is difficult. Structure of grains is radial, fibrous, spongy, granular, lamellar.

Marked by the grain shape and dirty olive-green color and aggregate polarization. Developed from biotite. As alteration proceeds the index declines, the birefringence decreases, the pleochroism diminishes, and aggregate polarization develops. Least altered biotite grains give rise to lamellar glauconite, which shows biaxial negative sign and marked pleochroism. On further alteration the "cryptocrystalline" variety which has only aggregate polarization forms. May occur in foraminiferal shells. Alters to limonite. X-ray data also indicate the affinity of glauconite and biotite.

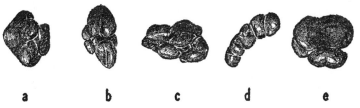

a   b   c   d   e

FIG. 222.—Glauconite: *a-e*, greensand, New Jersey (in reflected light).

GLAUCOPHANE (*see* Amphiboles)

---

[1] E. Wayne Galliher, Glauconite genesis: *Bull. Geol. Soc. Amer.*, vol. 46, pp. 1351-1366, 1935. John W. Gruner, The structural relationship of glauconite and mica: *Am. Mineralogist*, vol. 20, pp. 699-714, 1935.

Gypsum (Selenite)

$CaSO_4 \cdot 2H_2O$

Monoclinic; tabular (010); twinning and comp. pl. (100).
Cleavage: (010) perfect; (111)(100) imperfect

$G = 2.44$
$Dc = 6\text{-}7$
$Mag = IV$

Colorless
$\alpha = 1.520, \beta = 1.523, \gamma = 1.530.$ $B = .010$
Biaxial, positive. $Y = b$. $X \wedge c\ 37\frac{1}{2}°$
$2V = 58°$. Negative elongation.

Subrounded, euhedral, rhombic-shaped (010) grains common, especially in clays. Dusty included matter (probably clay) gathered into streaks parallel or less commonly radially or zonally arranged. Wind-blown gypsum sands distinctly rounded, circular to oval in outline. Inclined extinction. Twinning observed in a few grains; marked by sharp line of demarcation between the twinned halves which extinguish $37\frac{1}{2}°$ from line of contact.

Characterized by low specific gravity, low refractive index, crystal form and solution in hot hydrochloric acid. Probably authigenic in clays.

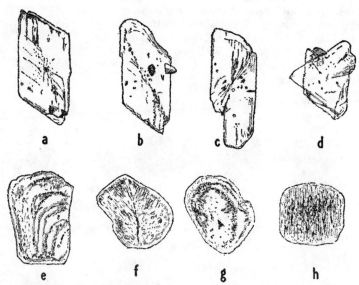

Fig. 223.—Gypsum: *a-d*, London clay, Surrey; *e-h*, dune sand, Tularosa Basin, New Mexico.

Hematite (*see* Opaque Minerals)

DESCRIPTION OF MINERALS

HYPERSTHENE and Enstatite
$(Mg, Fe)SiO_3$
Orthorhombic; prismatic, flattened $||$ to (100) or (010)

$G = 3.18\text{-}3.49$
$Dc = 6\text{-}7$ (enstatite $= 8\text{-}9$)

Perfect (110) at about 87-88°; fair $||$ to (100) and (010)   Mag = III and IV

Colorless (enstatite) to pale pink and green (hypersthene)
Pleochroism very faint to marked. X = clear red or pink,
    Y = yellowish, Z = green
$\alpha = 1.665\text{-}1.715$, $\beta = 1.669\text{-}1.728$, $\gamma = 1.674\text{-}1.731$. B = .009-.016
Biaxial, positive (enstatite) to negative (hypersthene) Optic plane (010),
    Z = c, Y = b. Parallel extinction.
2V = high, (63-90°)
Elongation positive (parallel to cleavage).

Worn elongate to stubby cleavage fragments usually hypersthene. Conspicuous striations parallel to cleavage with marked cross fractures. Other grains irregular and marked only by conchoidal fracture. Contains highly colored (brown) thin, plate-like inclusions (schiller-structure) which, if numerous, give bronze luster (bronzite).

Marked by high relief, low birefringence, parallel extinction, and striking pleochroism. Enstatite and hypersthene are members of an isomorphous orthorhombic pyroxene group. They grade into one another. With increase in content of iron goes increase in specific gravity, index, birefringence, the appearance of color and pleochroism, and change of optic sign from positive to negative. The distinction between them is arbitrary. The variety marked by negative sign, distinct color, and pleochroism may be called hypersthene and the colorless positive species enstatite. Derived from basic and ultrabasic igneous rocks.

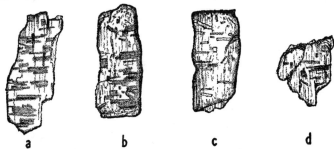

FIG. 224.—Hypersthene: *a-d,* bronzite variety, Lake Erie shore sand, Cedar Point, Ohio.

ILMENITE (*see* Opaque Minerals)
INDICOLITE (*see* Tourmaline)
KAOLINITE (*see* Clay Minerals)

436 SEDIMENTARY PETROGRAPHY

Kyanite (Disthene)
Al_2SiO_5
Triclinic; bladed
Perfect pinacoid (100) and (010) (001) parting. Nearly at right angles

$G = 3.6$
$Dc = 7-8$
$Mag = IV$

Colorless. Rarely pale blue
Pleochroism faint. X = colorless, Y = violet blue, Z = cobalt-blue
$α = 1.712, β = 1.720, γ = 1.728. B = .016$
Biaxial, negative. X nearly perpendicular to (100). Extinction on (100) $Z \wedge c$ $-30°$
$2V = 82°$
Elongation generally positive

Usually colorless, decidedly elongate grains of marked rectangular outline to short, moderately rounded elliptical grains. Conspicuous cross cleavage associated with step-like changes in order of interference color. Nearly all fragments yield centered acute bisectrix figure. Carbonaceous inclusions frequent. Grains may be altered along edges to micaceous material.

Marked by cleavage-controlled rectangular detrital form, by low birefringence, inclined extinction, centered interference figure, and step-like changes in order of interference color. An index of current velocity. Rounded kyanite denotes low velocity; angular kyanite, high velocity. Derived from schists and gneisses.

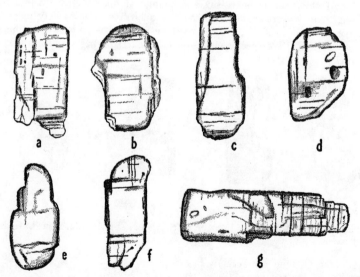

Fig. 225.—Kyanite: *a-f*, Lower Greensand, Surrey; *g*, Corallian sand, Oxford, England.

DESCRIPTION OF MINERALS

LEUCOXENE (*see* Opaque Minerals)
LIMONITE (*see* Opaque Minerals)
MAGNETITE (*see* Opaque Minerals)
MARCASITE (*see* Opaque Minerals)
MICROCLINE (*see* Orthoclase)
MONTMORILLONITE (*see* Clay Minerals)

MONAZITE
$(Ce, La, Nd, Pr)_2O_3 \cdot P_2O_5$ $G = 5.1\text{-}5.2$
Monoclinic; tabular parallel to (100) $Dc = 7\text{-}8$
Perfect basal (001) cleavage; good (100) and   Mag = III
   imperfect (010)

Yellow, brown, red
Pleochroism faint. X = light yellow, Y = dark yellow, Z = greenish yellow.
$\alpha = 1.786\text{-}1.800$, $\beta = 1.788\text{-}1.801$, $\gamma = 1.837\text{-}1.849$. $B = .049\text{-}.051$
Biaxial, positive. Y or X = b, $Z \wedge c$ $2°\text{-}10°$
2V small (11-14°); dispersion weak.

Grains rounded, equidimensional, often lying on (001). Such grains yield good interference figure. Euhedra rare. They exhibit the same color between crossed nicols as in ordinary light owing to their high birefringence.

Marked by color between crossed nicols, high relief (in balsam), and characteristic light yellow color. Differs from titanite in its very slight dispersion (and therefore complete extinction in white light), distinguished from zircon by its color and biaxial character; and differs from epidote in its very weak pleochroism and its small optic angle. Derived from granites.

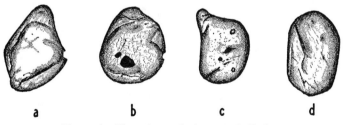

a   b   c   d

FIG. 226.—Monazite: *a-d*, shore sand, Ceylon.

Muscovite

$2H_2O \cdot K_2O, 3(Al,Fe)_2O_3 \cdot 6SiO_2$
Monoclinic; pseudo-hexagonal plates
Perfect basal (001) cleavage

$G = 2.80\text{-}2.88$
$Dc = 10\text{-}12$
$Mag = IV$

Colorless
$\alpha = 1.551\text{-}1.572, \beta = 1.581\text{-}1.611, \gamma = 1.587\text{-}1.615.$ $B = .036\text{-}.046$
Biaxial, negative. $Z = b$, $X \wedge c = 0°$. Bx_a nearly $\perp$ (001)
$2V = 29\text{-}42°$

Detrital muscovite occurs in thin transparent flakes (001) marked by low relief (in C.B.), and low bluish gray interference color. These grains yield perfectly centered acute bisectrix figure with low optic angle. In thin-section muscovite is marked by single cleavage and by brilliance of interference colors.

Characterized by cleavage, well-centered biaxial figure and negative sign. Distinguished from biotite by its lack of color. Derived from acid igneous rocks and from gneisses and schists.

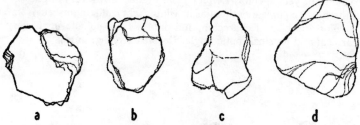

Fig. 227.—Muscovite: *a-d*, Kellaway's Rock, Yorkshire.

Nacrite (*see* Clay Minerals)
Nontronite (*see* Clay Minerals)

DESCRIPTION OF MINERALS 439

OLIVINE
$(Mg, Fe)_2SiO_4$
Orthorhombic; flattened $||$ to (100) or (010)
Cleavage: Good $||$ to (010) and (100)

$G = 3.5$
$Dc = 6-7$
$Mag = III$

Colorless; pale yellow. Non-pleochroic.
$\alpha = 1.662$, $\beta = 1.680$, $\gamma = 1.699$. $B = .037$
Biaxial, positive. Op. pl. $||$ to (001) $Z \perp$ to (010) (−) or $\perp$ 100 (+).

Fragments irregular, colorless, and with bright interference colors. Olivine usually occurs in sediments as irregular and much-fractured grains, showing traces of decomposition and giving good axial bar interference figures.

Rare as a detrital. Found as shore or dune sand near source. Derived from basic and ultrabasic igneous rocks.

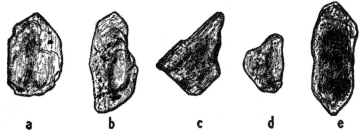

FIG. 228.—Olivine: *a-e,* shore sand, Red Sea.

OPAQUE MINERALS
Hematite:
Fe_2O_3; Hexagonal-rhombohedral
$G = 5.2$; $Dc = > 81$; $Mag = II$
Metallic to earthy luster. Indian red to black in reflected light.

Common as microscopic inclusions; also as irregular powdery aggregates; as oölites and pseudomorphs after fossils; rarely minutely botryoidal; also as grain coatings.

Ilmenite:
$FeTiO_3$; Orthorhombic
$G = 4.6-4.9$; $Dc = > 33.7, < 81$; $Mag = II$
Metallic luster. Brownish to purplish black in reflected light.

Common in sediments as irregular to well-rounded grains; often in part altered to leucoxene.

Leucoxene:
Composition and crystallization uncertain
$G = 3.5\text{-}4.5$; $Dc = ?$; $Mag = IV$
Dull luster. Dead white in reflected light.

Appears as rounded grains, sometimes with unaltered core of ilmenite. Mat surface sometimes minutely pitted.

Limonite:
$Fe_2O_3 \cdot nH_2O$; Amorphous
$G = 3.8$ (approx.); $Dc = 6\text{-}7$; $Mag = $ variable
Earthy to metallic luster; ochre yellow, brown, to brownish black in reflected light.

As rounded granules or as powdery aggregates and coatings. Also as pseudomorphs and decomposition product of pyrite, marcasite, and glauconite.

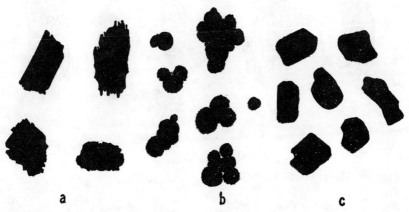

FIG. 229.—Opaque minerals: *a*, pyrite, Wadhurst clay, Kent; *b*, spherulitic pyrite from clay-filled pocket in Niagaran limestone, Joliet, Illinois; *c*, magnetite, Bagshot sand, Essex.

DESCRIPTION OF MINERALS 441

Magnetite:
$FeO.Fe_2O_3$; Isometric, octahedra; octahedral parting.
$G = 5.17$; $Dc = > 33.7, < 81$; $Mag = I$
Metallic luster. Bluish black in reflected light.

Angular and well-rounded grains abundant; crystal facets noted on some grains. Distinguished from ilmenite with difficulty. Marked by strong magnetic character and crystal form.

Marcasite:
FeS_2; Orthorhombic; crystals poor, commonly radiating, fibrous.
$G = 4.887$; $Dc = > 33.7, < 81$; $Mag = III$
Metallic luster. Pale yellow in reflected light, often tarnished dull.

Small grains, irregular, readily decomposed; radial structure common. Authigenic.

Pyrite:
FeS_2; Isometric, cubes and pyritohedra, often striated; conchoidal fracture.
$G = 5.02$; $Dc = > 33.7, < 81$; $Mag = IV$
Metallic luster. Pale brass-yellow in reflected light, readily tarnished.

Commonly well crystallized, though detrital euhedra rare; in sediments often as globular to irregular aggregates, or clusters of globules; also as nodules and small concretions. Authigenic for the most part.

ORTHOCLASE AND MICROCLINE
$K_2O \cdot Al_2O_3 \cdot 6SiO_2$
Monoclinic
Perfect (010) (001) cleavage

G = 2.56
Dc = 6-7
Mag = IV

Colorless
$\alpha = 1.518$, $\beta = 1.524$, $\gamma = 1.526$. B = .008
Biaxial, negative. Y or Z = b, X $\wedge$ $a = 5°$
2V = 2-70°. Dispersion weak

Orthoclase is found as angular to rounded detrital grains in sandstones and as authigenic euhedra in dolomites and limestones. Grains flattened parallel to (001) most abundant. A few rectangular grains showing Carlsbad twinning occasionally seen. Orthoclase is commonly much altered and rendered nearly opaque by alteration products (kaolin and sericite). Such grains appear white in transmitted light and show aggregate polarization between crossed nicols. Secondary enlargement known. Observed as "rim" of limpid feldspar added in optical continuity to clouded detrital grain. The nucleus may be microcline.

Marked by low specific gravity and low index. Distinguished from quartz by lower index of refraction, by cloudy alteration products (dusty appearance), by cleavage, and by biaxial figure. Allogenic feldspar derived from acid igneous rocks and gneisses.

PLAGIOCLASE FELDSPARS
$Na_2O \cdot Al_2O_3 \cdot 6SiO_2$ (albite)
$CaO \cdot Al_2O_3 \cdot 2SiO_2$ (anorthite)
Triclinic; polysynthetic twinning on (010) almost universal
Perfect (001) and less perfect (010) cleavage

G = 2.60 (albite)-
2.76 (anorthite)
Dc = 6-7
Mag = IV

Colorless
$\alpha = 1.525$, $\beta = 1.529$, $\gamma = 1.536$ (albite); $\alpha = 1.576$, $\beta = 1.584$, $\gamma = 1.588$ (anorthite). B = .011-.012
Biaxial, positive (albite), negative (anorthite)
Extinction, oblique, varies from 4° (albite) through 0° to 37° (anorthite) on grains lying on (001) cleavage and from 18° (albite) through 0° to 36° (anorthite) on grains lying on (010) faces
2V = 74-88°

Plagioclase grains generally basal (001). Marked by low relief (in Canada balsam). Usually show conspicuous multiple twinning bands under crossed nicols. Sometimes zoned. May be cloudy with alteration products mainly kaolin and mica; such products are usually most conspicuous along cleavage cracks. Albite may be authigenic, as a "rim" of limpid feldspar around a detrital grain or as minute euhedra in dolomites and limestones.

DESCRIPTION OF MINERALS 443

Characterized by twinning bands, low index, and low specific gravity. Distinguished from quartz by twinning, cleavage, and ease of alteration. The sodic members of this group are more common than the calcic members in sediments, but distinction between the several subspecies is difficult. Derived from crystalline igneous rocks and metamorphic rocks.

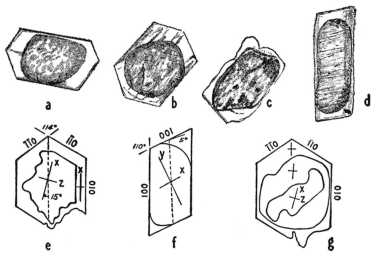

FIG. 230.—Feldspar: *a-c*, secondary enlargement of detrital feldspars, New Richmond sandstone, Minnesota (drawn from photographs of S. S. Goldich), microcline nucleus with orthoclase rim; *d*, secondarily enlarged feldspar, Bonneterre dolomite, Missouri; *e-g*, diagrams showing optical relations in enlarged feldspar grains (after Goldich); *e*, (001) section; *f*, (010) section; *g*, (001) section of a grain with two zones of secondary enlargement.

PLEONASTE (*see* Spinel)
PICOTITE (*see* Spinel)
PYRITE (*see* Opaque Minerals)

PYROXENES (Monoclinic) (*for orthorhombic pyroxenes, see* Hypersthene)
Augite and Diallage:
$CaO.2(Mg,Fe)O,(Al,Fe)_2O_3.3SiO_2$ $G = 3.4$
Monoclinic $Dc = 6-8$
Perfect prismatic (110) cleavage at about   Mag = II or III
  right angles; rarely ‖ to (100)

Color usually pale brownish gray; also pale grayish green
Non-pleochroic except lavender variety (titanaugite) or soda-rich species (ægirinaugite)
$\alpha = 1.696$-1.700, $\beta = 1.702$-1.718, $\gamma = 1.714$-1.742. $B = .018$-$.043$
Biaxial, positive. Optic plane (010). $Y = b$. $Z \wedge c$ 44-48°.
$2V = 59$-$67°$

Grains usually elongate worn cleavage fragments, sometimes with dentate ends Poorly rounded or irregular. Marked by dark platy inclusions or by cloudy alteration products. Some augite is marked by fine ruling structure ("herringbone structure") (variety diallage), which is commonly at right angles or diagonal to the elongation of the grain. In rare cases zoned; also multiple-twinned.

High index, high birefringence, high extinction angle, and brownish color are characteristic. Rare except in glacial sediments. Varieties: Soda-augite (ægirinaugite) decidedly pleochroic and green. Titanaugite (titaniferous) also decidedly pleochroic, of violet color, and of higher index. Augite derived mainly from intermediate and basic igneous rocks.

Diopside:
A pyroxene closely resembling augite and not always certainly distinguished from it. Has lower index ($\alpha = 1.664$, $\beta = 1.671$, and $\gamma = 1.694$, $B = .030$) and lower extinction angle (38½°). Diopside is seemingly more resistant than augite, and moderately well-rounded clear grains are observable. Grains prismatic, sometimes with dentate ends. Marked by lack of color or very pale green appearance in detrital grains. Derived mainly from metamorphic contact rocks and schists.

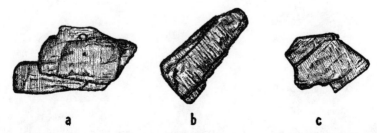

FIG. 231.—Pyroxene: *a-c,* augite, variety diallage, Lake Michigan shore sand, Illinois.

DESCRIPTION OF MINERALS 445

Quartz, Chalcedony, Chert $\quad$ G = 2.66
SiO$_2$ $\quad$ Dc = 6-7 (chalcedony, 8-9)
Trigonal: Hexagonal prisms and pyramids
Cleavage lacking. Poor rhombohedral parting $\quad$ Mag = IV

Colorless
$\epsilon = 1.553$, $\omega = 1.544$. B = .009
Uniaxial, positive. Rarely anomalously biaxial
Extinction sometimes wavy

Ordinary quartz occurs in brilliant, sharply angular grains marked by conchoidal fracture to beautifully rounded grains with a smooth dull or frosted surface. "Platy quartz," composed of several parallel or subparallel lamellæ of slightly different optical orientation and grains possessing parting, closely resemble grains of plagioclase lying on the 010 cleavage. Mackie classified the detrital quartz grains on the basis of contained inclusions: (a) with automorphic mineral inclusions, (b) with liquid and gas inclusions, usually in well-defined planes, (c) with acicular (rutile?) inclusions, (d) with no inclusions.

Commonly subject to secondary enlargement yielding well-terminated euhedra with detrital grain as nucleus. Continued secondary growth results in quartzitic mosaic of interlocking quartz grains in each of which the original grain is preserved as a dusty outline. Observed best in thin section.

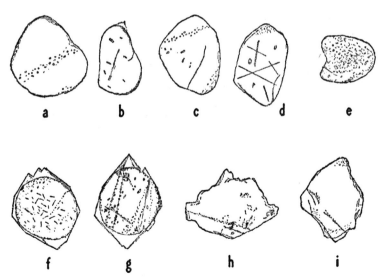

Fig. 232.—Quartz: *a-e*, rounded grains from St. Peter sandstone, St. Paul, Minnesota; *f-g*, secondarily enlarged grains, Klondike, Missouri; *h-i*, angular grains, glacial outwash, Marchington Lake, District of Kenora, Ontario.

Cryptocrystalline quartz (chalcedony), especially chert, also commonly detrital. Chert appears as a very fine granular mosaic which shows aggregate polarization under crossed nicols. May be variously colored owing to pigmentation, yellow (organic), gray (pyrite or graphite), red (hematite). Often traversed by minute quartz veinlets.

In thin section quartz exhibits low interference colors of the first order, principally white and gray, but in detrital well-worn grains the quartz shows brilliant colors of higher orders concentrically arranged.

Distinguished by low relief (in Canada balsam), low specific gravity, abundance, planes of fluid and gas inclusions, uniaxial positive character. An ubiquitous mineral derived from acid igneous rocks, especially granite, from metamorphic schists, gneisses, and quartzites, from veins and from older sediments. At times difficult to distinguish from orthoclase.[1]

Rutile
TiO_2

Tetragonal: Prismatic; geniculate twins
Distinct (100) (110) cleavage; also (111)

$G = 4.24$
$Dc = 5\text{-}6$
$Mag = IV$

Yellow, reddish brown, red
Pleochroism faint; maximum absorption parallel to principal axis
$\epsilon = 2.903$, $\omega = 2.616$. $B = .287$
Unixial, positive. Parallel extinction. Positive elongation

Grains irregular, generally elongate, prismatic forms with rounded pyramidal ends common. Twinned crystals rare. Longitudinal and oblique striations common. Inclusions abundant. Rutile shows the same color under crossed nicols as in ordinary light owing to its extreme birefringence. Deep red-brown varieties nearly opaque. Has very high relief in any medium and therefore broad, dark borders around the grains.

Characterized by form, high relief, high birefringence, deep color, and striæ. May be authigenic, but worn character of most rutile suggests detrital origin. Derived from acid igneous rocks and crystalline metamorphics.

Serpentine (*see* Chlorite)

---

[1] For means of distinction see R. Dana Russell, Frequency percentage determination of detrital quartz and feldspar: *Jour. Sed. Petrology*, vol. 5, pp. 109-114, 1935.

DESCRIPTION OF MINERALS 447

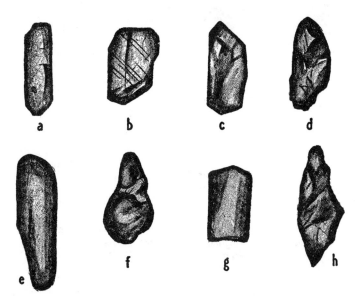

Fig. 233.—Rutile: *a-e*, Tertiary, New South Wales; *f-g*, glacial sand, Sudbury, Suffolk; *h*, Lancashire sand (Pleistocene), Lancashire.

Siderite
$FeCO_3$
Hexagonal-rhombohedral; rhombohedra
Perfect rhombohedral ($10\bar{1}1$) cleavage

G = 3.89
Dc = 3.5-4 (Berg)
Mag = II

Brown, yellow, gray
$\omega = 1.875$, $\epsilon = 1.633$. B = .242
Uniaxial, negative

Found as subangular to rounded rhombohedra. Also occurs as spherulitic or slightly ovate grains with radial structure. Soluble in acid.

Rare as detrital. Constituent of some iron ores, especially "clay ironstone."

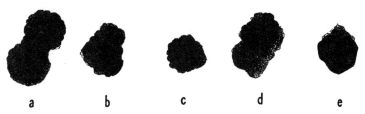

Fig. 234.—Siderite: *a-e*, spherulitic siderite, Fairlight clay, Sussex.

SILLIMANITE
$Al_2O_3 \cdot SiO_2$
Orthorhombic. Acicular parallel to c
Perfect (100) cleavage

$G = 3.23$
$Dc = 4\text{-}5$ (Berg)
$Mag = IV$

Colorless
$\alpha = 1.659$, $\beta = 1.660$, $\gamma = 1.680$. $B = .021$
Biaxial, positive. $X = b$, $Z = c$. Optic plane (100)
Extinction parallel
$2V = 20°$. Dispersion weak
Elongation positive

Grains irregular to short prismatic. Marked by longitudinal splitting and striæ parallel to length. Good interference figures rare, since plane of optic axes and cleavage are same. Fibrolite, or fibrous variety, common.

Distinguished from kyanite by straight extinction and from andalusite by positive elongation. Distinguished from both by its higher birefringence. Derived from metamorphosed argillaceous rocks.

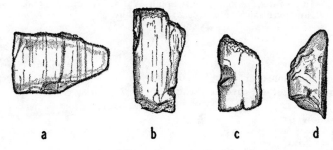

Fig. 235.—Sillimanite: *a-d,* alluvials, Ceylon.

DESCRIPTION OF MINERALS 449

SPINEL (*including* Pleonaste and Picotite)
(Mg,Fe)O.Al$_2$O$_3$ (spinel, pleonaste) G = 3.6-3.8 (spinel)
(Mg,Fe)O.(Al,Cr)$_2$O$_3$ (picotite) G 4.08 (pleonaste)
Isometric; octahedral Dc = 6-7
Cleavage: (111) imperfect to none; Mag = III-IV
  conchoidal fracture

Colorless (pure spinel), grass-green (pleonaste), coffee-brown (picotite)
$n =$ 1.718-1.75, spinel; 1.77, pleonaste (ceylonite); 2.05, picotite
Isotropic

As slightly worn octahedra to well-rounded grains. Grain surface marked by conchoidal fracture and pitting. Inclusions rare.

Common spinel (colorless to faint pink) is very similar to garnet (almandite), and unless crystal form is observable it is difficult to distinguish one from the other. Spinel, however, has refractive index lower than methylene iodide (1.74) while that of garnet is usually higher. Pleonaste and picotite characterized by their color. Spinel derived from metamorphic schists and limestone; pleonaste and picotite derived from basic and ultrabasic igneous rocks as well as metamorphic sources.

FIG. 236.—Spinel: *a-c*, variety ceylonite, Tertiary, New South Wales.

Staurolite

$2FeO.5Al_2O_3.4SiO_2.H_2O$ $G = 3.7$
Orthorhombic: Short prisms parallel to c. $Dc = 6\text{-}7$
Cruciform twins
Distinct (010) cleavage; (110) imperfect. $Mag = III$
Hackly fracture.

Yellow, gold, brown
Pleochroism marked, $X < Y < Z$. X = colorless, Y = pale yellow, Z = golden yellow. Also straw-yellow to russet brown
$\alpha = 1.736$, $\beta = 1.741$, $\gamma = 1.746$. $B = .010$
Biaxial, negative. $Z = b$, $X \wedge c = 0°$. Bx_a nearly | (001)
$2V = 88°$. Dispersion weak

Irregular, somewhat platy grains, determined by cleavage, marked by hackly to subconchoidal fracture. Well-formed crystals rare. Inclusions numerous, sometimes imparting a porous ("Swiss cheese") appearance to the grains. Minerals found as inclusions are garnet, tourmaline, rutile, biotite, and carbonaceous matter. Inclusions more abundant in deeper-colored varieties. Exhibits bright interference colors.

Characterized by color and pleochroism, high index, porous appearance of some varieties, conchoidal fracture. Derived from crystalline schists.

Titanite

$CaO.TiO_2.SiO_2$ $G = 3.5$
Monoclinic: wedge-shaped, flattened ‖ to (001), etc. $Dc = 5\text{-}6$
(110) cleavage rather distinct; difficult ‖ to (110) and $Mag = IV$
(112); (111) rare

Colorless, pale yellow, light brown
Pleochroism weak. X = nearly colorless, Y = pale greenish, etc., Z = yellow, reddish, brownish
$\alpha = 1.900$, $\beta = 1.907\pm$, $\gamma = 2.034$. $B = .134$
Biaxial, positive. $Y = b$, $Z \wedge c = 51°$. Optic plane (010)
$2V = 27°\pm$
Dispersion very strong, $r > v$; elongation negative

Often in diamond-shaped euhedral grains. More commonly as irregular grains of subangular aspect. Grains chipped and marked by conchoidal fracture. Exhibits same color under crossed nicols as in ordinary light owing to high birefringence. Many grains fail to show complete extinction in white light due to high dispersion. The grain turns bluish as the extinction position is reached. Such grains yield good interference figures. The figure is marked by a large number of color bands, very small optic angle, and by color fringes along the brushes (due to dispersion). Decomposition products produce dusky interior in some grains. Network of cracks due to fracture and poor cleavage mark some titanite.

Titanite is marked by its color, high index, extreme birefringence, and incomplete extinction in white light and by its interference figure. Derived from acid and intermediate plutonic igneous rocks. Also from schists and gneisses.

DESCRIPTION OF MINERALS 451

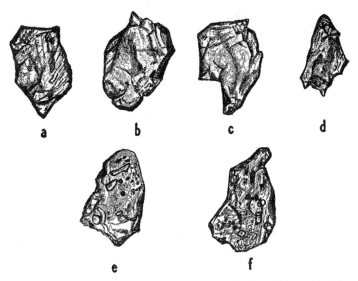

FIG. 237.—Staurolite: *a-d,* Trias, Yorkshire; *e-f,* Corallian sand (Jurassic), Oxford.

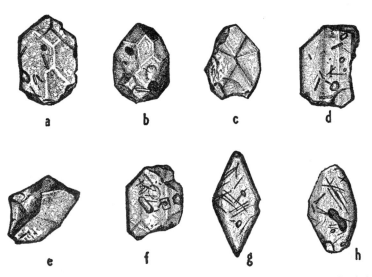

FIG. 238.—Titanite: *a-f,* Aptian, Surrey; *g-h,* Nutfield buffstone, England (after Newton).

Topaz
2(AlF)O.SiO$_2$
Orthorhombic: Elongation c, often vertically striated
Perfect (001) cleavage; imperfect ‖ to (201) and (021)

G = 3.58
Dc = 6-7
Mag = IV

Colorless
$\alpha = 1.619$, $\beta = 1.620$, $\gamma = 1.627$. B = .008
Biaxial, positive. X = a, Z = c. Optic plane (010)
2V = 49-66°
Dispersion distinct $r > v$

Irregular fractured grains. Basal grains (due to perfect basal cleavage) common on which well-centered interference figures are readily obtained. Interference colors bright.

Marked by high relief (in balsam), fracture, and optic character. Resembles some andalusite, from which it is distinguished by optic sign, lack of pleochroism, basal cleavage, and general freedom from cloudy alteration products. Derived from pegmatitic granites and greisen.

Tourmaline (Indicolite)
(Na,Ca)R_3(Al,Fe)$_6$B$_3$Si$_6$O$_{27}$(O,OH,F)$_4$ with
R = Mg,Fe″, Fe‴, Al,Li,Mn and Cr
Hexagonal-rhombohedral; striated prismatic
Cleavage lacking or poor (11$\bar{2}$0) and (10$\bar{1}$0); basal parting

G = 3.0-3.3
Dc = 5-6
Mag = III

Yellow-brown, dark brown, indigo, to black
Pleochroism strong. $\omega > \epsilon$. Common tourmaline: ω = dark brown, ϵ = honey-yellow. Variety indicolite: ω = indigo blue to black, ϵ = pale violet to colorless
$\epsilon = 1.621$-1.658, $\omega = 1.636$-1.698. B = .019-.032
Uniaxial, negative
Extinction parallel to length (and to striations)
Elongation negative

Detrital tourmaline occurs as (a) elongate prismatic grains, (b) more commonly irregular fractured pieces, and (c) well-rounded oval grains. The prismatic grains generally with fractured ends and well-marked striations and markedly pleochroic. Pleochroism of irregular to well-rounded grains variable to lacking. In the latter case a well-centered uniaxial figure may be seen. Inclusions common.

Characterized by color and strong pleochroism, negative uniaxial figure. Distinguished from biotite in detrital form by marked pleochroism (biotite flakes are basal and therefore non-pleochroic) and by lack of cleavage. Derived from pneumatolytic rocks, pegmatites, schists, gneisses, marbles.

DESCRIPTION OF MINERALS 453

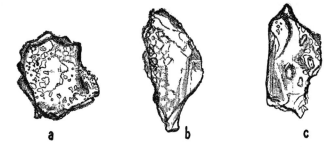

Fig. 239.—Topaz: *a-c*, alluvials, Nigeria.

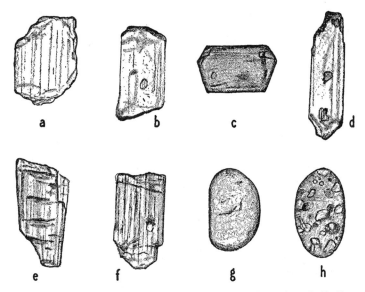

Fig. 240.—Tourmaline: *a-d,* Pliocene, Cornwall; *e-f,* variety indicolite, shore sand, Cornwall; *g-h,* St. Peter sandstone, St. Paul, Minnesota.

TREMOLITE (*see* Amphiboles)

VOLCANIC GLASS
Variable composition, from that of rhyolite to basalt $G = 2.2\text{-}2.7$
Amorphous
Conchoidal fracture

Colorless, green, yellow to brown
$n = 1.48\text{-}1.61$ (increases with decrease of SiO_2)
Isotropic

Splinters, fibers, bubble-walls and cellular pumice fragments of volcanic glass are common constituents of some sediments, both medium- and fine-grained. Such fragments are collectively called shards. Usually associated with euhedral crystals or crystal fragments of igneous rock-making minerals.

All gradations exist between normal sediments wholly free of volcanic admixtures and true tuffs. Tuffaceous sediments marked by presence of glass shards or pseudomorphs thereof. Glass alters or devitrifies readily. Beidellite and montmorillonite are two clay minerals resulting from such alteration. Bentonite is a clay derived by alteration of an ash. Volcanic glass characterized by low refractive index, low specific gravity, isotropic character, and shard structures.

ZIRCON
$ZrO_2.SiO_2$ $G = 4.6\text{-}4.7$
Tetragonal: Short prisms with pyramids $Dc = 6\text{-}7$
(110) cleavage rare; (111) poor $Mag = IV$

Usually colorless; some grains are mauve, yellow to brown. Strongly colored varieties are pleochroic.
$\epsilon = 1.985\text{-}1.991$, $\omega = 1.926\text{-}1.936$. $B = .055\text{-}.059$
Uniaxial, positive. Anomalously biaxial, small $2V$
Extinction parallel to c
Elongation positive

Zircon often shows euhedral form even in far-traveled sands with well-marked crystal facets. Commonly prisms (100) or (010) or both with pyramid terminations (111), (331), (311). Basal grains rare. Well-worn zircons usually elongate, elliptical to globular. Contains large rod-shaped inclusions of other minerals and large liquid or gas inclusions. Many grains show well-defined zoning; others marked by "dusky" appearance due to numerous inclusions crowded together.

Characterized by crystal form, high index and inclusions and at times by its zoning. Differs from titanite by its parallel and complete extinction in white light and by its crystal habit. Closely resembles xenotime. Derived from acid to intermediate igneous rocks.

DESCRIPTION OF MINERALS

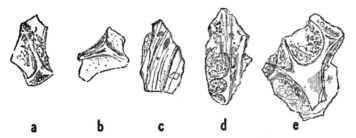

Fig. 241.—Volcanic glass fragments: *a-e,* volcanic ash, from Brule clay, Morrill County, Nebraska.

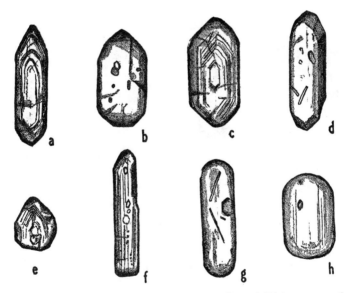

Fig. 242.—Zircon: *a-c,* Burke County, North Carolina; *d,* Pleistocene sand, Middlesex; *e-f,* glacial sand, Sudbury, Suffolk; *g-h,* St. Peter sandstone, St. Paul, Minnesota.

ZOISITE
$4CaO.3Al_2O_3.6SiO_2.H_2O$
Orthorhombic: Prismatic (c) grains
Cleavage: (010) very perfect; also ∥ to (100)

$G = 3.3$
$Dc = 8-9$ (clinozoisite)

Colorless to shades of rose, green, brown. Pleochroism faint
    Variety thulite shows distinct pleochroism, pink to deep rose or yellow
$\alpha = 1.700$, $\beta = 1.702$, $\gamma = 1.706$. $B = .006$
Biaxial, positive. $X = c$ and $Y = b$, or $X = b$ and $Y = c$
$2V = 0\text{-}60°$. Abnormal, ultra-blue interference colors

Generally observed as colorless grains ∥ to (010). Contains inclusions of amphibole microlites.

Characterized by high index, abnormal interference colors. Derived from crystalline schists, metamorphosed basic rocks.

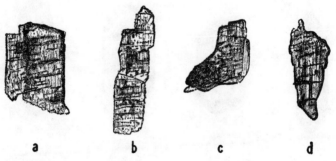

    a   b   c   d

FIG. 243.—Zoisite: *a-d*, crushed fragments.

DETERMINATIVE MINERAL TABLES

USE OF THE TABLES

The common minerals of the sedimentary rocks, both detrital and authigenic, are tabulated and arranged according to their properties. The arrangement is largely on the basis of optical characteristics inasmuch as these are the most useful for purposes of identification.

The tables are constructed so that the more easily determined characteristics form the main subdivisions. Entering the table with a given set of properties leads ultimately to a small group of minerals or a single

DESCRIPTION OF MINERALS 457

mineral. As a rule, however, a group rather than a single species is the end of the search. The student will then have to turn to the more detailed description and tabulation of properties of the minerals in question given in the preceding section of this book.

A dichotomous classification has been found by experience, both of the authors and of their students, to fail because of the indeterminable nature of some of the minor features by means of which one mineral species is finally separated from all others. A careful study of a group of properties and the figured drawings of the grains should lead the student to a final determination rather than a single property. The optic sign, for example, is sometimes not determinable by reason of an aggregate structure of the grains, or because 2V approximates 90° or no satisfactory interference figure is available owing to unfavorable orientation or cleavage.

To facilitate identification the student should keep a careful record of the various optical properties of an unknown species. Haphazard and unsystematic use of the tables is to be discouraged, as it often leads to wrong results and to a waste of time.

The student is cautioned against a too restricted use of the tables given here, since these tables include only the more common minerals of sediments. Locally other minerals not here tabulated or described will appear, and recourse must be had to larger works in which more complete tables and data are available.

For teaching purposes, however, such larger tables without grain drawings are difficult for the beginner to handle. For this reason we have included the abbreviated tables given here. Yet we believe that they are full enough so that for routine work they will prove adequate.

At the end of this chapter we have added a few supplementary tables for reference purposes summarizing the principal modes of occurrence of the common minerals of the sediments and their relative stability. Sample form sheets are added on which systematic records of sediments examined may be kept.

TABLES OF DETRITAL MINERALS

TABLE NO.	MINERAL GROUP
I.	Opaque Minerals
II.	Isotropic Minerals
III.	Anisotropic Colorless Minerals
IV.	Anisotropic Colored Minerals
V.	Aggregate Grains and Rock Fragments

TABLE I
Opaque Minerals

Color in Reflected Light	Mineral
White	Leucoxene
Black	Magnetite, Ilmenite, Picotite
Yellow	Pyrite, Marcasite
Red	Hematite
Yellow-brown	Limonite

NOTE.—Large grains of some deeply colored minerals, notably glauconite, hornblende, and tourmaline, may appear opaque. See table of colored minerals.

TABLE II
Isotropic Minerals

Relief in C.B.	Very Low	Moderate	Very High
Colorless	Opaline Silica, Volcanic Glass	Collophane	Fluorite ($n = 1.43$) Spinel ($n = 1.71$-1.75) Garnet ($n = 1.70$-2.0)
Colored	Volcanic Glass	Collophane	Spinel (pink) Pleonaste (Ceylonite—green) Picotite (coffee-brown) Garnet (pink, orange, yellow, green)

NOTE.—Basal sections of minerals of the tetragonal and hexagonal systems are isotropic. Such grains, however, yield well-centered uniaxial interference figures and are thus distinguished from the minerals listed above.

DESCRIPTION OF MINERALS

TABLE III
COLORLESS ANISOTROPIC MINERALS

	MEAN REFRACTIVE INDEX					
	< 1.54 (C.B.) Relief Low in C. B.		> 1.54 < 1.68 Relief Low to Moderate in C.B.		> 1.68 (piperine) Relief High in C.B.	
UNIAXIAL						
	....		* Quartz	+	* Zircon	+
	** Gypsum	+	** Calcite	—	** Siderite	—
			** Dolomite	—		
BIAXIAL						
	** Calcite	—	....		Clinozoisite	+
	** Dolomite	—			** Diopside	++
					Titanite	++
			* Anhydrite	++		
			Barite	++		
			Celestite	++		
			** Diopside	++		
			* Sillimanite	++		
			Topaz	++		
			Zoisite	++		
	Plagioclase	±	Plagioclase	±	Olivine	±
			Kaolin Group	±		
	Nontronite	±				
			Muscovite	—		
	Microcline	—	* Andalusite	—	** Kyanite	—
	Montmorillonite	—	Aragonite	—		
	Beidellite	—	Muscovite	—		
	Orthoclase	—	** Tremolite	—		
SPECIFIC GRAVITY	> 2.85		< 2.85 (bromoform)		> 2.85	

NOTES.— * notable cleavage or elongation and parallel extinction.
** notable cleavage or elongation with inclined extinction. +, ±, and − refer to optic sign.

TABLE IV

COLORED ANISOTROPIC MINERALS

	MEAN REFRACTIVE INDEX		
	< 1.54 (C.B.) *Relief Low in C.B.*	> 1.54 < 1.68 (piperine) *Relief Low to Moderate in C.B.*	> 1.68 (piperine) *Relief High in C.B.*
PLEOCHROIC UNIAXIAL	....	TOURMALINE (br., y.) INDICOLITE (indigo)	RUTILE (br.-r., y.) CASSITERITE (c., y., r.-br.)

PLEOCHROIC BIAXIAL	....	AMPHIBOLES
		  Tremolite (c., pale g.)
		  Actinolite (g.)
		  Hornblende (bl.-g., br., y.-g.)
		  Glaucophane (c., bl., violet)
		ANDALUSITE (rose, c.)
		DUMORTIERITE (bl., violet)
		THULITE (var. zoisite) (rose)
			BROOKITE (y., o., br.)
			EPIDOTE (c., lemon-y.)
			HYPERSTHENE (pink, y., g.)
			MONAZITE (dirty y., br.)
			STAUROLITE (golden y.)
			TITANITE (c., y., br.)
			DUMORTIERITE (bl., violet)

NON-PLEOCHROIC UNIAXIAL	DOLOMITE (pale br., c.)	DOLOMITE (pale br., c.) *BIOTITE (br., g.) *CHLORITE (g., y.)	ANATASE (y., bl.) ZIRCON (c., pale y., br.) CORUNDUM (c., bl.)
NON-PLEOCHROIC BIAXIAL	....	OLIVINE (pale g., y., c.) SERPENTINE (y., g.) DIOPSIDE (pale g., c.)	AUGITE (pale br., c.) DIOPSIDE (pale g., c.) OLIVINE (pale g., y., c.)

NOTE.—Basal sections of uniaxial minerals and optic axis sections of biaxial minerals are non-pleochroic.
* Pseudo-uniaxial and non-pleochroic in detrital grains (owing to very perfect basal cleavage).
bl.—blue; br.—brown; c.—colorless; g.—green; o.—orange; r.—red; y.—yellow.

DESCRIPTION OF MINERALS

TABLE V

COMPOUND GRAINS AND AGGREGATES (SHOWING AGGREGATE POLARIZATION)

Rock fragments	Chalcedony Chert Quartzite	Low relief, transparent, coarse to very fine mosaic under crossed nicols. Usually colorless; sometimes pale yellow, light to dark brown.
	Shale Slate	Marked by semi-opaque, "dirty" appearance. May be streaked or laminated. Gray to black.
	Felsitic fragments	Contain feldspar microlites, some isotropic glass, occasional biotite flakes, and small quartz grains. Larger fragments may show flow structure.
	Basaltic fragments	Somewhat similar to felsitic material, except much darker and often altered to green or yellow chloritic material.
	Limestone and dolomite grains	In recent sediments, especially glacial, small aggregates of calcite and/or dolomite may be present. Marked by high-order interference tints.
Compound grains	Glauconite	A semi-opaque mineral; grass-green in transmitted light; greenish black in reflected light. Grains globular, ovoid, compound pellets.
	Microperthite	Orthoclase feldspar in which are thin lamellæ of plagioclase. Plagioclase bands marked by twinning.
	Leucoxene	A semi-opaque mineral of high relief. Brilliant white in reflected light; dull luster. May be attached to core of ilmenite from which it is derived.
	Serpentine	Pale yellow to dark green aggregate; fibrous, semi-radiating to felted mass. Black granular inclusions.
	Pinite	Aggregate of mica flakes derived by alteration of cordierite. Brilliant polarization of sericite characteristic. Associated with iron ores.
Minerals with alteration coating		Certain easily altered minerals, notably the amphiboles, pyroxenes, and feldspars, may be so completely altered as to be unidentifiable. The identity of such grains is best established by finding partially altered grains with all degrees of transition to the fresh unaltered mineral.

TABLE 44
Detrital and Associated Minerals
(After Milner, *Mining Mag.*, vol. 28, p. 84, 1923)

*** Anatase (c)
** Andalusite (l)
* Apatite (r)
** Augite (r)
*** Barite (r)
** Biotite (l)
*** Brookite (c)
† Calcite (l) or (A)
*** Cassiterite (l)
*** Chalcedony (c)
† Chlorite (c)
*** Chromite (r)
*** Columbite (vr)
* Cordierite (l)
*** Corundum (l)
*** Diamond (vr)
** Epidote (c) or (A)
*** Fluorite (l)

*** Garnet (c)
* Glauconite (c)
* Glaucophane (r)
*** Gold (vr)
** Gypsum (l) or (A)
† Hematite (A)
** Hornblende (c)
** Hypersthene (r)
** Ilmenite (c)
† Kaolinite (A)
*** Kyanite (c)
*** Leucoxene (c)
† Limonite (A)
** Marcasite (l)
*** Magnetite (c)
** Microcline (l)
*** Monazite (r)
*** Muscovite (c)

* Olivine (r)
** Orthoclase (c)
** Plagioclase (c)
* Pyrite (c)
** Pyrolusite (r)
* Pyrrhotite (l)
*** Quartz (c)
*** Rutile (c)
† Siderite (l) or (A)
*** Sillimanite (l)
*** Spinel (r)
*** Staurolite (c)
*** Titanite (r)
*** Topaz (c)
*** Tourmaline (c)
*** Wolframite (l)
*** Xenotime ? (vr)
*** Zircon (c)

(c)—common
(l)—local
(r)—rare
(vr)—very rare
(A)—alteration product (secondary)

*** stable
** moderately stable
* unstable
† stable as a secondary product

TABLE 45
Detrital Mineral Suites Characteristic of Source Rock Types

Acid Igneous Rocks

Apatite
Biotite
Hornblende
Monazite
Muscovite
Titanite
Zircon (euhedra)

Pegmatites

Cassiterite
Fluorite
Topaz
Tourmaline
Wolframite

Ultrabasic Igneous Rocks

Anatase
Augite
Brookite
Chromite
Hypersthene
Ilmenite
Leucoxene
Olivine
Rutile

Dynamic Metamorphic Rocks

Andalusite
Garnet
Glaucophane
Hornblende (blue-green var.)
Kyanite
Sillimanite
Staurolite

Reworked Sediments

Glauconite
Iron ores
Quartz
Rutile
Tourmaline
Zircon (rounded)

TABLE 46
Common Authigenic Minerals

Albite
Anatase
Anhydrite
Aragonite
Barite
Brookite
Calcite
Celestite

Clay minerals
Collophane
Dolomite
Glauconite
Gypsum
Leucoxene
Limonite
Hematite

Orthoclase
Pyrite
Quartz
Rutile
Siderite

SAMPLE FORM FOR RECORDING MINERALOGICAL ANALYSIS OF SEDIMENT

Sample No. ...... Collection date.............. Study ....
Locality ....
Laboratory treatment:
  Sample weight............
  Acid-soluble............(weight)............per cent
  Grade size........................

LIGHT SEPARATE

S. G. ...... Weight...... Per cent...... No. grains counted.......

Species †	Per Cent	Color and Pleochroism	Form *	Remarks **

Etc.

HEAVY SEPARATE

S. G. ...... Weight...... Per cent...... No. grains counted.......

Species †	Per Cent	Color and Pleochroism	Form *	Remarks **
Zircon	22	Colorless	Euhedra; prism and bipyramid.	Acicular inclusions abundant. Zoning common.
Hornblende	14	X = y, Y = bl.-g, Z = g.-bl.	Worn prismatic cleavage frag.	Remarkably fresh.

Etc.   ....Petrographer

** Striations, zoning, twinning, optical anomalies, inclusions, etching, alteration.

* Degree of rounding, crystal habit, colloform, etc.

† Record varieties as separate species. Distinguish, for example, between brown tourmaline and blue tourmaline; between rounded zircon and euhedral zircon.

CHAPTER 18

MINERAL FREQUENCIES AND COMPUTATION

PEBBLE COUNTS

SINCE the coarsest clastics consist of rock fragments, the mineral composition is generally of less interest than, or subordinate to, the composition as expressed in terms of the rock units. Hence "pebble counts" are usually made, the nature of the rock being determined by megascopic study except in a few cases where positive identification must be made, in which case a thin section must be used. The composition of the breccia or gravel is thus determined by sorting over by hand and counting the various rock types represented. This is probably best done from a large sample collected and screened through a coarse sieve. The fines are washed through the sieves and either discarded or reserved for mineralogical analysis. The results can be expressed in percentage by number, or the separated fractions can be weighed and the composition expressed in percentage by weight. Such pebble counts are commonly made as a basis of discrimination between drift sheets of different glacial epochs.[1] The mineral nature of the sand portion has been used for the same purpose and is usually related to the coarser constituents.[2]

THIN-SECTION ANALYSIS

In thin or polished sections the relative volumes of the various minerals are measured. Usually these volume measurements are converted to weight and the final results expressed in weight-percentage. This is the method most commonly used in the study of the igneous rocks. It can be used for some sediments, as, for example, the crystalline dolomitic limestones stained to differentiate between calcite and dolomite. The method is based on the assumption that the sum of the areas of each of

[1] Example: H. R. Wanless, Nebraskan till in Fulton County, Illinois: *Trans. Ill. Acad. Sci.*, vol. 21, pp. 273-282, 1929.

[2] Examples: A. Raistrick. The petrology of some Yorkshire boulder clays: *Geol. Magazine*, vol. 66, pp. 337-344, 1929. Viktor Leinz, Ein Versuch, Geschiebemergel nach dem Schwermineraliengehalt stratigraphisch zu gliedern: *Zeits. f. Geschiebeforschung*, vol. 9, pp. 156-168, 1933.

the components in a random section of a uniform rock is proportional to the volume of that constituent in the rock. The actual areas are rarely estimated; generally linear measurements along two mutually perpendicular directions are made. This is the so-called "Rosiwal method."[1] The linear measurements are reduced to 100, or percentages, the values representing the relative volume of each constituent. These values are then multiplied by the specific gravity of the respective minerals and the whole is again reduced to parts per hundred or percentage by weight. It is generally said that the total linear measurement should exceed 1,000 times the length of the largest grain in order to achieve a fair degree of accuracy. A mechanical stage on the microscope will facilitate this work. Shand[2] developed a recording micrometer to reduce the labor and eye-strain involved in this kind of work. Devices of a similar sort have since been developed by Wentworth[3] and by the Leitz company[4] (Figure 244). An integrating stage based on principles rather different from those employed in the stage of Wentworth is that recently described by Dollar.[5] This stage is a modified mechanical stage with six revolving drums, one for each constituent present in the rock. All of these devices may be attached to the stage of an ordinary petrographic microscope and may be used not only for the measurements of the constituents of thin sections, but also for crushed and screened mill products with particles as small as .006 mm. In the latter case the grains are embedded in sealing wax or other mountant and a polished section is made. Other methods[6] of quantitative mineral analysis of thin sections include the measurement of areas by means of an ocular with ruled micrometer grid or a camera lucida drawing or photograph on paper with ruled grid, by use of a

[1] August Rosiwal, Ueber geometrische Gesteinsanalysen. Ein einfacher Weg zuziffernmässigen Festellung des Quantitätsverhältnisses der Mineralbestandtheile gemengter Gesteine: Verhandl. K. K. Geol. Reichsanstalt, Wien, p. 143, 1898.
[2] S. J. Shand, A recording micrometer for geometric rock analysis: *Jour. Geology*, vol. 24, pp. 394-404, 1916.
[3] Chester K. Wentworth, An improved recording micrometer for rock analysis: *Jour. Geology*, vol. 31, pp. 228-232, 1923.
[4] F. E. Thackwell, Quantitative microscopic methods with integrating stage: *Econ. Geology*, vol. 28, pp. 178-182, 1933.
[5] A. T. J. Dollar, An integrating micrometer for the geometrical analysis of rocks: *Mineral. Mag.*, vol. 24, pp. 577-594, 1937.
[6] For descriptions of several other methods of measuring relative proportions of minerals in thin sections or rock specimens see: Ira A. Williams, The comparative accuracy of the methods for determining the percentage of the several components of an igneous rock: *Am. Geol.*, vol. 35, pp. 34-46, 1905. F. C. Lincoln and H. J. Reitz, The determination of the relative volumes of the components of rocks by mensuration methods: *Econ. Geology*, vol. 8, pp. 120-139, 1913. Albert Johannsen, A planimeter method for the determination of the percentage compositions of rocks: *Jour. Geology*, vol. 27, pp. 276-285, 1919. H. L. Allin and Wilbur G. Valentine, Quantitative microscopic analysis: *Am. Jour. Sci.*, vol. 14, pp. 50-65, 1927.

FREQUENCIES AND COMPUTATION

planimeter on such a drawing or photograph, and by cutting up a drawing or photograph and weighing the cut-out grain images.

These methods work very well for rocks of medium or coarse grain, which are of uniform grain size and without banding. Numerous difficulties are encountered when these requisite conditions are not fulfilled. This is especially

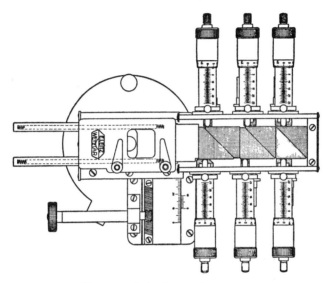

FIG. 244.—Leitz integrating stage.

the case with sediments which are often of too small a grain to permit easy recognition of the minerals, such as the shales, or contain, in addition to large and easily recognized components, a fine-grained paste of ill-defined composition, such as the graywackes. Sections should be cut at right angles to the bedding, otherwise biased results will be obtained.

MINERAL FREQUENCIES

Estimation. Study of published papers on the minerals of sedimentary rocks show no agreement as to the method to be used to express abundance of the several mineral components. The earlier workers simply listed the minerals in order of abundance or used such descriptive terms as "flood," "rare," "abundant," etc., and in some cases tabulated their results, using the initial letters of such adjectives or capital letters or italics to denote degrees of rarity or abundance. Others have devised a "scale of abundance," arranging the descriptive terms or initial letters in order of increasing rarity or abundance. As an example of such a scale we have that given by Milner.[1] He concluded that it is possible after considerable practice to arrive at fairly consistent results by simple estimation by eye after inspection of several random fields. Accuracy was less in residues with small grains. Milner admitted, however, that rarely did two estimations by different persons agree, and hence in order to achieve results that could be duplicated by other investigators or that can be put on record and used by other workers for comparative purposes a more accurate method of frequency determination is necessary.

Term	Symbol	Number
Flood	F	9
Very abundant	A	8
Abundant	a	7
Very common	C	6
Common	c	5
Scarce	s	4
Very scarce	S	3
Rare	r	2
Very rare	R	1

Table after Milner.

Number scales, such as that included with the descriptive scale in the above table, have no particular advantage over those based on descriptive terms, since both are based on estimation rather than actual count. Number scales are of several kinds. One is that proposed by Watts (see table above) which uses the numbers 1 to 9, 1 representing "very rare," 2 representing "rare," etc., and 9 representing "flood." It is interesting to note that Artini [2] apparently first proposed a number scale, but this scale employed the numbers 1 to 10 and used them in the reverse sense to that of Watts, namely, 1 was most copious and 10 represented most rare. Salmojraghi [3] found difficulty in applying this scale and at first reverted to a set of six descriptive terms, namely, *dominant, abundant, frequent, scarce, rare,* and *most rare.* He later adopted a numerical scale from 1 to 10, using 1 for the most rare grains and 10 for the most abundant. This is the scale used by Boswell.[4]

[1] H. B. Milner, *Sedimentary Petrography,* 2nd ed. (1929), p. 386.
[2] E. Artini, Intorno alla composizione mineralogica della sabbie di Alcuni fiumi del Venete, con applicazione terreni di trasporte: *Riv. di Min. Crist. ital.,* Padova, vol. 19, pp. 33-94, 1898.
[3] F. Salmojraghi, Sullo studio mineralogico delle sabbie e sopra un modo di rappresentarne i risultati: *Atti. soc. ital. sci. nat.,* vol. 43, pp. 54-89, 1904.
[4] P. G. H. Boswell, The petrography of the Cretaceous and Tertiary outliers of the west of England: *Quart. Jour. Geol. Soc. London,* vol. 79, p. 226, 1923.

FREQUENCIES AND COMPUTATION

Ultra-dominant	10
Dominant	9
Very abundant	8
Abundant	7
Very frequent	6
Frequent	5
Scarce	4
Very scarce	3
Rare	2
Exceedingly rare	1

As Dryden[1] has pointed out, the frequency numbers used are not numbers in the mathematical sense. The numbers 1 to 5 of a simplified scale are not directly proportional to percentages. "One" does not stand for 20 per cent, "two" for 40 per cent, etc. Instead one means only that a few grains were present in the slide examined, two denotes, perhaps, 5-10 grains, three means about 10 per cent, four some 30-50 per cent and five something over 50 per cent. The numbers do not have any simple relation to each other. Two is not twice one, etc. The numbers are symbols only and are not amenable to the ordinary laws of mathematics.

Counting. In most of the papers that have been published on the petrography of sedimentary rocks the authors reported the abundance of the different minerals by the use of descriptive terms only, though the use of a numerical scale was not uncommon. Since 1920, however, there have been an increasing number of studies in which the method of counting grains and calculating percentage was used. W. F. Fleet[2] exemplifies the worker desiring quantitative results. He mounted the *entire* heavy mineral residue. He concluded that "...to obtain accurate results that may readily be compared with those of other workers actual counting of the grains is considered essential...at first sight, a tedious task, but it has been found that, using 0.4 objective with an eyepiece giving a large field, and by employing a mechanical stage, it is not difficult to traverse the whole width of a slide several times, counting every grain. According to the closeness with which the minerals appear on the mount it may be necessary to count the grains in anything from twelve to fifty fields, and the number of grains may range from 500 to 1,000 or even more." Fleet counted *all* the grains in the entire heavy-mineral residue. In all probability such completeness is seldom necessary and counting a part of the grains, about 300, in part of the residue will

---

[1] A. L. Dryden, Accuracy in percentage representation of heavy mineral frequencies: *Proc. Nat. Acad. Sci.,* vol. 17, pp. 233-238, 1931.
[2] W. F. Fleet, Petrological notes on the Old Red Sandstone of the West Midlands: *Geol. Magazine,* vol. 63, pp. 505-516, 1926.

be satisfactory. Otto[1] has shown that it is possible to quarter down a large heavy-mineral residue without greatly changing the proportions of the minerals of that residue by the use of some such device as his Miscrosplit. Wentworth, Wilgus, and Koch,[2] because of the inherent limitations of the Jones-type sample splitter, devised a rotary type, but this did not materially improve the results in so far as splitting of heavy mineral separates was concerned. Appel[3] and others have also developed microsplitting devices.

It is common practice not only to mount just a fraction of the heavy residue but also to count only a part of the material mounted—a couple of microscope fields, or a hundred grains. At once the question arises, what error will be made in the case of each mineral species if some number n, less than N, the total number of grains, be counted? Dryden[4] has written on this question.

The chance that any grain will belong to a certain species is p, the probability, while the chance that it will not belong to such a category is q. $q = 1 - p$. $p = a/N$ where a is the number of grains of a given species actually present and N is the total number of grains. $q = b/N$, where b is the number of grains not of the species in question and N is the total number of grains as before. For example, if a heavy mineral residue contains 2,000 grains of which 500 are zircons, the chance that any random grain will be zircon is 500/2,000 or $\frac{1}{4}$, while the chance that it will not be zircon is 1-500/2,000 or $\frac{3}{4}$.

The probable error (P.E.) is given by the formula:

$$\text{P.E. (in no. of grains)} = 0.6745\sqrt{npq}$$

where n is the number of grains counted, p is the probability of a given grain being a particular mineral and q is the chance that such a random grain is not the mineral in question. It is clear that for a species of some given frequency, p and q will not vary and that the P.E. will be a function of n, or more correctly $\sqrt{n}$. The probable error in percentage, however, is $(\text{P.E.}/a)100$ in which a is the "true" frequency of the species present. For example, if the heavy mineral content of a sand actually consists of garnet 40 per cent, tourmaline 30 per cent, staurolite 15 per cent, and kyanite 15 per cent, and only 50 grains are counted, what will be the P.E. (in percentage) of the value obtained for garnet?

[1] George Otto, Comparative tests of several methods of sampling heavy mineral concentrates: *Jour. Sed. Petrology*, vol. 3, pp. 30-39, 1933.
[2] C. K. Wentworth, W. L. Wilgus, and H. L. Koch, A rotary type of sample splitter: *Jour. Sed. Petrology*, vol. 4, pp. 127-138, 1934.
[3] J. E. Appel, unpublished research at the University of Chicago.
[4] A. L. Dryden, *loc. cit.*

In this case $p = 40/100$ or $\frac{2}{5}$ and $q = 1 - \frac{2}{5} = \frac{3}{5}$. Hence P.E. (in number of grains) $= .6745 \sqrt{50} \times \frac{2}{5} \times \frac{3}{5} = .6745 \sqrt{12} = 2.33$ grains and $(2.33/20)100 = 11.6$ per cent. This means that the chances are even that the observed frequency of garnet will show a deviation up to 11.6 per cent either way from the "true" frequency of 40 per cent or from 35.4 per cent to 44.6 per cent. Had 200 grains been counted, the chances are even that the observed percentage would lie between 37.7 and 42.3. It is evident that the larger the number of grains counted, the smaller the probable error. In fact, if the number counted is increased by four, the error is reduced to one half. Moreover, the probable error is greatest for the rarer constituents and lowest for the abundant components.

These relations are graphically summarized in Figure 245. Here the relation of the total number of grains counted to the probable error in percentage is shown for several different frequencies. For example, if one wants the probable error to be 10 per cent for those constituents making up 10 per cent of the sample, some 400 grains must be counted. As seen from the diagram, the accuracy increases very slowly after the sharp "bends" of the curves are passed and their nearly straight parts reached. Each worker will have to decide just what accuracy is demanded by each problem. As Dryden points out, the "law of diminishing returns" probably will make us count but a few hundred grains. Perhaps 300 will be satisfactory for most work. Obviously statement of the quantity of a constituent to the nearest whole per cent figure is all that is justified on the basis of the probable error involved. The worker should also state the number of grains counted. Such a statement enables the reader to determine for himself the probable accuracy of the results.

It should be added that the curves shown in Figure 245 and the examples given above on the calculation of the probable error assume that one knows the true frequency value of the mineral in question. Actually this is what one is trying to find. It is not possible to take the percentage determined experimentally and say what its probable error is.

Since Dryden showed that not more than two significant figures are justified when even as many as 4,000 grains are counted, the authors recommend that for research work the entire residue or a *carefully* quartered portion be mounted, which will contain about 300 grains, and the mineral frequencies be determined by count. In oil-field laboratories where the number of samples is large and where time is an important factor and comparison of the results of several workers is not to be made, counting grains may not be desirable or necessary. Even in this

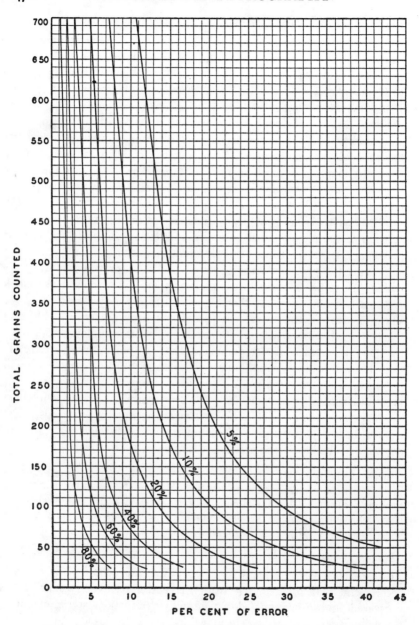

Fig. 245.—Curves for estimation of probable error (after Dryden).

case certain workers[1] have found it "desirable for most work to count all the grains in a slide, and work strictly according to percentages."

Some workers prefer to express the frequency directly in percentage by number of the total number counted. Others have combined the percentage by number with some scale of abundance. W. A. P. Graham,[2] for example, counted 100 heavy mineral grains and expressed his results as follows:

Very rare	V R	1- 2%
Rare	R	3- 4%
Common	C	5- 6%
Very common	V C	7- 8%
Abundant	A	9-10%
Very abundant	V A	10-11%

Evans, Hayman, and Majeed[3] used a series of frequency numbers which bear a logarithmic relation to the actual frequency as determined by counting. Their scale is given below:

Percentage	Frequency Number	Proportion
80	8	Very abundant
40	7	Abundant
20	6	Fairly abundant
10	5	Very common
5	4	Common
2-3	3	Fairly common
1-2	2	Scarce
1/2-1	1	Rare
0-1/2	1*	One grain per slide
0	0	Absent

This scale is straightforward down to frequency 3, but for lower numbers it is slightly arbitrary. This scheme was also used by the same authors in modified form by using + and − so that further subdivisions of the scale were possible.

Since it is well known that mineral frequencies in the different grade sizes of the same sediment are not alike,[4] it is necessary for scientific work to study the mineral composition of each grade size by itself, or in

[1] P. Evans, R. J. Hayman and M. A. Majeed, The graphical representation of heavy mineral analyses: *World Petrol. Congress, Proc. 1933* (London), vol. 1, pp. 251-256.
[2] W. A. P. Graham, A textural and petrographic study of the Cambrian sandstones of Minnesota: *Jour. Geology*, vol. 38, pp. 696-716, 1930.
[3] Evans, Hayman, and Majeed, loc. cit., p. 254.
[4] See for example: L. Hawkes and J. A. Smythe, Garnet-bearing sands of the Northumberland coast: *Geol. Magazine*, vol. 68, pp. 345-361, 1931.

the case of a comparative study of many samples, to study the same grade size in all of the investigated samples (see Figures 246 and 247). Rubey,[1] who attempted to evaluate some of the factors that determine the variations in frequency in the different grade sizes, concluded that the "heavy minerals from at least two size fractions of each sample should be examined. These two fractions should be so chosen that one of them has the same *actual* limits in all of the samples compared and the other represents the same *relative* size within the distribution curves of the different samples." Russell,[2] as a result of a careful study of sands

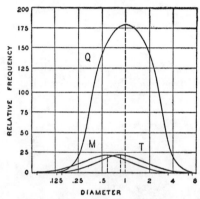

Fig. 246.—Diagram showing the theoretical proportions of quartz (Q), magnetite (M), and tourmaline (T) in the various grade sizes of a water-laid sandstone (after Rubey).

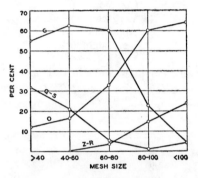

Fig. 247.—Observed relations of mineral frequencies and grade sizes in sample of beach sand (after Hawkes and Smythe). G, garnet; Q-S, quartz and shell; O, orthoclase; Z-R, zircon and rutile.

transported by the lower Mississippi, came to the conclusion that in the case of the light mineral the difficulty arising from the different proportions of minerals in the different grades can largely be overcome by analyzing a single size grade from each sample, providing the grade chosen represents the average composition of the light portion and is near the central portion of the average sample. With the heavy minerals, satisfactory results may be secured only by averaging the percentages of the heavy minerals in two size grades of each sample. One of these grades should be the same for all samples, the other should occupy a definite position with respect to the size frequency distribution of the sample.

[1] William W. Rubey, The size-distribution of heavy minerals within a water-laid sandstone: *Jour. Sed. Petrology*, vol. 3, pp. 3-29, 1933.
[2] R. Dana Russell, The size distribution of minerals in Mississippi River sands: *Jour. Sed. Petrology*, vol. 6, pp. 125-142, 1936.

FREQUENCIES AND COMPUTATION

Presentation of results. The data may be given in tables with mineral species listed against stratigraphic horizon and the quantity of any one species indicated by letter or other symbol or by frequency number or even directly as a percentage. It is difficult readily to visualize all the

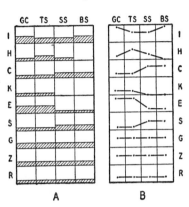

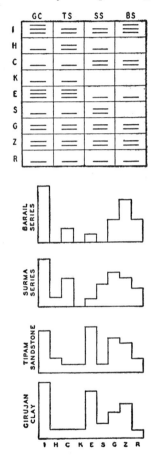

FIG. 248.—Neaverson chart (after Evans, Hayman, and Majeed). GC, Girujan clay; TS, Tipam sandstone; SS, Surma series; BS, Barail series. I, Ilmenite-magnetite; H, hornblende; C, chloritoid; K, kyanite; E, epidote; S, staurolite; G, garnet; Z, zircon; R, rutile. (Upper left figure.)

FIG. 249.—Bar chart or histogram of mineral frequencies. (Abbreviations as in Figure 248.) (Lower left figure.)

FIG. 250.—Other types of charts to show mineral frequencies (after Evans, Hayman, and Majeed). (Abbreviations as in Figure 248.) (Upper right figure.)

data in such a table and still more difficult to compare with other tables, hence some workers have used graphical methods.

E. Neaverson [1] used a chart in which rare minerals are shown by a single dash (—), common minerals by two dashes (=), and abundant minerals by three dashes (≡); the minerals are tabulated horizontally and the localities vertically (Figure 248).

[1] E. Neaverson, The petrology of the Upper Kimmeridge Clay and Portland sand in Dorset, Wiltshire, Oxfordshire and Buckinghamshire: *Proc. Geo. Assoc.,* vol. 36, pp. 240-256, esp. p. 252, 1925.

Some workers have used the conventional bar diagram for presenting results, while others have used other diagrams or graphs (Figures 249 and 250). The Burmah Oil Company [1] has developed a method which appeals most to the authors. As indicated above, the geologists for this company represent the frequency of the mineral species by a series of frequency numbers which are, in a general way, logarithms of the frequency percentage determined by counting. This is done to emphasize the lower frequency values. It is clear that a change in frequency from 6 to 11 per cent is more significant than a change from 76 to 81 per cent, hence a logarithmic scheme of presentation is employed. The actual frequency scale, however, as used by these geologists departs somewhat from true logarithmic plan. It is given below:

TABLE 47

Frequency	Approximate Percentage
8+	90-100%
8	75-89%
8—	60-74%
7+	45-59%
7	35-44%
7—	28-34%
6+	23-27%
6	18-22%
6—	13-17%
5	7-13%
4	4-6%
3	2-3%
2	1-2%
1	½-1%
1*	1 grain only

In plotting, 8+ is plotted as 8⅓ squares and 7— as 6⅔ squares, 1* is plotted as a half-square. The scale of frequencies of ⅟₁₀ in. per unit has proved very suitable and a convenient scale of thickness is 1,000 ft. to the inch. In order to avoid an arbitrary arrangement of minerals it was found convenient to place them in order of refractive index. A "range table" prepared by these workers for an area in Upper Assam is here given (Figure 251). By comparison with the other graphic methods given it can be seen that this method of plotting has the advantage that the data from each sample of a series of samples are individually shown. Averaging the results by formations obscures the

[1] Evans, Hayman, and Majeed, *loc. cit.*, p. 254.

lesser fluctuations in frequency which make correlation possible. (Compare, for example, Figure 250 and Figure 251.)

When the data on mineral content are to be used for other purposes than correlation, other graphic methods of presentation may be used (see chapter on graphical presentation).

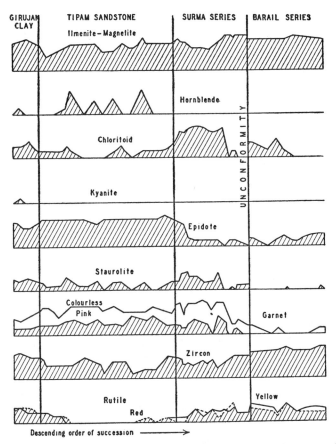

FIG. 251.—Profile chart of heavy-mineral frequencies and relation to stratigraphic horizons (after Evans, Hayman, and Majeed). Horizontal scale: 1 in. equals 2,000 ft. Vertical scale: .05 in. equals 1 frequency unit.

Calculation of mineral frequencies in sample based on frequencies determined in the fractions separately. A separation scheme involving one or more methods which would divide the sediments into fractions each of which was but a single mineral would perhaps be ideal. Then each fraction could be weighed and the composition of the sediment as a whole expressed in percentage by weight of each species in a fashion analogous to that used in chemical analyses in which the quantity of each oxide is expressed in weight percentage.

This ideal is not yet attainable. The fractions obtained by any separation method consist usually of several species. The mineral content of each fraction is determined microscopically by counting the grains, and their frequency is expressed in percentage based on their *number*. Inasmuch as the quantity of each fraction is usually expressed by percentage based on the *weight* of the fraction, difficulties appear when an attempt is made to combine all the data into a single tabulated analysis. These difficulties are due to size and specific gravity differences among the mineral grains. If, for example, one chooses to express his results on a weight basis, it is obvious that even though a given fraction is composed of but two minerals in equal numbers, 50 per cent each, the two may not make just half of the fraction by weight if they are of different specific gravities. It is equally clear that if the two species are of different sizes a similar difference in proportion by weight will appear. These difficulties are overcome if correction is made in computations for specific gravity difference and if the fractions studied are closely sized, as by screening, prior to separation magnetically or in any other way.

If the investigator wishes to express his results in number frequency rather than by weight, difficulties due to size and specific gravity will also appear. For example, even though two fractions are of equal weight, they may contain very different numbers of grains. The problems of computing are simplified if all the grains in each separate (or in some known portion of each separate, such as one half, or one fourth) are counted and the absolute frequencies are added together before computing percentages. This is the method recommended. The final results, of course, will not be comparable to those obtained by recalculation on a weight basis. It has not yet been shown, however, that in the case of the sediments a weight percentage is more significant than a number percentage.

> One would not, for example, recommend a statement of the proportion of racial groups in the city of Chicago on a weight basis. Why then, should the mineral proportions of a sand grain population be placed on a weight basis? On the other hand, in the study of the accessory minerals of the crystalline rocks, where the several species are first released by crushing and then separated from one another by various means, the true number frequency is materially affected by the crushing process. It is clear that in this case weight is the desirable basis for expression of abundance.
> Example: The heavy minerals of a sediment are divided into several frac-

tions by means of an electromagnet. The non-magnetic fraction is too large to mount entire, so it is reduced by a Microsplit to one-half in size. The tabulated results are as follows:

TABLE 48

SAMPLE RECOMBINATION OF MINERAL FREQUENCIES FROM MAGNETIC AND NON-MAGNETIC FRACTIONS

(Based on count of entire amount of each fraction)

MINERAL	NUMBER OF GRAINS IN EACH FRACTION				TOTAL	PER CENT
	Strongly Magnetic	*Moderately Magnetic*	*Weakly Magnetic*	*Non-magnetic* *		
Magnetite ...	81	0	0	0	81	20
Garnet	0	28	65	4	101	25
Zircon	0	0	0	73	146	37
Hornblende ..	0	12	48	5	70	18
Totals	81	40	113	82	398	100

* Since the non-magnetic fraction was large, it was split in two parts. It is therefore necessary to multiply the frequencies of any species in that fraction by two in order to get a correct total for the entire sediment.

Several workers have noted the discrepancies involved in combination of weight and number percentages. Alan Stuart,[1] for example, criticized the combination of weight and number data in the same analysis as is commonly done when the quantity of heavy minerals is given in percentage by weight while the various mineral species that make up the heavy fraction are expressed by number.[2] Salmojraghi[3] also discussed the problem of whether results should be expressed on a volume-weight or a number basis. He too pointed out that, if the grain dimensions and form were uniform, weight percentage could be computed if allowance

[1] Alan Stuart, Heavy mineral frequencies: *Geol. Magazine*, vol. 64, p. 143, 1927.
[2] Alan Stuart, On a black sand from south-east Iceland: *Geol. Magazine*, vol. 64, pp. 540-545, 1927. Stuart gives the mineral composition of a sand which he studied in percentage by weight, which he obtained by recalculation of the percentage frequencies of the minerals as determined by count in several fields (in this case fifty). Such calculations are based on the known specific gravity of the sand as a whole and the approximate specific gravity of the several separates obtained by the use of heavy liquids (taking into account the frequencies of the minerals in each separate and approximating by using the known specific gravities of each of the minerals), and assuming the same to be perfectly graded.
[3] F. Salmojraghi, Sullo studio mineralogico delle sabbie e sopra un modo di rappresentarne i risultati: *Atti. soc. ital. sci. nat.*, vol. 43, pp. 54-89, 1904.

was made for specific gravity differences. He concluded that, since such uniformity does not exist, one would have to be content with numerical percentages.

STATISTICAL METHODS

Invariably the percentage composition of one sample will differ in some degree from that of a second sample even though both were taken from the same stratum or deposit. To some extent such differences are due to errors of sampling and laboratory analysis, but even when such errors are allowed for, variations remain which must be taken into

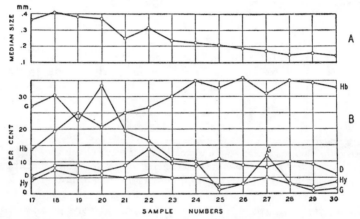

FIG. 252.—Relation of certain mineral frequencies to distance of travel along shore of Cedar Point (Lake Erie), Ohio. A, median size-distance relations; B, mineral frequency-distance relations. Hb, hornblende; G, garnet; Hy, hypersthene; D, diopside.

account. If the samples were collected in vertical sequence the variations are related to time and may be of correlative value. If the variations are observed in synchronous samples they are related to place and may be called *spatial variations*. The change (increase or decrease) in percentage of a given mineral from place to place may be termed a *mineral gradient*. A regular decline in content of garnet, for example, in the direction of littoral drift along a beach or along a stream course may be spoken of as a garnet gradient (see Figure 252). Not all variations are regular and progressive.

The main known causes of systematic mineral changes are (1) contamination, (2) selective abrasion, and (3) selective transportation. The principle of contamination has been outlined by Cayeux[1] and applied to

[1] L. Cayeux, *Les roches sédimentaires de France* (Paris, 1929).

FREQUENCIES AND COMPUTATION 481

heavy minerals by Krynine.[1] Contamination results in the increase or decrease in percentage of certain constituents due to influx of new material (as at the junction of two streams draining petrographically dissimilar areas) of different petrographic character. If of totally different character, the percentage of all minerals present will be depressed. If of partially different nature, the percentage of certain minerals will be depressed; the percentage of other minerals, common to both sediments involved, may be depressed little or even augmented.

But as postulated by Krynine,[2] the several grades are differently affected. The heavy minerals of the far-traveled sediment are, by wear and breakage, reduced in size and confined largely to the fine fraction, whereas the little-traveled local material is of larger size and therefore added mainly to the medium sand fraction.[3] Krynine, having the concept of contamination in mind, pointed out that the ratios between minerals of the original sediment are unchanged by any amount of additions of different petrographic make-up. A hornblende-kyanite ratio, for example, remains unchanged if the newer heavy minerals added are garnet and monazite. If, however, appreciable additions of kyanite were made to the sediment, the hornblende-kyanite ratio would obviously be altered.

Correlation and differentiation of deposits on the basis of certain significant mineral ratios thus becomes possible. A younger deposit, for example, made up largely of material reworked from subjacent beds and having therefore much the same absolute mineral frequencies as the older deposit can readily be distinguished on the basis of these ratios.

Krynine illustrates the principle involved by an example. (See Figure 253.) Sand A, characterized by a heavy residue containing 60 per cent of hornblende, 20 per cent of kyanite, and 20 per cent of other minerals, begins its journey down stream at point one, where $\frac{2}{3}$ of the heavies are in the medium sand fraction (Am) and $\frac{1}{3}$ are in the fine fraction (Af). At point two, some distance downstream, a portion of the grains in the medium sand fraction will be reduced in size by attrition and breakage and will have passed into the fine fraction (Af_2). At this point the medium portion (Am_2) will contain but half the total number of grains and the fine fraction ($Af + Af_2$) the other half, but the frequencies of the minerals in each part remain unchanged. At a point three, the medium fraction (Am_3) contains only $\frac{1}{3}$ the total number of grains and the fine fraction ($Af + Af_3$) contains $\frac{2}{3}$. At this point a short tributary enters carrying the same load as the main stream but

[1] Paul D. Krynine, Glacial sedimentology of the Quinnipiac-Pequabuck lowland in southern Connecticut: *Am. Jour. Sci.* (5), vol. 33, pp. 111-139, 1937.
[2] See also Paul D. Krynine, Age of till on "Palouse Soil" from Washington: *Am. Jour. Sci.*, vol. 33, pp. 205-216, 1937.
[3] As Krynine points out, this is true if the local sediment itself is not derived from a preëxisting deposit of fine grain. In that event the contamination is described as "abnormal."

consisting of sand B, of local origin, characterized by 70 per cent garnet, 20 per cent monazite, and 10 per cent of other constituents. The short distance of travel and lack of wear results in ⅔ of the heavy-mineral grains being present in the medium fraction of sand B and only ⅓ in the fine fraction. As a result of the mixing in equal amounts of A and B, the new sand C contains in its medium fraction (Am₃ + Bm) ⅔ of material from B and

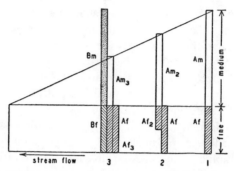

Fig. 253.—Diagram illustrating differential contamination (after Krynine).

⅓ from A, and in its fine fraction (Af + Af₃ + Bf) ⅔ from sand A and but ⅓ from sand B. The following table depicts the resulting mineral frequencies:

TABLE 49

MINERAL	SAND A (FAR-TRAVELED) Medium and Fine Fraction	SAND B (LOCAL) Medium and Fine Fraction	SAND C (EQUAL MIXTURE OF A AND B)	
			Medium Fraction	Fine Fraction
Garnet	..	70	47	23
Hornblende .	60	..	20	40
Kyanite	20	..	7	13
Monazite ..	..	20	13	7
All others ..	20	10	13	17

An examination of the table shows that though the absolute frequencies are entirely different, Sand A and both the medium and fine fractions of Sand C have in common the same ratio of hornblende to kyanite (3 to 1). Likewise Sand B and the two fractions of Sand C have a common ratio of garnet to monazite (3½ to 1).

Not all minerals are equally resistant to abrasion or solution to which they are subjected during transportation. As a result, less stable species suffer more rapid size reduction due to abrasion or breakage than do other species. The percentage of a less resistant mineral will show a decline in the coarse grades with respect to more stable minerals present (which appear, therefore, in increased percentage). Conceivably a complementary increase in abundance of the unstable species might take place in the finer grades, though, on the other hand, gains in a particular grade (by breakage of larger sizes) might be balanced by losses, so that certain equilibrium proportions would be maintained.

Analysis of the results of differential breakage is not simple. Let us take a specific example. Assume a heavy mineral assemblage of garnet (unbreakable) and hornblende (breakable). Assume moreover the same percentage of each in all grades, namely, 50-50. What happens to a thousand grains of each in the coarsest grade if the sediment travels from point A to point B? If half of the hornblende grains broke into two equal parts, 1,000 grains would result which will drop into the next lower grade. This would leave 500 hornblende grains and 1,000 garnet grains. The ratio would now be about 33 to 67. If in the next lower size grade, similar breakage takes place, 500 hornblende grains will become 1,000 in number and will drop into a still smaller grade. To the 500 grains of hornblende remaining, however, we must add the 1,000 grains contributed by the coarsest grade. We have therefore 1,500 hornblende grains and, as before, 1,000 garnet grains. The ratio hornblende to garnet becomes 60 to 40. It is clear from our simplified hypothetical case that differential breakage would result in changing ratios and that the change would not always be in the same direction in each grade.

Some breakage must occur, and what does take place is probably differential; that is, all minerals are not equally affected. The question might be raised, however, as to whether breakage is important or significant. Little direct evidence is available on this point. Experiments by Anderson [1] and Marshall [2] make it seem likely that, barring the presence of gravel, abrasion of sand is a very slow process and that transportation for hundreds of miles is necessary to achieve perceptible effects. Pettijohn and Ridge,[3] however, in their study of the sands of the Cedar Point spit in Lake Erie, noted marked systematic variations in the proportions of the several heavy mineral species from sample to sample. They suggest that the progressive change from sample to sample is due to selective sorting. Selective sorting by currents depends on the size, shape, and specific gravity of the mineral grains. Small grains tend to outrun coarse; grains of low sphericity will outrun those with high

[1] G. E. Anderson, Experiments on the rate of wear of sand grains: *Jour. Geology*, vol. 34, pp. 144-158, 1926.
[2] P. Marshall, The wearing of beach gravels: *Trans. New Zealand Instit.*, vol. 58, pp. 507-532, 1927.
[3] F. J. Pettijohn and J. D. Ridge, A mineral variation series of beach sands from Cedar Point, Ohio: *Jour. Sed. Petrology*, vol. 3, pp. 92-94, 1933.

sphericity; and minerals of low specific gravity outrun those of high specific gravity. Pettijohn and Ridge found, for example, a marked increase (in the .177-.125 mm. grade) in the percentage of hornblende and a decrease in the percentage of garnet in the direction of travel over a distance of less than seven miles (Figure 252). Abrasion certainly is not responsible in this situation, as the grains fail to become rounded. Breakage is perhaps responsible, but does not seem to be a wholly satisfactory explanation when such experimental data as exist are taken into account. MacCarthy[1] observed, for example, that the sand on a beach grows regularly finer in the direction of travel, and it also has been observed to grow more "angular" in the same way. This has been explained as due to a tendency of the finer and more angular (larger surface area) grains to outrun the larger and more rounded grains in the "down current" direction of travel.

To these geologic factors must be added one other. A given mineral may be removed or otherwise altered in an ancient deposit through solution or diagenetic change after deposition. Loss of one species changes the absolute percentages of those remaining.

Russell,[2] for example, studied the sands of the Mississippi River between Cairo, Illinois, and the Gulf of Mexico. He found that in over 1,000 miles of travel there was little progressive change in mineral content of the investigated sands. The slight changes noted were attributed to "selective destruction by abrasion and chemical weathering" during transport. On the whole, however, the destruction of the "less resistant" minerals was of such slight consequence that Russell was forced to the conclusion that the absence of the less resistant minerals in sediments derived from older formations is due to decomposition and solution of the minerals in question *after* the deposition of the sediment, followed perhaps by reworking and deposition of the material now free from the less stable species.

The writers have felt it necessary to discuss the geological factors involved in the study of mineral variations, since any statistical methods invoked to handle numerical data derived from mineral analyses must be appropriate to the problem involved. The indiscriminate use of statistical procedures to none too clearly thought out situations leads only to confusion. Moreover, to correctly interpret mineral gradients one must have a clear picture of the underlying geological factors involved. In any event it does not follow that an increase in the percentage of a particular

[1] Gerald R. MacCarthy, Coastal sands of the eastern United States: *Am. Jour. Sci.* (5), vol. 22, pp. 35-50, 1931; The rounding of beach sands: *Am. Jour. Sci.* (5), vol. 25, pp. 205-224, 1933.
[2] R. Dana Russell, Mineral composition of Mississippi River sands: *Geol. Soc. America, Bulletin,* vol. 48, pp. 1307-1348, 1937.

FREQUENCIES AND COMPUTATION

TABLE 50
ANALYSIS OF HEAVY-MINERAL SAMPLES ON THE BASIS OF A REFERENCE SAMPLE (AFTER COGEN)

NUMBER OF GRAINS IN 20 G. OF ——— MM. GRADE SIZE*

Sample No.	A	B	C	D	E	Other Minerals	
						F	G
Reference sample	150	200	50	25	100		
1	305	435	100	55	210		
2	100	140	35	16	66		
3	145	210	50	27	98		
4	155	204	125	28	95		
5	a	b	c	d	e	f	g

PERCENTAGE BY NUMBER OF GRAINS

Sample No.	A	B	C	D	E	Other Minerals	
						F	G
Reference sample	100	100	100	100	100		
1	203	218	200	220	210		
2	67	70	70	64	66		
3	97	105	100	108	98		
4	103	102	250	112	95		
5	$100\frac{a}{150}$	$100\frac{b}{200}$	$100\frac{c}{50}$	$100\frac{d}{25}$	$100\frac{e}{100}$	$100\frac{f}{k}$	$100\frac{g}{k}$

* Actually the number of grains in a carefully split-off portion of the heavy residue is determined and the results are recalculated on a 20-g. basis.

mineral means that there is a real increase in the quantity of that mineral. One should not, therefore, conclude that, because a particular species showed an increase in some given direction, the cause was closer proximity to the source area.

To get around difficulties due to variations from the causes mentioned several schemes have been proposed. One of the writers several years ago, took the ratio between any mineral, whose variation was to be studied, and garnet. Garnet was chosen in the case studied because it is a relatively stable detrital, one of the last to be removed by solution or abrasion, it is easily recognized, and it was present in all samples in sufficient quantity as not to be greatly in error in percentage determination. The formula $A : B = 100 : x$ was used, where A is the percentage of garnet, B is the percentage of any other constituent and x is the relative value of this other constituent where garnet is taken as 100.

Cogen[1] later tried a similar scheme. Cogen, however, took the ratio of the frequency of each component in the sample to the frequency of the same components in a "reference" sample—a composite sample collected from the stratum with which it is desired to compare materials of unknown age—according to the formula $P = 100 \frac{n}{t}$, where P is percentage, n is the number of grains of a mineral species in a given amount of a selected grade size of sand, and f is the number of grains of a mineral species in the same amount of a selected grade size of sand from the reference sample. Cogen's scheme lends itself conveniently to graphical representation when the mineral percentages are plotted on a logarithmic scale. The norms or reference sample will always be represented by a horizontal line cutting the ordinate at 100 per cent. Curves of other samples may lie either above or below the norm, but as long as they remain nearly horizontal they suggest close correspondence to the reference sample. Wide deviations from the horizontal are indications of significant fluctuations in the mineral percentages in the sample (Figure 254).

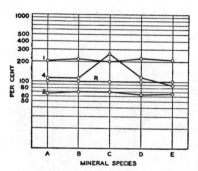

FIG. 254.—Diagram of composition of several samples in terms of a reference sample (after Cogen).

Without an objective measure of similarity the comparison between heavy mineral suites is a matter of personal judgment. For example, consider the three samples analyzed in Table 51 on page 487.

By inspection one can see that #24 and #18 are very much alike, while #53 is somewhat dissimilar. Study of numerous samples and attempts to state the degree of similarity involved is both awkward and inconvenient. Dryden[2] used the "coefficient of correlation" of the statistician to avoid such difficulties.

[1] William M. Cogen, Some suggestions for heavy mineral investigation of sediments: *Jour. Sed. Petrology*, vol. 5, pp. 3-8, 1935.
[2] Lincoln Dryden, A statistical method for the comparison of heavy mineral suites: *Am. Jour. Sci.* (5), vol. 29, pp. 393-408, 1935.

TABLE 51

Mineral	Samples		
	#24	#18	#53
Apatite	2	4	9
Augite	0	0	7
Biotite	1	0	1
Epidote	21	19	30
Garnet	7	5	0
Hornblende	27	34	10
*Oxides	11	11	10
**Ores	15	17	21
Tourmaline	7	4	1
Zircon	5	3	9
Pyrite	0	0	3

Data taken from Frederick C. Kruger, A sedimentary and petrographic study of certain glacial drifts of Minnesota: *Am. Jour. Sci.* (5), vol. 34, pp. 345-363, 1937. Percentage by number, $1/8$-$1/16$ mm. grade. #24 and #18, Wisconsin gray drift; #53, Wisconsin red drift.
 * Oxides = hematite, limonite, leucoxene
 ** Ores = magnetite, ilmenite

The term *correlation* must not be confused with correlation in the geological sense. Statistical correlation implies no time relationship nor any other causal relation. It only states objectively similarity or mathematical dependence of one set of data upon another set.

The use of the coefficient of correlation for such purposes, however, has been objected to by Eisenhart,[1] and the chi-square test has been proposed in its stead.[2] Helson,[3] on the other hand, sanctions the use of the coefficient of correlation for the purpose mentioned.

If one chooses to use the coefficient of correlation, the following example will illustrate the necessary calculations. Samples #24 and #18 above will be compared.

$$r = \frac{\Sigma(XY) - nM_x M_y}{\sqrt{(\Sigma(X^2) - (nM_x^2))(\Sigma(Y^2) - nM_y^2)}} \quad \ldots \quad (1)^4$$

[1] Churchill Eisenhart, A note on "A statistical method for the comparison of heavy mineral suites": *Am. Jour. Sci.*, vol. 30, pp. 549-553, 1935.
[2] Churchill Eisenhart, A test for the significance of lithological variations: *Jour. Sed. Petrology*, vol. 5, pp. 137-145, 1935.
[3] Harry Helson, On statistical methods of comparing heavy mineral suites: *Am. Jour. Sci.*, vol. 32, pp. 392-395, 1936.
[4] For theory of see page 260.

TABLE 52

Mineral	Sample #24		Sample #18		
	X	X²	Y	Y²	XY
Ap	2	4	4	16	8
Bio	1	1	0	0	0
E	21	441	19	361	399
G	7	49	5	25	35
Hb	27	729	34	1,156	918
Ox	11	121	11	121	121
Ores	15	225	17	289	255
T	7	49	4	16	28
Z	5	25	3	9	15
	9|96 10.7	1,644	9|97 10.8	1,993	1,774

where r is the coefficient of correlation, Σ denotes "the sum of," X and Y are the items to be compared, i.e., the frequency in percentage of the mineral constituents in each sample, n is the number of items (number of minerals in the suite), and M_x, and M_y are the mean values of the X and Y items respectively.

Substituting in (1) the values from the above table, we have

$$r = \frac{1774 - (9 \times 10.7 \times 10.8)}{\sqrt{(1644 - 9 \times 10.7^2)(1993 - 9 \times 10.8^2)}} \quad \ldots \quad (2)$$

In a similar manner samples #24 and #53 may be compared. The coefficient in this case is .53.

The coefficient should not, perhaps, be called the coefficient of correlation. Eisenhart points out that the coefficient of correlation requires that the paired values compared be independent of each other. This is not wholly true inasmuch as the percentages must total 100; consequently when all but one pair are known, the remaining one is uniquely determined—it is not independent. Eisenhart recommended the use of frequencies instead of percentages because, in addition, results are more reliable for large samples than for small. The use of percentage instead of frequency obscures the sample size, which is important.

Eisenhart, moreover, states that the coefficient of correlation cannot be used even if the true frequencies are taken because the "mineral composition is not a measurable characteristic." A measurable characteristic would be, for example, sphericity. Sphericity could be compared to grain size by use of the coefficient of correlation. There exists a continuous measurable variation from one sphericity value to another. There is, however, no such continuity from a given hornblende frequency to some stated frequency of garnet. A mineral suite is discontinuous and *qualitative*. What really is required is a coefficient of association, which like the correlation coefficient would vary from 0 to 1. To a certain extent the values computed above (equations 1 and 2) seem to meet such requirements, for it is clear that the values obtained show what is evident on inspection, namely that samples #24 and #18 are quite similar, while #24 and #53 are much less so. Identical samples yield a value of 1, while highly dissimilar samples give a very low value.

Eisenhart, therefore, proposed the chi-square method of comparing samples or testing their homogeneity. Consider but a single sample. In respect to a chosen attribute the sample is or is not homogeneous. For example, the pebble content of a gravel sample is under consideration. If all of the pebbles are limestone, the sample is homogeneous; if they are mostly limestone, but include also some shale pebbles, then it is non-homogeneous.

In the case of two gravel samples, if both contain limestone and shale pebbles (and not others), under what conditions can the two samples be considered lithologically homogeneous? If they are not identical, some criterion must be used in deciding whether or not the two samples are homogeneous in the sense that they have both come from the same parent deposit, whose mineral proportions are approximated by the proportions in the samples, the variations being due to sampling fluctuations. The chi-square test is such a criterion.[1]

The student is cautioned against indiscriminate application of any statistical measure of correlation or homogeneity to data collected without a full understanding of the underlying requirements which must be met before the particular statistical device under consideration can be applied. A larger background of statistical training than is possessed by the average geologist is required to handle with assurance the more complex statistical problems which arise.

[1] For both the development of the theory underlying the chi-square test and its use the reader is referred to p. 264.

CHAPTER 19

CHEMICAL METHODS OF STUDY

A PARTIAL or complete chemical analysis is often of great value to a sedimentologist. Such analyses are of special value in the study of fine-grained sediments, the mineral composition of which is difficult if not impossible to determine. Such materials include the clays (and slates), the fine-grained limestones, phosphate rock, coal, volcanic ash, bentonite, etc. From such chemical analyses the mineral composition may be quantitatively determined by computation if certain simplifying assumptions are made. Chemical analyses of certain sediments are also made for commercial purposes where even minor amounts of certain materials may be deleterious. Analyses of limestones are made for the MgO content; sedimentary iron ores are analyzed for P_2O_5 and S content. Analyses of cement rock and added materials are made for control of the manufacture of Portland cement. Sediments differ from igneous rocks in a number of important ways, and analyses of metamorphic rocks, therefore, may be of value in the determination of the sedimentary or igneous origin of these rocks. Such a method presupposes a knowledge of the chemical composition of the sediments themselves. The changes that a rock undergoes on weathering are better understood if the analyses of both the weathered and unweathered materials are available. Similarly, a knowledge of the chemical composition of rocks and minerals in general aids in understanding of many other processes—kaolinization, phosphatization, dolomitization, etc.

It is clear, then, that chemical analyses are an aid to identification of minerals and are of value in determination of commercial ore and industrial treatment and in study of the geologic origin and changes that rocks undergo. Both the geologist and the technologist have need to make and use chemical analyses.

QUANTITATIVE ANALYSIS

Methods. To make a good quantitative analysis requires considerable skill and experience. It is not within the scope of this book to give instructions in this technique. The reader is referred to the larger works

CHEMICAL METHODS OF STUDY

on this subject and particularly to those works which deal with the problem of analysis of rock and mineral materials.[1]

Computations based on quantitative analyses. Let us assume that one has a satisfactory analysis of a rock. It is desired to compute the probable mineral composition of the rock. Either the "molecular ratio method" or a geological slide rule [2] may be used. If the thin section shows certain minerals to be present, the proportions of these minerals may be determined by calculation. It is only necessary to know the chemical composition of the minerals themselves. The molecular ratios are quantities which indicate the relative numbers of the several molecules in the rock, and in that respect they are more significant than the percentages. They are simply obtained by dividing the percentage of the component present by the molecular weight of that component. For example, if the percentage of SiO_2 is 76.84, the molecular ratio is 76.84 divided by the molecular weight of SiO_2 or 60, which is 1.531. Since the molecular composition of the common rock-making minerals is now understood, it is therefore possible to calculate from the molecular proportions the proportions of the minerals known to be present.

TABLE 53

TABLE OF CALCULATIONS

Compound	Analysis	Mol. Wt.	Mol. Ratio	An	Ab	Or	Mag	Ferant	At	Kaolin	Qz
SiO_2	76.84	60	1.531	3.48	14.76	6.12	....	1.44	0.72	1.62	48.70
Al_2O_3	11.76	102	0.105	1.22	4.18	1.73	....	....	1.84	2.79	....
Fe_2O_3	0.55	160	0.005	....	....	....	0.55	....	....	....	....
FeO	2.88	72	0.040	....	....	....	0.15	2.52	....	....	....
MgO	1.39	40	0.036	....	....	....	....	....	1.40	....	....
CaO	0.70	56	0.012	0.70	....	....	....	....	....	....	....
Na_2O	2.57	62	0.041	....	2.57	....	....	....	....	....	....
K_2O	1.62	94	0.017	....	....	1.62	....	....	....	....	....
H_2O	1.87	18	0.103	....	....	....	....	0.43	0.63	0.97	....
	100.18			5.40	21.51	9.47	0.70	4.39	4.59	5.38	48.70

Also MnO: Trace. Calculated total 100.14; error 0.04 per cent.

[1] H. S. Washington, *The Chemical Analysis of Rocks* (New York, 1930). W. F. Hillebrand, The analysis of silicate and carbonate rocks: *U. S. Geol. Survey, Bull. 700,* 1919. A. W. Groves, *Silicate Analysis* (London, 1937). W. Van Tongeren, *Gravimetric Analysis* (Amsterdam, 1937). A laboratory manual with special reference to the analysis of natural minerals and rocks.
[2] W. J. Mead, Some geologic short cuts: *Econ. Geology*, vol. 7, pp. 136-144, 1912. J. H. Hance, Use of the slide rule in the computation of rock analyses: *Jour. Geology*, vol. 23, pp. 560-568, 1915.

Example. Graywacke, Hurley, Wisconsin. Analysis by H. N. Stokes. Described by W. S. Bayley, *U. S. Geol. Survey, Bull. 150*, p. 84, 1898. Contains quartz, feldspars, iron oxides, and probably kaolin. In the cement are chlorite, quartz, magnetite, pyrite, rutile, and either muscovite or kaolin.

The principal minerals and their assumed composition are given in the table below:

Quartz (Qz)	SiO_2
Feldspar	
Orthoclase (Or)	$K_2O, Al_2O_3, 6SiO_2$
Albite (Ab)	$Na_2O, Al_2O_3, 6SiO_2$
Anorthite (An)	$CaO, Al_2O_3, 4SiO_2$
Kaolin	$2H_2O, Al_2O_3, 2SiO_2$
Chlorite	
Ferroantigorite (Ferant)	$2H_2O, 3FeO, 2SiO_2$
Amesite (At)	$2H_2O, 2MgO, Al_2O_3, SiO_2$
Magnetite (Mag)	FeO, Fe_2O_3

By inspection it is clear that all of the K_2O is in orthoclase, all the Na_2O is in albite, and all the CaO is in anorthite. Each of these, therefore, may be calculated at once. The percentage of orthoclase, for example, is made up of the appropriate percentages of K_2O, Al_2O_3, and SiO_2. Since the molecular proportions in orthoclase of these constituents are 1:1:6 respectively, the percentage to be assigned to each is computed by multiplying the molecular ratio by the molecular weight of each. K_2O, $0.17 \times 94 = 1.62$; Al_2O_3, $.017 \times 102 = 1.73$; $6SiO_2$, $.017 \times 6 = .102$ and $.102 \times 60 = 6.12$. Adding $1.62 + 1.73 + 6.12$ we find 9.47 or the percentage of orthoclase. In like manner the percentage of albite and anorthite are computed. The Fe_2O_3 content may be assigned to magnetite; the unused FeO is assigned to ferriferous chlorite or ferroantigorite. The MgO may be allotted to a magnesium chlorite, amesite, and the percentage of this mineral determined. Unused Al_2O_3 may be computed as kaolin and unused SiO_2 as quartz. The computations made show that the rock is about half quartz (48.70), more than one-third feldspar (36.38) and the balance chlorite (8.98), kaolin (5.38), and magnetite (0.70).

The variable and complex composition of some minerals (such as chlorite) make exact computation impossible. In those cases where the minerals present are not known, as in the case of the very fine clays, the computations are no more accurate than the assumptions underlying them.[1]

[1] For further details see: J. F. Kemp, The recalculation of the chemical analyses of rocks: *School of Mines Quarterly*, vol. 22, p. 75, 1901. Whitman Cross, J. P. Iddings, L. V. Pirsson and H. S. Washington, *Quantitative Classification of Igneous Rocks* (1903). A. Osann, *Beitrage zur chemischen Petrographie: I. Molekularquotienten zur Berechnung von Gesteinsanalyses* (Stuttgart, 1903).

MICROCHEMICAL METHODS

Chemical tests may be applied to minerals as an aid to their identification. It is necessary, in the case of the sediments, that the tests be suitable for very small amounts of material—be applicable to single mineral grains. This requirement is best met by microchemical methods. These methods, in addition to being suitable for very small amounts of material, have the advantage of requiring very little accessory material other than a set of reagents. The more usual tests can be carried out on a glass slide on the stage of the microscope. The resulting crystals formed are examined optically in a manner familiar to petrographers. The tests are simple and straightforward. On the other hand, while they work very well for simple compounds, they are often indecisive in the case of complex mixtures or at best give non-characteristic precipitates. For certain types of minerals, particularly the readily soluble opaque sulphides and sulpho-salts, these methods are notably successful. But for analysis of the difficultly soluble silicates they are less satisfactory than the straight optical methods more often used. Since it is the latter type of material that is most usually involved in the study of the sediments, the writers do not feel justified in giving any considerable space to chemical microscopy. Cayeux, on the other hand, relied heavily on these methods for identification. The reader is referred to Cayeux's monographic work [1] on the sediments as well as to the standard works on chemical microscopy.

ORGANIC CONTENT

The organic content of sediments is important in the recognition of source beds of petroleum and in researches on the biological aspects of sedimentation. Trask and Hammar [2] have reviewed the problems involved in the determination of the organic content of sediments and its significance. These workers determined the carbon content, the nitrogen con-

---

[1] L. Cayeux, *Introduction a l'étude pétrographique des roches sédimentaires* (Paris, Imprimerie Nationale, 1916), quarto, vol. 1 (text) and vol. 2 (plates). See chapter II, "Analyse microchimique," pp. 95-170. M. N. Short, Microscopic determination of the ore minerals: *U. S. Geol. Survey, Bull. 825*, part. 4, "Microchemical Methods," pp. 115-201, 1931. Lloyd W. Staples, Mineral determination by microchemical methods: *Am. Mineralogist*, vol. 21, pp. 613-634, 1936. Emile Chamot and C. W. Mason, *Handbook of Chemical Microscopy*, vol. II, *Chemical Methods and Inorganic Qualitative Analysis* (New York, 1931). H. Behrens and P. D. C. Kley, *Mikrochemische Analyse*, 4th ed. (Leipzig, 1921).

[2] Parker D. Trask and Harold E. Hammar, Organic content of sediments: *Preprint Amer. Petrol. Inst.*, 15th Ann. Meeting, Dallas, Texas, 1934.

tent, the volatility, and the degree of oxidation of the sediments.[1] For the analytical methods used in the determination of carbon and of nitrogen the standard works should be consulted.

It may be of value to determine the bituminous content of a sandstone. Usually some solvent is used in conjunction with an extraction device. The Soxhlet extractor is such a device. The section on the preparation of samples for mineral studies has further details.

INSOLUBLE RESIDUES

The study of insoluble residues in limestones and dolomites has become popular in recent years. While the study of the residues is not strictly a problem in chemical analysis, the method of preparing the residues involves the use of a chemically active solvent. As long ago as 1888, Wethered[2] examined the residues obtained from the solution of limestone in hydrochloric acid. McQueen[3] in 1931 revived interest in the study of residues and demonstrated their value in stratigraphic studies, particularly in the case of well cuttings. Many workers have extended the use of the residues.

The method of obtaining the residues is simple and as developed by various workers differs only in details. In general the broken or crushed rock is placed in dilute HCl until action ceases; the mud and excess acid are then carefully decanted. The residue is washed several times and the washings are decanted. The residue is then dried and examined under a binocular microscope. Commercial HCI (muriatic acid) may be used, but the formation of gypsum (due to the presence of sulphate ions in the acid) should be guarded against. Precipitated gypsum will obscure the true nature of the residue from the limestone investigated.

The character of the residue varies considerably with the strength of the acid used. Rapid evolution of carbon dioxide gas upon addition of strong acid destroys delicate arenaceous foraminifera and other delicate structures—notably dolocasts. Dunn,[4] therefore, recommends placing the sample in water and adding acid slowly. He recommends a strength of

[1] Parker D. Trask and Harold E. Hammar, The degrees of reduction and volatility as indices of source beds: *Preprint Amer. Petrol. Inst.*, 16th Ann. Meeting, 1935.
[2] E. Wethered, Insoluble residues obtained from the Carboniferous-Limestone series at Clifton: *Quart. Jour. Geol. Soc. London*, vol. 44, pp. 186-198, 1888.
[3] H. S. McQueen, Insoluble residues as a guide in stratigraphic studies: *Missouri Bur. Geol. and Mines, 56th Bien. Rept.* Appendix, pp. 104-107, 1931.
[4] Paul H. Dunn, Microfaunal technique in the study of older Paleozoics: *Trans. Ill. Acad. Sci.*, vol. 25, pp. 140-141, 1933.

CHEMICAL METHODS OF STUDY 495

40 parts water to one part acid. St. Clair[1] used acetic acid in place of hydrochloric acid and found that he obtained a better quality of residue. Acetic acid did not flocculate the mud released and so decantation was rendered easier. Moreover delicate structures were preserved, as were certain mineral species ordinarily dissolved by HCl.

The results of laboratory study may be graphically presented. Ireland[2] and also Hills[3] used a modified bar graph, with the width of each bar representing thickness of beds sampled and the length of each bar representing the approximate amount of residue present. Each bar was subdivided, and each subdivision was appropriately colored or shaded to indicate the different kinds of material composing the residue. The bars were plotted in their appropriate stratigraphic position on long paper strips (Figure 255). Strips representing different sections or different wells can then be placed side by side and moved until a match or correlation between them is obtained.

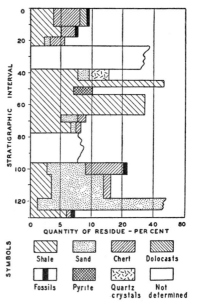

Fig. 255.—Insoluble residue diagram. (Modified from Ireland.)

STAINING METHODS

For purposes of identification or for facilitating quantitative estimate of several minerals of otherwise similar appearance, staining methods have been used. Such methods are generally of limited interest, but in the case of the sediments, two situations arise to which a staining technique may be profitably applied.

Orthoclase and quartz are often difficult to distinguish with certainty and rapidity. Where accurate count of the proportions of these two minerals in the "light" separate of a sand is desired, a stain selectively

---

[1] Donald St. Clair, The use of acetic acid to obtain insoluble residues: *Jour. Sed. Petrology*, vol. 5, pp. 146-149, 1935.
[2] H. A. Ireland, Use of insoluble residues for correlation in Oklahoma: *Bull. Am. Assoc. Petrol. Geol.*, vol. 20, pp. 1086-1121, 1936.
[3] John M. Hills, The insoluble residues of the Cambro-Ordovician limestones of the Lehigh Valley, Pennsylvania: *Jour. Sed. Petrology*, vol. 5, pp. 123-132, 1935.

adsorbed may be applied. Russell,[1] after a review of the problem, recommends malachite green. The grains to be studied were sprinkled uniformly over a glass slide previously coated with a film of kollolith (see page 360). The slide is heated until the grains sink in but leave their upper portions exposed. After cooling, a drop or two of hydrofluoric acid is allowed to remain on for about one minute and is then washed off. After washing, the grains are covered with a solution of malachite green (or any other water-soluble organic dye). The gelatinous aluminum fluosilicate film produced on the feldspars adsorbs the dye, while the quartz remains clear.

Calcite and dolomite are also difficult to tell apart. A number of methods in which a stain is employed have been recommended for use with these minerals.[2] Otto,[3] who investigated all of the known methods, concluded that the silver nitrate-potassium chromate method was most reliable. For this method a 10 per cent solution of silver nitrate is required. The mineral powder (or thin section) is treated with this reagent for 2 to 5 min. and then carefully washed with distilled water to remove any trace of the nitrate; it is then immersed in a neutral solution of potassium chromate (K_2CrO_4) for about 1 min. and then washed with water. Calcite will be colored a brownish red, while dolomite (or aragonite, if present) is without color (though with longer treatment some irregular coloration will be noted).

Various staining and microchemical methods have been proposed for other minerals in addition to those noted above, notably calcite and

[1] R. Dana Russell, Frequency percentage determinations of detrital quartz and feldspar: *Jour. Sed. Petrology*, vol. 5, pp. 109-114, 1935.

[2] J. Lemberg, Zur microchemischen Untersuchung von Calcit, Dolomit und Predazzit: *Zeitschr. deutsch. geol. Gesellsch.*, vol. 39, pp. 489-492, 1887; also, vol. 40, pp. 357-359, 1888. Zur microchemischen Untersuchung einiger Minerale: *Zeitschr. deutsch. geol. Gesellsch.*, vol. 44, pp. 224-242, 1892. E. E. Fairbanks, A modification of Lemberg's staining method: *Am. Mineralogist*, vol. 10, pp. 126-127, 1925. G. G. Suffel, Dolomites of western Oklahoma: *Okla. Geol. Survey, Bull. 49*, pp. 9-11 (Lemberg method), 1930. W. Heeger, Ueber die mikrochemische Untersuchung fein verteilter Carbonate im Gesteinsschliff: *Centralbl. f. Min., etc.*, pp. 44-51, 1913. Fritz Hinden, Neue Reaktionen zur Unterschiedung von Calcit und Dolomit: *Verhandlung der naturforschritte Gesellsch. in Basel*, vol. 15, p. 201, 1903. Lloyd G. Henbest, The use of selective stains in paleontology: *Jour. Paleon.*, vol. 5, pp. 355-364, 1931. Lucien Cayeux, *Introduction a l'étude pétrographique des roches sédimentaires* (Paris, Imprimerie Nationale, 1916). Chap. III, "Analyse chromatique," pp. 171-187, Texte. F. Cornu, Eine neue Reaktion zur Unterschiedung von Dolomit und Calcit: *Zentralbl. f. Min., etc.*, p. 550, 1906. M. Dominikirwicz, A microchemical method for the analysis of carbonates: *Roczniki Chemji*, vol. 3, pp. 165-176, 1923. K. Spangenberg, On the copper sulphate test for dolomite and calcite: *Zeits. f. Kryst. Min.*, vol. 52, p. 529, 1913.

[3] Personal communication. The writers are indebted to Mr. Otto for many of the references on staining methods.

CHEMICAL METHODS OF STUDY

aragonite [1] and anhydrite and gypsum.[2] Ordinarily, however, these minerals are better distinguished optically than chemically. For this reason no further details are given here.[3]

[1] J. Johnston, H. E. Merwin and E. D. Williamson, On tests for calcite and aragonite: *Am. Jour. Sci.*, vol. 41, p. 473, 1916. W. Meigen, Eine einfach Reaktion zur Unterscheidung von Aragonit und Kalkspath: *Centralbl. f. Min., etc.*, pp. 577-578, 1901. St. J. Thugutt, Ueber chromatische Reaktionen auf Calcit und Aragonit: *Kosmos (Radziszewski-Festband)*, vol. 35, p. 506, 1910 (*Abst. Chemical. Zentralbl.*, 1910, II, p. 1084).

[2] G. Berg, A quick determination of the presence of anhydrite in rocks and artificial formation of microscopic anhydrite crystals: *Centralbl. f. Min., etc.*, pp. 688-690, 1907. H. F. Gardner, Notes on the chemical and microscopic determination of gypsum and anhydrite: (*Appendix C-11*) *Proc. Amer. Soc. Testing Materials*, vol. 26, part I, pp. 296-301, 1926. Fran Tucan, Microchemical reactions of gypsum and anhydrite: Zagreb (Croatia). *Centralbl. f. Min., etc.*, pp. 134-136, 1908.

[3] See also: E. Steidtman, Origin of dolomite as disclosed by stains and other methods: *Bull. Geol. Soc. Am.*, vol. 28, pp. 431-450, 1924. Otto Mahler, On the use of $FeCl_3$ and $CuSO_4$ for distinguishing between calcite and dolomite: *Inaugural dissertation, Freiburg*, 1906.

CHAPTER 20

MASS PROPERTIES OF SEDIMENTS

INTRODUCTION

Just as the behavior and functions of a complex organism are the sum of the behavior and functions of the component cells, so also is the character of the aggregate sediment (or any other particulate substance) a summation of the characters of the individual particles or grains of which it is made.

We have seen that a study of the properties of a clastic sediment is ultimately a study of the fundamental properties of its component grains, namely, size, shape, surface texture, mineral composition, and grain orientation. Dependent upon these basic attributes are such properties of the whole aggregate as porosity, permeability, color, etc. These latter secondary characteristics the writers have chosen to call the "mass properties" or the properties of the whole.

Many of the mass properties are of greater technological than geological interest, and since this volume is written primarily for the geologist, the treatment of those properties that have proved to be of greatest geological worth is given greatest space. For the sake of completeness and for the benefit of the non-geological reader or of the geologist who is obliged to make certain simple tests to determine the suitability of a substance for some technological use, a little has been included on methods of measurement of these properties together with a few leading references to the literature. Moreover, since these secondary properties are dependent on the fundamental attributes of the grains, and since the latter are themselves determined by the geological history of the deposit, the mass properties are also different for deposits of different origin and similar for those materials with a common origin. Sauramo,[1] for example, found that hygroscopicity, a function of grain size, was in turn related to the flocculation or non-flocculation of the clays studied. It was thus possible to tell whether the clay had been deposited in brackish or in fresh water.

---

[1] Matti Sauramo, Studies on the Quaternary varve sediments in southern Finland: *Bull. Commission Geol. de Finlande*, vol. 11, No. 60, 1923.

As the interrelationship between the fundamental properties and the mass properties becomes better known, it should be possible to compute the latter from the former. When the relation of these characters to the geological origin is understood, the location and exploitation of the materials of economic value will become more certain and less difficult.

COLOR OF SEDIMENTARY ROCKS

The color of a sediment is a property which depends on the color of the component particles, or on the presence of some pigment, or upon the fineness of the materials. Small quantities of limonite or hematite may, for example, give a marked color to the sediment. Sauramo [1] noted that in the case of *very* fine-grained sediments the color itself is a function of particle size. The smaller the particle size the darker the color. Trask and Patnode [2] have shown that the color of sediments from oil-fields varies more or less directly with the organic content. Light sediments, in general, contain little organic matter, and dark sediments ordinarily contain much. It is apparent, therefore, that an accurate statement of color is desirable.

The accurate naming and determination of colors is a matter of considerable difficulty owing both to subjective errors and to the lack of any generally recognized scale of colors.

Several methods of recording colors of sediments are open to the worker. He may compare the color of the sediment with a color chart or atlas such as those of Ridgway or Ostwald.

Ridgway's book has 1,115 colors arranged in systematic manner for comparative purposes. Ridgway has selected from the spectrum 59 hues or colors which he has named and numbered consecutively from 1 red to 59 violet. Thus, 11 is orange, 23 yellow, 35 green, and 49 blue. Intermediate hues are named by a combination of the names of the hues between which they lie, as for example yellow-orange, orange-yellow, etc., the last term of the combination designating the dominant hue. Ridgway has arranged the hues in horizontal succession across the middle of the page. Upward from the middle, he made his colors lighter by adding white, designating the increasing whiteness by letters *a* to *g*, while downward the colors are darkened and are marked by the letters *h* to *n*. In addition to this set of modified pure colors there are five sets of colors each made grayer than the preceding by the addition of increasing amounts of neutral gray. The progressive graying is designated in Ridgway's book by prime signs running from one to five, thus: ' — ''''' . Each hue thus grayed is both lightened and darkened like the pure colors. To

---

[1] Matti Sauramo, *ibid.*, pp. 19-20.
[2] Parker D. Trask and H. Whitman Patnode, Means of recognizing source beds: *Am. Petrol. Inst.*, 17th Ann. Meeting, 1936. Preprint.

complete his work, Ridgway has added a column of neutral grays which are gradations between black and white.[1]

The color may be matched with a composite color obtained by mixing standard colors in various proportions. Hutton,[2] for example, obtained the desired color by combination, in a rapidly rotating disc, of varying proportions of four standard colors, namely, white (neutral 9), black (neutral 1), red (red 4/9) and yellow (yellow 8/8). These colors were obtained from the Munsell Color Company of Baltimore, Maryland, and were certified to by the United States Bureau of Standards.

The comparison of color is facilitated if the streak of the rock fragments is studied. Comparison in the usual way by drawing fragments of the material over an unglazed porcelain plate enables smaller color differences to be detected than is possible with the use of chips. Grawe[3] suggested an improvement in this technique by grinding the rock, sieving it, and mounting the material passing 250-mesh ($\frac{1}{16}$-mm.) screen in a cardboard frame (a cardboard microslide, for example), protecting it with an ordinary thin cover glass.

For greater accuracy in color determination Grawe used a color photometer. Such an instrument permits quantitative analysis and synthesis of colors. Information on the construction and use of color photometers may be obtained from manufacturers.

SPECIFIC GRAVITY OF MINERAL GRAINS AND OF SEDIMENTARY ROCKS

The specific gravity may be determined by direct measurement of the weight and volume. The gravity of fragments of the size of cobbles or pebbles is thus determined. Generally the weight can be determined with a fair degree of accuracy. The accuracy of the method, therefore, depends primarily upon the measurement of volume, which is determined by noting the difference between the levels of a liquid partly filling a graduate before and after immersion of the fragment in the liquid.

---

[1] Ridgway's book is obtainable from A. Hoen & Company, Chester, Chase and Biddle Streets, Baltimore, Maryland, for $12 plus postage (shipping weight, 2 lb.). Following the Ridgway system, the Subcommittee on Color Chart of the Committee on Sedimentation of the Division of Geology and Geography of the National Research Council prepared and published a color chart of 114 colors. The colors chosen were those believed to be most applicable to the sediments. The charts are obtainable from the National Research Council, 2101 Constitution Avenue, Washington, D. C., at a cost of 75c each.

[2] J. G. Hutton, Soil colours; their nomenclature and description: *Proc. 1st. Int. Congr. Soil Sci.*, vol. IV, pp. 164-172, 1928.

[3] Oliver R. Grawe, Quantitative determination of rock color: *Science* (n.s.), vol. 66, pp. 61-62, 1927.

Accuracy in reading the volume is facilitated by the use of a cylindrical graduate of the smallest diameter which will admit the object whose volume is to be determined. With the weight and volume known, the specific gravity is numerically equal to the density [1] when expressed as weight in grams divided by volume in cubic centimeters. The direct method of volume measurement outlined above is often much in error due to air-filled pore spaces in the object.

There are various methods by which the weight of a body is directly measured and the volume is determined by subsequent weighing in water and noting the resultant loss in weight from which the weight of the water displaced is known according to the well-known Archimedes principle. One of these methods is the pycnometer method. The pycnometer is essentially a small container so constructed that it may be filled to a definite volume with a very high degree of accuracy (Figure 256).[2] It is especially useful for the determination of the specific gravity of a rock, since the material is first crushed to a powder, thereby eliminating all pore space and the error therefrom.

FIG. 256.—Pycnometer.

Weigh the pycnometer full of water at 15° C. Call this weight a. Remove most of the water and insert 1 g. of powder of the substance whose gravity is to be determined. Take pains to remove air bubbles after putting the powder in the bottle. Use suction or boiling. Fill again with water and be careful to have water of same temperature as before. Weigh and record this weight as b. The weight of the water displaced by the gram of rock is $(a + 1g.) - b$. The specific gravity is, therefore, $= \dfrac{1}{(a+1)-b}$. Sometimes some organic liquid, such as benzene or carbon tetrachloride, is used instead of water in order to avoid difficulty with air bubbles adhering to the sand or crushed material. In this case the computation must take into account the specific gravity of the liquid used.

The use of the Jolly spring balance and various steelyard balances involves weighing the material whose specific gravity is desired in air and again in water, although standard units of weight are not used. All that

---

[1] The density of a substance is the *weight per unit volume*, whereas the specific gravity is the *ratio* between the weight of a body and the weight of an equal volume of water (at 4° C). In the metric system the weight in grams per cubic centimeter is numerically the same as specific gravity.

[2] H. V. Ellsworth, A simple and accurate constant-volume pyknometer for specific gravity determinations: *Mineral. Mag.*, vol. 21, pp. 431-436, 1928.

is necessary is that the units of weight for one weighing shall be equal to those for the other. The specific gravity of the material in question is then found by dividing the weight in air by the loss of weight in water, S.G. $=\dfrac{W_a}{W_a - W_w}$, where W_a is the weight in air and W_w is the weight in water. The Jolly spring balance is best suited for small masses, while some of the steelyard balances are better adapted for larger masses. In either case errors in specific gravity appear due to air-filled pore spaces.

Direct comparison with heavy liquids is the best method for individual grains of small size. This may be done either with a series of liquids differing by some small interval in specific gravity [1] or by means of a "diffusion column" such as that used by Sollas.[2] A set of heavy liquids for specific gravity determination has been described by Spencer and by Landes.[3] The set made by Landes with Clerici's solution ranged from 2.0 to 4.1. The liquids are kept in small stoppered tubes arranged in a rack designed for the purpose. The mineral fragment to be tested is placed in a special holder consisting of a small perforated cup soldered on the end of a rigid wire handle. The perforated cup containing the mineral is lowered to the bottom of the test-tube and the action of the mineral noted. This float-sink test is performed until the density is narrowed down to the interval between two test liquids. It is necessary to rinse and dry the mineral-holder after each test in order to avoid dilution of the test solution. The process of immersing, washing, and drying does not take over 30 sec. to the tube. The density of the mineral is then compared to published densities or to suitable charts.[4]

The diffusion column for determination of specific gravity is prepared by half-filling a graduate or glass tube with a heavy liquid and then filling the upper part with a lighter miscible liquid, and allowing the tube to stand until the two liquids have diffused so that an even gradation from maximum density at the bottom to minimum density at the top is developed. The diffusion may be hastened also by preparing a series of

---

[1] Kenneth K. Landes, Rapid specific gravity determinations with Clerici's solution: *Am. Mineralogist*, vol. 15, pp. 159-162, 1930.

[2] W. J. Sollas, On the physical characters of calcareous and siliceous sponge-spicules and other structures: *Jour. Roy. Geol. Soc. Ireland*, vol. 7, p. 35, 1886; A method of determining specific gravity: *Nature*, vol. 43, pp. 404-405, 1891.

[3] L. J. Spencer, Specific gravities of minerals: an index of some recent determinations: *Mineral. Mag.*, vol. 21, pp. 337-365, 1927. Kenneth K. Landes, *ibid.*, pp. 159-162.

[4] Kenneth K. Landes, A mineral specific gravity chart: *Am. Mineralogist*, vol. 15, pp. 534-535, 1930. Joseph L. Rosenholtz and Dudley T. Smith, Tables and charts of specific gravity and hardness for use in the determination of minerals: *Rensselaer Polytechnic Inst., Eng. and Sci. Series*, no. 34, 1931.

equal fractions of the total volume of the column, the fractions forming a series ranging from 100 per cent heavy liquid and 0 per cent light liquid to 100 per cent light and 0 per cent heavy. If these fractions are then put in the column in their proper order the diffusion process is speeded up. When the diffusion column is ready, several particles of known specific gravity are carefully dropped into the column. These mark definite specific gravity levels in the column. The unknown minerals can then be dropped into the column and their specific gravities determined by comparison with the position of the indicators. This method provides a rapid and fairly accurate determination of specific gravity of several different mineral grains at one time.

Graham[1] placed the unknown mineral grain in a test-tube together with several indicators, or mineral grains of known specific gravity. The heavy liquid to be used was then poured in. A burette was filled with a suitable diluent. The diluent was run into the test-tube with constant stirring until the first indicator remained suspended, when the burette was read. Further readings were taken as equilibrium was reached with the other indicator minerals and with the unknown grain. The burette readings were finally plotted against the specific gravities of the known indicators, and the value for the unknown mineral was found from the point at which its burette reading intersected the curve.

The reader is referred to the section dealing with heavy liquids for the properties, preparation, standardization, and recovery of these liquids.[2]

The reader is also referred to the section dealing with mineral separations based on specific gravity.

POROSITY

DEFINITIONS AND GEOLOGICAL SIGNIFICANCE

The porosity of a rock is the percentage of pore space in the total volume of the rock—that is, the space not occupied by solid mineral matter. This percentage expresses practically the volume that can be occupied by water.

[1] R. P. D. Graham, The determination of specific gravity of mineral fragments by heavy liquids: *Roy. Soc. Canada, Proc. and Trans.*, ser. 3, vol. 11, sec. iii, pp. 51-53, 1917.
[2] Harold H. Hawkins, Procedure for restandardizing Clerici's solution: *Am. Mineralogist*, vol. 17, pp. 157-159, 1932.

The porosity as here defined is the *total* pore space as contrasted with the *effective* or available pore space. Total pore space includes all interstices or voids whether connecting or not and is larger than the limiting value of the effective pore space.

Porosity (and permeability) is studied most especially in conjunction with problems of petroleum geology and hydrology. The rate of petroleum production and the ultimate amount of oil and gas in a sand are items of economic import related to both porosity and permeability of sediments. They are important also in connection with foundations, dam sites, and filter beds.

The porosity of a sediment is governed by (1) uniformity of grain size; (2) shape of the grains; (3) method of deposition and manner of packing of the sediment; and (4) compaction during and after deposition. For a thorough discussion of the factors which affect both porosity and permeability the reader is referred to the paper by H. J. Fraser.[1]

COLLECTION OF UNCONSOLIDATED MATERIALS FOR POROSITY DETERMINATION

The procedure in determining porosity usually consists of two parts: (1) to obtain the bulk volume of the sample, (2) to obtain the aggre-

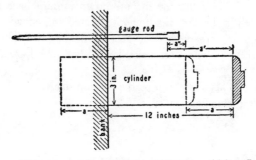

FIG. 257.—Meinzer sampling cylinder and gauge. (After Stearns.)

gate volume of the grains. The total volume of the sample minus the aggregate volume of the grains gives the volume of the pore space. The porosity is computed by the formula $P = 100\,(V-v/V)$, where P is the porosity in per cent, V is the bulk volume, and v is the aggregate volume of the grains.

The kind of method used to determine the volume of the sample depends on whether the sample is unconsolidated or consolidated. If it is

[1] H. J. Fraser, Experimental study of the porosity and permeability of clastic sediments: *Jour. Geology*, vol. 43, pp. 910-1010, 1935.

unconsolidated, the volume may have been determined in the field by means of a special sampling cylinder and gage rod such as that devised by Meinzer (see Figure 257).[1] This consists of a heavy brass cylinder, 3 in. in diameter and about 1 ft. long, closed at one end and having a cutting edge at the other. The gage rod is a steel rod about 2 ft. long, sharpened at the front end and having a definite reference mark at the rear end. To use this device for unconsolidated materials, a smooth face of the deposit is exposed and the gage rod is inserted to a marked depth. The cylinder is then pushed or driven in parallel to the rod. By means of the gage rod the distance the cylinder is inserted is known and hence the volume of the sample can readily be determined.

Fraser[2] has also described a method of sampling incoherent sands. For distinctly wet sands, such as beach sand, which are feebly coherent, Fraser found it possible to collect a sample and dip the same in paraffin at the sampling locality. Sand is removed from around the four sides of a column measuring some 2 in. square and 2 in. high by means of a trowel. The trowel is then carefully passed through the base of the column and the damp sand lifted out as a roughly cubical block. This block, carefully trimmed in order to remove adhering grains, is then transferred to an L-shaped dipper. The dipper is made by bending a strip of galvanized iron 2 in. wide and 6 in. long into the L-form. The sample is immersed in melted paraffin, temperature about 55° C., and immediately withdrawn. It is thereafter treated as a coherent sample would be. It is necessary, of course, later to dissolve away the paraffin coat and weigh the dried sand in order to obtain the water-free weight of the original sample.

In the event the sand is wholly dry in the first place, a different sampling technique is required. Fraser recommends either saturating the sand with water and treating as outlined above, or saturating with hot melted paraffin.

If the volume was not determined in the field, it is necessary to compact the sample artificially in the laboratory in a container of known volume. After such compacting by jarring or tamping is complete, the volume is estimated.[3] For purposes of determining the minimum porosity, Lamar devised an apparatus consisting of a metal tube 1¼ in. in diam-

---

[1] Norah Dowell Stearns, Laboratory tests on physical properties of water-bearing materials: *U. S. Geol. Survey, Water Supply Paper 596-F*, pp. 121-176, 1927.

[2] H. J. Fraser, Sampling incoherent sands for porosity determinations: *Am. Jour. Sci.* (5), vol. 22, pp. 9-17, 1931.

[3] J. E. Lamar, Geology and economic resources of the St. Peter sandstone of Illinois: *Ill. State Geol. Survey, Bull. 53*, pp. 149-150, 1928.

eter, working in two guide sleeves, which was raised about half an inch from below by a plunger operating on an eccentric and allowed to drop. The raising and dropping was repeated at about the rate of 100 times a minute. The cylinder struck a piece of felt of such thickness as to produce a "dead" fall, thus reducing to a minimum the amount of rebound. The machine was motor-driven. Insertion of a graduated rod into the cylinder enabled the operator to determine the depth of the material therein and to determine the endpoint of compaction. Such artificial packing does not duplicate the natural condition of the sand and is not generally to be recommended.

Preparation of the Sample

In all cases it is necessary to examine closely the specimen used for analysis. The piece should be broken in two, and one half should be saved for reference or for duplicate analysis. If the rock contains bituminous or oily material, this should be removed. Heating the sample in a gas oven over a Bunsen flame (200°-300° C.) for 1 to 4 hrs. is recommended by Sutton.[1] Care should be taken not to break down the carbonates and not to dehydrate the minerals. Errors due to dehydration are likely to be small except in the case of shales. Specimens with cracks should be rejected.

Methods of Measurement

Meinzer[2] gives the following methods for determining porosity: (1) measuring the quantity of water required to saturate a known volume of dry material, (2) comparing the volume of a sample with the aggregate volume of its constituent grains, (3) comparing the specific gravity of a sample with the weighted average of the known specific gravities of its constituent materials, (4) comparing the specific gravity of a dry sample with that of a saturated sample of the same material, (5) obtaining the uniformity coefficient and estimating the porosity on the basis of the observed relation between porosity and uniformity coefficient, and (6) producing a partial vacuum in a vessel that contains a dry sample and observing the change in air pressure when this vessel is connected with another that contains air under atmospheric pressure, the volume of each vessel and of the sample being known.

---

[1] Chase E. Sutton, Use of the acetylene tetrachloride method of porosity determination in petroleum engineering field studies: *U. S. Bureau of Mines, Repts. Investigations No. 2876,* 1928.

[2] O. E. Meinzer, The occurrence of ground water in the United States: *U. S. Geol. Survey, Water Supply Paper 489,* pp. 11-17, 1923.

Most of the methods of porosity determination proposed and used are based on method 2 above, particularly Melcher's and Russell's methods, though Nutting used method 3 and more recently Washburn and Bunting[1] proposed method 6.

Hirschwald method. Hirschwald[2] determined the absolute pore space of a rock by dipping in paraffin, determining the lump volume by means of a volumeter, and computing the lump specific gravity. The pore space was computed from the specific gravity of the grains and that of the whole fragment.

Melcher method. Melcher[3] was about the first to make an extended study of porosity. Melcher determined the volume of consolidated materials by coating the specimen with paraffin and weighing the rock so coated in both air and water.

The specimen is first cleaned of adhering loose grains and weighed. It is then dipped in melted paraffin—which is just a little above the melting point. The coating is then examined for pinholes and air bubbles, which are removed with a heated wire. The dipping process is as short as possible to prevent paraffin from entering the rock pores. When the paraffin coat is hard, the specimen is again weighed to determine the weight of the coating acquired. The coated sample is next suspended by a fine wire and weighed immersed in water. The temperature of the water is also recorded. The sample is dried and again weighed in air. A check against the original air weight is thus obtained to see if any water was absorbed. If an appreciable quantity of water was absorbed, its weight is added to the weight of the water displaced. (The apparent loss of weight or weight in air minus the weight in water is the weight of the water displaced according to the well-known principle of Archimedes.)

From the weight of the water displaced, its temperature, and its density, the volume of the sample plus paraffin is obtained. From a previous determination of the density of the paraffin and weight of the paraffin covering the sample, the volume of the paraffin is computed. This volume is subtracted from the total volume and thus the volume of the fragment (plus voids) is obtained.

The volume of the grains (less voids) was determined by Melcher by means of a pycnometer.[4]

[1] E. W. Washburn and E. N. Bunting, Porosity: VI, Determination of porosity by the method of gas expansion: *Jour. Am. Ceram. Soc.,* vol. 5, pp. 113-129, 1922.
[2] Julius Hirschwald, *Die Prüfung der Natürlichen Bausteine auf ihre Wetterbeständigkeit* (W. Ernst und Sohn, Berlin, 1908).
[3] A. F. Melcher, Determination of pore space of oil and gas sands: *Mining and Metallurgy,* No. 160, sec. 5, 1920.
[4] Melcher preferred a Johnston and Adams plane-joint pycnometer instead of a common pycnometer.

A sample of the rock, crushed to fine size, weighing about 5 g. is dried at 110° C. for 30 min. to 1 hr. and cooled in a desiccator and weighed. It is then exposed to the air until constant weight is reached and placed in a pycnometer bottle of known weight. The pycnometer is filled with distilled water and the adhering air is removed by an aspirator.[1] After reaching a constant temperature, the pycnometer and contents were weighed. From a previous calibration of the pycnometer, which gives the weight of the water necessary to fill it, the weight of the water displaced by the grains is found. The volume of the particles is then computed from the weight of the water displaced and the table of densities of water at the temperature recorded.

The derivation of the formula (after Melcher) on which the computations of porosity are based is given below with an example:

Let V = volume of the fragment
V_{tg} = volume of grains of that fragment free from moisture
V_p = volume of pore space

$$V_p = V - V_{tg} \qquad (1)$$

The percentage of pore space is

$$P = \frac{V - V_{tg}}{V} 100 = 100 \left(1 - \frac{V_{tg}}{V}\right) \qquad (2)$$

To find V and V_{tg}, break the sample in two parts, one to be dipped in paraffin and the other to be used for the determination of the grains.

Let W and W_1 = weights, respectively of the two pieces
W_p = weight of one of the fragments dipped in paraffin
then $W_p - W$ = weight of the paraffin

The volume of the paraffin is

$$V_{p1} = \frac{W_p - W}{0.906}$$

where 0.906 = density of paraffin at $20.4°$ C.

Let W_{p1} = weight of fragment dipped in paraffin plus wire carrier in boiled distilled water
W_c = weight of wire carrier immersed an equal distance in water as when fragment was attached

The weight of the water displaced by the fragment plus its coating of paraffin is $W_p - (W_{p1} - W_c)$.

[1] Melcher, and also Stearns, employed the aspirator to first evacuate the air from the sample and then, while still in vacuum, added boiled distilled water. But if acetylene tetrachloride or tetralin (tetrahydronaphthalene) is used, there is no need of the aspirator. These liquids have greater "wetting power" than water. Of these two Nutting preferred tetralin. The ideal pycnometer fluid is one which gives nothing to the grains and dissolves from them any oil or moisture without increase of its own volume. A good pycnometer liquid should completely wet the grains without being adsorbed.

$$V = \frac{W_p - (W_{p1} - W_c)}{D_t} - V_{p1}$$

where D_t = density of water taken from the density tables at the temperature of the water when weighing was made. Substituting for V_{p1} its value,

$$V = \frac{W_p - (W_{p1} - W_c)}{D_t} - \frac{W_p - W}{0.906} \quad \ldots \ldots \quad (3)$$

If W_g = weight of grains after sample is crushed, $W_1 - W_g$ = weight lost or gained by breaking-up of the sample. The difference ($W_1 - W_g$) is usually quite small and may be neglected, if precautions be taken.

Let W_{g1} = weight of crushed sample, oven-dried
W_k = weight of pycnometer
W_{k1} = weight of water content of pycnometer at temperature $t1$
W_{k2} = weight of pycnometer with crushed sample filled with water
D_{t1} = density of water at temperature $t1$

The weight of water displaced by the crushed sample is

$$W_{k3} = W_{k1} - [W_{k2} - (W_k + W_{g1})]$$

and the volume of the grains is

$$V_g = \frac{W_{k3}}{D_{t1}} = \frac{W_{k1} - [W_{k2} - (W_k + W_{g1})]}{D_{t1}} \quad \ldots \ldots \quad (4)$$

The total volume of the grains, V_{tg}, in the fragment that was coated with paraffin is found from the proportion

$$V_{tg} = \frac{W V_g}{W_g}$$

Substituting for V_g its value in equation (4),

$$V_{tg} = \frac{W \{W_{k1} - [W_{k2} - (W_k + W_{g1})]\}}{W_g D_{t1}}$$

Substituting in equation (2) for V and V_{tg} their values

$$P = 100 \left[1 - \frac{D_t \, W \, \{W_{k1} - [W_{k2} - (W_k + W_{g1})]\}}{D_{t1} \, W_g \left[W_p - (W_{p1} - W_c) - \left(\frac{W_p - W}{0.906}\right) D_t \right]} \right]$$

where W_k, W_{k1} and W_c are experimental constants and D_t and D_{t1} are constants found from the tables on density of water free from air. These leave six quantities W, W_g, W_{k2}, W_{g1}, W_p, and W_{p1} to be found by weighing.

Russell method. Russell[1] in 1926 described a new and short method of porosity determination. He devised an apparatus[2] (Figure 258) con-

---

[1] W. L. Russell, A quick method for determining porosity: *Am. Assoc. Petrol. Geol., Bull. 10*, pp. 931-938, 1926.
[2] Obtainable from Eimer and Amend, New York.

sisting of two graduated glass tubes which at one end are enlarged into the space A and at the other end, B, and ground to fit tightly into the stopper C. The apparatus is so made that when the stopper is in place the volume below the base of the lowest graduation on the tubes, marked "zero point," is equal, or nearly equal, to the volume above the top of the highest graduation of the graduated tubes, marked Z'. Hence if the level of the liquid in the tubes is at Z' when the apparatus is inverted, it will be at the zero point, or close to it, when the upright position is resumed. It is almost impossible to achieve this relation exactly, hence a small correction for the displacement of the zero point must be made.

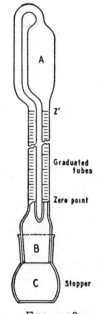

FIG. 258.—Russell porosity apparatus.

To use the apparatus [1] the stopper is taken off and acetylene tetrachloride is poured in until space A is filled and the liquid rises to the first graduation, Z', or a little above. A reading is then taken. Meanwhile the rock specimen has been immersed in acetylene tetrachloride until bubbles cease to rise. The specimen is removed with tweezers, held momentarily with one corner downward while the excess of liquid, which runs to the lower end of the specimen, is quickly blotted up with a suitable cloth or paper, and then placed in the stopper which is fitted over the ground glass joint (previously rubbed with stopcock grease) of the tubes. The apparatus is then turned upright and a second reading is made (and correction applied for the zero-point error). The difference between the two readings gives the volume of the solid specimen, both grains and pores (which were filled with acetylene tetrachloride from previous immersion). The volume of the grains is determined in like manner by grinding the specimen of the rock in question and placing a weighed quantity of the dried powder in the Russell apparatus. Rotating the apparatus in an inclined position will free adhering air bubbles. Correction must be made for any loss of material during the crushing. The difference between the volume of the solid fragment and the original volume of the grains, multiplied by 100, gives the porosity of the rock in percentage.

The principal sources of error with this apparatus are those due to changes in temperature and to loss due to grinding prior to grain volume measurement. The first is appreciable only if the air temperature changes rapidly or if the instrument has an undue amount of handling. The loss on grinding can be corrected for by weighing the original specimen and

[1] See also description of details of the method by C. E. Sutton, *loc. cit.*, 1928.

weighing the dried ground powder. The volume difference resulting from loss is proportional to the observed weight difference.

Gealy's method. Gealy[1] used the Russell apparatus, but used mercury to measure the volume of the rock chunk. A wire basket was added in which the rock fragment was placed in order to insure complete immersion of the specimen in mercury. It is not necessary, of course, to immerse the fragment prior to volume measurement as in the case of the acetylene tetrachloride method. The mercury does not penetrate the rock pores at all. Care should be used not to break the apparatus with the mercury, and the detachable cap should be held firmly in place when the mercury is in this end, otherwise it is likely to come off under the pressure of the liquid. The mercury becomes dirty after usage and may be cleaned by squeezing it through two or three thicknesses of clean cloth or by spraying into dilute nitric acid. The grain volume, however, must be measured with a pycnometer or with the Russell apparatus in the usual way.

Nutting's method. Nutting,[2] in 1930, modified the Melcher method of porosity determination. Nutting coated his specimens with paraffin in the usual way and determined their density by putting them in a low, wide cylindrical pycnometer vessel some 30 mm. in diameter and 20 mm. deep (Figure 259). This pycnometer is provided with a flange 4 mm. wide on which rests a circular plate-glass cover some 2 mm. thick and 38 mm. in diameter. The top of the flange is ground and polished optically flat, so that the cover defines a precise volume. The volume is determined from the weight of water displaced by the lump. The same pycnometer is utilized in the determination of the grain densities or volume, but tetralin (tetrahydronapthalene) is used instead of water.

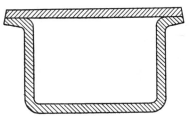

FIG. 259.—Nutting pycnometer.

Method of Washburn and Bunting. In the sixth method outlined, the method of Washburn and Bunting, an air-tight vessel, B, is joined by a capillary tube to a second air-tight vessel, A (Figure 260). There is a stopcock on each vessel and in the tube that connects the two vessels. There is also a manometer connected with vessel A. A dry sample of

[1] W. B. Gealy, The use of mercury for determination of volume of rocks with the Russell porosity apparatus: *Am. Assoc. Petrol. Geol., Bull. 13,* pp. 677-682, 1929.
[2] P. G. Nutting, Physical analysis of oil sands: *Am. Assoc. Petrol. Geol., Bull. 14,* pp. 1342-1347, 1930.

known volume of the material to be tested is placed in vessel A. Most of the air in vessel A is then pumped out, and the pressure of the remaining air is observed. The air in vessel B is at the atmospheric pressure. The stopcock in the connecting tube is then opened, and the

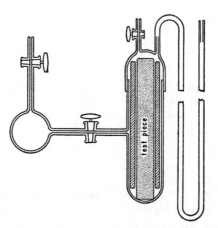

FIG. 260.—Washburn and Bunting expansion porosimeter.

resulting pressure is observed. After each operation the apparatus must be allowed to stand until temperature equilibrium is attained. The porosity is computed from the formula:

$$P = 100 \left[\frac{(p_1 - p_3)v_1}{(p_3 - p_2)v_3} - \frac{v_2 - v_3}{v_3} \right]$$

where P = porosity, in percentage by volume
p_1 = initial pressure in vessel B, or atmospheric pressure
p_2 = initial pressure in vessel A
p_3 = pressure in both vessels after the stopcock between them has been opened
v_1 = volume of vessel B
v_2 = volume of vessel A
v_3 = volume of sample

The volume of the sample may be measured either by the method of Melcher or by one of the several types of volumeters developed for the purpose.[1] Several types of gas-expansion porosimeters were also devel-

[1] Ernest Goodner, A mercury volumeter: *Jour. Am. Ceram. Soc.*, vol. 4, pp. 288-300, 1921. A. E. R. Westman, The mercury balance—an apparatus for measuring the bulk volume of brick: *Jour. Am. Ceram. Soc.*, vol. 9, pp. 311-318, 1926. H. G. Schurecht, A direct reading overflow volumeter: *Jour. Am. Ceram. Soc.*, vol. 3, p. 731, 1920.

oped by MacGee,[1] including a direct reading type to simplify calculations.

The principle of the Washburn-Bunting method has been applied to a porosimeter of the McLeod gage type. In this instrument the air is drawn from the rock pores by means of a mercury piston and subsequently compressed into a burette at atmospheric pressure. Fancher, Lewis, and Barnes[2] recommend it as the most desirable method of measuring porosity where accuracy, cost, and rapidity of test are all involved. It should be noted, however, that only the *effective* pore volume is measured by the porosimeter.

The porosimeter consists of a sample chamber connected in series to an expansion chamber and to a calibrated heavy glass capillary tube and valve. This unit is set in a fixed position, with the capillary tube uppermost. The bottom of the sample chamber is connected by a heavy rubber tube to a glass leveling bulb containing mercury. The leveling bulb is arranged so that it can be readily raised or lowered. Lowering the bulb lowers the mercury and thus reduces the pressure so that the air is drawn from the pores of the test piece. The bulb is then raised until the mercury, in which the test piece is still immersed, stands at the same level in the capillary tube as in the leveling bulb itself. The volume of air collected (at atmospheric pressure) is then read to the nearest 0.01 ml. Certain corrections, for the air remaining in the sample under reduced pressure and for adsorption on the walls of the porosimeter, need to the made. The bulk volume of the specimen is obtained by means of a volumeter. The reader is referred to the bulletin by Fancher, Lewis, and Barnes for details.

The Barnes method of saturation with tetrachlorethane under reduced pressure is also described by the same authors.

PERMEABILITY

Permeability is defined as the faculty of allowing passage of fluids without impairment of structure or displacement of parts. In geology a rock is termed "permeable" when it permits an appreciable quantity of fluid to pass through in a given time; and "impermeable" when the rate of passage is negligible. Very obviously permeability depends not only

[1] A. Ernest MacGee, Several gas expansion porosimeters: *Jour. Am. Ceram. Soc.*, vol. 9, pp. 814-822, 1926.
[2] G. H. Fancher, J. A. Lewis, and K. B. Barnes, Some physical characteristics of oil sands: *Penna. State College, Bull. 12*, pp. 64-167, 1933. This paper contains the most complete review of the methods of porosity and permeability measurement known to the authors.

on the rock itself but on the fluid involved and the hydraulic head or pressure.

Permeability is governed by (1) grain size, (2) grain shape, (3) porosity, (4) uniformity of grain size, and (5) packing or arrangement of the grains. It is also modified by the nature of the fluid involved, especially viscosity, and by the hydraulic gradient.

A full discussion of the interrelations between porosity and permeability as well as a more complete discussion of the latter is found in the paper by Fraser.[1]

The method to be used in measuring permeability will depend on whether the material is consolidated or unconsolidated.

Unconsolidated materials. Permeability in unconsolidated materials is usually determined in the laboratory by measurement of the rate of flow of a fluid, usually water, through a column of the sediment in question contained in some kind of a percolation cylinder. The general formula for permeability is written as follows:

$$P = \frac{qlt}{Tah}$$

where P is the coefficient of permeability, q is the quantity of water, l is the length of the column of sample, t is the correction for temperature, T is the time, a is the cross-section area of the column, and h is the head or hydraulic gradient. This formula is based on Darcy's law that the rate of flow varies in direct proportion to the hydraulic gradient.

Stearns[2] describes the test devised by O. E. Meinzer. Two types of apparatus were used. These differed only in the length of the percolation cylinders. One such cylinder, constructed of copper, was 48 in. high; the other was but 8 in. long. These have a diameter of 3 in. There is one opening near the bottom for the inflow of water and one near the top for discharge of water that has risen through the column of sediment. The difference in head of the water at the top and bottom is regulated by an adjustable supply and determined by reading of two pressure gages which connect to the top and the bottom of the cylinder (Figure 261). The overflow is caught in a graduated cylinder.

Prior to operation of the permeability test several rubber stoppers are placed in the bottom of the percolation cylinder and on these is placed a circular piece of fine copper gauze to keep the sand from sifting through. The sample is placed in the cylinder and jarred or tamped so

[1] H. J. Fraser, *loc. cit.*, 1935.
[2] Norah Dowell Stearns, *loc. cit.*, pp. 144-158.

MASS PROPERTIES OF SEDIMENTS

that it occupies a minimum volume. Water is allowed to enter slowly, and after uniform discharge appears at the top of the cylinder the test is begun. The temperature is taken, the difference in pressure in the two gages is read, and the rate of discharge is determined by stopwatch for either a 30- or a 60-sec. period. Several tests with different hydraulic heads are recorded.

Fraser [1] used a similar apparatus only slightly modified and conducted his tests in much the same manner.

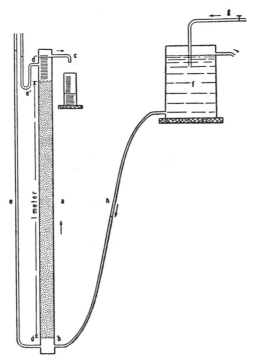

FIG. 261.—Meinzer long tube permeameter. (After Stearns.)

It is possible, on the basis of certain simplifying assumptions, to compute the coefficient of permeability from the known properties of both the liquid and the sediment involved in the system in question, such as porosity, effective grain size, coefficient of viscosity, hydraulic gradient, area of cross-section of the column, and height of the column. Fraser[2] has reviewed the formulas of Slichter, Terzaghi, and others and has compared the values with those experimentally derived from col-

[1] H. J. Fraser, *loc. cit.*, pp. 947-950.
[2] H. J. Fraser, *loc. cit.*, pp. 950-957.

umns of uniformly sized spheres. The equations of Slichter and Terzaghi were found to give reasonably satisfactory values for small, uniformly sized spheres, but when they were applied to sediments variably sized and irregularly shaped grains, discrepancies appeared. Fraser further discusses the causes of these discrepancies, but a review of this problem is out of place in a book dealing primarily with analytical methods.

Consolidated rocks. A method of measurement of permeability of consolidated rocks is described by Nutting.[1] A small test plate is cut from the rock to be studied. A standard size is about ½ in. in diameter and 5 mm. thick. This plate is carefully coated around the edge with sealing wax and then cemented over a 10-mm. round hole in an ordinary pipe cap—the cast-iron bowl used by plumbers to cap blind pipes. This cap is then screwed onto a "T" attached to a sink tap; the other side of the T is for attaching a pressure gage. Flow is measured by catching the drip for a few minutes in a graduate. Because the rate of flow falls off rapidly (ordinarily to about one half in an hour) the measurement is begun as soon as the flow is full and steady. Nutting prepared the test disc from a core sample by working down to dimensions on a coarse carborundum stone.

Nevin,[2] following Botset,[3] depreciates the determination of permeability by measuring the water flow through the rock or sand. The rate of flow of water through a sandstone will diminish with time. Such diminution is apparently due, according to Botset, to the hydrolysis of silica by water and the production of silicic acid. Nevin therefore recommends the measurement of the flow of air through the sand. To make this measurement air was fed from a large storage tank, where it was under a pressure of 200 lbs., through a reducing valve, thence through a pressure gage to the sample. The purpose of the valve is to deliver air at a specified pressure to the sample. The time required for 10,000 c.c. to pass through the sample was recorded with a stopwatch. The volume was measured with a gas gage or by displacement of a like volume of water from a bell-jar.

The sample itself, cut to disc shape, was mounted in brass fittings sealed with sealing wax or with modeling clay.[4] Samples must be freed

---

[1] P. G. Nutting, *loc. cit.*, pp. 1347-1349.
[2] Charles Merrick Nevin, Permeability, its measurement and value: *Bull. Am. Assoc. Petrol. Geol.*, vol. 16, pp. 373-384, 1932.
[3] H. G. Botset, The measurement of permeabilities of porous alundum discs for water and oils: *Review of Scientific Instruments*, vol. 2, No. 2, pp. 84-95, 1931.
[4] A. F. Melcher, Apparatus for determining the absorption and the permeability of oil and gas sands for certain liquids and gases under pressure: *Bull. Am. Assoc. Petrol. Geol.*, vol. 9, pp. 447-450, 1925.

of oil (with carbon tetrachloride in a reflux condenser) before permeability tests can be made.

Nevin computed the permeability from the usual formula (page 514). The permeability of two sediments may be compared by this formula provided that the units of measurements in each case are the same.[1] Nevin found that there was a direct relationship between the permeability to air and that to water, so that for comparative purposes the permeability to air is all that is required.

A complete discussion of the theory and methods of permeability measurement for consolidated sediments is given by Fancher, Lewis, and Barnes.[2] A small core of the sandstone, cut to proper dimensions is inserted into a bored No. 7 rubber stopper, which in turn is tightly wedged into a core holder. Air or liquid is passed through the test piece, and the volume delivered in a given time is measured. Suitable gages measure both the pressure of the fluid delivered and the back pressure. The coefficient of permeability is computed from Darcy's law. For full details of technique, precautions to be taken, and computations, the reader is referred to the original paper.

PLASTICITY

DEFINITIONS

Plasticity is defined by Seger[3] as the property which solid bodies show of absorbing and holding a liquid in their pores and forming a mass that can be pressed or kneaded into any desired shape, which it retains when the pressure ceases, and that, on the withdrawal of the water by evaporation, changes into a hard mass.

The definition as given above is essentially the technical or "practical" definition. Physically a substance is said to be plastic when it exhibits plastic flow. Pure plastic flow, essentially that shown by clay, is expressed by the equation

$$D = 1/\eta \; (\tau - f)$$

where D is the velocity gradient, η is the cofficient of viscosity, τ is the shearing stress, and f is the yield value. By velocity gradient is meant the

[1] Nevin let the viscosity of water equal unity. In that case the viscosity factor for air is approximately 60 times less. A viscosity factor of 60 was therefore used.
[2] G. H. Fancher, J. A. Lewis, and K. B. Barnes, loc. cit. (1933).
[3] H. Seger, Beziehungen zwischen Plasticität und Feuerfestigheit der Thone: Tonindustrie-Zeitung, vol. 14, pp. 201-202, 1890.

dv/dy relations, where dv is the velocity increment of one lamina of flow with respect to another over a distance dy. As may be seen by inspection of Figure 262, plastic flow is a laminar motion essentially the same as viscous flow except that a plastic substance is not deformed until stresses greater than the yield value, f, accumulate. The curves of viscous flow and plastic flow are essentially the same except that the curve of viscous flow passes through the origin, whereas that of plastic flow cuts the X-axis at the yield point, f. The rate of yield is directly proportional to the shearing stress. The slope of the curves in both cases is expressed by the tan a or $1/\eta$.

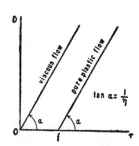

FIG. 262.—Viscous and plastic flow. D, velocity gradient; f, yield point; r, shearing stress.

METHODS OF MEASUREMENT

Owing to the importance of plasticity in certain industries, notably the ceramic industry, many practical methods for the measurement of this property have been devised. These methods, however, do not all give results consistent with one another nor do they give results which can be interpreted in terms of the physical constants, η and f, that is, the coefficient of viscosity and yield value.

The "practical methods" are sometimes tied up with more than one physical constant, and it is not always easy to interpret the results obtained by these methods. Nevertheless, since the methods are well established it seems advisable to review in brief fashion their main features.[1]

The plasticity of a clay is influenced by (1) the past history of the clay, (2) the type of clay, (3) the amount of working the clay mixture has received, (4) the time of contact of the clay and water, (5) the temperature of the mixture, and (6) the amount of water. Comparative studies of plasticity should, therefore, be made under similar conditions.

A statement of plasticity based on a determination of yield value and viscosity is perhaps the most sound. If the rate of flow (viscosity) be held constant, the plasticity will vary directly as the yield value. Accordingly a plastometer, of the Bingham type,[2] which measures the rate of flow through a capillary tube under constant measurable pressure, may be used. The flow at several pressures is determined, and the volume

[1] For a full discussion of the theory of plasticity and the "scientific" methods of measuring this property, the reader is referred to the volume by R. Houwink, *Elasticity, Plasticity and Structure of Matter* (Cambridge, England, 1937). Also: Hind, S. R., The plasticity of clay: *Trans. Ceram. Soc.*, vol. 29, pp. 177-207, 1930.
[2] F. P. Hall, The plasticity of clays: *Jour. Am. Ceram. Soc.*, vol. 5, pp. 346-354, 1922.

MASS PROPERTIES OF SEDIMENTS 519

discharged per second is plotted against the pressure producing the flow. The slope of the line shows the mobility of the clay, and the intercept on the pressure axis shows the yield value.

Some of the "practical methods" that have been used are the determination of the length of a pencil extruded from a cylindrical die before breaking under its own weight; the product of the weight required to force a Vicat needle of given dimensions into the clay a given distance in a given time;[1] the increase in area of a clay cylinder, 2 cm. x 5 cm., under compression at the point of fracture, the product of the deformability or the pressure required to produce cracks on clay cylinders or spheres of specific dimensions and the tensile strength;[2] the difference between the amount of water required to achieve the lower limit of fluidity (at which two portions of the clay will not flow together if jarred) and the amount of water at the rolling-out limit (at which the clay can no longer be rolled into thin cylinders);[3] and the difference between the amounts of water required to make the clay sticky and that needed to make it just wet enough to be rolled, or the loss of color of a dye solution when mixed with clay.[4]

HYGROSCOPICITY

The amount (percentage) of water absorbed and held in equilibrium by a sediment in a saturated atmosphere is the *hygroscopic coefficient,* and the water so held, as a film of molecular dimensions, is termed hygroscopic water. The amount of hygroscopic water is dependent on the total surface area of the grains, which is a function of their size or state of division, and the chemical nature of the sediment. In the case of a suite of clay samples, for example, of about the same chemical composition, the hygroscopicity is inversely proportional to the coarseness of the grain, or increases as the grain size decreases. As Sauramo noted, the determination of hygroscopicity is very sensitive and its results are good characteristics for the study of the mode of origin of the sediment.

The procedure followed by Sauramo is given herewith: About 20 g. of the material are ground fine and kept for 5 hr. in a steam bath in vacuum with phosphoric pentoxide. The sample is thus made completely dry. It is then weighed and placed under reduced pressure over a 10 per cent

[1] F. F. Grout, The plasticity of clays: *Jour. Am. Chem. Soc.,* vol. 27, pp. 1037-1049, 1905.
[2] B. Zschokke, Untersuchungen über die Bildsamkeit der Tone: *Tonindustrie-Zeitung,* vol. 29, pp. 1657-1662, 1905. J. W. Mellor, On the plasticity of clays: *Trans. Faraday Soc.,* vol. 17, pp. 354-365, 1921.
[3] Albert Atterberg, Die Plastizität der Tone: *Int. Mitt. für Bodenkunde,* vol. 1, pp. 10-43, 1911.
[4] Harrison Everett Ashley, The colloid matter of clay and its measurement: *U. S. Geol. Survey, Bull. 388,* pp. 42-47, 1909.

aqueous solution of sulphuric acid and there retained for 5 or 6 days. The material will then have taken up its maximum amount of water and is weighed once more. The difference between this and the first weighing gives the amount of water absorbed, which is expressed as a percentage of dry substance. The figure thus found expresses the relative hygroscopicity and is termed the hygroscopic coefficient.[1]

Wallace and Maynard [2] also measured the hygroscopicity of clays. Ten-gram samples were placed in a desiccator in the bottom of which a wet sponge had been placed to insure a saturated atmosphere. Weighings of the samples were made every few hours until constant weight was reached. The percentage of water absorbed or hygroscopic moisture was then calculated.

The relation between grain size and hygroscopicity was further studied by Maynard.[3]

MISCELLANEOUS MASS PROPERTIES

Adsorption. Adsorption is the power a substance (particularly clay) has of removing solid substances from solutions which are in contact with it.

The power of adsorption varies with different clays and has been used to characterize certain clays. The adsorption of sodium carbonate by a clay is an index to the colloid content of a clay. Wallace and Maynard [4] weighed out samples, dried at 110° C. to constant weight, and placed these in 500-c.c. Erlenmeyer flasks. To each sample was added 100 c.c. of standardized normal sodium carbonate solution. At the end of 112 hours 25 c.c. of the sodium carbonate solution were withdrawn by means of a pipette and titrated against a previously standardized solution of HCl. The decrease in strength of the sodium carbonate solution represented the amount of sodium carbonate adsorbed by the clay. The percentage adsorbed was calculated.

Moisture equivalent. The moisture equivalent is a measure of the water-retention capacity of a soil. It is expressed by the formula

$$M_w = \frac{w}{100W}$$

where M_w is the moisture equivalent in per cent by weight, w is the weight of the moisture, and W is the weight of the dry sample. Experimentally, w is the weight of the moisture retained in the saturated specimen subject to

---

[1] Matti Sauramo, Studies on the Quaternary varve sediments in southern Finland: *Bull. Comm. Geol. de Finlande*, vol. 11, No. 60, pp. 17-20, 1923. Sven Oden, Note on the hygroscopicity of clay and the quantity of water absorbed per surface unit: *Trans. Faraday Soc.*, vol. 17, pp. 244-248, 1921.

[2] R. C. Wallace and J. E. Maynard, The clays of the Lake Agassiz basin: *Roy. Soc. Canada, Proc. and Trans.*, 3rd ser., vol. 18, sec. iv, 1924.

[3] J. E. Maynard, The clays of the Lake Agassiz basin: *Roy. Soc. Canada, Proc. and Trans.*, 3rd ser., vol. 19, sec. iv, pp. 103-114, 1925.

[4] R. C. Wallace, and J. E. Maynard, *loc. cit.*, pp. 15-18, 1924.

constant centrifugal force. The moisture retained is determined by loss in weight after drying in an oven at 110° C. For further details the interested reader is referred to the paper by Stearns.[1]

Shrinkage. The linear shrinkage on air- or oven- (105° C.) drying is often measured. The linear shrinkage was one of several physical properties used by Sauramo [2] to distinguish between Quaternary fresh-water and brackish-water clays.

Tensile strength. Tensile strength is the resistance to rupture exhibited by a clay after air-drying. The tensile strength and shrinkage are a measure of the bonding power of a clay or molding sand and are of importance in foundry use.

Crushing strength. The crushing strength is the resistance to compression after air-drying. It is of but slight importance.

Fusibility. The fusion point of a clay is the temperature at which fusion or melting takes place. It depends largely on the chemical composition of the clay. It is a matter of great importance in the ceramic industries.

Slaking. Slaking is the ease with which a piece of dried clay falls apart upon immersion in water.

Cohesiveness. Cohesion is that form of attraction by which the particles of a body are united throughout the mass, whether like or unlike.

Shrinkage, tensile strength, crushing strength, cohesiveness, fusibility, and slaking are properties which, for the most part, are not known to have any geologic significance. They are, however, important properties from the technological point of view and must be measured in order to know the economic worth of a clay deposit. For these various tests one must refer to books and articles dealing with the technology of clay and similar materials.[3]

[1] Norah Dowell Stearns, Laboratory tests on physical properties of water-bearing materials: *U. S. Geol. Survey, Water Supply Paper, 596F,* pp. 134-137, 1927.
[2] Matti Sauramo, *loc. cit.,* p. 19, 1923.
[3] Heinrich Ries, *Clays, Their Occurrence, Properties, and Uses,* 3rd ed. (New York, 1927).

CHAPTER 21

THE LABORATORY, EQUIPMENT, AND ORGANIZATION OF WORK

THE LABORATORY

IN DESIGNING and equipping a laboratory for the study of sediments, special consideration has to be given to the necessity of simultaneous handling of a great many samples. Speed as well as accuracy are especially called for in an oil-field laboratory where the laboratory worker is in direct touch with the drilling. Work should be kept up to date so that as the producing horizon is approached no delay in completion of the well will take place. To be of practical use, therefore, the laboratory and equipment should be arranged so that many samples may be examined for their fossil and mineral content and the information so obtained may be communicated to those in charge of oil-field operations.

The requirements of a good laboratory are (1) adequate space, (2) good lighting, (3) good ventilation, and (4) special facilities for work and storage (Figure 263). Basement laboratories are generally unsatisfactory in respect to both lighting and ventilation. Good artificial lighting and a motor-driven ejector fan may in part overcome these difficulties. A basement laboratory, however, has the advantage of constant temperature which is so necessary in methods of analysis dependent upon the settling of grains in a column of liquid. For color determination daylight is required. A northern exposure unobstructed by trees and buildings is most desirable. The special facilities required include an acid-proof sink, preferably soapstone, acid-proof drainboard, hot and cold water, gas, and electricity. Both alternating and direct current are desirable. The latter is needed for operation of an arc-light and an electromagnetic separator. Otherwise a battery and battery charger are required for the magnetic work. Numerous convenient outlets are required for microscope lamps, shaking machine, Ro-Tap sieve shaker, hot-plate, and other similar apparatus. A steam table for evaporations is desirable, though a *large* electric hot-plate is satisfactory. The power need for such heating equipment usually requires a special line.

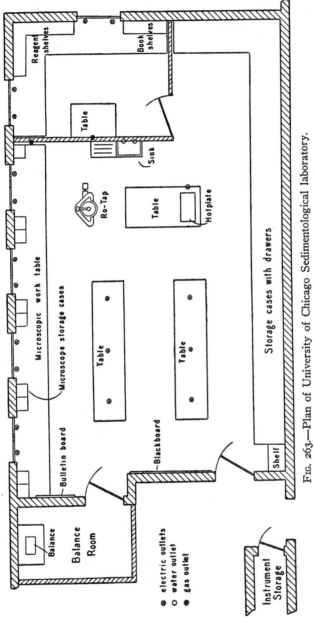

FIG. 263.—Plan of University of Chicago Sedimentological laboratory.

Work space should be provided on tables, slate-covered for chemical work and linoleum-covered for other work. These tables should be against a wall with electric outlets or should themselves be wired for electricity. The tables should be of such height and supplied with stools of such height that the worker can readily look down the vertical tube of a microscope. It is necessary that the microscope tube be vertical and the stage horizontal for both immersion methods and microchemical work. Improper table height is very inconvenient.

Storage facilities are important. Samples may be taken care of by suitable cases, provided with drawers, set against one wall. Cabinets with deep drawers and cupboards provide space for microscopic and other special equipment. Numerous shelves are needed for chemical reagents and for books. Odd corners or other waste space may be utilized for storage of screens and sample splitter or other bulky equipment.

Special arrangements need to be made, moreover, in oil-field laboratories for keeping extensive files of data on various wells and for storage of reference specimens and well samples. For storage of this collection provision should be made with drawers built into cabinets in the laboratory. Each drawer should be made to hold five tiers of the standard sample envelopes.

Hanna and Driver [1] recommend the use of the original envelopes, since much labor and annoyance is avoided in the transfer to glass bottles or similar containers. They also recommend a linear system of filing. All samples from one well, arranged according to depth, are given an accession or lot number, and the lot is filed at the end of the collection irrespective of what precedes or may follow. Drawers are thus filled from the beginning and no shifting of lots ever becomes necessary.

It is desirable to have a small separate balance room, free from dirt and drafts, for weighing, a small closet for storage of valuable instruments, and a special laboratory for chemical work equipped with a fume chamber. Acid fumes in the general laboratory damage and corrode delicate apparatus. Preliminary disaggregation of the sample and splitting and similar operations which create a great deal of dust should not be carried out in any of the laboratories mentioned but should be done in another place.

If the laboratory is to be used for teaching purposes, a blackboard and a bulletin board, for posting assignments and instructions, are quite necessary. Even if the laboratory is used solely for research such additions are also useful.

[1] G. D. Hanna, and H. L. Driver, The Study of subsurface formations in California oil field development: *Summary of Operations, California Oil Fields*, vol. 10, pp. 5-26, 1924.

LABORATORY AND EQUIPMENT

It should be added that if no special laboratory is available most of the analyses described in this volume can be carried out either in a good chemical laboratory or in a well-equipped laboratory for mineralogical and petrographic research.

APPARATUS

A great part of the analytical work on sediments may be carried out with the apparatus found in most laboratories for chemical or petrographic research. Lack of special equipment, however, often renders the work laborious. To facilitate setting up a laboratory or to aid the worker about to start on a research project on sedimentary materials, an annotated list of equipment has been prepared.

This list is broken down into three parts. List "A" is the minimum essential equipment. The items contained therein, excepting the chemical balance and the petrographic microscope, are all inexpensive and easily procured. Since most laboratories have both the balance and the microscope, work may generally be carried forward without large appropriations. List "B" contains highly desirable items many of which are to be found in an up-to-date laboratory. Others are special items, though, excepting for a few, they may be obtained from scientific supply houses. Most of the items of this list have been figured and described in the preceding chapters of this book. List "C" is of special apparatus needed for some research projects.

List A: Essential Equipment

Chemical balance
Balance weights
Testing sieves, cover and pan
    Wentworth intervals (.0625-16 mm.)
Petrographic microscope and case
Microscopic accessories
    Stage micrometer
    Micrometer eyepiece
    Objective lenses
      (4-, 16-, and 40-mm.)
    Gypsum plate
    Quartz wedge
    Camera lucida
Immersion liquids, bottles and case
    (30 liquids, 1.45-1.75, interval of .010)

Small agate mortar and pestle
Glassware
    Petrographic slides
    Cover glasses (7/8 in. square)
    Beakers, 50-, 250-, 500-, and 1000-c.c.
    Erlenmeyer flasks, 250- and 500-c.c.
    Florence flasks, 1-liter
    Graduates, 25-, 50-, 100-, 500-, and 1,000-c.c
    Separatory funnel
    Funnels, 3-in. and 6-in. diam.
    Evaporating dishes, 6-in.
    Graduated volumetric flask
    Watch-glasses, 3-in.
    Test-tubes
    Pipettes, 10-c.c., 20-c.c.

List A: Essential Equipment (*Continued*)

Burettes
Pycnometer
Balsam bottle
250-c.c. reagent bottles
Large bottles (4-liter capacity)
Glass tubing (3/16- or 1/4-in.)
Glass rodding (1/8-in.)
1 gross 3 x 1 shell vials
Thermometers
   — 10° C. to 200° C. and — 10° to 60° C.
Hardware
  Test-tube holder
  Test-tube rack
  Ring-stand and rings
  Clamps
  Tripod
  Wire gauze
  Bunsen burners
  Tongs
  Pinch-clamps
  File (triangular)
  Cork-borer
  Wing-top for burner
Chemical reagents
  Sodium oxalate
  Sodium carbonate
  Hydrochloric acid
  Bromoform
  Ethyl alcohol
Xylol
Canada balsam
Potassium acid sulphate
Sodium hydroxide
Sodium hyposulphite
Calcium chloride (anhyd.)
Hot-plate, electric
Large horse-shoe magnet
Sample containers and miscellaneous
  Cylindrical cardboard cartons, 1/2-pint, pint, and quart sizes
  Boxes for microscope slides
  Counter brush
  Test-tube brush
  "Bon Ami"
  Corks, assorted
  Rubber stoppers, assorted
  Rubber tubing, several sizes
  Matches
  Stone laboratory jars
Paper and printed forms
  Graph paper
    Arithmetic
    Semi-log
  Filter-paper
  Form sheets
    Screen analysis
    Pipette analysis
    Mineral analysis
  Ledger book

List B: Desirable Equipment

Inexpensive triple-beam balance
Ro-Tap shaking machine for screens, stop-clock and switch
Accessory sieves ($\sqrt{2}$ and $\sqrt[4]{2}$ set)
Binocular microscope with case and several objectives
Mechanical stage attachable to rotating stage
Oil immersion lens
Microscope lamp
Refractometer, Abbé type, 1.45-1.81 range.
One-circle goniometer
Slide-rule
Planimeter, polar type, reading in mm.^{2}
Microsplitter
Gage for pebble diameters
Roundness circle gage

LABORATORY AND EQUIPMENT

List B: Desirable Equipment (*Continued*)

Drawing equipment
  T-square, board, triangles, protractor, scale, etc.
Overflow volumeter (Schurecht type)
Aspirator
Electromagnet, with resistance or with rectifier and transformer, and with adjustable pole pieces
Machine for shaking clay suspensions
Special stirring device for clay suspensions
Pasteur-Chamberland filter
Russell porosity apparatus
Drying oven, with thermostat control
Steep-sided funnel with valve in stem
Heavy-liquid hydrometer or Westphal balance
Large iron mortar and pestle
Diamond mortar
Jones riffle-type sample splitter with pans and scoop
Copper still, tin-lined, for distilled water
Abrasives
Chemicals
  Acetylene tetrabromide
  Piperine
  Bakelite varnish
  Nitrocellulose solution

List C: Special Equipment (Not Used in Routine Work)

Optical equipment
  Microprojector—Promar type (Clay-Adams) or Bausch and Lomb type
  Universal stage (Bausch and Lomb 5-circle)
  Integrating stage (Leitz)
  Temperature control cell for double variation work
  Monochromator
  Binocular eyepiece for petrographic microscope

Photomicrographic equipment

Special reagents
  Microchemical set
  Clerici's solution
  Furfural
  Benzene

Additional equipment for quantitative analysis of silicate and carbonate rocks
  Microburner
  Meeker burner

Platinum crucible and platinum-tipped tongs
Shapometer of Tester and Bay
Wentworth flat-type convexity gage

528 SEDIMENTARY PETROGRAPHY

List C: Special Equipment (*Continued*)

Pointolite
Soxhlet extractor
Centrifuge and centrifuge tubes
Berg centrifuge pipette
Crook electrostatic plates
Bingham plastometer
Viscosimeter
Odén balance or Wiegner tube
Sedimentation hydrometer
Washburn-Bunting porosimeter
Permeameter
Computing machine
Grinding laps

REFERENCE BOOKS

In addition to the laboratory and the apparatus listed above, certain standard reference books are desirable for the research worker and student. We have prepared a partial list of such larger works both in the field of sedimentology and in the related sciences. These works might well be placed on the book-shelves of the laboratory conveniently located for the worker.

Sedimentology and Sedimentation

Twenhofel, W. H., *Treatise on Sedimentation*, 2nd ed. (Baltimore, 1932).
Milner, H. B., *Sedimentary Petrography*, 2nd ed. (London, 1929).
Boswell, P. G. H., *On the Mineralogy of Sedimentary Rocks* (London, 1933).
Gessner, Hermann, *Die Schlämmanalyse* (Leipzig, 1931).
Hahn, F.-V. von, *Dispersoidanalyse* (Leipzig, 1927).
Cayeux, Lucien, *Introduction à l'étude pétrographique des roches sédimentaires*, Texte et Atlas (Imprimerie Nationale, Paris, 1931, re-impression).
Hatch, F. H., and Rastall, R. H., *Petrology of the Sedimentary Rocks*, rev. ed. (London, 1923).
Ries, Heinrich, *Clays, Their Occurrence, Properties and Uses*, 3rd ed. (New York, 1927).
Tickell, Frederick G., *The Examination of Fragmental rocks* (Stanford University, 1931).

Statistics

Mills, F. C., *Statistical Methods* (New York, 1924).
Arkin, H., and Colton, R. R., *An Outline of Statistical Methods*, in College Outline Series (New York, 1934).
Camp, B. H., *The Mathematical Part of Elementary Statistics* (New York, 1931).

LABORATORY AND EQUIPMENT

Fisher, R. A., *Statistical Methods for Research Workers* (London, 1932).

Microscopy and Mineralogy

Johannsen, Albert, *Manual of Petrographic Methods*, 2nd ed. (New York, 1918).

Larsen, E. S., and Berman, Harry, The microscopic determination of the nonopaque minerals: *U. S. Geol. Survey, Bull. 848, 1934.*

Winchell, Alexander N., *Elements of Optical Mineralogy, Part I, Principles and Methods*, 5th ed. (New York, 1937); *Part II, Descriptions of Minerals*, 3rd ed. (1933); *Part III, Determinative Tables*, 2nd ed. (1929).

Hartshorne, N. H., and Stuart, A., *Crystals and the Polarising Microscope* (London, 1934).

Wright, F. E., *The Methods of Petrographic-microscopic Research, Their Relative Accuracy and Range of Application:* Carnegie Inst. Wash. Public. 158, 1911.

Fry, W. H., Petrographic methods for soil laboratories: *U. S. Dept. Agric., Tech. Bull. 344, 1933.*

Chemistry and Microchemistry

Clark, F. W., The data of geochemistry: 5th ed., *U. S. Geol. Survey, Bull. 770, 1924.*

Hillebrand, W. F., The analysis of silicate and carbonate rocks: *U. S. Geol. Survey, Bull. 700, 1919.*

Groves, A. W., *Silicate Analysis* (London, 1937).

Washington, H. S., *The Chemical Analysis of Rocks* (New York, 1930).

Van Tongeren, W., *Gravimetric Analysis* (Amsterdam, 1937).

Miscellaneous

Handbook of Chemistry and Physics, 22nd ed. (Cleveland, 1937).

Journal of Sedimentary Petrology, vols. 1-8, 1931-1938. Published by the Society of Economic Paleontologists and Mineralogists, Tulsa, Oklahoma.

Reports of the Committee on Sedimentation, published by the National Research Council, 2101 Constitution Avenue, Washington, D. C.

ORGANIZATION

The plan of work in effect in the laboratory will depend on the objective of that work. If the objective is economic, an abbreviated, rapid, efficient routine will be established. If, for example, the laboratory exists as an adjunct to oil-field operations, large numbers of samples will need to be examined for both their fossil and their mineral content.

Operational problems that require in part routine laboratory control are (1) where to cement off the water, (2) how far the drill is from the main producing horizon, (3) whether the zone of production has been passed.

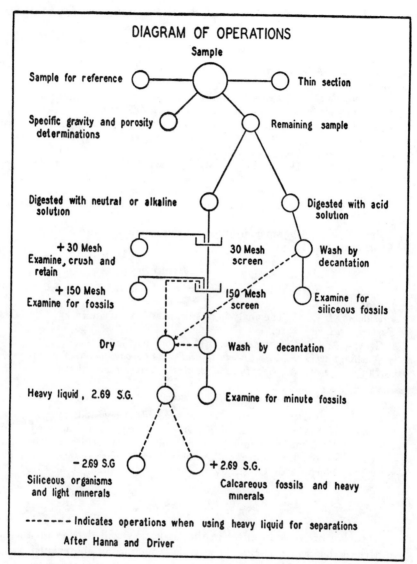

Fig. 264.—Diagram of operations (for an oil-field laboratory). (After Hanna and Driver.)

Where the oil is produced from several thin sand layers, failure to locate accurately the position of the uppermost one may result in shutting off this layer with the water. It may also happen that the water is cemented off too high, which will lead to difficulties of water incursion. Some of these difficulties are overcome by careful stratigraphic control based on laboratory study of core samples.

Where the exact structure is unknown but the general stratigraphy is understood, it is often desirable to know how far the drill is from the zone of production. Paleontologic and petrographic work on the well cuttings and cores may lead to an answer to this question. In like manner laboratory studies indicate whether the producing sand has been passed or not.

No plan of work that the writers might prepare would be found satisfactory in all situations. That shown in Figure 264, taken from Hanna and Driver, is to be regarded only as an example of a plan found workable in one oil-field laboratory. It is perhaps more complete than will be required in other places. For further description of current practice in oil-field laboratories, the reader is referred to other papers.[1]

For laboratories with different economic objectives, as for example the investigation of molding sands or the study of the clays of some region, a very different plan would be followed. Each worker will, as he becomes acquainted with the problems involved and the methods available, devise his own scheme.

The appended "flow sheet" of operations (Figure 265, next page) is the authors' attempt to summarize graphically the various interrelated techniques of study of the sediments. It is not to be taken literally as a plan of operations.

[1] C. B. Claypool and W. V. Howard, Method of examining calcareous well cuttings: *Bull. Am. Assoc. Petrol. Geol.,* vol. 12, pp. 1147-1152, 1928. Earl A. Trager, A laboratory method for the examination of well cuttings: *Econ. Geology,* vol. 15, pp. 170-178, 1920. R. D. Reed, Microscopic subsurface work in oil fields of United States: *Bull. Am. Assoc. Petrol. Geol.,* vol. 15, pp. 731-754, 1931 (with bibliography). Robt. M. Whiteside, Geologic interpretation from rotary well cuttings: *Bull. Am. Assoc. Petrol. Geol.,* vol. 16, pp. 653-674, 1932.

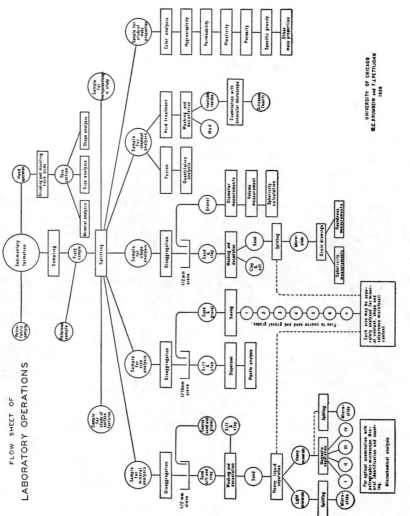

Fig. 265.

AUTHOR INDEX

Ahrens, W., 364
Alexander, A. E., 360
Alexander, L. T., 53, 63, 67, 164, 172, 175, 181
Allen, H. S., 100
Allin, H. L., 466
Alty, S. W., 376
Anderson, G. E., 277, 483
Andreason, A. H. M., 97, 140, 152, 164, 171, 181
Andrews, L., 151
Antevs, E., 11
Appel, J. E., 46, 145, 359, 365, 470
Appiani, G., 149
Arkin, H., 528
Arnold, H. D., 97, 99, 100
Artini, E., 468
Ashley, H. E., 519
Atterberg, A., 77, 78, 98, 149, 519

Baker, H. A., 196, 255
Barnes, K. B., 143, 513, 517
Barnette, R. M., 55
Barth, T., 384
Bay, H. X., 282
Beam, W., 61
Behrens, H., 493
Berek, M., 366
Berg, E., 342
Berg, G. A., 348, 497
Berman, H., 359, 413, 529
Blake, L. I., 352, 353
Boswell, P. G. H., 98, 468, 528
Botset, H. G., 516
Bouyoucos, G. J., 54, 172, 173, 181
Bracewell, S., 322
Bradfield, R., 123, 164
Brammal, A., 379, 422
Brauns, R., 326
Bray, R. H., 21, 411
Briggs, L. J., 53, 61, 123, 149
Brögger, W. C., 336
Brown, I. C., 341
Buerger, M. J., 384

Bunting, E. N., 507
Burri, C., 196, 197

Calkins, F. C., 389
Cameron, E. N., 360, 364
Camp, B. H., 217, 249, 250, 251, 252, 253, 265, 528
Carnes, A., 181
Casagrande, A., 173
Cayeux, L., 311, 337, 480, 493, 496, 528
Chamot, E. M., 127, 366, 367, 385, 386, 401, 493
Choate, S. P., 222, 239
Christiansen, J. E., 104, 105, 109
Church, A. H., 336
Church, C. C., 312
Clark, C. L., 68
Clarke, F. W., 529
Claypool, C. B., 342, 532
Clemmer, J. B., 325, 326
Clerici, E., 327, 328, 385
Cogen, W. M., 486
Cohee, G. V., 201, 322
Cohen, E., 333
Colton, R. R., 528
Cornu, F., 496
Correns, C. W., 62, 67, 70, 77, 79, 119, 181, 222, 342
Coutts, J. R. H., 118, 157
Cox, E. P., 281
Crook, T., 153, 344, 346, 352
Cross, W., 492
Crowther, E. M., 118, 157, 162, 171
Cushman, A. S., 154

Dana, E. S., 49, 366
Darby, G. M., 180
Davies, G. M., 321
Davis, R. O. E., 54
Delesse, A., 345
Derby, A., 319
Deverin, L., 311
Diller, J. S., 338
Doelter, C., 315, 346

AUTHOR INDEX

Dollar, A. T. J., 466
Dominikirwicz, M., 496
Dragan, I. C., 52
Dreilbrodt, C., 338
Driver, H. L., 312, 524
Drosdoff, M., 314, 411
Dryden, L., 86, 263, 469, 470, 486
Dunn, P. H., 494

Eakin, H. M., 31
Edson, F. C., 360
Eichenberg, W., 316
Eisenhart, C., 40, 264, 487
Ekman, V. W., 28
Ellsworth, H. V., 501
Emery, K. O., 157
Emmons, R. C., 331, 369, 383, 389
Evans, P., 473, 476
Ewing, C. J. C., 319

Fairbanks, E. E., 496
Fairburn, H. W., 271
Fancher, G. H., 143, 513, 517
Fischer, G., 77, 81, 129, 289
Fisher, A., 266
Fisher, R. A., 42, 112, 217, 265, 529
Fleet, W. F., 469
Földvari, A., 64, 70
Foque, F., 345
Ford, W. E., 366
Fraser, H. J., 270, 278, 339, 504, 505, 514, 515
Fry, W. H., 366, 409, 411, 529

Gallay, R., 67
Galliher, E. W., 433
Gardner, H. F., 497
Gardner, W., 163, 191
Gary, M., 154
Gealy, W. B., 511
Georgesen, N. C., 49, 313
Gessner, H., 67, 70, 108, 118, 122, 135, 142, 152, 153, 156, 160, 164, 172, 202, 528
Gilligan, A., 376
Goldich, S. S., 376
Goldman, M. I., 8, 312
Goldschmidt, V., 330, 332
Goldstein, S., 108
Gollan, J., 153
Gonell, H. W., 155

Goodner, E., 512
Graham, R. P. D., 503
Graham, W. A. P., 473
Grawe, O. R., 500
Green, H., 128
Gregory, H. E., 277
Griffith, J. S., 202
Grim, R. E., 21, 409, 411
Gripenberg, S., 28, 187, 220, 229
Gross, J., 152
Grout, F., 22, 519
Groves, A. W., 491, 529
Gruner, J. W., 433
Guinard, M., 312

Hagerman, T. H., 129, 288, 289
Hall, F. P., 518
Hallimond, A. F., 347
Hammar, H. E., 493, 494
Hance, J. H., 491
Hanna, G. D., 312, 524
Hanna, M. A., 322, 323
Harland, M. B., 63
Harrington, V. F., 384
Harris, S., 206
Hartshorne, N. H., 366, 529
Hatch, F. H., 346, 356, 528
Hatch, T., 129, 222, 226, 239
Hatfield, H. S., 348
Hauenschild, A., 337
Hawkes, L., 473
Hawkins, H. H., 503
Hayman, R. J., 473, 476
Head, R. E., 363
Heeger, W., 496
Helson, H., 487
Henbest, L. G., 496
Hendricks, S. B., 409
Hervot, L., 153
Hewett, D. F., 8
Heyward, F., 24
Hilgard, E. W., 98, 151
Hillebrand, W. F., 491, 529
Hills, J. M., 495
Hind, S. R., 518
Hinden, F., 496
Hirschwald, J., 507
Hirst, P., 364
Hissink, D. J., 53, 70
Hoernes, R., 277
Holman, B., 348
Holmes, A., 356
Hopkins, C. G., 78

AUTHOR INDEX

Hough, J. L., 298
Howard, A. D., 389
Howard, W. V., 342, 532
Hubbard, P., 154
Hutton, J. G., 500

Iddings, J. P., 366, 492
Ireland, H. A., 495

Jennings, D. S., 163, 191
Johannsen, A., 12, 37, 177, 363, 366, 367, 466, 529
Johnson, W. H., 158
Johnston, C. M., 174
Johnston, J., 497
Joseph, A. F., 52, 172, 180

Keen, B. A., 53, 62, 157, 158, 162, 164, 172
Keller, W. D., 360
Kelley, T. L., 238
Kelley, W. J., 161
Kemp, J. F., 205, 492
Kerr, P. F., 408, 411, 427
Keyes, M. G., 363
Kinney, S. P., 128
Klein, D., 330
Kley, P. D. C., 493
Knapp, R. T., 161
Knight, Jr., F. P., 33
Knopf, A., 354
Knopf, E. B., 268
Koch, H. L., 46, 358, 470
Kohler, R., 63
Köhn, M., 119, 149, 164, 181
Kopecky, J., 152
Krauss, G., 163
Kreider, D. A., 340
Krumbein, W. C., 40, 46, 54, 63, 70, 84, 85, 91, 129, 170, 193, 197, 202, 209, 229, 239, 252, 266, 300, 306
Kruyt, H. R., 57
Krynine, P. D., 481
Ksanda, C. J., 346
Kuhn, A., 181
Kunitz, W., 341

Ladenburg, R., 99
Lakin, H. W., 172, 175, 181
Lamar, J. E., 280, 310, 505

Landes, K. K., 502
Larsen, E. S., 359, 380, 384, 413, 529
Laspeyres, H., 337
Lasson, M. L., 64
Leggette, M., 364
Leighton, M. M., 21
Leinz, V., 465
Lemberg, J., 496
Levy, H., 191
Lewis, J. A., 143, 513, 517
Lincoln, F. C., 466
Löber, H., 156
Locke, C. E., 98
Loebe, R., 63
Lorentz, H. A., 98
Luedecke, O., 338
Lugn, A. L., 28
Lundberg, J. J. V., 97

MacCarthy, G. R., 484
MacClintock, P., 21
MacGee, A. E., 513
Mackey, C. O., 206, 211
Mackie, W., 277, 314, 378
McQueen, H. S., 494
Mahler, O., 497
Majeed, M. A., 473, 476
Mann, A., 50, 313
Mann, P., 345
Marshall, C. E., 408, 410, 411, 427
Marshall, P., 483
Martens, J. H. C., 361
Martin, F. J., 65, 180
Martin, F. O., 61, 123, 149
Mason, C. W., 127, 366, 367, 385, 386, 401, 492
Maynard, J. E., 520
Mead, W. J., 491
Meinzer, O. E., 506, 514
Melcher, A. F., 507, 516
Mellor, J. W., 519
Merwin, H. E., 8, 332, 384, 389, 497
Middleton, H. E., 53, 63, 67, 164
Mills, F. C., 187, 197, 217, 229, 230, 238, 240, 243, 247, 248, 252, 261, 262, 263, 266, 528
Milner, H. B., 5, 12, 38, 315, 324, 363, 364, 468, 528
Miner, N. A., 269
Mitscherlich, E. A., 24, 124
Moore, H. W., 180
Morris, M., 50
Mortimore, M. E., 278, 280

AUTHOR INDEX

Müller, H., 59, 341
Murray, J., 27, 28
Muthmann, W., 325

Neaverson, E., 475
Negreano, D., 352
Neumaier, F., 50, 69, 70
Nevin, C. M., 516
Nichols, J. B., 123
Nicollier, V., 153
Niggli, P., 256
Nikitin, W., 369
Nolte, O., 52, 55
Novak, W., 69, 180
Noyes, A. A., 315
Nutting, P. G., 511, 516

Odén, S., 55, 62, 94, 112, 115, 118, 157, 160, 161, 191, 520
Oebbeke, K., 336, 339
Olmstead, L. B., 53, 55, 63, 67, 164, 172, 175, 181
O'Meara, R. G., 325, 326
Osann, A., 492
Oseen, C. W., 107
Otto, G. H., 14, 20, 39, 40, 43, 44, 45, 46, 265, 355, 357, 470, 496
Owens, J. S., 98

Parmelee, C. W., 180
Part, G. M., 364
Partridge, F. C., 389
Patnode, H. W., 499
Pearce, J. R., 61, 123, 149
Pearson, J. C., 155
Pearson, R. W., 49, 150
Pedigo, J., 206
Pellegrini, N., 180
Penfield, S. L., 340
Pentland, A., 281
Perrin, J., 100
Perrot, G. St. J., 128
Pettijohn, F. J., 44, 196, 198, 204, 240, 271, 483
Piggot, C. S., 29
Pirsson, L. V., 492
Plummer, F. B., 206
Postel, A. W., 56
Pratje, O., 16, 30, 181
Probert, A., 152
Puchner, H., 180

Purdy, R. C., 258
Puri, A. N., 53, 62, 65, 75, 173

Raeburn, C., 12
Raistrick, A., 465
Rankama, K., 328
Rastall, R. H., 346, 356, 528
Rauterberg, E., 153
Redwine, L. E., 30
Reed, R. D., 314, 360, 532
Reitz, H. J., 466
Retgers, J. W., 330
Rham, W. L., 156
Rhoades, H. F., 153
Riboni, P., 352
Richards, R. H., 98
Richardson, E. G., 175, 176
Richter, G., 52, 55, 70, 149
Richter, K., 269
Ridge, J. D., 196, 240, 483
Ries, H., 521, 528
Rietz, H. L., 41
Rinne, F., 366
Rittenhouse, G., 167, 168
Robeson, F. A., 174
Robinson, G. W., 65, 67, 83, 94, 163
Rohrbach, C., 331
Roller, P. S., 122, 128, 147, 155, 258
Rosenholtz, J. L., 351, 502
Rosiwal, A., 466
Ross, C. S., 321, 326, 335, 364, 389, 408, 411, 427
Roth, R., 307
Rouse, H., 105, 171
Rubey, W. W., 52, 83, 102, 474
Running, T. R., 206
Russell, R. D., 446, 474, 484, 496
Russell, W. L., 509
Rutherford, R. L., 386

St. Clair, D., 495
Salmojraghi, F., 320, 468, 479
Sander, B., 271
Sauramo, M., 498, 499, 520, 521
Schaffer, R. J., 364
Schöne, E., 94, 98, 121, 150
Schott, W., 62, 67, 70, 181
Schoutens, Wa., 175
Schramm, E., 159
Schroeder, F., 341
Schroeder van der Kolk, J. L. C., 321
Schucht, F., 156

AUTHOR INDEX

Schurecht, H. C., 175, 292, 512
Scott, W., 119
Scripture, E. W., 159
Seger, H., 517
Sexton, H. D., 181
Shand, S. J., 466
Shaw, C. F., 118
Shaw, T. M., 164
Shepard, F. P., 201
Shepherd, St. J. R. C., 348
Sherzer, W. H., 303
Shewart, W. A., 265
Short, M. N., 493
Simonson, R. W., 49, 150
Simpson, D., 25
Sligh, W. H., 155
Smeeth, W. F., 338
Smith, D. T., 351, 502
Smithson, F., 319, 322, 345
Smythe, J. A., 473
Snow, O. W., 52, 65
Sollas, W. J., 332, 502
Sommers, R. E., 411
Sondstadt, E., 330
Sorby, H. C., 278
Soule, F. M., 27, 30
Spangenberg, K., 496
Spencer, L. J., 502
Spieker, E. M., 49, 316
Stach, E., 317
Staples, Lloyd W., 493
Stearns, N. D., 505, 514, 521
Steele, J. G., 123, 164
Steidtman, E., 497
Stokes, G. G., 95
Stow, M. H., 316
Stuart, A., 366, 479, 529
Suffel, G. G., 496
Sullivan, J. D., 321, 323, 325, 327, 331
Sutton, Chase E., 506, 510
Svedberg, T., 123
Szadeczky-Kardoss, E. V., 119, 158, 286

Taylor, G. L., 49, 313, 340
Taylor, J. R., 49, 150
Tester, A. C., 48, 282, 334
Thackwell, F. E., 369, 466
Thomas, E. T., 362
Thomas, M. D., 64, 163, 191
Thoreen, R. C., 174
Thoulet, J., 330, 338
Tickell, F. G., 159, 282, 313, 348, 528
Tolmachoff, I., 50, 312

Tomlinson, C. W., 356
Trager, E. A., 532
Trask, P. D., 28, 30, 123, 149, 159, 201, 222, 229, 230, 236, 493, 494, 499
Trowbridge, A. C., 278, 280
Truog, E., 49, 150, 314
Tucan, Fran, 497
Tuorila, P., 59, 73
Twenhofel, W. H., 6, 528
Tyler, S. A., 379
Tyrrell, G. W., 5

Udden, J. A., 77, 214
Ulbricht, R., 49
Ungerer, E., 63
Urbain, P., 411

Valentine, W. G., 466
Van Niewenberg, C. J., 175
Van Orstrand, C. E., 191, 222, 239
Van Tongeren, W., 491, 529
Van Werveke, L., 335
Van Zyl, J. P., 149
Varney, F. M., 30
Vassar, H. E., 328, 332
Veatch, A. C., 25
Vendl, M., 119, 158
Vhay, J. A., 328
Vinther, E. H., 64
von Hahn, F.-V., 55, 101, 127, 135, 161, 175, 176, 528
Von Sanden, H., 191
von 'Sigmond, A. A. J., 70, 149
von Smoluchowski, M., 58
von Wolff, F., 342

Wadell, H., 23, 93, 94, 104, 105, 127, 178, 188, 268, 270, 281, 283, 284, 292, 295, 296, 298
Wahnschaffe, F., 156
Walker, T. L., 330
Wallace, R. C., 520
Wanless, H. R., 465
Washburn, E. W., 507
Washington, H. S., 491, 492, 529
Weeks, M. E., 49, 150
Wentworth, C. K., 31, 33, 36, 37, 45, 46, 77, 80, 126, 136, 140, 147, 222, 239, 279, 290, 303, 306, 358, 466, 470

Werner, D., 158
Westman, A. E. R., 512
Wethered, E., 494
Wetzel, W., 312
Weyland, H., 364
Weymouth, A. A., 363
Weyssenhoff, J., 99
Whiteside, R. M., 27, 532
Whittles, C. L., 52, 54, 73, 165
Wicksell, S. D., 129
Wiegner, G., 55, 59, 66, 159
Wiley, H. W., 148
Wilgus, W. L., 46, 358, 470
Williams, I. A., 466
Williams, L., 304
Williamson, A. T., 328

Williamson, E. D., 497
Willis, E. A., 174
Winchell, A. N., 366, 529
Winterer, E. V., 118
Wintermyer, A. M., 174
Winters, Jr., E., 63
Wolff, E., 148
Woodford, A. O., 339
Wright, F. E., 366, 382, 386, 529
Wülfung, E. A., 337

Zimmerley, S. R., 152
Zingg, Th., 81, 257, 288
Zschokke, B., 519
Zunker, F., 161

SUBJECT INDEX

Acetylene tetrabromide, 325
Actinolite, 414
Acute bisectrix, 396
Adsorption, 520
Aegirinaugite, 444
Aggregate analysis, 153
Aggregate, properties of, 7
Aggregates, compound grains and, table of, 461
Air elutriation, 122, 154
Albite, 442
Allophane, 427
Amphibole group, 414
Analysis:
    graphic data, 205
    orientation, 268
    preparation for, 43
    variance, 42
Anatase, 416
Anauxite, 426
Andalusite, 417
Angularity, 280, 286
Anhydrite, 418
Anisotropism, 390
Anorthite, 442
Apatite, 418
Apparatus, list of, 525
Aragonite, 420
Arithmetic mean, 217, 224, 240
Auger samples, collection of, 24
Augite, 444
Average, *see* Mean
Average size, significance of, 218
Average values, comparison of, 245

Bags, for samples, 34
Bakelite, 364
Baker's equivalent grade, 255
Baker's grading factor, 255
Bar charts, 205
Barite, 420
Barium mercuric nitrate, 331
Base exchange, 64, 75, 411
Beaches, sampling of, 15, 19

Beach pebbles, orientation of, 270
Becke line, 379, 387
Becke method, 386
Bedding, 8
Beidellite, 427
Bentonite, 454
Biaxial crystal, 396
Biaxial figure, 403
Biaxial minerals, optic sign of, 405
Biotite, 421
Birefringence, 392
    factors controlling, 393
Bisectrix, acute, 396
    obtuse, 396
Bisectrix figure, 403
Bitumen, removal of, 49, **315**
Boulder, defined, 78
Bromoform, 320
    clarification of, 321
    objection to, 324
    procedure for recovery, **343**
    properties of, 321
    recovery of, 321
Bronzite, 435
Brookite, 422

Cadmium borotungstate, **330**
Calcite, 423
Camera lucida, 372
Canada balsam, 360
    vs. piperine, 360
Cartons, for samples, **36**
Cassiterite, 424
Celestite, 420
Cement:
    calcite, 48
    iron oxide, 48
    organic, 49
    removal of, 48
    silica, 48
Central tendency, measures of, 217
Centrifuge:
    tube, Kunitz, 341; Schröeder, 341;
        Taylor, 341

540 SUBJECT INDEX

Centrifuge (*continued*)
  use with heavy liquid, 340
Ceylonite, 449
Chalcedony, 445
Chamberland filter, 67
Channel sample, collection of, 17
Chatter marks, 306
Chemical methods of analysis, 490
Chert, 445
Chiastolite, 417
Chi-square, 264, 489
Chlorite, 425
Christiansen effect, 387
Circle scale, 298
Circularity, degree of, 285
Clarification of grains, 314
Class intervals, unequal, 77, 85
Clay:
  defined, 78
  potash-bearing, 427
  preparation for mineral analysis, 316, 408
  special methods of study, 407
Clay ironstone, 447
Clay minerals:
  base exchange capacity, 411
  chemical analysis, 410
  dehydration tests, 410
  optical methods, 409
  X-ray methods, 410
Cleavage, 376
  table of, 377
Clerici's solution, 328
Clino-zoisite, 456
Coagulant, 58
Coagulation, 57
  limits of, 60
  orthokinetic, 59
  perikinetic, 58
  tests for, 73
Coal, preparation for mineral analysis, 317
Cobble, defined, 78
Coefficient:
  of correlation, 260, 486
  of mineral association, 264
  of resistance, 104
  of pebble size, 210
Cohesiveness, 521
Colloidal suspensions, 57
Collophane, 428
Cologarithm, 241
Color:
  interference, 393

  of minerals, 400
  of rocks, 499
Commercial analysis, 12
Compass needle figure, 403
Composite sediments, analysis of, 142
Concretions, 9
Contamination, mineral, 481
Control, theory of, 265
Convergent light, 401
Correlation, coefficient of, 260, 486
Corundum, 429
Crowther's sedimentation tube, 163
Crushing, 313
Crushing strength, 521
Crystal:
  biaxial, 396
  form, 375
  negative, 395
  positive, 395
  striation, 379
  uniaxial, 395
Crystalline rocks, preparation for analysis, 317
Cumulative curves, 188, 215
  differentiation of, 190
Curves, *see* Cumulative, Frequency, Mathematical analysis, Sedimentation curves

Darcy's law, 514
Decantation:
  centrifugal, 149
  methods, 119, 147
Delta, pebbles in, 268
Dependent variable, 182
Detrital minerals:
  description of, 412
  tables of, 458
Diallage, 444
Diamagnetic substances, 344
Diameter:
  mean, 146, 279
  measurement of, 145
  nominal sectional, 296
  true nominal, 94
Dickite, 426
Dielectric constant, 348
Differential contamination, concept of, 481
Diffusion column, 502
Digest:
  acid, 312
  alkaline, 312
Diopside, 444

SUBJECT INDEX 541

Disaggregation, 51, 310, 318
Disperse systems, 91, 92
Dispersion:
  acid treatment in, 65
  boiling in water, 55, 66
  breakage during, 69
  critique of, 68
  effect of base exchange on, 64
  effect of boiling on, 66
  effect of electrolytes on, 62
  factor, 53
  Földvári's theory of, 64
  function of time, 53
  ignition, 55
  international method, 65
  of particles, 51
  optical, 407
  physical vs. chemical, 69
  procedures, chemical, 56
  procedures, physical, 52
  routine, generalized, 70
  rubbing in water, 52
  shaking in water, 52
  soaking in water, 52
  statistical, see Standard deviation, Quartile deviation
  stirring in water, 54
  tests for, 73, 75
  use of beads in, 54
  vibration in water, 54
Disthene, 436
Dolomite, 423, 424
Double refraction, 390
Drilling:
  diamond core, 26
  percussion, 26
  rotary, 27
Drive-pipe samples, collection of, 25
Dumortierite, 430

Electrolytes, see Peptizer, Salt
Electromagnet, 345
  specifications, 346
Electrostatic plates, Crook, 352
Elongation, sign of, 394
Elutriation, 119
  rising-current, 121, 150
  theory of, 121
Elutriator:
  Andreason, 153
  Andrews' kinetic, 151
  Crook, 153

Cushman and Hubbard, 154
Gary, 154
Gollan, 153
Gonell, 155
Hilgard, 151
Kopecky, 152
Roller, 155
Schöne, 150
Enstatite, 435
Epidote, 430
Equivalent grade, 255
Error function, 40
Error:
  independent, 266
  of the mean, 41
  probable, 266
  sampling, 40
Esker, pebbles in, 268
Exponential function, 208
Extinction, 391
  angle of, 396
  explanation, 392
  incomplete, 394
  oblique, 397
  parallel, 395
  symmetrically inclined, 395
Extraordinary ray, 395

Feldspar, 442
Field schedule, 8
Fineness factor, 258
Flatness ratio, 279
Flocculation, 57; see also Coagulation
"Flow-sheet" of operations, 532
Fluorite, 431
Fracture, 376, 378
Frequency, curve, 190, 218
  mineral, see Mineral frequency
  polygon, 186
  pyramid, 184
  weight vs. number, 226
Frequency distribution, 213, 221
Friable material, thin sections of, 364
Functions, properties of, 210
Funnel, for heavy-liquid separation, 335
Fusibility, 521

Gage, convexity, 280
Garnet, 432
Geometric mean, 217, 240, 243
  graphic computation of, 254

SUBJECT INDEX

Glacial till:
  pebbles, 268, 273
  sampling of, 13
Glass, volcanic, 454
Glauconite, 433
Glaucophane, 415
Gloss, defined, 304
Goniometer, for refractive index work, 384
Grade scale, 76
  A. S. T. M., 81
  Atterberg, 78
  bureau of soils, 78
  choice of, 88
  engineering, 81
  fourth root of 2, 137
  functions of, 86
  geometrical, 77
  Hopkins, 78
  I. M. M., 82
  Krumbein, 84
  logarithmic, 83, 89
  Phi, 84
  Robinson, 83
  Rubey, 83
  square root of 2, 137
  Udden, 77
  Wentworth, 80
  Zeta, 84
Grading factor, 255
Grain shape, 375
Grains, manipulation of, 389
Gram-Charlier series, 253
Granule, defined, 80
Graphic computation of geometric mean, 254
Graphic differentiation, 190
Graphic presentation, 182
Graph:
  paper, 183
  principles of, 182
  three variables, 200
  two variables, 184
Gravel, defined, 78
Grid for sampling, 15
Gypsum, 372, 434

Halloysite, 427
Harmonic mean diameter, 128
Heavy liquid, 320
  choice of, 330
  fractionation with centrifuge, 340
  ideal, 320
  imperfect separations of, 331
  nomenclature, 321
  separation apparatus, 335
  standardization, 331
  table, 329
Heavy minerals:
  concentration by sieving, 319
  enrichment by dry agitation, 320
  graphic presentation, 477
  panning, 319
  rough concentration, 319
  separation procedure, 343
Hematite, 439
Histogram, 184, 214
  arithmetic, 185
  logarithmic, 185
  uses of, 187
Hornblende, 414
Hydraulic value, 94
Hydrometer, 334
  Blake, 334
  Bouyoucos, 172
  Casagrande, 173
  heavy-liquid, 334
  Puri, 173
  size analysis by, 172
  Tester, 334
Hygroscopicity, 519
Hypersthene, 435
Hyrax, 360

Ignition, dispersion by, 55
Illumination:
  bright-field, 409
  dark-field, 409
Ilmenite, 439
Image, 285
Immersion liquids:
  bottles for, 380
  choice of, 381
  of low index, 384
  mixing of, 382
  preparation of, 380
  refractive indices, 381
  standardization of, 384, 385, 386
  table of, 383
Immersion media, of high index, 384
Immersion method, 379
Inclusions, 378
Independent variable, 182
Index liquids, see Immersion liquids
Index of refraction:
  anisotropic substances, 398

SUBJECT INDEX 543

Becke method, 386
  inclined illumination, 388
  liquids, table of, 381, 383
  minimum deviation method, 389
  relation to specific gravity, 333
  variation methods, 385
  vertical incidence, 384
Index, sorting, 258
Indicolite, 452
Indurated samples, size analysis of, 129
Inflection point, 192
Insoluble residues, 494
  graphs of, 495
Interference figure, 401
  biaxial, 403
  bisectrix, 403
  centered uniaxial, 402
  optic axis, 403
Interference colors, 393
Integrating stage, 369, 466
Iron oxide, removal of, 314
Isoaxial crystals, 396
Isogyres, 403
Iso-megathy map, 201
Isopleth map, 201
Isotropism, 390

Jolly balance, 501

Kaolin group, 426
Klein solution, 330
Kollolith, 360
Kurtosis, 220, 250
  quartile, 238
  significance of, 220
Kyanite, 436

Laboratory, 522
  plan of, 523
  requirements of, 522
  schedule of operations, 9, 530, 531
Lens:
  objective, 370
  ocular, 370
Leucoxene, 440
Leverrierite, 426
Light, polarized, 390
Limestone, preparation for mineral analysis, 317
Limonite, 440
Linear function, 206
Loess, sampling of, 17

Logarithmic mean, 242
Luster, defined, 304, 306

Magnetic permeability, 344
Magnetite, 441
Magnets, permanent, 344
Map, isopleth, 201
Marcasite, 441
Mass properties of sediments, 498
Mathematical analysis, 205
Mean:
  deviation, 219, 247
  error of, 41
  see also Arithmetic, Geometric, Logarithmic; Median, Mode
Measurement, microscopic, 177
Mechanical analysis, 91
  "break" in, 136, 189
  centrifugal force, 123
  comparisons of, 180
  flow-sheet, 171
  inherent error, 118
  methods of, 135
  microscopic methods, 126, 176, 302
  modern methods, 117
  preparation for, 47
  principles of, 91
  size of vessel, 99
  summary of, 134
  thin-section, 129
Mechanical stage, 368, 373
Median, 217, 223, 229
Median map, 201
Methylene iodide, 326
Microchemical methods, 493
Microcline, 442
Micrometer:
  eyepiece, 177, 373
  stage, 373
Micrometric analysis, 126, 176, 302
Microprojector, 179, 374
Microscope, 366
  centering procedure, 402
  measurements with, 373
  mechanical analysis with, 176
  mechanical components of, 367
  nicol prisms, 371
  objective lenses, 370
  ocular lenses, 371
  optical components, 369
  polarizing, 366
  uses of, 366
Microscopic particles, "size" of, 127

Microsplit, 46, 357
Microstriations, 304
Mineral analysis, preparation for, 309
Mineral frequency, 465, 467
  by count, 469
  calculation of, 477
  estimation, 467
  histogram of, 475
  log scale of, 473
  presentation of results, 475
  probable error of, 470
  relation to grade size, 474
  scale of abundance, 467, 473
  statistical methods, 480
  variations of, 480
  weight vs. number, 478
Mineral gradient, 480
Mineral grains, specific gravity of, 500
Mineralogical analysis:
  preparation for, 48
  sample form sheet, 464
Minerals:
  biaxial, 401
  colored anisotropic, table of, 460
  colorless anisotropic, table of, 459
  common authigenic, table of, 463
  description of, 412
  detrital and associated, table of, 462
  dielectric constants of, 350, 351
  identification of, 366
  isotropic, table of, 458
  magnetic classes of, 346
  mounting for microscopic study, 357
  opaque, 439; table of, 458
  solubility, table of, 355
  uniaxial, 401
Mineral suites, table of, 463
Mineral tables, determinative, 456
Modal class, 214
Mode, 217, 243, 273
Moisture equivalent, 520
Moment:
  analysis, 223, 239
  fourth, 252
  kurtosis, 250
  product, 262
  skewness, 250
  third, 251, 253
Monazite, 437
Monodisperse systems, 92
  sedimentation of, 111
Montmorillonite, 427
Mortar, diamond, 313
Mountant, 360

Mounting, 359
  permanent, 360
  procedure for Canada balsam, 360
  procedure for piperine, 361
  slips, 362
  temporary, 359
Moving average, 198
Muscovite, 438

Nacrite, 426
Neaverson chart, 475
Nicol prism, 390
Niggli's statistical method, 256
Nitrocellulose peels, 365
Nominal diameter, 143, 279, 292
Nominal sectional diameter, 128, 297
Nontronite, 427
Normal curve, 252
Normal phi curve, 252
Number vs. weight frequency, 226

Objective lens, 370
Obtuse bisectrix, 396
Octahedrite, 416
Ocular lens, 371
  micrometer scale, 371
Odén curve, 113, 160
Odén method, 157
Odén sedimentation balance, 157, 158
Odén theory of sedimentation, 112, 115
Olivine, 439
Opaque minerals, 439
Optical constants, fundamental, 374
Optical methods, 366
Optic angle, estimation of, 406
Optic axes, 393
Optic axial angle, 396
Optic axis figure, 403
Optic normal, 396
Optic orientation, 395
Optic plane, 396
Optic sign:
  biaxial minerals, 405
  determination of, 404
  uniaxial minerals, 404
Ordinary ray, 395
Organic content of sediments, 493
Orientation analysis, 268
  esker pebbles, 268
  laboratory procedure, 271
  statistics of, 272
  till pebbles, 269, 273

SUBJECT INDEX 545

use of goniometer in, 272
Orientation, optic, 395
  diagrams, 397
Oriented samples, collection of, 22, 270, 273
Orthoclase, 442
Orthokinetic coagulation, 59
Outcrop samples, 13

Panning, 319
Parallel extinction, 395
Paramagnetic substances, 344
Parting, 376, 378
Pebble:
  defined, 78
  shape, geological meaning, 277
Pebble counts, 465
Pebbles:
  collection of, 32
  diameters of, 143, 145, 279
  orientation of, 269, 273
  surface textures of, 304
  table of surface textures, 305
  volume measurement, 292
Peptization, 56
  procedures, 61
Peptizer, 58
  ammonium carbonate as, 66
  ammonium hydroxide as, 62, 67
  gum arabic as, 152
  lithium chloride as, 63
  sodium carbonate as, 62, 72
  sodium hydroxide as, 65
  sodium oxalate as, 63, 72
  sodium pyrophosphate as, 64
  sodium silicate as, 64
Percussion marks, 306
Perikinetic coagulation, 58
Permeability:
  coefficient of, 514
  consolidated rocks, 516
  defined, 513
  factors influencing, 514
  formula for, 514
  unconsolidated materials, 514
Permeameter, Meinzer, 515
Petrofabric analysis, *see* Orientation analysis
Phi curve, normal, 252
Phi mean, 242
Phi scale, 84, 88, 90, 233
Phi values, conversion chart for, 244

Photocell method of mechanical analysis, 175
Picotite, 449
Pie diagrams, 205
Piperine, 360, 361
Pipette:
  Berg, 342
  Jennings, Thomas, and Gardner, 164
  Krauss, 165
  Robinson, 164
Pipette method, 162, 167
  error in, 164
  procedure for, 166
  theory of, 163
Plagioclase feldspar, 442
Planimeter, 296
Plastic flow, 518
Plasticity, 517
  factors governing, 518
  methods of measurement, 518
Plastometer, 518
Pleochroic formula, determination of, 401
Pleochroism, 400
Pleonaste, 449
Plummet method of mechanical analysis, 174
Pointolite, 299
Polarization, anomalous, 394
Polarized light, 390
Polaroid, 372
Polish, defined, 304, 306
Polydisperse systems, 92
  sedimentation of, 112
Porosimeter:
  Fancher, Lewis and Barnes, 513
  Russell, 510
  Washburn and Bunting, 512
Porosity:
  Barnes method, 513
  defined, 503
  effective, 513
  factors governing, 504
  Gealy's method, 511
  geological significance of, 503
  Hirschwald method, 507
  Melcher method, 507
  method of Washburn and Bunting, 511
  methods of measurement, 506
  Nutting's method, 511
  preparation of the sample, 506
  Russell method, 509
  unconsolidated materials, 504
Potash-bearing clays, 427

Potassium mercuric iodide, 330
Power function, 207
Pressure-chamber disaggregation, 50, 313
Prism, nicol, 371, 390
  passage of light through, 391
Probability graph paper, 189
Probable error, 18, 266, 470
  curves for estimation, 472
Product moment, 262
Projection apparatus, microscope, 178
Pycnometer, 501
  Nutting, 511
Pyrite, 441
Pyrophyllite, 427
Pyroxenes, monoclinic, 444

Quantitative analysis, chemical, 490
  computations based on, 491
Quartering, see Splitting
Quartile, defined, 230
Quartile deviation, 219, 230
  arithmetic, 230
  geometric, 230
  logarithmic, 231
  phi, 234
Quartile kurtosis, 238
Quartile measures, 223, 229, 232
Quartile skewness, 235
  phi, 237
Quartz, 445
Quartz wedge, 372

Radius:
  equivalent, 94
  measurement of, see Diameter
  sedimentation, 94
Range, statistical, 219
Reference books, 528
Reference sample, 486
Refractive index, see Index of refraction
Refractometer, 384
Relief, optical, 387
Resistance, coefficient of, 104
Retgers' liquids, 330
Reynolds number, 104
Rising-current elutriation, 121, 150
Rock specimens, disruption of, 49
Rohrbach's solution, 331
Rosiwal method, 466
Rounding, scale of, 283

Roundness:
  choice of method, 291
  classes, Szadeczky-Kardoss, 287
  defined, 277; see also Shape
  factors controlling, 278
  Fischer method, 289
  ratio, 279
  restricted meaning, 284
  Szadeczky-Kardoss method, 286
  Tester method, 282
  visual method, 280
  vs. shape, 283
  Wadell formulas, 285
  Wadell's concept of, 285
  Wadell method, 295
  Wentworth method, 279
Rutile, 446

Sailboat, diagrams, 205
Salt, soluble, removal of, 66
Sample:
  containers, capacities of, 34
  splitter, 44, 357; Appel, 46, 358, 359; Jones, 45; knife edge, 44; oscillatory, 46, 358, 359; Otto, 45, 357; rotary, 46, 358; statistical tests of, 46; Wentworth, Wilgus, and Koch, 46, 358
Sampler:
  clam-shell, 30
  drag bucket, 27
  Ekman, 28
  hand auger, 23
  Heyward, 24
  hydraulic core, 30
  Knudsen, 30
  Lugn, 28
  Mitscherlich, 24
  Piggot, 29
  Simpson, 25
  Veatch, 25
Samples:
  bottom, 27
  catalogue for, 37
  channel, 16
  collection of, 11
  composite, 18, 42
  compound, 18
  containers, 33, 35
  core, 26
  disaggregation of, 47
  discrete, 13
  dispersion of, 72

SUBJECT INDEX 547

display, 11
disruption of, 49
drive-pipe, 25
for commercial analysis, 12
grab, 13
hand auger, 23
hand specimen, 12
indurated, 22
labeling and numbering, 36
linear series, 14
percussion drill, 26
quartering, 44, 357
random, 40
rotary drill, 27
serial, 14
single, vs. composite, 20
size of, 31
spot, 13
sub-surface, 23
weathering, 21
well, 26
wet vs. dry, 69
Sampling:
  complexity of, 12
  error, 40, 220
  field observations during, 10
  grids, 15
  purposes of, 11
  theory of, 38
Sampling tubes, 28; see also Samplers
Sand:
  defined, 78
  unbedded, sampling of, 13
Sand grains:
  "roundness" of, 281
  shape of, 299
  Sorby's classification, 278
  surface textures, 306
  table of surface textures, 307
Scale:
  arithmetic, 183
  logarithmic, 183
Scale units, choice of, 183
Scatter diagram, 199
Secondary enlargement, 376
Sedimentary analysis, organization of, 529
Sedimentary petrography, defined, 4
Sedimentary petrology, defined, 5
Sedimentation:
  balance, Odén, 157
  curves, 113, 160, 195
  defined, 6
  glass, Spaeth, 339

methods, 119
Odén's theory, 112
tubes, see Settling tubes
unit, 20
Sedimenting systems, theory of, 110
Sediments:
  constituents of, 311
  degree of induration, 311
  organic constituents of, 9
Selective abrasion, 483
Selective sorting, 483
Selenite, 434
Semi-log graph paper, 183
Separation apparatus, mineral, see Vessel; size, see Settling tube
Separation (mineral separation):
  chemical methods, 354
  dielectric, 348; procedure, 350, 351
  electromagnetic, 345
  electrostatic, 352
  errors in, 356
  flotation, 354
  hand sorting, 353
  magnetic methods, 344; procedure for, 346
  methods, 319
  on the basis of shape, 353
  systematic schemes of, 356
  systematic, table of, 349
Separator:
  electromagnetic, 347
  magnetic, Davis, 347; Hallimond, 347
Series, discrete, 213
Serpentine, 425
Settling tube:
  Appiani, 148
  Atterberg, 149
  Clausen, 156
  Crowther, 163
  Emery, 157
  Kelley, 161
  Kühn, 148
  Trask, 149
  von Hahn, 162
  Werner, 159
  Wiegner, 160
Settling velocity, 95
  effect of vessel on, 99
  Goldstein's law, 108
  nomograph for, 170
  Oseen's law, 107
  of spheres, 110
  Stokes' law, 95

Settling velocity (*continued*)
  summary of laws, 109
  table of, 111, 112, 166
  time chart, 169
  Wadell's law, 104
Shaker:
  end-over-end, 53
  reciprocating, 53
  Ro-Tap, 139
  rotary, 53
Shape:
  analysis, preparation for, 47
  choice of method, 291
  classes, Zingg, 288
  Cox method, 281
  defined, 277; *see also* Roundness
  factors controlling, 278
  Hagerman method, 288
  Lamar method, 280
  minimum porosity method, 280
  of mineral grains, 375
  Pentland method, 281
  relation to crystal habit, 375
  relation to mineral cleavage, 376
  of sand grains, 299
  short method, 299
  Tickell method, 282
  verbal description, 290
  vs. roundness, 283
  Zingg method, 288
Shapometer, 282
Shrinkage, 521
Siderite, 447
Sieve analysis, 137
  report form, 141
  vs. micrometric analysis, 302
Sieves, Tyler, 138
Sieving, 137
  as function of time, 125
  coarse-particle, 142
  completeness of, 140
  disadvantages of, 124
  theory of, 124
  timer, 140
Sign of elongation, 394
Sillimanite, 448
Silt, defined, 78
Silt flask, Bennigsen, 156
Size analysis, 91
  methods of, 135
Size, concept of, 93
Size distribution, microscopic, 128
Size frequency distribution, 87
Skewness, 215, 219

Skewness measures:
  comparison of, 236
  quartile, 235
  significance of, 220
Slaking, 521
Slides for mounting loose grains, 362
Sondstadt's solution, 330
Sorting:
  coefficient, Trask, 230
  factor, 214
  indices, 258
  significance of, 219
Soxhlet extractor, 315
Spatial distribution, 213
Spatial variations, 480
Specific gravity:
  of mineral grains, 500
  relation to refractive index of liquid, 333
Specific surface, 128
Sphericity:
  defined, 292
  degree of, 283
  formulas for, 284
  histogram, 300
  nomograph computation, 293
  Wadell's method, 295
Spinel, 449
Splitter, *see* Sample splitter
Splitting, 44, 357
Spot samples, collection of, 13
Sprengel tube, 332
Stage:
  integrating, 466
  mechanical, 373
Stage micrometer, 372
Staining methods, 495
Standard deviation, 219, 224
  arithmetic, 248
  geometric, 250
  logarithmic, 249
Standard size, 285, 296
Star diagrams, 205
Statistical analysis, elements of, 212
Statistical correlation, 260
Statistical devices, choice of, 259
Statistical dispersion, measures of, 247
Statistical measures, 217
  application of, 228
Statistical methods, summary of, 267
Statistics:
  defined, 212
  of particle orientation, 273
Staurolite, 450

SUBJECT INDEX 549

Stokes' law, 95
  assumptions of, 96
  experimental, 98
  limits of, 100
  summary of, 101
  time chart, 169
Striation, crystal, 379
Striations:
  defined, 305
  pattern on pebbles, 305
  relations to pebble axis, 306
Surface factor, 258
Surface mean diameter, 258
Surface (particle surface):
  dull, 304, 306
  etched, 307
  faceted, 307
  frosted, 307
  furrowed, 306
  mat, 307
  pitted, 306
  ridged, 306
  rough, 305, 306
  scratched, 305
  smooth, 305, 306
Surface texture:
  analysis, preparation for, 47
  defined, 303
  pebbles, 304
  sand grains, 306
  significance of, 303
Suspensions, colloidal, 57

Tensile strength, 521
Tetrabrom-ethane, 325
Thallium formate, 327
Thallium formate-malonate, 328
Theory of control, 265
Thin-section analysis, 128, 178, 465
Thin sections, preparation of, 363
Thoulet solution, 330
Three-variable surface, 200
Time graph, 197
Time series, 213
  analysis, 198
Titanaugite, 444
Titanite, 450
Topaz, 452
Tourmaline, 452
Tremolite, 414
Tremolite-actinolite, 414
Triangle diagrams, 202

Tribrom-methane, 320

Uniaxial crystal, 395
Uniaxial figure, 402
Uniaxial mineral, optic sign of, 404
Universal stage, 369

Variable, choice of, 182
Variation area, 255
Varietal characters, 378
Varves, sampling of, 12
Vessel:
  Brögger, 336
  Church, 336
  Fraser, 339
  Harada, 336
  Hauenschild, 337
  Laspeyres, 337
  Luedecke, 338
  Penfield, 340
  Smeeth, 338
  Thoulet, 338
  Wülfung, 337
Vibration axes, 396
Vibration directions, 396
Vibration, dispersion by, 54
Viscosity, 101
Viscous flow, 518
Volcanic glass, 454
Volume, Wadell's method, 295
Volumeter:
  Gealy, 511
  Schurecht, 292

Water-soluble salts, removal of, 56
Water, viscosity of, 102
Weight vs. number frequency, 226
Westphal balance, 333
Wet sieving, 142
Whipple disk, 373
Wiegner curve, 160
Wiegner sedimentation tube, 160
  modified, 161

Yield point, 518

Zeta scale, 84, 90
Zircon, 454
Zoisite, 456
Zoning, mineral, 379

(3)